Mechanics and Mechanisms of Fracture: An Introduction

Alan F. Liu

ASM International®
Materials Park, Ohio 44073-0002
www.asminternational.org

Prepared under the direction of the ASM International Technical Books Committee (2004–2005), Yip-Wah Chung, FASM, Chair.

ASM International staff who worked on this project include Scott Henry, Senior Manager of Product and Service Development; Bonnie Sanders, Manager of Production; and Madrid Tramble, Senior Production Coordinator.

Library of Congress Cataloging-in-Publication Data

Liu, A.F.
    Mechanics and mechanisms of fracture: an introduction / A.F. Liu.
        p. cm.
    Includes bibliographical references and index.
    1. Fracture mechanics. I. Title.

TA409.L58 2005
620.1′126—dc22   2005042107

ISBN: 0-87170-802-7
SAN: 204-7586

ASM International®
Materials Park, OH 44073-0002
www.asminternational.org

Printed in the United States of America

Contents

Preface

Failure of a machine part (or structural part) could be attributed to many known or unknown factors. The most fundamental one is due to static overload of a structural member. The term overload might mean the structural member is subjected to a load that exceeds the ultimate strength of the material. Overload might also mean the applied load is higher than the anticipated design load of the structure, the value of which is normally set to lower than the material's ultimate strength. Incidentally, structural parts often fail at load levels much lower than the design load. An initial defect (or damage) in the part, cyclic load-induced cracking, or time-dependent/environmentally-assisted cracking all can cause a machine part to fail prematurely. The studies of the problems and the remedies for handling the failure modes in the first two types of failures mentioned are called fracture mechanics and fatigue; high-temperature creep, stress-corrosion, corrosion-fatigue, or hydrogen-embrittlement are the sources for the rest.

Fracture mechanics is the study of the influence of loading, crack size, and structural geometry on the fracture resistance of materials containing natural flaws and cracks. When applied to design, the objective of the fracture mechanics analysis is to limit the operating stress level so that a preexisting crack would not grow to a critical size during the service life of the structure. Fatigue is the study of the effects of repeated loading on a machine part that was initially defect free, and how those loads (in combination with other factors) may shorten its anticipated life. Understanding the mechanisms of creep and stress-corrosion/hydrogen-embrittlement is important when a machine part is intended for use in an extreme environment.

Both the metallurgical and mechanical elements in each type of these failure modes will be discussed. In the metallurgical field, the model that offers explanations to a phenomenon is often case specific. Experience gained may or may not be applicable to another case, however similar it may seem. Revised or updated concepts or solutions appear in the literature from time to time. It is impossible to discuss all the concepts and methods in a single book. Only those relevant to the intended specific themes are included here. In contrast, the mechanical causes of failure can be analyzed by means of solid mechanics. Nevertheless, the solution to a problem is valid only for the set-up and assumptions as intended; thus it is also somewhat case specific.

The main focus of this book is to explicate how materials respond to applied forces, and relate them to design analysis, material evaluation, and failure prevention. Metals occupy the main part of the book; also included are non-metallic materials such as ceramics, plastics, and fiber-reinforced polymer-matrix composites. Both the fundamental and practical concepts of fracture are described in terms of stress analysis and the mechanical behavior of materials. The metallurgical aspects of deformation and fracture in metals also are discussed. The first two chapters of this book can be regarded as the fundamentals of stress analysis and mechanical behavior of materials. These chapters provide necessary knowledge for the understanding and appreciation of the contents in chapters that follow.

With regard to structural design and analysis, a strong emphasis is placed on showing how fracture mode is influenced by the state of stress in the part. The stress analysis section in Chapter 1 serves as a crash course (or a refresher course) in the strength of materials and prepares the reader with the basic analytical tools for the remainder of the book. A progressive approach is taken to show the effect of structural geometry and loading conditions to the resulting stresses—first to show the stresses in undamaged structural members; next to show the stresses in structures that contain a geometric discontinuity; then finally, to show the stress field in the vicinity of a crack. Stress analysis of cracks requires the use of a new analytical tool called fracture mechanics. This is discussed in Chapter 1 for primary elastic with small scale yielding at the crack tip, and in Chapter 6 for large scale yielding at the crack tip. Application of fracture mechanics to design/analysis and the related tasks is presented in Chapters 4 through 6. Numerous examples are given throughout this book to illustrate the elastic and plastic behavior of materials at a stress raiser, and how the static, fatigue, and residual strengths of a machine part might be affected by it. Finally, the structural analysis methods, as well as the damage tolerant aspects of fiber-reinforced composites, are discussed in Chapter 8.

Strangely, in the last couple years, "101" has become a household term. It penetrates into our homes by way of television, radio, and newspaper. Politicians, entertainers, the media, and talk show hosts and their guests, would freely use " 'Whatever' 101" to link to the topic or event that is being discussed. Therefore this book could, in that sense, qualify to be titled "Mechanics and Mechanical Behavior of Materials 101."

To reiterate: This book is about how machine (or structural) parts fail, why one piece fails in a certain way and another piece fails differently; and will provide engineering tools for analyzing these failures, and ultimately, preventing failure. This book can be used as a desktop reference book or as a self-study book, and can be used by engineering students and practicing engineers with or without some prior training in solid mechanics and/or mechanical metallurgy.

I want to sincerely thank the reviewers at ASM for their constructive comments. A special thanks to Steve Lampman for his effort in integrating handbook content in the area of brittle and ductile fractures, and the mechanism of intergranular fracture, into Sections 2.4.3 and 2.4.4. Last but not least, I wish to acknowledge the excellent work that Kathy Dragolich put in the coordination of this book.

A.F. Liu
West Hills, California

CHAPTER 1

Solid Mechanics of Homogeneous Materials

ENGINEERING MATERIALS can be conveniently grouped into five categories: metals and alloys, intermetallics, ceramics and glasses, polymers (plastics), and composites. General advantages of these materials are summarized in Fig. 1.1 along with the predominant type of bonding and spatial arrangement of their constituents. The bond type is related to the sharing of electrons due to differences in the number of electrons in the outer electron shells of individual atoms. This results in different types of bonding and bond strength (Table 1.1), including strong primary bonds (covalent and ionic bonds), intermediate-strength metallic bonds, and weaker secondary bonds (van der Waal bonds and hydrogen bonds). The type of bonding is an essential factor that influences the physical and mechanical properties of a substance, and makes each class of materials unique. The exception may be composites, which are not homogeneous materials. A composite is a kind of engineered material that consists of particulate fillers or strong/stiff fibers in a soft matrix.

Table 1.1 Bond energies for various materials

Bond type	Material	Bond energy	
		kJ/mol	kcal/mol
Ionic	NaCl	640	153
	MgO	1000	239
Covalent	Si	450	108
	C (diamond)	713	170
Metallic	Hg	68	16
	Al	324	77
	Fe	406	97
	W	849	203
van der Waals	Ar	7.8	1.8
	Cl_2	31	7.4
Hydrogen	NH_3	35	8.4
	H_2O	51	12.2

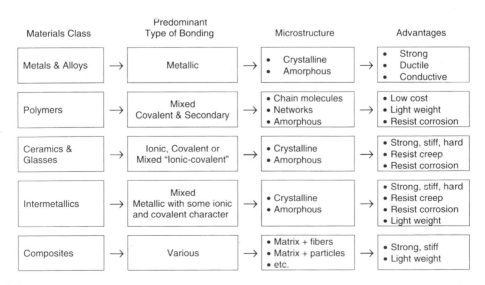

Fig. 1.1 General characteristics of major classes of engineering materials

The bonds between atoms also occur in various spatial orientations. The spatial arrangement of atoms may be random ("amorphous"), or the atoms may be periodically arranged either in a two-dimensional array (as in the case of a graphite sheet structure) or in a three-dimensional array of a crystal lattice. Over still greater distances, spatial arrangements may be completely amorphous, a single crystal, or polycrystalline (as is the case in most metals and alloys, where many individual crystalline grains are rotated with respect to one another). Microstructures may also be mixtures of crystalline and noncrystalline regions or have regions of oriented structures (as in the case of aligned carbon backbone chains in polymers).

Mechanical behavior depends on both the type of bonding and the long-range spatial arrangement of constituents in a material. Metals represent the majority of the pure elements and form the basis for the majority of the structural materials. The metallic bonds typically form crystalline grains, and the most common crystal structures in most engineering metals include face-centered cubic (fcc), body-centered cubic (bcc), or hexagonal close-packed (hcp) structures. Deformation and fracture mechanisms of metals are described in Appendix 1.

Ceramics and glasses, which have strong ionic-covalent chemical bonds, are very strong, stiff, and brittle. Crystal structures of ceramic materials are often more complex than those of metallic materials. High packing density cubic structures do exist in some ceramic materials (rock salt and cesium chloride structures), but the presence of more than one type of atom together with ionic bonding causes differences in critical stresses for ductile (slip) and brittle (cleavage) behavior compared to metallic materials. Qualitatively, the critical stresses for slip are so high relative to those required to propagate cleavage that these materials are considered to be inherently brittle.

The physical and mechanical properties of intermetallics fall in between ceramics and metals, but their bonding system is a combination of metallic, ionic, and/or covalent types. Intermetallics are also crystalline, and so the mechanisms of deformation and fracture can be discussed within the same framework as metals (e.g., the role of antishear twinning, microtwinning, etc.*). The microstructure of these materials controls microscale fracture mechanisms and fracture site initiation. In this book, intermetallics, such as TiAl, Ti_3Al, NiAl, and Ni_3Al, are treated the same as metals because they are used for high-temperature applications the same way as high-temperature superalloys.

Polymers, on the other hand, are long-chain molecules that may bond together by the weak van der Waals force, the stronger hydrogen bond, or the much stronger covalent bond. The polymer itself is a long chain of monomer units (typically made from compounds of carbon, but also from inorganic chemicals, such as silicates and silicones) that form a chain from strong covalent bonds between the monomer units. These long polymer chains may also bond together in various ways. For example, weak van der Waals bonds occur between polymer chains in the case of thermoplastic polymers, while stronger hydrogen or covalent bonds occur between the polymer chains in cross-linked thermosetting plastics.

Crystallinity is seldom complete in a polymeric material, although some limited crystallinity can occur in polymers. Crystallinity occurs when the polymer chains arrange themselves into an orderly structure. In general, simple polymers (with little or no side branching) crystallize very easily. Crystallization is inhibited in heavily cross-linked (thermoset) polymers and in polymers containing bulky side groups. The ability of polymeric materials to form crystalline solids depends in part on the complexity of the pendant side groups of atoms attached to the covalently bonded carbon atom backbone. Polymeric materials containing mixtures of crystalline and noncrystalline regions show decreased ductility and increased strength and stiffness as the degree of crystallinity increases. Conversely, polymers exhibit low strength, low stiffness, and susceptibility to creep when the temperature is above the glass-transition temperature (i.e., when the polymer structure becomes amorphous with pronounced viscoelastic behavior).

The aim of this book is to introduce the various types of mechanical behavior/properties of these materials and provide analytical tools to assess the strength and life of a structural design in terms of mechanical properties. The first task of designing a mechanical/structural component is to select a group of materials that can withstand the in-service environment. For instance, among all the materials, one type of alloy is known to have good resistance to corrosion; another type exhibits excellent high-temperature

---

*See Appendix 1 for a description of micromechanisms in the deformation and fracture of metals.

performance; a third group is specially developed for low-temperature application, and so on. Depending on the service environment, as well as the stiffness, strength, and life requirements, there is a little leeway left for material selection. Structural mechanics is an essential tool for design. It helps to understand what leads a material to respond to stresses in a certain way. An optimum design is the one that turns the initially unfavorable failure mode to a favorable one. This can be done by cleverly combining stress analysis and material selection. Much of this book will focus on various aspects of stress analysis and related topics.

1.1 Key Types of Mechanical Behavior

Most engineering materials of concern in this book can be regarded as deformable solids. They respond to applied loads in three different ways: elastic, plastic, and viscoelastic. Elastic and plastic behaviors deal with deformation (when the body is subjected to external force). Viscoelastic behavior actually is a combination of elastic and viscous behavior. A material having this property is considered to combine the features of a perfectly elastic solid and a perfect fluid. To honor those who first introduced the three types of ideal substances, the perfectly elastic solid is commonly called the Hooke solid; the perfectly plastic solid is known as the St. Venant solid. The perfect fluid, which is considered a viscous material, is called the Newtonian liquid.

Consider a cylindrical bar or rod fixed at one end and with an external force P pulling at the other end, as shown in Fig. 1.2(a). The two gage marks placed on the surface of the rod in its unstrained state will move apart as the length of the rod is elongated under the applied load. The deformation δ, i.e., the change in gage length to the original gage length ratio, is called the average strain and is represented by:

$$\varepsilon = \delta/L_0 \qquad \text{(Eq 1.1a)}$$

As shown in Fig. 1.2(b), the external load P is balanced by the internal resisting force $\sigma \cdot A$, where σ is the stress normal to the cutting plane and A is the cross-sectional area of the rod. Therefore, in equilibrium:

$$\sigma = P/A \qquad \text{(Eq 1.1b)}$$

If we gradually increase the strain of a Hooke solid, the stress is also increased proportionally. The relation of traction to strain is a linear line as shown in Fig. 1.2(c). This behavior is commonly known as Hooke's law. However, it need not be a perfectly straight line; the definition of elasticity only requires that on gradually diminishing strain the same curve is retraced, and that on complete release of the stress no permanent deformation remains.

For the St. Venant solid, we assume the force per unit of area that causes an extension or contraction must reach a certain finite value for the initiation of permanent distortion. When this characteristic load is reached, the substance starts to deform permanently, or to "yield," and continues so at this load. We further assume that the change in shape during increasing load is small. This case is then represented by drawing a horizontal straight line in the diagram in which the permanent extensions are plotted as abscissas and the forces per unit area as ordinates, as shown in Fig. 1.3(a). This behavior is commonly known as "rigid–perfectly plastic."

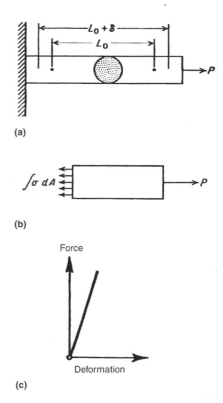

Fig. 1.2 Cylindrical bar subjected to axial load. (a) Linear elongation along the loading direction. (b) Free body diagram. (c) Linear relationship between force and elastic deformation (or stress and elastic strain)

Very few substances exhibit only one form of deformation. They usually behave in combination of any two of the three, and in a few cases all three. For example, viscoelasticity is a combination of viscosity and elasticity. However, the most common type of substance combines elasticity and plasticity. The theory of plasticity considers three types of substances that involve plastic flow, presented in idealized forms as: rigid ideal plastic material (Fig. 1.3a), ideal plastic material with an elastic region (Fig. 1.3b), and piecewise linear (strain-hardenable) material (Fig. 1.3c). As the assumption in material behavior changes from (a) toward (c), the complexity in mathematical representation of each of these models increases.

The curve in Fig. 1.3(b) represents an elastic–perfectly plastic material. It exhibits elastic behavior up to the yield stress, then plastic flow follows without further increase in stress. Upon release of stress (at some point), that part of the total deformation is regained. When a body is stressed, and lets the strain increase indefinitely, two things may happen. If the material behavior is plastic, after the yield stress is reached it deforms endlessly. If the material is elastic and brittle, it will break at a stress level corresponding to its breaking (fracture) stress. A more realistic representation of an elastic-plastic material is presented in Fig. 1.3(c). This stress-deformation diagram shows that at first the material deforms elastically. After the material reaches its yield point, increase in strain continues when the material is subjected to higher stresses. Still, Fig. 1.3(c) represents only a simplified version of a real material. A typical stress-strain curve of most metallics does not have a break between the elastic and the plastic regions; the transition is rather smooth (Fig. 1.3d). The inelastic portion of the curve may be very long or very short, depending on the ductility of the material. It may even become nil if the material is very brittle. Ductility and brittleness are not just the inherent characteristics of a material. Depending on many factors, such as the ambient environment (including temperature) and the active stress system, a normally ductile material may become (or act) brittle, and

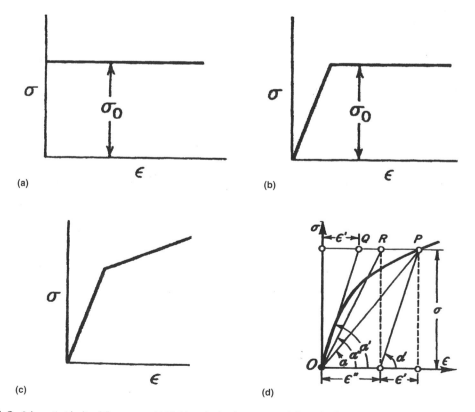

Fig. 1.3 Schematic idealized flow curves (a) Rigid–perfectly plastic material. (b) Perfectly plastic material with an elastic region. (c) Piecewise linear (strain-hardening) material. (d) Typical of engineering alloys

vice versa. These factors are discussed at great length throughout this book.

In Chapter 2, a more detailed account is given on the stress-strain curve, the information obtained from a tensile stress-strain curve, and what that information means in design analysis and failure prevention. Before presenting the stress-strain relations or getting into the characteristic details of the stress-strain curve, it is necessary to introduce some fundamental elastic constants that link the stresses and strains. First notice the initial elastic behavior of the tensile stress-strain curve in Fig. 1.3(d). The slope of the straight line OQ that relates ε and σ is called the modulus of elasticity in tension, or simply the Young's modulus, designated by the symbol E. In the case of compression, E is the modulus of elasticity in compression. Within the elastic limit, the modulus of elasticity in compression is the same as in tension. The linear relationship is known as Hooke's law. This law states that for a homogeneous and isotropic material, stress is proportional to strain as long as the stress does not exceed a limiting value. If the material is subjected to shear deformation, as in torsion, the ratio of the shearing stress to the shearing strain within the elastic range of stress is called the modulus of rigidity (or modulus of elasticity in shear), designated by the symbol G. The units of E and G are the same as stress. The strain and stress are related as follows:

$$\sigma = \varepsilon \cdot E \qquad \text{(Eq 1.1c)}$$

for tension or compression, and

$$\tau = \gamma \cdot G \qquad \text{(Eq 1.1d)}$$

for torsion, with τ and γ being the torsional stress and strain, respectively.

When a piece of structural member (or test bar) is loaded in tension, the axial elongation is accompanied by a lateral contraction. The absolute value of the ratio of the lateral strain to the axial strain is known as the Poisson ratio, and is designated by the symbol ν (or μ as it also appears in the literature). For a perfectly isotropic material, $\nu = 0.25$. For some metals ν can be as high as 0.33. For many ductile metals, $\nu = 0.3$. Within the elastic limit, the Poisson ratio in compression is the same as in tension. Physically, the material expands in the lateral direction during compression. The volume change in a material during deformation is given by:

$$\Delta V = (\varepsilon_x + \varepsilon_y + \varepsilon_z)V$$
$$= \left[\frac{1 - 2\nu}{E}(\sigma_x + \sigma_y + \sigma_z)\right]V \qquad \text{(Eq 1.2a)}$$

where V is the original volume and ΔV is the volume change. For many applications/models, constancy of volume is assumed. In other words, the numerical sum of strains in a three-dimensional solid is zero. That is:*

$$\varepsilon_x + \varepsilon_y + \varepsilon_z = 0 \qquad \text{(Eq 1.2b)}$$

It has been shown that the elastic constants E, G, and ν are not independent but are related by:

$$G = \frac{E}{2(1 + \nu)} \qquad \text{(Eq 1.3)}$$

This equation is derived on the basis of the two-dimensional, pure shear condition. The magnitude of the stress in one direction is equal, and opposite, of stress in the other direction. The two stresses are acted normal to each other.

It is also shown that in the plastic region ν increases with strain. When the plastic strain is very large, the value of ν approaches a limiting value of 0.5 for an incompressible material (Ref 1.2). Substituting 0.5 for ν in Eq 1.2(a), $\Delta V = 0$. Therefore, Eq 1.2(a) is reduced to Eq 1.2(b) in the large plastic region. The intermediate values of ν is given by:

$$\nu = \nu'' - (\nu'' - \nu')\frac{\cot \alpha'}{\cot \alpha} \qquad \text{(Eq 1.4)}$$

Here, α' is a constant angle, associated with E and the elastic strain ε'. The angle α is associated with the total strain (the elastic strain ε' plus the plastic strain ε''). The value of the angle α starts with α' and becomes smaller as the strain increases. When

$$\alpha = \alpha' \qquad \nu = \nu'$$
$$\alpha \to 0 \qquad \nu \to \nu''$$

Here, ν' is the elastic Poisson's ratio, and ν'' is a limiting value, which is 0.5 (see Fig. 1.3d).

To describe the deformation behavior of a viscous substance, we first need to describe the vis-

---

*Unless otherwise noted, derivation and proof of all the equations presented in this and subsequent sections are available in Ref 1.1 to 1.3, among other textbooks on solid mechanics.

cosity behavior of a Newtonian liquid. Consider a viscous substance, such as a heavy lubricating oil, sandwiched between two closely spaced parallel plates as shown in Fig. 1.4. One plate is moved relative to the second one in a direction parallel to the plates. It will be found that the force per unit area of the contact surface τ is proportional to the relative sliding velocity u and inversely proportional to the distance of the two plates. Thus, a fluid exerts an internal resistance against this motion. This shearing stress is expressed by:

$$\tau = \mu \cdot u/a \qquad \text{(Eq 1.5a)}$$

where a is the distance between the two plates. The material constant μ is called the coefficient of viscosity. Equation 1.5(a) can be expressed in a more general manner by letting rectangular coordinates X in the direction of motion and Y in the direction perpendicular to the plates:

$$\tau = \mu \cdot (\partial u/\partial y) \qquad \text{(Eq 1.5b)}$$

Now we can plot the tangential force per unit area τ versus u/a, or $\partial u/\partial y$. The quantity u/a, or $\partial u/\partial y$, is called "rate of shear." A typical plot of this behavior is shown in Fig. 1.5.

Now consider a viscous substance in the form of a rod (Fig. 1.2), but hang the rod vertically and attach a dead weight at its bottom end. The rod will be seen to stretch continuously very slowly at a constant velocity that is proportional to the attached weight. Let u be the displacement and v be the velocity in a point x at time t, assuming that $u = 0$ at $t = 0$. The permanent unit strain ε_x will increase proportionally with the time t with constant $\partial \varepsilon_x/\partial t$, where the subscript x denotes the direction of deformation parallel to the axial (length) direction of the rod. Therefore:

$$v = \partial u/\partial t, \qquad \varepsilon_x = \partial u/\partial x, \qquad \partial \varepsilon_x/\partial t = \partial v/\partial x \qquad \text{(Eq 1.5c)}$$

Figures 1.2 and 1.5 are analogies, as are Eq 1.1(d) and 1.4(b). Thus, we can write an equation for the viscous substance in analogy to Eq 1.1(c) as:

$$\sigma_x = \zeta \cdot (\partial v/\partial x) = \zeta \cdot (\partial \varepsilon_x/\partial t) = \zeta \cdot \dot{\varepsilon}_x \qquad \text{(Eq 1.5d)}$$

where the proportionality factor ζ may be called the coefficient of viscosity for tension. It is apparent that that ζ is in analogy to E, μ to G, and $\dot{\varepsilon}_x$ to ε_x. Since E is related to G, ζ is related to μ in the same way. If a value of $v = 0.5$ is substituted into Eq 1.2, $E = 3G$; likewise, for a viscous substance, $\zeta = 3\mu$. That is, the coefficient of viscosity for tension is equal to three times the coefficient of viscosity for shear.

As shown in more detail in this chapter, the stress and strain system in a three-dimensional elastic solid consists six components of stresses (three tension, three shear) and six components of strains. Also, because the stresses are the same in an elastic solid and a viscous substance, we know that the rate of deformation in a viscous material consists of six quantities. This means that both materials have the same form of equations. One group of the equations can be transformed into the other group if the quantities appearing in the stress-strain relations (ε_x, ε_y, ε_z, γ_{yz}, γ_{zx}, γ_{xy}, E, and G) are replaced for quantities of strain-rate relations in a viscous material ($\dot{\varepsilon}_x$, $\dot{\varepsilon}_y$, $\dot{\varepsilon}_z$, $\dot{\gamma}_{yz}$, $\dot{\gamma}_{zx}$, $\dot{\gamma}_{xy}$, ζ, and μ). An example of such substitution has just been shown using the analogies between Eq 1.1(c) and 1.5(d). Similarly, we can transform Eq 1.2 to suit a viscous material as:

$$\dot{\varepsilon}_x + \dot{\varepsilon}_y + \dot{\varepsilon}_z = 0 \qquad \text{(Eq 1.5e)}$$

The strength of a crystalline solid, such as metal or ceramic, is deformation dependent. The

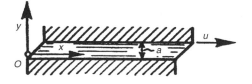

Fig. 1.4 Viscous flow between two parallel plates. Source: Ref 1.2

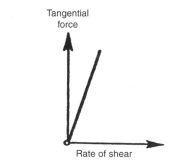

Fig. 1.5 Schematic of viscous deformation

strengths of viscoelastic polymers are deformation rate dependent. Crystalline solids may become deformation rate sensitive when stressed at high temperature for a period. Time-dependent creep is an important phenomenon in metals and ceramics at high temperatures. The viscoelastic behavior of many polymers at room temperature is well known, and hence its time dependence is a rule rather than an exception. This type of transformation (from strain to strain-rate behavior) is discussed later in the sections for creep, creep crack growth, and polymers.

1.2 Stress and Strain

1.2.1 Normal and Shear Stresses

Consider a cylindrical bar fixed at one end and with an external force P pulling at the other end, as shown in Fig. 1.2(a). If X is the loading direction, the stress over the circular cross section that is perpendicular to the axial load P will be:

$$\sigma_x - P/A \qquad \text{(Eq 1.6a)}$$

where A is cross-sectional area. Now consider the cross section pq in Fig. 1.6, which is inclined to the axis by an angle φ. As shown in Fig. 1.6, there is a normal stress component σ_n and a

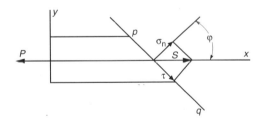

Fig. 1.6 Normal and shear stresses acting on a plane inclined to the loading direction

shear stress component τ acting on the pq plane. The normal stress component is:

$$\begin{aligned} \sigma_n &= P'/A' = (P \cdot \cos \varphi)/(A/\cos \varphi) \\ &= P \cdot \cos^2 \varphi/A \end{aligned} \qquad \text{(Eq 1.6b)}$$

The shear stress component is:

$$\tau = P \cdot \sin \varphi \cdot \cos \varphi/A \qquad \text{(Eq 1.6c)}$$

or

$$\tau = P \cdot \sin 2\varphi/2A \qquad \text{(Eq 1.6d)}$$

Note that this group of equations can also be used in the case of compression. Tensile stress is assumed positive and compression stress negative (see Fig. 1.7a and b). Therefore, we may only change the signs for σ_x and σ_n to indicate compression. The sign convention for the shearing stress would be that shown in Fig. 1.7(c) and (d). Thus, positive sign for shearing stress is taken when they form a couple in the clockwise direction and negative sign for the opposite direction. Similarly, applying this sign convention to a biaxial stress case, the shearing stress τ in Fig. 1.7(e) would be positive whereas τ_1 would be negative.

In another situation, when an out-of-plane traction force is applied at an arbitrary angle with respect to a plane, as shown in Fig. 1.8, the applied force is resolved into two components: one normal to the plane, another acting on the plane. The stress corresponding to the force acting normal to the plane is given by:

$$\sigma = \frac{P}{A} \cos \theta \qquad \text{(Eq 1.6e)}$$

The shear stress lying in the plane along the line OC, directly projected from P, will be:

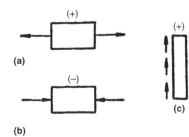

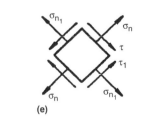

Fig. 1.7 Sign conventions for normal and shear stresses

$$\tau = \frac{P}{A} \sin \theta \qquad \text{(Eq 1.6f)}$$

This shear stress can be resolved into two components, along the X and Y directions, respectively.

$$X \text{ direction:} \quad \tau = \frac{P}{A} \sin \theta \cdot \sin \phi \qquad \text{(Eq 1.6g)}$$

$$Y \text{ direction:} \quad \tau = \frac{P}{A} \sin \theta \cdot \cos \phi \qquad \text{(Eq 1.6h)}$$

Thus, in general, a given plane may have one normal stress and two shear stresses acting on it.

Now consider a cube (Fig. 1.9), its size is infinitesimally small. For equilibrium calculations, assume the traction is applied through a point at the centroid of the cube. As shown in Fig. 1.9, one normal stress and two shearing stresses act on each face of the cube. The notations of these stresses are as depicted: σ for the normal stresses and τ for the shear stresses. The subscript for σ denotes the direction of the stress. The first subscript for τ indicates the plane on which the shear stress lies and the axis to which this plane is normal. The second subscript for τ denotes the direction. Two subscripts for the normal stresses follow the same notation system. For example, σ_y or σ_{yy} is the stress perpendicular to the plane normal to the Y-axis, acting in the Y-direction. In the one-subscript system, the "plane normal to the Y-axis" is implied. The component τ_{yx} is the shear stress lying on the plane normal to the Y-axis, acting in the X-direction.

Again, as shown in Fig. 1.9, one normal and two shearing stresses act on each face of the cube. The notations and magnitude of each pair of components (acting on the back-to-back faces) are equal. However, each pair of these stresses acts on opposite directions. Thus, the following nine stress components are parallel to the three coordinate axes:

$$\begin{bmatrix} \sigma_x & \tau_{xy} & \tau_{xz} \\ \tau_{yx} & \sigma_y & \tau_{yz} \\ \tau_{zx} & \tau_{zy} & \sigma_z \end{bmatrix}$$

Their positive directions are shown as solid arrows in Fig. 1.9, for the three faces of the element the external normals of which point in the directions of the positive X, Y, and Z axes.

From the condition of equilibrium with respect to moments of the forces transmitted through the six faces of the cube, the six components of the shearing stresses must satisfy the equalities:

$$\tau_{xy} = \tau_{yx}, \qquad \tau_{yz} = \tau_{zy}, \qquad \tau_{zx} = \tau_{xz} \qquad \text{(Eq 1.7)}$$

Thus, the state of stress at a point is completely described by six components: three normal stresses and three shear stresses (Ref 1.2, 1.3).

Many engineering problems are two dimensional rather than three dimensional. For example, stresses may occur on the X-Y plane of a thin plate when forces are applied at the boundary, parallel to the plane of the plate. The stress components σ_z, τ_{xz}, and τ_{yz} are zero on both plate faces. For now, we can also assume that the Z-directional stresses are zero within the plate. Thus, the state of stress is then specified by σ_x, σ_y, and τ_{xy} only, and is called the "plane stress" condition. However, as will be shown later, the

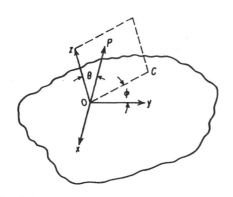

Fig. 1.8 Resolution of traction force into stress components

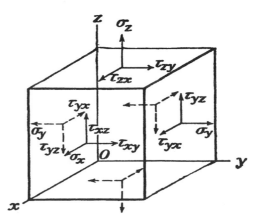

Fig. 1.9 Components of stresses

strain in the Z-direction is a nonzero quantity. Also, the plane stress assumption requires the nonzero stresses to be only functions of x and y.

Consider an element lying on the X-Y plane subjected to stresses σ_x and σ_y, and shear stress $\tau_{xy}(=\tau_{yx})$. Figure 1.10 shows a part of this element with an oblique plane intersecting the X and Y axes. The oblique plane can be at any orientation with respect to the X and Y axes. In Fig. 1.10, it is shown at an angle α with the Y-axis. The stresses acting on the oblique plane would be σ and τ, as shown. These stresses are given by:

$$\sigma = \frac{1}{2}(\sigma_x + \sigma_y) + \frac{1}{2}(\sigma_x - \sigma_y)\cos(2\alpha)$$
$$+ \tau_{xy}\sin(2\alpha) \qquad \text{(Eq 1.8a)}$$

$$\tau = -\frac{1}{2}(\sigma_x - \sigma_y)\sin(2\alpha) + \tau_{xy}\cos(2\alpha)$$
$$\text{(Eq 1.8b)}$$

The characteristics of these equations can be summarized as:

- The maximum and minimum values of normal stress on the oblique plane through the center of the element occur when the shear stress is zero.
- The maximum and minimum values of both normal stress and shear stress occur at angles that are 90° apart.
- The maximum shear stress occurs at an angle halfway between the maximum and minimum normal stresses.
- The variation of normal stress and shear stress occurs in the form of a sine wave, with a period of $\alpha = 180°$.

1.2.2 Linear and Shear Strains

The circular bar shown in Fig. 1.2(a) will deform linearly corresponding to the applied axial load. The average linear strain, which is a di-

mensionless quantity, is the ratio of the change in length to the original length (Eq 1.1a). This is regarded as the linear strain. In a two-dimensional problem, consider a rectangle on the X-Y plane (Fig. 1.11). When the rectangle is subjected to in-plane loading, the shear stresses will deform the rectangle as indicated by the dotted lines. This shearing action is called the "simple shear." The angular change in a right angle is known as "shear strain." As illustrated in Fig. 1.11, the angle at point A, which is originally a right angle, is reduced by a small amount θ. The shear strain γ is equal to the displacement a divided by the distance between the planes h. The ratio a/h is also the tangent of the angle through which the element has been rotated. That is:

$$\gamma = a/h = \tan\theta \qquad \text{(Eq 1.9a)}$$

For the small angles usually involved, the tangent of the angle and the angle (in radius) are equal. Therefore, shear strains are often expressed as angles of rotation:

$$\gamma = \theta \qquad \text{(Eq 1.9b)}$$

This representation of shear strain is also applicable to the three-dimensional problem, because the X-Y plane can be considered one face of the cube.

The entire strain system for the three-dimensional case is the same as the stress system. That is, six components of strain are sufficient to define the strains at a point. The components of strains corresponding to the component of stresses are, similarly: ε_x, ε_y, ε_z, γ_{xy}, γ_{yz}, and γ_{zx}.

1.2.3 Stress-Strain Relations

In a one-dimensional problem, and referring to an experimentally established phenomenon

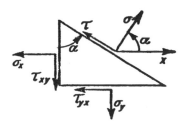

Fig. 1.10 Stresses on an oblique plane in plane stress

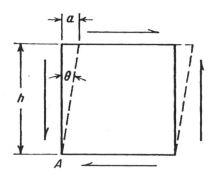

Fig. 1.11 Shear strain

(Fig. 1.2, 1.3), elastic stress and elastic strain are related simply by:

$$\varepsilon_x = \sigma_x/E \qquad \text{(Eq 1.10a)}$$

where E is the modulus of elasticity in tension. Extension of the element in the X-direction is accompanied by lateral contractions in the Y and Z directions. Therefore:

$$\varepsilon_y = \varepsilon_z = -v \cdot \varepsilon_x = -v \cdot \sigma_x/E \qquad \text{(Eq 1.10b)}$$

where v is Poisson's ratio. Equations 1.10(a) and 1.10(b) can be used also for simple compression. Within the elastic limit, the modulus of elasticity and Poisson's ratio in compression are the same as in tension. Physically, the material expands in the lateral direction during compression.

For a three-dimensional problem, that is, when the stress system includes σ_x, σ_y, and σ_z:

$$\varepsilon_x = \frac{1}{E}[\sigma_x - v(\sigma_y + \sigma_z)]$$

$$\varepsilon_y = \frac{1}{E}[\sigma_y - v(\sigma_z + \sigma_x)]$$

$$\varepsilon_z = \frac{1}{E}[\sigma_z - v(\sigma_x + \sigma_y)] \qquad \text{(Eq 1.11a)}$$

In many practical problems, all displacements can be considered to be limited to the X-Y plane, so that strains in the Z-direction can be neglected in the analysis. Examples for such a system are a very thick plate, or constraints applied to restrict the plastic flows in the Z-direction. In analogy to plane stress, plane strain is mathematically defined as $\varepsilon_z = \gamma_{xz} = \gamma_{yz} = 0$, which ensures that $\tau_{xz} = \tau_{yz} = 0$ and $\sigma_z = v(\sigma_x + \sigma_y)$. Therefore, similar to plane stress we need consider only the same three stress components: σ_x, σ_y, and τ_{xy}. The strain-stress relationships for the condition of plane strain thus become:

$$\varepsilon_x = \frac{1 + v}{E}[(1 - v)\sigma_x - v\sigma_y]$$

$$\varepsilon_y = \frac{1 + v}{E}[(1 - v)\sigma_y - v\sigma_x] \qquad \text{(Eq 1.11b)}$$

The modulus of rigidity (G) has been defined as the ratio of the shearing stress to the shearing strain. According to Timoshenko and Goodier (Ref 1.3), the relations between the shearing strain components and shearing stress components then are:

$$\gamma_{xy} = \tau_{xy}/G, \qquad \gamma_{yz} = \tau_{yz}/G, \qquad \gamma_{zx} = \tau_{zx}/G \qquad \text{(Eq 1.12)}$$

The elongations (Eq 1.11a) and the distortions (Eq 1.12) are independent of each other. Hence the general case of strain, produced by three normal and three shearing components of stress, can be obtained by superposition. Addition of the three equations in Eq 1.11(a) results in:

$$\varepsilon_x + \varepsilon_y + \varepsilon_z = \frac{1 - 2v}{E}(\sigma_x + \sigma_y + \sigma_z) \qquad \text{(Eq 1.13)}$$

The left-hand term of Eq 1.13 is called the volume strain, designated by the symbol Λ. Using Eq 1.13 and solving Eq 1.11(a) for σ_x, σ_y, and σ_z, we find the stress in terms of strain:

$$\sigma_x = \frac{vE}{(1 + v)(1 - 2v)}\Lambda + \frac{E}{1 + v}\varepsilon_x$$

$$\sigma_y = \frac{vE}{(1 + v)(1 - 2v)}\Lambda + \frac{E}{1 + v}\varepsilon_y$$

$$\sigma_z = \frac{vE}{(1 + v)(1 - 2v)}\Lambda + \frac{E}{1 + v}\varepsilon_z \qquad \text{(Eq 1.14)}$$

For the case of plane stress ($\sigma_z = 0$), this group of equations reduces to:

$$\sigma_x = \frac{E}{1 - v^2}(\varepsilon_x + v\varepsilon_y) \qquad \text{(Eq 1.15a)}$$

$$\sigma_y = \frac{E}{1 - v^2}(\varepsilon_y + v\varepsilon_x) \qquad \text{(Eq 1.15b)}$$

1.3 Principal Stresses and Principal Strains

In the vast majority of practical applications, a two-or three-dimensional state of stress exists rather than the simple one-dimensional or uniaxial type shown in Fig. 1.12(a). The 2D and 3D states of stress are shown in Fig. 1.12(b) and (c). These latter systems, called "state of combined stress," greatly affect the strength and ductility

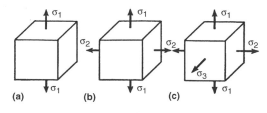

(a)   (b)   (c)

Fig. 1.12 State of stress

of materials. In general, each plane passing through a point of the body will be acted on by both normal and shearing stresses. The problem is, however, greatly simplified if the planes cutting out the element of material are so rotated that only normal stresses appear on the faces of the element.

This has been done in Fig. 1.12. In other words, the stresses shown in Fig. 1.12 coincide with the X, Y, or Z axis. It is these normal stresses acting on planes on which no shearing stresses exist that are of greatest significance in stress analysis problems and in the relation of stress to mechanical properties. Those normal stresses acting on planes containing no shearing stresses are called "principal stresses." At a point in a stressed body, there are three principal stresses acting on mutually perpendicular planes. They are designated by σ_1, σ_2, and σ_3, where the subscripts refer to the directions of the stresses. Customarily, these symbols also are used for labeling the magnitudes of the principal stresses. The subscript 1 indicates the largest principal stress, and the subscripts 2 and 3 indicate the intermediate and the smallest, respectively. The graphical method described below is the simplest way to derive a set of equations for the principal stresses in terms of the normal stresses and shearing stresses of the Cartesian coordinate. This method is best known as the Mohr's circle. The next section describes the method with a two-dimensional plate (biaxial stress) problem, demonstrating how to construct a Mohr's circle to determine σ_1, σ_2, and τ_{max} provided that σ_x, σ_y, and τ_{xy} are given.

1.3.1 Mohr's Circle for Stresses

Consider σ_x and σ_y acting normal to the edges of a rectangular plane element as shown in Fig. 1.13(a). Shearing stresses also act around the edges of this element so that, by definition, none of these stresses is a principal stress. To find the magnitudes and the directions of the principal stresses σ_1 and σ_2, construct a Mohr's circle as follows. Referring to Fig. 1.13(b), select a scale for stresses and measuring normal components along the horizontal axis and shearing components along the vertical axis. Then, take point O as the origin and plot the positive (tensile) stresses to the right, negative (compressive) stresses to the left. The first step is to set the stresses σ_x and σ_y to scale at points E and E_1.. The second step is to let ED and E_1D_1 equal the positive and negative values of τ_{xy}. The third step is to connect DD_1 and use DD_1 as the diameter of the Mohr's circle. The intersection of this diameter with the X-axis gives the center C of the circle, so that the circle can be readily constructed. The magnitudes of the principal stresses are defined by points A and B because the shear stresses are zero at these points. The maximum shearing stress is equal to the radius of the circle. Using the Mohr's circle, the formulas for calculating σ_1, σ_2, and τ_{max} can be obtained:

$$\sigma_1 = \overline{OA} = \frac{\sigma_x + \sigma_y}{2}$$
$$+ \sqrt{\left(\frac{\sigma_x - \sigma_y}{2}\right)^2 + \tau_{xy}^2} \qquad \text{(Eq 1.16a)}$$

$$\sigma_2 = \overline{OB} = \frac{\sigma_x + \sigma_y}{2}$$
$$- \sqrt{\left(\frac{\sigma_x - \sigma_y}{2}\right)^2 + \tau_{xy}^2} \qquad \text{(Eq 1.16b)}$$

$$\tau_{max} = \frac{\sigma_1 - \sigma_2}{2}$$
$$= \sqrt{\left(\frac{\sigma_x - \sigma_y}{2}\right)^2 + \tau_{xy}^2} \qquad \text{(Eq 1.16c)}$$

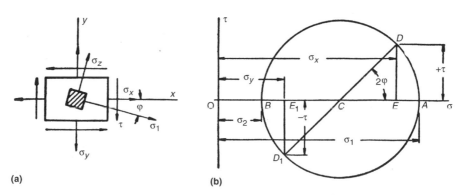

(a) (b)

Fig. 1.13 Two-dimensional Mohr's circle. Note that τ in the figure is the same as τ_{xy}.

The directions of the principal stresses can also be obtained from the Mohr's circle. The angle DCA, labeled 2φ, is the double angle between one of the principal axes with respect to the axis of reference. The X-axis is usually taken as the reference axis. For the calculation of the numerical value of the angle φ we have, from Fig. 1.13:

$$|\tan 2\varphi| = \frac{\overline{DE}}{\overline{CE}} = \frac{\tau_{xy}}{(\sigma_x - \sigma_y)/2} \qquad \text{(Eq 1.17)}$$

This equation defines two mutually perpendicular directions for which the shear stress is zero (the shear stresses along φ and $90° + \varphi$ are zero). These directions are called the "principal directions of stress." The normal stresses corresponding to these directions are the "principal stresses." The sign convention for φ is explained as follows: When we constructed the Mohr's circle, we took the τ-axis positive in the upward direction and considered shearing stresses as positive when they gave a couple in the clockwise direction. So, we set τ_{xy} upward at point E. However, by definition the lower half-circle of Fig. 1.13(b) represents the stress variation for all values of 2φ between 0° and 180° (or, simply, $0 < \varphi < \pi/2$). The upper half-circle gives stress for $0 > \varphi > -\pi/2$.* Therefore, the angle between the X-axis and σ_1 in this case (Fig. 1.13a) is negative; that is, φ moves clockwise from σ_x.

In summary, when we know σ_x, σ_y, and τ_{xy} we can use Eq 1.16 and 1.17, or construct a Mohr's circle, to determine the magnitude of the principal and the maximum shear stresses, and the direction of these stresses. On the other hand, if we know the quantities of the latter, we can construct a Mohr's circle in a reverse manner to determine σ_x, σ_y, and τ_{xy}. The 2D Mohr's circle can be drawn the same way for the one-dimensional stress system. In that case, σ_x is the same as σ_1. The 1D Mohr's circle would then be used for determining σ_n and τ on oblique planes respective to the loading direction.

1.3.2 Mohr's Circle for Strains

Instead of using computational methods, stresses on the surface of a structural piece can be determined on the basis of experimentally measured strains. The strain gage method is the most popular experimental technique. However, data reduction is not as simple as one might think. Rather than detailing the strain gage technique, this section touches on a few key points regarding conversion of measured strains to actual strains in a multidirectional strain field.

A Mohr's circle for strains can be constructed using the same method for constructing a Mohr's circle for stresses. In a two-dimensional plane problem, the Mohr's circle for strains is largely useful for determining principal strains, or strains along any desired orientation, from experimentally measured strains. Then these strains can be converted to stresses through the stress-strain relationships described earlier.

In general, the state of strain is completely determined if ε_x, ε_y, and γ_{xy} can be measured. However, strain gages can make only direct readings of linear strain. Shear strains must be determined indirectly. Thus, three independent measurements of linear strain in different directions are needed in order to calculate the magnitudes and directions of the two principal stresses. Strain gages are usually made up as a set in the form of a "rosette." The most common type consists of three gages positioned at relative directions. The one with its legs placed 45° apart from one another is called the rectangular rosette. Figure 1.14(a) shows a general arrangement of a three-gage rosette. The three axes of its legs (A, B, and C) are placed at arbitrary angles in relation to a pair of reference axes, OX and OY (90° apart). If corresponding linear strains ε_a, ε_b, and ε_c are measured in their respective directions, one can calculate the linear and shearing strains, ε_x, ε_y, and γ_{xy}, corresponding to the OX and OY axes of reference. The values of ε_x, ε_y, and γ_{xy} are calculated in terms of the measured strains ε_a, ε_b, and ε_c from the following set of simultaneous equations (Ref 1.4):

$$\varepsilon_a = \varepsilon_x \cos^2 \theta_a + \varepsilon_y \sin^2 \theta_a + \gamma_{xy} \sin \theta_a \cos \theta_a$$
$$\varepsilon_b = \varepsilon_x \cos^2 \theta_b + \varepsilon_y \sin^2 \theta_b + \gamma_{xy} \sin \theta_b \cos \theta_b$$
$$\varepsilon_c = \varepsilon_x \cos^2 \theta_c + \varepsilon_y \sin^2 \theta_c + \gamma_{xy} \sin \theta_c \cos \theta_c$$
$$\text{(Eq 1.18)}$$

The principal strains would then be calculated from the expressions:

$$\varepsilon_1 = \frac{\varepsilon_x + \varepsilon_y}{2}$$
$$+ \sqrt{\left(\frac{\varepsilon_x - \varepsilon_y}{2}\right)^2 + \frac{1}{4}\varepsilon_{xy}^2} \qquad \text{(Eq 1.19a)}$$

---

*As stated in Ref 1.3, this rule is used only in the construction of Mohr's circle. Otherwise, the rule given in Fig. 1.10 still holds.

$$\varepsilon_2 = \frac{\varepsilon_x + \varepsilon_y}{2}$$
$$- \sqrt{\left(\frac{\varepsilon_x - \varepsilon_y}{2}\right)^2 + \frac{1}{4}\varepsilon_{xy}^2} \qquad \text{(Eq 1.19b)}$$

By changing designation subscripts x and y to 1 and 2, respectively (Eq 1.15), the magnitudes of the principal stresses σ_1 and σ_2 can be calculated. The directions can be determined from the following relationship:

$$\tan(2\varphi) = \frac{\gamma_{xy}}{\varepsilon_x - \varepsilon_y} \qquad \text{(Eq 1.20)}$$

A Mohr's circle for strains can be constructed the same way as that in Fig. 1.13. This time, the linear strains are plotted as the abscissa and $\gamma_{xy}/2$ is plotted as the ordinate.

The following demonstrates how to determine ε_1 and ε_2 (as well as σ_1 and σ_2) using the directly measured strains ε_a, ε_b, and ε_c. Because the demonstration must be case specific, only an example for the rectangular rosette with three gages is given. In Fig. 1.14(b), the three legs for the rectangular rosette are OA, OB, and OC, 45° apart between OA and OB, and between OB and OC. If the OA axis of the rosette (Fig. 1.14b) is taken as the reference, then OA is considered coincident with OX. For the arrangement of the strain gage axes in the rectangular rosette with $\theta_a = 0°$, $\theta_b = 45°$, and $\theta_c = 90°$, Eq 1.18 reduces to:

$$\varepsilon_x = \varepsilon_a \qquad \varepsilon_y = \varepsilon_c \qquad \text{(Eq 1.21a)}$$

and

$$\varepsilon_b = \varepsilon_a/2 + \varepsilon_c/2 + \gamma_{xy}/2 \qquad \text{(Eq 1.21b)}$$

or

$$\gamma_{xy} = 2\varepsilon_b - (\varepsilon_a + \varepsilon_c) \qquad \text{(Eq 1.21c)}$$

Subsequently, Eq 1.19(a) and (b) can be expressed as:

$$\varepsilon_1 = \frac{\varepsilon_a + \varepsilon_c}{2}$$
$$+ \frac{1}{2}\sqrt{(\varepsilon_a - \varepsilon_c)^2 + [2\varepsilon_b - (\varepsilon_a + \varepsilon_c)]^2}$$
$$\text{(Eq 1.22a)}$$

$$\varepsilon_2 = \frac{\varepsilon_a + \varepsilon_c}{2}$$
$$- \frac{1}{2}\sqrt{(\varepsilon_a - \varepsilon_c)^2 + [2\varepsilon_b - (\varepsilon_a + \varepsilon_c)]^2}$$
$$\text{(Eq 1.22b)}$$

Comparing Eq 1.19(a) and (b) to Eq 1.22(a) and (b), and considering the fact that OA is perpendicular to OC and OB lies in the middle and forms 45° angles with the other gages, a Mohr's circle is constructed as shown in Fig. 1.15. The dimensions A, B, C, D, and E are defined as:

$$A = \frac{\varepsilon_a + \varepsilon_c}{2} \qquad \text{(Eq 1.23a)}$$

$$B = \frac{1}{2}\sqrt{(\varepsilon_a - \varepsilon_c)^2 + [2\varepsilon_b - (\varepsilon_a + \varepsilon_c)]^2} \qquad \text{(Eq 1.23b)}$$

$$C = \frac{\varepsilon_a - \varepsilon_c}{2} \qquad \text{(Eq 1.23c)}$$

$$D = \varepsilon_b - (\varepsilon_a + \varepsilon_c)/2 = [2\varepsilon_b - (\varepsilon_a + \varepsilon_c)]/2 \qquad \text{(Eq 1.23d)}$$

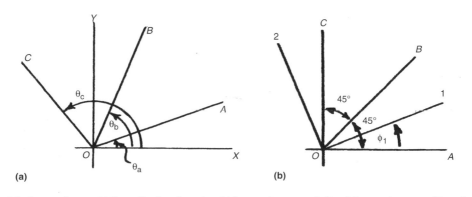

Fig. 1.14 Rosette diagram. (a) Generalized configuration. (b) Rectangular rosette. A, B, and C are strain gage positions; 1 and 2 are principal stress directions

$$E = \varepsilon_a - \varepsilon_c \qquad \text{(Eq 1.23e)}$$

Substituting Eq 1.21(a) and (c) into Eq 1.20 yields the expression:

$$\tan(2\varphi) = \frac{2\varepsilon_b - (\varepsilon_a + \varepsilon_c)}{\varepsilon_a - \varepsilon_c} \qquad \text{(Eq 1.24)}$$

which yields two values of φ. Each value relates to a principal stress axis. That is, φ_1 corresponds to σ_1, and φ_2 corresponds to σ_2. The following rules make up the sign convention for determining φ_1 and φ_2:

- When $\varepsilon_b > (\varepsilon_a + \varepsilon_c)/2$, φ_1 lies between $0°$ and $+90°$, measured positive in the counter-clockwise direction from the positive OA axis of the strain rosette to the positive $O1$ axis, which corresponds to the direction of σ_1.
- When $\varepsilon_b < (\varepsilon_a + \varepsilon_c)/2$, φ_1 lies between $0°$ and $-90°$.
- When $\varepsilon_b = (\varepsilon_a + \varepsilon_c)/2$ and $\varepsilon_a > \varepsilon_c$, $\varepsilon_a = \varepsilon_1$ and $\varphi_1 = 0$.

- When $\varepsilon_b = (\varepsilon_a + \varepsilon_c)/2$ and $\varepsilon_a < \varepsilon_c$, $\varepsilon_a = \varepsilon_2$ and $\varphi_1 = \pm 90°$.
- $\varphi_2 = 90° + \varphi_1$.

The proofs of these rules are given in Ref 1.4.

The principal stress can now be determined. Actually, in most cases, one does not need to know the numerical values of the principal strains; therefore, a little time and effort can be saved by using the directly measured strains. The expressions for σ_1, σ_2, and $\tau_{\max}$ have been given by Murray and Stein (Ref 1.4) as:

$$\sigma_1 = E\left\{ \frac{\varepsilon_a + \varepsilon_c}{2(1 - \nu)} + \frac{1}{2(1 + \nu)} \\ \cdot \sqrt{(\varepsilon_a - \varepsilon_c)^2 + [2\varepsilon_b - (\varepsilon_a + \varepsilon_c)]^2} \right\}$$

(Eq 1.25a)

$$\sigma_2 = E\left\{ \frac{\varepsilon_a + \varepsilon_c}{2(1 - \nu)} - \frac{1}{2(1 + \nu)} \\ \cdot \sqrt{(\varepsilon_a - \varepsilon_c)^2 + [2\varepsilon_b - (\varepsilon_a + \varepsilon_c)]^2} \right\}$$

(Eq 1.25b)

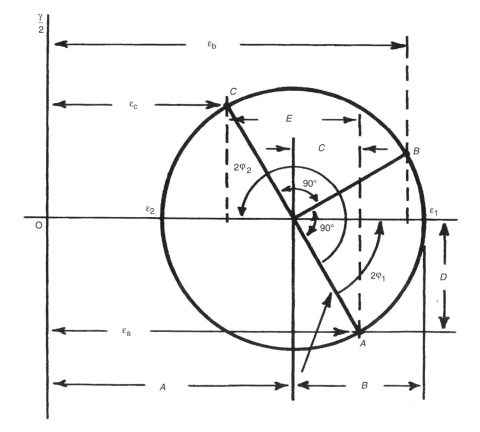

Fig. 1.15 Mohr's circle for the rectangular rosette with three observations of strain. Source: Ref 1.4

$$\tau_{max} = \frac{E}{2(1 + v)}$$
$$\cdot \sqrt{(\varepsilon_a - \varepsilon_c)^2 + [2\varepsilon_b - (\varepsilon_a + \varepsilon_c)]^2}$$
(Eq 1.26)

More equations like these, for the other types of rosette configurations, are available in the literature (Ref 1.4, 1.5).

1.4 Equivalent Stress and Equivalent Strain

In design and analysis, extensive use is made of test data developed from laboratory coupons under uniaxial loading condition. Structural/ machine components operate under multiaxial stress conditions. It is often helpful to convert data derived under one state of stress to another state of stress. This can be accomplished by use of the so-called tensile equivalent stresses and strains. When a structural piece is subjected to a set of combined loads, biaxial or triaxial, it must have a stress level of which the equivalent stress will take into account the effect of each stress component. The term "equivalent" simply means the combined stress is equivalent to the situation of simple tension so that the severity of the stress system can be evaluated by means of the material's tensile test data. Among the many proposed methods, the von Mises relationship is most appealing from the standpoint of continuum calculation and experimental data treatment, and is almost universally used by engineers. The von Mises relationship for equivalent stress is represented by (Ref 1.2):

$$\bar{\sigma} = \sqrt{\frac{1}{2}[(\sigma_1 - \sigma_2)^2 + (\sigma_2 - \sigma_3)^2 + (\sigma_3 - \sigma_1)^2]}$$
(Eq 1.27a)

The stress components in Eq 1.27(a) are principal stresses. For the condition of biaxial stresses (e.g., $\sigma_3 = 0$) Eq 1.27(a) is reduced to:

$$\bar{\sigma} = \sqrt{\sigma_1^2 - \sigma_1\sigma_2 + \sigma_2^2}$$
(Eq 1.27b)

In simple tension, the value of $\bar{\sigma}$ in these equations reduces to σ_1. Alternatively, these equations can also be expressed using the rectangular coordinate system:

$$\bar{\sigma} = \sqrt{\frac{1}{2}[(\sigma_x - \sigma_y)^2 + (\sigma_y - \sigma_z)^2 + (\sigma_z - \sigma_x)^2 + 6(\tau_{xy}^2 + \tau_{yz}^2 + \tau_{zx}^2)]}$$
(Eq 1.28a)

and

$$\bar{\sigma} = \sqrt{\sigma_x^2 - \sigma_x\sigma_y + \sigma_y^2 + 3\tau_{xy}^2}$$
(Eq 1.28b)

The equivalent strain is given by:

$$\bar{\varepsilon} = \frac{1}{3} \cdot \sqrt{2[(\varepsilon_1 - \varepsilon_2)^2 + (\varepsilon_2 - \varepsilon_3)^2 + (\varepsilon_3 - \varepsilon_1)^2]}$$
(Eq 1.29a)

and

$$\bar{\varepsilon} = \frac{1}{3} \cdot \sqrt{2\left[(\varepsilon_x - \varepsilon_y)^2 + (\varepsilon_y - \varepsilon_z)^2 + (\varepsilon_z - \varepsilon_x)^2 + \frac{3}{2}(\gamma_{xy}^2 + \gamma_{yz}^2 + \gamma_{zx}^2)\right]}$$
(Eq 1.29b)

When the axial components are absent, as in the state of pure shear (in-plane loading or torsion), the equivalent stress and strain become:

$$\bar{\sigma} = \sqrt{3}\tau$$
(Eq 1.30a)

and

$$\bar{\varepsilon} = \gamma/\sqrt{3}$$
(Eq 1.30b)

1.5 Stress Analysis of Monolithic Load-Carrying Members

The force system acting at a given cross section of a structural member can be resolved into:

- A force normal to the section and acting through the centroid, producing axial loading
- A couple lying in a plane normal to the plane of the cross section, producing flexure or bending
- A force lying in the plane of the cross section (perpendicular to the longitudinal axis of the member), producing cross shear

- A couple lying in the plane of the cross section, producing torsion

Because material failure (or fracture) is governed by tensile and shear properties, any type of loading should be resolved to tensile and shear components.

This book is not a textbook that covers all aspects of structural analysis. For handbook solutions, a library of stress formulas is available in Ref 1.6. This section presents a few simple yet typical cases. The sections that follow present some advanced structural analysis concepts. Because stress analysis procedures are case specific, generalized solutions are rare. Therefore, several key steps that lead to the final solution will be shown as required. Otherwise, derivation of all the equations in this section can be found in Ref 1.2, 1.3, 1.7, and 1.8, as well as other textbooks that cover strength of materials.

1.5.1 Axial Loading

The distribution of stress across any transverse section of an axially loaded member is usually assumed uniform. For a monolithic member, say a rod, subjected to tension loads at both ends (Fig. 1.2a), the stress is assumed uniformly distributed over the cross section. If the material is homogeneous and if the rod has a constant cross section, all the adjacent cross sections remain plane and parallel under load. The cross-sectional stress is simply the applied load divided by the cross-sectional area:

$$\sigma = P/A \qquad \text{(Eq 1.31)}$$

If the rod is long and slender, and is subjected to axial compression instead of tension, the load may be sufficiently large to cause buckling of the member, as shown in Fig. 1.16(a). After buckling occurs, two adjacent plane sections such as AA and BB, although they remain plane, do not remain parallel. These planes actually experience a combination of axial load and bending load as schematically illustrated in Fig. 1.16(b). The critical stress (the critical P/A value in this case) for a rod of length L is given by the well-known Euler equation:

$$\sigma_{cr} = K \frac{\pi^2 E}{(L/k)^2} \qquad \text{(Eq 1.32)}$$

Here K is a factor accounting for loading condition. Columns of differing boundary condi-

tions have different values of K. Typical values for K range from 0.25 to 4, $K = 0.25$ for one end free and one end fixed, $K = 1$ for both ends free, $K = 4$ for both ends fixed. The parameter k is the radius of gyration with respect to the axis about which buckling will occur, given in textbooks of statics as:

$$k = \sqrt{I/A} \qquad \text{(Eq 1.33)}$$

where I is "moment of inertia" given in textbooks that cover strength of materials. It is seen that for a given material the value of the critical stress depends on the magnitude of the ratio L/k, which is called the slender ratio. Euler's equation for columns is not valid for small values of slender ratio because of its inherent limitation that stresses must be below the proportional limit. A number of empirical and semiempirical formulas have been developed for a wide range of slender ratios. They can be found in textbooks that cover strength of materials.

1.5.2 Simple Bending

The stress analysis procedures covered in this chapter deal with deformation in the elastic range only, with one exception: bending in the plastic range, which is included due to its importance in structural design/analysis.

Elastic Bending. A beam or a rod can be bent by a couple of forces/moment applied at the ends of the beam, or by lateral forces acting on the beam (Fig. 1.17a). The axial stresses acting normal to the cross section of the beam resulting from the bending moment are schematically

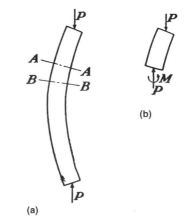

(b)

(a)

Fig. 1.16 Axial compression causing buckling

shown in Fig. 1.17(b). The conventional formula for the bending stress is:

$$\sigma_x = M \cdot y/I_Z \qquad\qquad \text{(Eq 1.34)}$$

where the subscript x indicates that the stress is in the X-direction, M is the bending moment (to be defined later), y is the distance from the neutral axis of the beam, and I_z is the cross-sectional moment of inertia about the Z-axis. Note that the geometry of the cross section shown in Fig. 1.17(b) is rectangular; its neutral axis is in the middle of the cross section. The resulting axial stresses along the depth direction are tension on one side and compression on the other side about the neutral axis, and are symmetrically balanced. For a nonsymmetric cross section, the neutral axis would be shifted and the axial stress distribution would be unbalanced.

The idealized loading conditions required by the bending formula are essentially realized by the center portion of a beam supported at the ends and carrying two equal and symmetrically placed loads. However, Eq 1.34 is universally applicable to any loading condition that produces bending, as long as the resulting stresses do not exceed the material's proportional limit.

Equation 1.34 is also based on several assumptions:

Statics

a. The resultant of the external forces is a couple that lies in, or is perpendicular to, a plane of symmetry of the cross section.
b. The beam is in equilibrium.

Geometry

c. The longitudinal axis of the beam is straight.
d. The beam has a constant cross section throughout its length.
e. A plane section that is normal to the longitudinal axis of the beam before the beam is bent remains plane after the beam is bent.
f. The beam bends without twisting.

Material Properties

g. The beam material is homogeneous and isotropic.
h. The stresses do not exceed the proportional limit of the material.

The most common types of bending problems encountered by engineers are the three-point bend and the four-point bend configurations shown in Fig. 1.18 and 1.19, respectively. These figures simply show that a simple supported beam is subject to a single force or a pair of forces acting opposite to the supporting forces. The bending moment M and the shearing force V acting at a given location of the beam can be determined as follows.

Consider, for example, a simply supported beam with a single concentrated load P (see Fig. 1.18). To calculate the reaction forces R_1 and R_2 at the supports, choose the point of support at

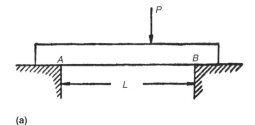

(a)

(b)

Fig. 1.17 Bending of a beam. (a) A couple of forces applied at the ends of the beam, or lateral force acting on the supports at the ends of the beam. (b) Axial stress distribution in a beam subjected to bending moment M

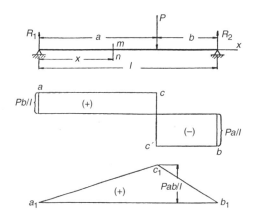

Fig. 1.18 Three-point bending

the right as the pivot point. All bending moments about that point should be balanced. Therefore:

$$R_1 \cdot l = P \cdot b \qquad \text{(Eq 1.35a)}$$

Thus

$$R_1 = Pb/l \qquad \text{(Eq 1.35b)}$$

and

$$R_2 = Pa/l \qquad \text{(Eq 1.35c)}$$

Taking a cross section mn to the left of P, it can be concluded that at such a cross section:

$$V = R_1 = Pb/l \qquad \text{(Eq 1.36)}$$

and

$$M = R_1 \cdot x = Pbx/l \qquad \text{(Eq 1.37)}$$

These equations imply that the shearing forces remain constant at locations to the left of the applied force. The bending moment varies in proportion to the distance x. The distributions of V and M are shown in Fig. 1.18, where V and M are represented by the straight lines ac and a_1c_1, respectively. The diagram for the shearing force is called the shear diagram, and the diagram for the bending moment is known as the moment diagram. For a cross section to the right of the applied load:

$$V = R_1 - P = -Pa/l \qquad \text{(Eq 1.38)}$$

and

$$M = R_1 \cdot x - P(x - a) = \frac{Pbx}{l} - P(x - a) \qquad \text{(Eq 1.39)}$$

with x always being the distance from the left end of the beam. The shearing force for this portion of the beam remains constant and negative. Note that the area on the left side of the shearing force diagram is balanced by the area on the right side.

Following this procedure, we can easily determine the shear and the moment diagrams for the four-point bend configuration (Fig. 1.19). This procedure can be applied to any other loading condition that involves only four variables: number of Ps, the magnitude for each P, the po-

sition of each P, and the span between the supports. Note that the moment diagram in Fig. 1.19 is a unique feature of the four-point bending condition. It always results in pure bending between the two applied forces. Pure bending means there is a constant bending moment acting on that segment of the bar. In this case, when the two forces are symmetrically placed with respect to the center of the bar, $R_1 = P$ and $R_2 = P$. A constant bending moment exists in between the two forces; its value is $(P \cdot a)$. No shearing force is associated with pure bending.

For bending of a cantilever beam, the same method described previously is used to construct the shearing force and bending moment diagrams (Fig. 1.20). Measuring x from the left end of the beam and considering the portion to the left of the load $P_2 (0 < x < a)$:

$$V = -P_1 \quad \text{and} \quad M = -P_1 \cdot x \qquad \text{(Eq 1.40)}$$

The minus signs in these expressions follow the rule depicted in Fig. 1.21(a) and (b) and 1.22(a) and (b). For the right portion of the beam $(a < x < l)$:

$$V = -P_1 - P_2 \text{ and } M = -P_1 \cdot x - P_2(x - a) \qquad \text{(Eq 1.41)}$$

The corresponding diagrams of shearing force and bending moment are shown in Fig. 1.20(b) and (c).

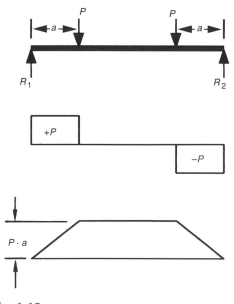

Fig. 1.19 Four-point bending

Plastic Bending. The bending formula for symmetrical bending (Eq 1.34) was derived on the basis of the assumption that stress and strain are proportional. If the proportional limit of the material is exceeded, assumption (h) is no longer valid. However, the relationship between the stress and the moment in a beam can be determined if we assume that the bending stress-strain characteristic of the beam is similar to the stress-strain characteristics in tension (or compression). Assumptions (a) through (g) can still be considered valid.

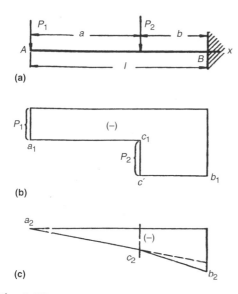

(a)

(b)

(c)

Fig. 1.20 Bending of a cantilever beam

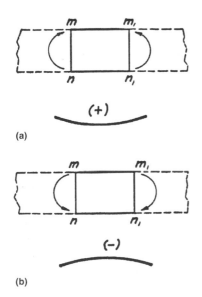

(a)

(b)

Fig. 1.21 Sign conventions for bending moment

When the material is loaded into the nonlinear range, the relationship between stress and normal strain can be represented by the general equation:

$$\sigma = f(\varepsilon) \qquad \text{(Eq 1.42)}$$

A possible stress-strain relationship that is presumably determined from a tension and a compression test is illustrated in Fig. 1.23. Since experimental evidence shows that a plane section before bending remains plane after bending, even after plastic action is taking place, the unit strain on a cross section of a beam is proportional to the distance from the neutral axis. Figure 1.24 presents a schematic illustration. Hence,

$$\varepsilon_y/\varepsilon_0 = y/c \qquad \text{(Eq 1.43)}$$

where ε_y is the unit strain at the point a distance y from the neutral axis, and ε_0 is the unit strain at a reference point located at a distance c from the neutral axis. In other words, the strain distribution across the beam is linear, as shown in Fig. 1.24(a). This relationship is valid for both elastic and plastic ranges. The stress distribution corresponding to the strain distribution of Fig. 1.24(a) is shown in Fig. 1.24(b), which is obtained by combining Fig. 1.23 and 1.24(a). It is seen that while the strain distribution is always linear, the stress distribution is nonlinear. However, nonlinearity starts from the outermost fibers of the beam, where the bending stress is the highest at a given load level. While the applied load level is not high enough to cause yielding across the entire cross section of the beam, there still is a linear region of stress in the middle of the beam. This can be called a stage of elastic-plastic bending. The distribution of bending stresses then become totally nonlinear when the stress-strain relationship is fully plastic.

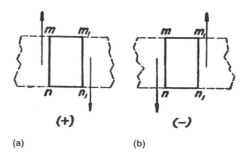

(a) **(b)**

Fig. 1.22 Sign conventions for shearing force

Here are a few key steps for obtaining an equation for the stresses in plastic bending. Begin with a beam that has a cross section of an arbitrary shape and thickness t (Fig. 1.25). As-

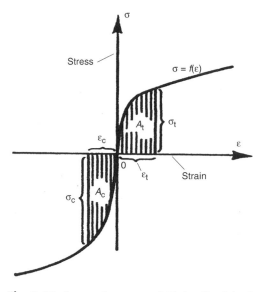

Fig. 1.23 Stress-strain curve $\sigma = f(\varepsilon)$ for bending. Subscripts c and t indicate compression and tension, respectively.

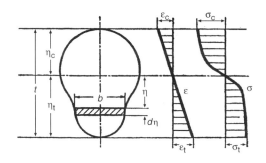

Fig. 1.24 Distribution of strain (a) and stress (b) in a beam above the proportional limit

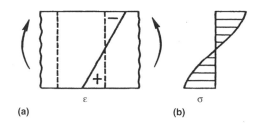

Fig. 1.25 Plastic bending stress-strain distribution in a beam with arbitrary cross section and thickness t. The stresses and strains are actually distributed along the length of the beam, as in Fig. 1.24.

sume that the radius of curvature of the neutral surface produced by the moments M is equal to ρ (Fig. 1.26). In this case, the unit elongation of a fiber a distance y from the neutral surface is:

$$\varepsilon = \eta/\rho \qquad (Eq\ 1.44)$$

Using η_c and η_t to denote the distances from the upper and lower beam surfaces respective to the neutral axis, the elongation in the outmost fibers is:

$$\varepsilon_t = \eta_t/\rho, \qquad \varepsilon_c = \eta_c/\rho, \qquad \eta_t + \eta_c = t \qquad (Eq\ 1.45)$$

where the subscripts t and c stand for tension and compression, respectively. It is seen that the elongation or contraction of any fiber is readily obtained, provided we know the position of the neutral axis, say η_t or η_c, and the radius of curvature, ρ. Since assumptions (a) and (b) are still valid, these two quantities can be found from the two equations of statics:

$$\int \sigma \cdot dA = 0 \qquad (Eq\ 1.46)$$

and

$$\int \sigma \cdot \eta \cdot dA = M \qquad (Eq\ 1.47)$$

From Fig. 1.25, $dA = b \cdot d\eta$, therefore:

$$\int_t \sigma \cdot b \cdot d\eta = \rho \int_t \sigma \cdot b \cdot d\varepsilon \qquad (Eq\ 1.48)$$

and

$$\int_t \sigma \cdot b \cdot \eta \cdot d\eta = \rho^2 \int_t \sigma \cdot b \cdot \varepsilon \cdot d\varepsilon = M \qquad (Eq\ 1.49)$$

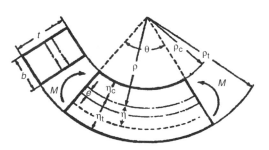

Fig. 1.26 Lengthwise view of a beam subjected to plastic bending (with a rectangular shape for simplicity). e is the distance between the neutral axis and the centroid (its value can be zero).

where the integral over thickness t is integrating from $-\eta_c$ to η_t, or from $-\varepsilon_c$ to ε_t wherever applicable.

Simplifying the problem for this demonstration, the cross section becomes rectangular; thus, b is no longer a variable, but is independent of η. Therefore, the cross section A is equal to $b \cdot t$. Inserting Eq 1.42 into Eq 1.46 and 1.47 yields:

$$\rho \cdot b \int_t f(\varepsilon) \cdot d\varepsilon = 0 \qquad \text{(Eq 1.50a)}$$

i.e.,

$$\int_t f(\varepsilon) \cdot d\varepsilon = 0 \qquad \text{(Eq 1.50b)}$$

and

$$\rho^2 \cdot b \int_t f(\varepsilon) \cdot \varepsilon \cdot d\varepsilon = M \qquad \text{(Eq 1.50c)}$$

From Fig. 1.26,

$$\theta = l/\rho \qquad \text{(Eq 1.51)}$$

where l is the length of the bar for which the angle θ was observed. From Eq 1.45, which relates ε, η, and ρ:

$$\frac{1}{\rho} = (\varepsilon_t + \varepsilon_c)/t \qquad \text{(Eq 1.52)}$$

and

$$\theta = \frac{l(\varepsilon_t + \varepsilon_c)}{t} \qquad \text{(Eq 1.53)}$$

Substituting Eq 1.52 into Eq 1.50(c) gives a formula for the bending moment as a function of the positive and negative strains:

$$M = \frac{t^2 \cdot b}{(\varepsilon_t + \varepsilon_c)^2} \int_t f(\varepsilon) \cdot \varepsilon \cdot d\varepsilon \qquad \text{(Eq 1.54)}$$

Now, further assume that we have tensile and compression stress-strain curves for the material in question. That is, the stress-strain relationship of Eq 1.42 is presented as those shown in Fig. 1.23. Introducing the areas under the respective stress-strain curves for compression and for tension, and on the basis of Eq 1.50(b), yields:

$$A_c = A_t \qquad \text{(Eq 1.55)}$$

which gives:

$$\sigma_c \cdot d\varepsilon_c = \sigma_t \cdot d\varepsilon_t \qquad \text{(Eq 1.56)}$$

After some manipulations, using Eq 1.50(b) to 1.56, we finally obtain the Herbert equation (Ref 1.2).

$$\frac{1}{\theta} \frac{d}{d\theta} (M\theta^2) = bt^2 \frac{\sigma_t \sigma_c}{\sigma_t + \sigma_c} \qquad \text{(Eq 1.57)}$$

For a material with the same stress-strain curve for tension and compression:

$$\sigma_c = \sigma_t, \qquad \varepsilon_c = \varepsilon_t, \qquad d\varepsilon_c = d\varepsilon_t \qquad \text{(Eq 1.58)}$$

Eq 1.57 becomes:

$$\frac{1}{\theta} \frac{d}{d\theta} (M\theta^2) = bt^2 \frac{\sigma_t}{2} \qquad \text{(Eq 1.59a)}$$

or

$$\sigma_t = \frac{2}{bt^2} \left(2M + \theta \frac{dM}{d\theta}\right) \qquad \text{(Eq 1.59b)}$$

where M is the bending moment applied at both ends of the bar. In the case of four-point bending (Fig. 1.19), $M = P \cdot a$. If θ or ρ can be measured by running an experiment, and a diagram for M versus θ (Fig. 1.27) is constructed, Eq 1.56(b) can be interpreted by a graphical solution as:

$$\sigma_t = \frac{2}{bt^2} (2\overline{PC} + \overline{PB}) \qquad \text{(Eq 1.59c)}$$

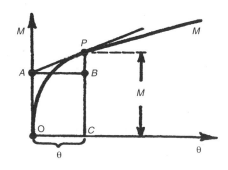

Fig. 1.27 Typical M vs. θ diagram in pure bending. Source: Ref 1.2

for any given θ. Substituting Eq 1.58 into Eq 1.53, we obtain the corresponding tensile strain as:

$$\varepsilon_t = \frac{t}{2l}\,\theta \qquad \text{(Eq 1.60)}$$

where l is the length of the bar for which the angle θ was observed, and t is the thickness of the beam. The subscript t indicates tension.

Cross Shear. It has been shown that during simple bending, a bending stress σ_x acts normal to the Y-Z plane of a cross section (see Fig. 1.17). In addition to σ_x (produced by the bending moment), a shearing stress τ_{xy} (produced by the shear force V, which is in a way also produced by the bending moment) also acts on the Y-Z plane. This shearing stress is called the "cross shear." The cross-shearing force V induces shearing stresses on the vertical transverse plane on which it acts and, in addition, on horizontal planes (see Fig. 1.28(a) and (b)). The shear stress that acts on the horizontal plane is call the "longitudinal shear." At any point in the member, the magnitudes of the shear stresses on the two perpendicular planes are equal, that is, $\tau_{yx} = \tau_{xy}$.

Let us begin with a simple configuration: a rectangular cross section. The equation for that is:

$$\tau_{xy} = \tau_{yx} = \frac{V}{2I_z}\left(\frac{h^2}{4} - y_1^2\right) \qquad \text{(Eq 1.61)}$$

where V is the total vertical shear from the shear diagram, I_z is the moment of inertia about the Z-axis, b is beam width, h is beam height, and y_1 is the vertical distance starting from the neutral axis toward the top or bottom of the beam. This equation imposes that the shear stress is maximum at the midheight of the beam and zero stress at both the top and bottom of the beam.

Because each cross-sectional geometry (e.g., an I-beam, a bar with a circular cross section) is unique, all the geometric variables in Eq 1.61 must be reformulated each time. No one single solution can cover all situations. To complicate the matter further, twisting may occur when the principal plane of the beam is not a plane of symmetry—for example, a channel, an angle, or a Z-section. It would require some effort of analysis to obtain a solution that is equivalently a simple bending without torsion. This type of analysis is beyond the scope of this book. Readers are referred to textbooks on solid mechanics or strength of materials.

1.5.3 Simple Torsion

Let a shaft of length L and radius a be subjected to pure torsion by torque. The torque forces M_T are applied at the shaft ends, as shown in Fig. 1.29. A straight line MA drawn on the surface of the shaft and parallel to the longitudinal axis when the shaft is unstrained will become a helix when the shaft is twisted. Point A on the circumference will move to point B. The movement of A relative to M (i.e., the arc AB) will be proportional to L. Now consider a small element that consists of points M and N on the surface of the shaft, as shown in the enlarged view at the left in Fig. 1.29. This element de-

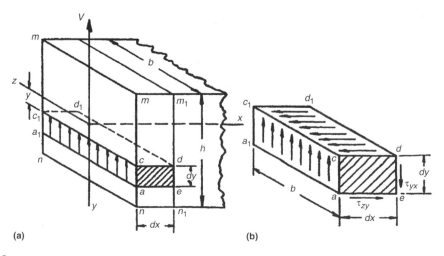

(a)   (b)

Fig. 1.28 Shearing stresses in a beam

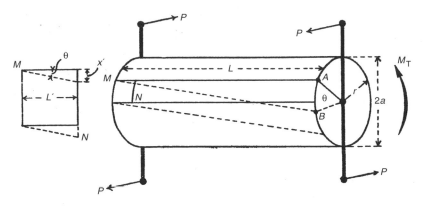

Fig. 1.29 Simple torsion

forms as shown by the dotted lines, and the shear strain on the outer perimeter is:

$$\gamma_a = x'/L' = \overline{AB}/L = \theta \cdot a/L \qquad \text{(Eq 1.62)}$$

where θ is the angle of twist (in radian), and a is the maximum radius of the shaft ($= d/2$). The shear stress is:

$$\tau = G \cdot \gamma, \qquad \tau_a = G \cdot \theta \cdot a/L \qquad \text{(Eq 1.63)}$$

The shearing strain and shearing stress at any point of the cross section is equal to:

$$\gamma = \theta \cdot r/L, \qquad \tau = G \cdot \theta \cdot r/L \qquad \text{(Eq 1.64)}$$

where G is the modulus of rigidity, and r is the radius at a point on the cross section. If the magnitude of the twisting moment is known, the shearing stress can be expressed as:

$$\tau = M_T \cdot r/J \qquad \text{(Eq 1.65)}$$

where M_T is the moment of twist (torque), and J is the polar moment of inertia of a plane area.

The above equations apply to elastic deformation only. Beyond the torsional yield strength, the shear stress over a cross section of the bar is no longer a linear function of the distance from the axis. The procedure for obtaining a solution is very similar to that for plastic bending that was demonstrated earlier. For a work-hardening material, the maximum shear stress at the outer fiber of the circular shaft (i.e., at $r = a$) is:

$$\tau_a = \frac{1}{2\pi a^3} \left(3M_T + \theta' \frac{dM_T}{d\theta'} \right) \qquad \text{(Eq 1.66a)}$$

where θ' is the angle of twist per unit length, that is, $\theta' = \theta/L$, or

$$\tau_a = \frac{1}{2\pi a^3} (3\overline{AP} + \overline{CP}) \qquad \text{(Eq 1.66b)}$$

by interpretation of the torque-twist diagram (i.e., Fig. 1.30). The corresponding shear strain at the outer fiber of the bar is:

$$\gamma_a = \theta \cdot a \qquad \text{(Eq 1.67)}$$

1.5.4 Pressurized Tubes

In the case of pressurized tubes, pipes, or vessels (Fig. 1.31), the initial elastic hoop stress σ_θ, the radial stress σ_r, and the stress in the axial direction σ_x are as follows:

$$\sigma_\theta = \frac{P \cdot r_i^2 \cdot (r_0^2 + r^2)}{r^2 \cdot (r_0^2 - r_i^2)} \qquad \text{(Eq 1.68a)}$$

$$\sigma_r = \frac{-P \cdot r_i^2 \cdot (r_0^2 - r^2)}{r^2 \cdot (r_0^2 - r_i^2)} \qquad \text{(Eq 1.68b)}$$

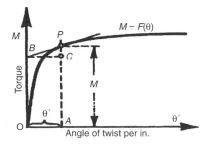

Fig. 1.30 Torque-twist diagram

$$\sigma_x = \frac{P \cdot r_i^2}{r_0^2 - r_i^2} \qquad \text{(Eq 1.68c)}$$

where P is the internal pressure, r is the radial distance, and the subscripts i and o stand for inner and outer, respectively. This set of equations is for a thick-wall cylinder; σ_r is always a compressive stress and σ_θ a tensile stress.

The maximum shearing stress acts at the inner surface of the cylinder:

$$\tau_{max} = \left(\frac{\sigma_\theta - \sigma_r}{2}\right)_{r=a} = \frac{P \cdot r_0^2}{r_0^2 - r_i^2} \qquad \text{(Eq 1.68d)}$$

By gradually increasing the internal pressure, we finally reach a point where the material at the inner surface begins to yield. This occurs when the maximum shearing stress reaches the yield point stress ($\tau_{y.p.}$). The tangential stress distribution across the tube wall is (Ref 1.2, 1.8):

$$\sigma_\theta = 2\tau_{y.p.}\left(1 + \log\frac{r}{b}\right) \qquad \text{(Eq 1.68e)}$$

Since shear stress is equal to one-half of the maximum stress in pure tension, we can equate this equation with the material tensile yield strength as:

$$\sigma_\theta = F_{ty}\left(1 + \log\frac{r}{b}\right) \qquad \text{(Eq 1.68f)}$$

As for a thin-wall cylinder, the stress that is going through the wall thickness (σ_r) does not exist, and its hoop and axial stresses become:

$$\sigma_\theta = P \cdot r_i/t \qquad \text{(Eq 1.69a)}$$

and

$$\sigma_x = P \cdot r_i/2t \qquad \text{(Eq 1.69b)}$$

where t is the wall thickness. No clear definition exists as to what constitutes a thick wall or thin wall; it takes some engineering judgment. For example, a copper pipe inside a residential house is a thin wall; an airplane fuselage is also a thin wall. On the basis of Timoshenko's calculation (Ref 1.8), the hoop stress is always maximum at the cylinder inner surface and minimum at the outer surface. The ratio of these two values is:

$$\frac{\sigma_{\theta,max}}{\sigma_{\theta,min}} = \frac{r_0^2 + r_i^2}{2r_i^2} \qquad \text{(Eq 1.70)}$$

So, for instance, if the outer diameter is 5% larger than the inner diameter, the hoop stress on the outer surface will be 5% higher than the hoop stress on the inner surface. Perhaps a 5% difference in hoop stress distributed over the wall thickness is reasonably uniform, so this cylinder is close enough to be considered a thin-wall cylinder. Otherwise, a more stringent criterion should be adopted.

1.5.5 Static Strength of a Pin-Loaded Lug

Pin-loaded attachment lugs have long been treated as a special class of structural member by engineers. Most textbooks on strength of materials offer no mention of stress analysis procedure for lugs. Up to now, methods for determining static strengths for lugs are covered in proprietary structural design/analysis manuals. Industry-developed methods are totally empirical, covering many possible failure modes, and are not discussed in this book. However, one published article (Ref 1.9) is available, and the method it proposes is discussed here.

Attachment lugs usually come as a set that includes a male part and a female part (Fig. 1.32). A pin is inserted to assemble the joint, as shown. A bushing may or may not be placed between the pin and the lug. The load may be applied axially, obliquely, or transversely in the plane of the lug. The allowable load for the joint is based on the minimum margins of safety obtained for either component in the joint. Failure mechanisms of the male lug are generally of

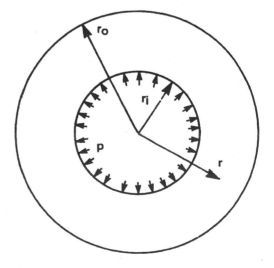

Fig. 1.31 Cross-sectional view of a pressurized thick-wall cylinder

greatest concern. Six types of failure modes are considered:

- Ultimate strength failure of the lug due to tension across the net section (Fig. 1.33a)
- Ultimate strength failure of the lug due to localized bearing failure, or shear tearout (Fig. 1.33b)
- Yielding of the lug
- Yielding of the bushing (if a bushing is used)
- Shearing of the pin
- Bending of the pin

Analytically speaking, tension or bearing failure of a male lug is primarily a stress concen-

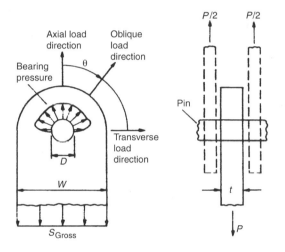

Fig. 1.32 Typical male and female lug set combination and loading conditions. The pin-lug contact pressure distribution shown is an example for axial loading. Source: Ref 1.10

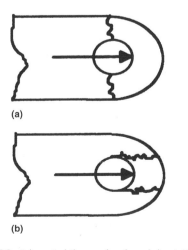

(a)

(b)

Fig. 1.33 Schematic failure modes of a male lug. (a) Tension. (b) Bearing (shear tearout)

tration problem. As discussed later in this chapter, any geometric discontinuity or cutout in a structural part raises the local stress level at the cutout. A lug hole is essentially a cutout. The peak tensile stress at the edge of the hole is responsible for initiating lug failure. Failure modes in lugs are functions of lug geometry and material mechanical properties. Tension mode failure usually occurs in materials of low ductility. In materials with high ductility, the failure mode of a lug can be either tensile or shear tearout, depending on the lug geometry. For narrow lugs, the fracture could propagate across the width of the lug in a tension failure mode. For wider lugs, shear tearout failure could be prominent.

Ekvall (Ref 1.9) has investigated the relationship between the elastic stress concentration factor of a lug geometry with the experimental failure load of the same lug. The lug geometries examined were either straight or tapered, with a uniform thickness (Fig. 1.34). After correlating data drawn from a huge database, a semiempirical equation was developed:

$$P = D \cdot t \cdot F_{tu} \cdot K_{BR} \qquad \text{(Eq 1.71)}$$

This equation can be used to determine failure loads whether the lug failed as a result of tension or shear tearout. The effect of pin bending has also been considered. The terms in Eq 1.71 are defined as follows: P = ultimate load at failure,

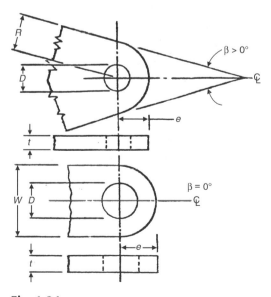

Fig. 1.34 Straight lug and tapered lug configurations

D = lug hole diameter, t = lug thickness, F_{tu} = material tensile ultimate strength (for the appropriate grain orientation), and K_{BR} = bearing efficiency factor, a function of the elastic stress concentration factor,* as shown in Fig. 1.35.

The accuracy of this analysis method has been demonstrated in Ref 1.9, showing good correlation between the predicted and the experimental failure loads for 263 data points. The tests were conducted on various types of lugs made of 25 different materials and loaded at five different angles. Complete details regarding lug material, geometry, and loading condition are given in Ref 1.9. High-strength pins were used in these tests. A majority of the test data developed had a pin for which the stiffness was three times greater than the stiffness of the lug material. According to Ref 1.9, variation in stiffness ratio of the pin and the lug had little effect on the ultimate strength of the lug. However, limited test results for lugs with a very low strength pin indicated that the ultimate strength of the lug could be reduced by as much as 14% due to excessive pin bending.

In summary, this method is applicable to both straight and tapered lugs of uniform thickness (Fig. 1.34). A lug may be subjected to axial, oblique, or transverse loading. Predicting the failure mode of a given lug is not an objective of this method. However, the predicted failure load is the same no matter which type of failure the lug experiences. Implementing this method first requires determining the stress concentration factor for the lug under consideration. The bearing efficiency factor is then determined from Fig. 1.35, and finally the ultimate strength using Eq 1.71.

1.6 Stress Analysis Using Finite Element Methods

Stress analysis by numerical techniques such as finite element analysis (FEA) is a useful engineering tool for solving problems that involve complex geometric and loading combinations. The complexity of the boundary conditions and the governing differential equations determine the possible methods of solution. In some cases, solutions may be adequately achieved by analytical (closed-form) equations, while in other

cases numerical techniques (such as FEA) may be required.

For example, consider the simple example of a uniformly loaded cantilever beam (Fig. 1.36a). In this case (assuming uniform loading and a rigid connection), closed-formed equations can provide analytical solutions for the deflection and bending stress over the length of the beam (Fig. 1.36a). However, if the boundary conditions are altered by the addition of simple supports (Fig. 1.36b), then the system becomes statically indeterminate (i.e., there is no longer a closed-formed solution that specifies deflection over the length of the beam). In this case, numerical techniques such as FEA are required to find approximate solutions of deflection and bending stresses.

The general methods of solving the underlying equations of a model depend on the complexity of boundary conditions and variables of stress, strain, and strain rate. Developing an analytical or a closed-form solution model may be advantageous in many instances, but numerical techniques may sometimes be required. In some cases analytic solutions can be obtained. However, this is only true of simple forms of the equations and in simple geometric regions. For many practical problems, computational or numerical solutions are needed.

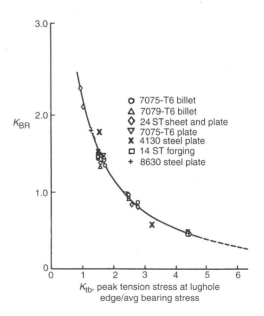

Fig. 1.35 Relationship between experimentally determined bearing efficiency factor and analytically determined elastic stress concentration factor. The lug may be straight or tapered. The straight lug may be loaded at 0° or 45°; the tapered lug may be loaded at 0°, 45°, or 90° to the lug axis. Source: Ref 1.9

---

*Stress concentration factor is discussed in sections that follow, which give stress concentration factors for various lug configurations.

Numerical methods provide approximate solutions by converting a complex continuum model into a discrete set of smaller problems with a finite number of degrees of freedom. The meshes are created by using structured elements like rectilinear blocks or unstructured meshes with variable-shaped elements (e.g., tetrahedra, bricks, hexahedral, prisms, and so forth) for better fidelity to the macroscopic conditions or boundaries. Once a discrete-element model has been created, mathematical techniques are used to obtain a set of equilibrium equations for each element and the entire model. By applying various boundary conditions and loads to the model, the solution of the simultaneous set of equations provides the resulting response anywhere in the model while still providing continuity and equilibrium. The process of solution is essentially a computer-based numerical method, where interpolation functions (polynomials) are used to reduce the behavior at an infinite field of points to a finite number of points. Generally, analysis falls into one of two categories:

- Closed-form analytical approximations
- Numerical methods using discrete elements

However, this division is not entirely distinct, as closed-form methods are often solved by discrete-element numerical methods as well. The general distinction between the two sets of analytical methods is that the first set is based on simplifying assumptions that permit closed-form solution as well as rapid numerical and graphical methods. Solving the system equations by numerical methods, in contrast, reduces the need for simplifying assumptions. Regardless of the solution method, continuum mechanics is the starting point. Each problem starts with the continuum equations and then, through various simplifying assumptions, leads to a solution method that may be closed form or numerical.

As more rapid and powerful computers help speed up the process of many computational tasks such as obtaining numerical solutions for problems of great complexity and structure size, the finite element method (FEM) helps in previously difficult (or impossible) system conditions. There are two basic types of finite element methods: the h-version and the p-version. The h-version is the conventional type. Computer codes such as NASTRAN belong to this category. The p-version computer code did not exist until the early 1980s. Both codes will be discussed here and briefly compared from the standpoint of the user.

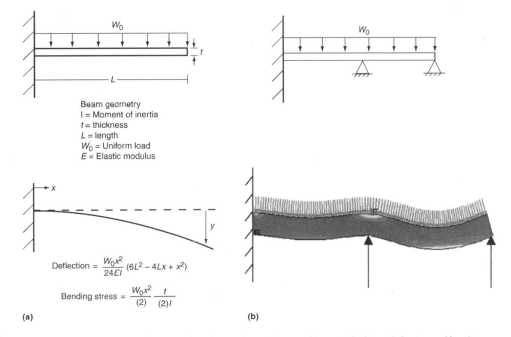

Fig. 1.36 Effect of boundary conditions on the solution of a cantilever problem. (a) The beam deflection and bending stresses for a uniformly loaded cantilever can be solved by a closed-form equation as shown. (b) A supported cantilever beam is statically indeterminate, and numerical methods are required to approximate deflection and bending-stress conditions that are consistent with the boundary conditions established by the additional supports.

The first step in using a finite element code to perform a stress analysis task is to construct a finite element model that closely resembles the shape of the part in question. A finite element code usually provides several types of elements that may be two-dimensional or three-dimensional, as shown in Fig. 1.37. The 2D elements are regarded as membrane elements, the 3D elements as solid. Both linear and quadratic types are available, as are special-purpose elements such as bar and spring elements, as well as elements designed to perform a specific task or have a special property. This provides great flexibility in structural modeling. By carefully choosing the elements and designing a network to connect them together through the nodes, any model can be constructed to simulate a complicated machine part, even an entire automobile or an aircraft, in as much detail as one can imagine.

To begin, a simple two-dimensional model is constructed using membrane elements (Fig. 1.38). The grid points are identified by numbers. The elements are also identified by numbers, although not shown in the figure. It is preferable, though not required, to number the nodes and the elements in an orderly manner. In this example, because the area at the bottom left corner is important to the analyst, a finer mesh is inserted there. A mesh much finer than this may be needed, depending on the purpose of the analysis. After a geometry is mapped, material properties and any other specific instructions regarding each element or node should be included in the data input, as required by the computer code. Loads are attached to selected grid points. For

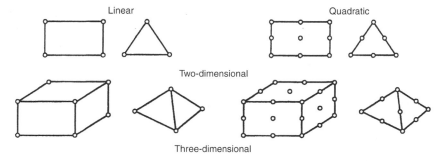

Fig. 1.37 Examples of element geometries available in an h-version finite element code

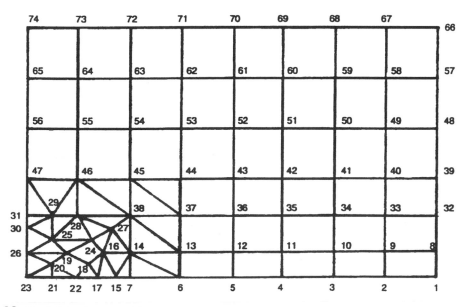

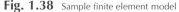

Fig. 1.38 Sample finite element model

example, vertical tension loads can be put on nodes 66 through 74. Proper constraints are then placed at the bottom of the model (along the bottom edge between nodes 1 and 23) so that the computed stresses in each element will be the result of the tension loads only. There are six degrees of freedom in each node: three directions in movement and three directions in rotation. It is very important that the boundary conditions be identified with proper constraints. The

same model can be used for other loading conditions by simply moving the loads to other grid points and changing the boundary constraints.

The next model illustrates modeling of a biaxially loaded test piece. The overall appearance of the specimen is shown in Fig. 1.39. Loads are applied at the ends of all four arms. Such a specimen has been used to study crack propagation behavior under various biaxial stress conditions, with different biaxial stress ratios (Ref 1.11). A finite element model simulating the actual (uncracked) specimen configuration has been constructed for determining the stresses σ_x and σ_y in the center of the specimen. The results, which are presented in Ref 1.11, will not be discussed here. However, it is strongly emphasized that conducting stress analysis (by using finite element or other means) is an essential step prior to mechanical testing. For an actual test, the required input load values (P_x and P_y) corresponding to any desirable σ_x and σ_y combinations should be predetermined through stress analysis.

Because the geometry and the loading condition are both symmetric, a model representing one-quarter of the specimen will be sufficient (Fig. 1.40). The thickness variation in different areas of the specimen is accounted for by using solid elements of different thicknesses. The thicknesses in various specimen regions are identified as t_1 to t_5; t_1 and t_2 represent the center region and the loading arms, respectively. The tapered region, which symmetrically connects

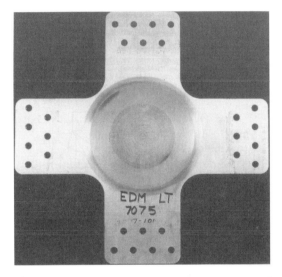

Fig. 1.39 Photograph of a cruciform specimen. Source: Ref 1.11

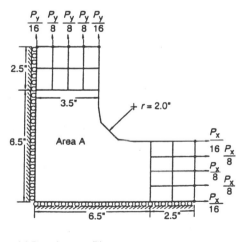

(a) Boundary conditions

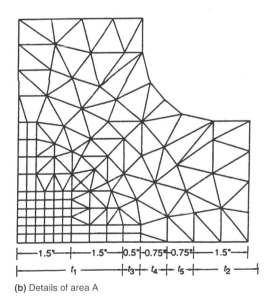

(b) Details of area A

Fig. 1.40 Finite element model of one-quarter of a cruciform specimen. Source: Ref 1.11

the center region to the top and bottom surfaces of the loading arms, is represented by three rings of triangular elements of different intermediate thicknesses (t_3, t_4, and t_5) to simulate the curvature connecting t_1 and t_2. As shown in Fig. 1.40, one-half of the load is evenly distributed at the end of each loading arm, because the model contains only one-half of a loading arm in each direction. Five grid points on each arm carry the loads. Take the vertical arm, for example. The node at the centerline shares the load with the other half of the same arm, which is not in the model. Therefore, the input load at this point is only one-half of the distributed load. Likewise, the node at the end of the free edge can only take one-half of the distributed loads to make up for the total. The rollers and the fixed foundation attached to the horizontal line at the bottom of the model are expressions for boundary constrains used in the older literature. They indicate that the nodes along that line are free to slide laterally but not vertically. Likewise, the movements of the nodes along the vertical boundary are constrained as indicated.

The major part of the computer output consists of stresses at the centroid of an element and the displacement of a grid point. Thus, the computed stresses are never on any free edge of the model. By placing very small elements along the edge, one might think that is close enough to obtain correct stress values. Depending on the geometry, it may not be a problem if the stress distribution is fairly uniform. However, extrapolation may be needed when the stress gradient is steep. Checking "load balance" is a very important step in post data reduction. This means that the total output loads should be the same as the input. For example, the total reaction load along the base (the bottom edge) of Fig. 1.38 should equal the total input loads distributed along the top edge (nodes 66 to 74). The magnitudes of the reactions are sometimes calculated by the computer and displayed in the printout. Otherwise, they can be easily converted from the stress output. The computed stresses may be in error if these loads are not in balance, most likely due to incorrect boundary constraints.

Most finite element codes also calculate the value of the von Mises stress (Eq 1.27a) in each element. By comparing the von Mises stress with the material proportional limit (or tensile yield strength if so chosen), a yield or no-yield decision can be made. A yield zone (a group of yielded elements) corresponding to a set of input loads can be identified. In other words, the load level that causes yielding (or wide spread of yielding) can be identified. During construction of the finite element model, a material tensile curve is entered into each selected element. The material tensile curve input may be an equation, or take the form of stepwise linear segments. The actual stresses in an element corresponding to an applied load level can be determined through the stress-strain relationship of the material tensile curve.

Commercially available p-version finite element codes (PFEC) are relatively small in size and can be run by a workstation instead of a mainframe for cost savings. Of course, the code can be merged into (or linked to) mainframe programs (e.g., NASTRAN, or some program for mesh generation) without difficulty. Characteristics of the p-version methodology include:

- Stress analysis can be done using a coarse mesh and/or odd-shaped elements. In the conventional h-version method, the shape of the element is either rectangular, square, or triangular. In any case, the element must have a reasonable aspect ratio (the height-to-width ratio). Using the p-version codes, the analyst can construct a model with odd-shaped elements (e.g., a triangle having a curved edge) and/or elements having an extreme aspect ratio.

- For data reduction, a PFEC allows the user to extract many data points of stress and displacement from each element. Stresses are obtained on any preselected line connecting two nodal points. An exact stress value can be obtained at the edge of a PFEM without extrapolation.

- When using the h-version method, the error of approximation is controlled by refining the mesh. In the p-version method, the element is described by a group of shape functions of high-degree polynomials (the range can be as high as a nine-degree polynomial, as compared with linear or quadratic in the h-version), with the flexibility of varying the p-level in any designated element.

- To verify the solution, the p-version allows creation of a series of solutions without the need to refine the mesh, as is necessary with the h-version. A full sequence of polynomial levels (from 1 to 9) can be run to allow examination of strain-energy convergence and other functionals of interest. In most cases, convergence can be reached within seven attempts. Beyond that, only insignificant improvements on minimizing the computational error can be achieved. The same mesh is used for all p-levels, requiring no additional effort by the user.

- The user can judge the accuracy of the solution based on converge on stress, or converge on displacement, or both.

In the following, an example PFEM is shown based on a PFEC called MECHANICA®–APPLIED STRUCTURE code (developed by the now-defunct RASNA Corporation, San Jose, California). The code contains both plane and solid elements and runs on an IBM workstation. This PFEC allows the user to extract 10 pairs of data points along any preselected line connecting any two nodal points of a given element. For example, one may select a line along an edge of the element, or a line connecting two diagonal nodal points, and so on. In the process of constructing a model,

the size and placement of elements are influenced by these code capabilities. Therefore, strategic deployment of elements is an important part of a modeling scheme.

Figure 1.41 shows the geometry of a T-beam, called a cap-web specimen because there is a big mass at the top and a very thin web hanging down from the cap. By adjusting the position of the pinhole along the Z-direction, the specimen will respond to the applied tension load with various combinations of tension and bending. The Z-dimension corresponding to pure tension is 66.827 mm (2.631 in.). Because of symmetry, a PFEM was constructed for only one-quarter of the specimen (Fig. 1.42). To load the specimen, uniformly distributed loads were applied to one-half of the pinhole, simulating the pin-to-hole contact pressure. The result, represented by a stress contour map, is shown in Fig. 1.43(a). Note that only the area near the midlength of the

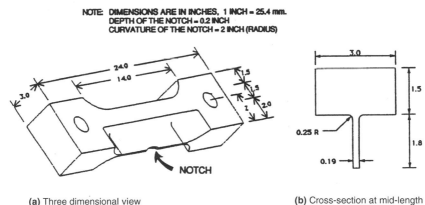

(a) Three dimensional view

(b) Cross-section at mid-length

Fig. 1.41   Cap-web specimen. Source: Ref 1.12

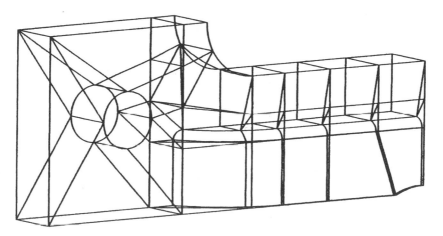

Fig. 1.42   Finite element mesh for one-quarter of a cap-web specimen

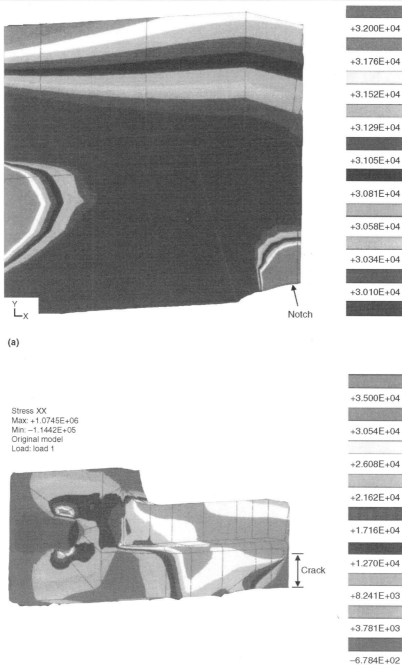

Fig. 1.43 Sample contour maps showing the highly stressed areas in the cap-web specimen. (a) Without crack. (b) With crack

specimen is shown. The highly stressed areas are at the notch and at locations where the web and the cap connect to the loading block. Shown in Fig. 1.43(b) is a stress contour map for the same cap-web specimen with a crack extending from the straight edge toward the top. The highest stress point is now at the crack tip, because stress concentration is highest there. The area along the crack line is unloaded. Figure 1.44 is a plot of the cross-sectional stress distribution for the notched specimen subjected to pure tension. The cross-sectional stress distribution in the area away from the notch root is fairly uniform, and the magnitude of the stress corresponding to pure tension is verified by the hand-calculated *P/A* value and by strain gage measurements. Also, the highest stress point is right at the notch root, where the local stress is 41% higher than the average *P/A* stress, due to stress concentration. In other words, the value of stress concentration is 1.41 at the notch root. The subject of stress concentration is discussed in the next section. Stress analysis of cracks will be covered at the end of this chapter and in Chapters 5 and 6.

1.7 Local Stress Distribution at a Geometric Discontinuity

Geometric discontinuities (i.e., cutouts or notches) are common in structural parts. A cutout in a structural piece acts as a stress raiser and

causes redistribution of stresses in its local vicinity, thereby altering the load-carrying characteristics. The severity of a stress raiser depends on its geometry. A cutout with a sharp corner is more detrimental than one that is round and smooth. A sharp notch is even worse, and a crack is most dangerous.

Stress analysis of cracks requires the use of fracture mechanics. The basic solutions of stresses at the crack tip are given in section 1.8 for approximately elastic conditions (i.e., with small-scale yielding at the crack tip). Chapter 6 describes stress analysis of cracks with large-scale yielding at the crack tip. Application of fracture mechanics to design/analysis and related tasks is presented in Chapters 4 to 6.

For distribution of stresses near a geometric discontinuity, consider the situation shown in Fig. 1.45. Figure 1.45(a) depicts an ordinary panel subjected to tension loads at both ends. Imagine the load paths are evenly distributed. Some of the (imaginary) flow lines cannot follow their initial paths after a portion of the panel is cut away. These flow lines must go around the cutout as shown in Fig. 1.45(b). In a way, the flow lines above the cutout are being pushed just enough to the right to go around the cutout. Near the tip of the cutout the flow lines are closely spaced, with more loads flowing through that packed region, thus producing a higher local stress level in this region. The local stress level will gradually decrease and eventually return to

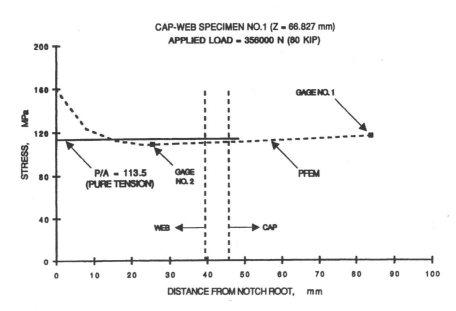

Fig. 1.44 Axial stress distribution across the midlength section of a cap-web specimen subjected to pure tension. Source: Ref 1.12

normal as normal flows are maintained in the area away from the cut. The stress distribution of the cap-web specimen (Fig. 1.44) shows just that. This phenomenon of load redistribution, which leads to local stress redistribution, is called the "stress concentration." The ratio of the highest stress (which is at the tip of and normal to the cutout) to the average stress is called the "stress concentration factor," designated by the symbol K_t. For the notch geometry shown in Fig. 1.41, $K_t = 1.41$. The average stress can be based on gross area stress (i.e., the A of the P/A stress is the original area ignoring the cutout) or on net section stress (i.e., the average stress is equal to P/A_{net}, and A_{net} is the original area minus the cutout). There is also a third type of K_t that is based on bearing stress. Remember, any type of K_t will lead to the same peak stress as long as the type of baseline stress is clear. The specific K_t for various case studies will be clearly noted throughout this book. Otherwise, a gross area K_t is implied.

Any cutout or discontinuity is considered a stress raiser, or site of stress concentration. The abruptness of the notch and fillets in Fig. 1.41 are relatively minor, whereas the notch in Fig. 1.45(b) is quite severe. Therefore, one would expect the K_t value for the latter to be much higher. Note that the mechanism of stress concentration in the fillets is not exactly the same as in the notch. This will be discussed in the next section.

Numerical solutions of K_t values for numerous geometric variations in combination with various loading conditions are available in the literature. Many stress concentration factors are compiled in Ref 1.13 to 1.17. With today's high-speed computers and more sophisticated finite element programs, the stress concentration factor and local stress distribution at a stress raiser can be easily determined. Problems that are not covered in handbooks can be solved by running a finite element program. However, the handbooks mentioned here help make an engineer's life easier. The next two sections discuss several common configurations and a few others that are not so common.

Thus far, discussion has involved a pure elastic analysis where everything obeys Hooke's law. When the resulting local stress level (i.e., $K_t \cdot P/A$) exceeds the material's proportional limit, the magnitude of K_t will be reduced in order for the actual stress to fall on the material's tensile test curve. This phenomenon is discussed in Chapter 2.

1.7.1 Elastic Stress Concentration Factors for Uniformly Loaded Members

Consider a circular hole in a plate, a case familiar to everyone. The equations below are the solution of G. Kirsch for a circular hole in an infinitely wide sheet (Ref 1.3). Let the width $m_1 n_1$ in Fig. 1.46 be infinitely wide, and the length infinitely long. The hole diameter is $2r$. Let R be the radius of any point on the sheet from the hole center. The Kirsch solutions for the radial, tangential, and shear stress around the hole are, respectively:

$$\sigma_r = \frac{S}{2}\left(1 - \frac{r^2}{R^2}\right) + \frac{S}{2}\left(1 + \frac{3r^4}{R^4} - \frac{4r^2}{R^2}\right)\cos 2\theta$$

$$\sigma_\theta = \frac{S}{2}\left(1 + \frac{r^2}{R^2}\right) - \frac{S}{2}\left(1 + \frac{3r^4}{R^4}\right)\cos 2\theta$$

$$\tau_{r\theta} = -\frac{S}{2}\left(1 - \frac{3r^4}{R^4} + \frac{2r^2}{R^2}\right)\sin 2\theta$$

$$\text{(Eq 1.72)}$$

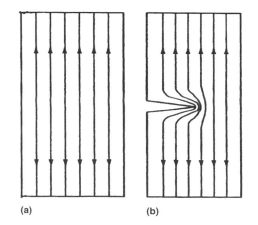

Fig. 1.45 Schematic flow lines in a flat panel. (a) Ordinary panel subjected to tension load. (b) With cutout

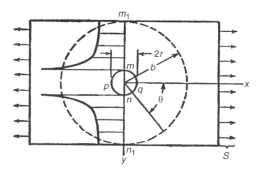

Fig. 1.46 Schematic σ_θ distribution in a sheet with a circular hole. Source: Ref 1.3

where S is the applied P/A stress, with A being the sheet thickness times the width $m_1 n_1$ (ignoring the presence of the hole), and θ starts from point q on the X-axis. A σ_θ-based stress concentration factor is customarily considered the stress concentration factor for a circular hole. A plot of the stress concentration factors around the hole (the ratios of σ_θ to S at $R = r$) is constructed as shown in Fig. 1.47. The point of highest tangential stress is at the edge of the hole, at $\theta = \pi/2$ and $\theta = 3\pi/2$ (i.e., points n and m), where

$$\sigma_\theta = 3S \qquad \sigma_r = \tau_{r\theta} = 0 \qquad \text{(Eq 1.73)}$$

Figure 1.47 also shows that the tangential stresses at points p and q, and in their vicinity,

are in compression. This is because the tension loads that are supposed to pass through the centerline of the panel were diverted and left behind an unloaded area. This vacuum area then feels pressure coming from the bulk of material in the neighborhood.

Equation 1.72 applies equally to a panel subjected to equibiaxial tension loads. For example, when tension loads of the same magnitude simultaneously act in both the X and Y directions, the stress concentration factor at each point along the hole circumference is obtained by superposition. At a given point, K_t is the sum of two separately computed stress concentration factors, one for each direction. For example, K_t equals 2 at point m because the sum of K_t at this point is $3 + (-1)$. For the same reason, K_t also equals 2 at point n, p, and q.

An equation for the local distribution of σ_θ above sections mm_1 and nn_1 is obtained by letting $\theta = \pi/2$ (or $\theta = 3\pi/2$), so that Eq 1.72 becomes:

$$\sigma_\theta = S\left(1 + \frac{1}{2}\frac{r^2}{R^2} + \frac{3}{2}\frac{r^4}{R^4}\right) \qquad \text{(Eq 1.74)}$$

A plot of Eq 1.74 is shown in Fig. 1.48. In cases where the sheet width is finite, numerical solutions of σ_θ for various hole diameter to width ratios have been given by Howland (Ref 1.18). The Howland solutions are plotted together with the Kirsch solution in Fig. 1.48. The stress concentration factors (those data points at $L/r = 0$)

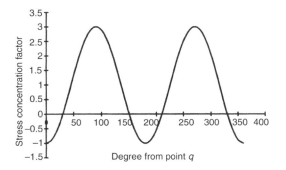

Fig. 1.47 Stress concentration factors along the circumference of a circular hole in an infinite sheet. Stresses are tangential stresses at the hole edge. See Fig. 1.46 for the location of points m, n, p, and q.

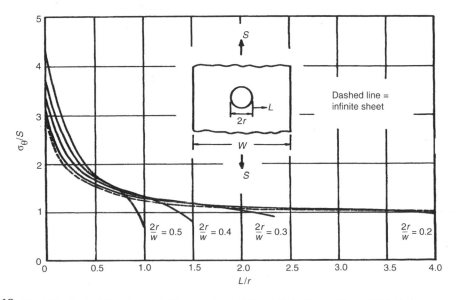

Fig. 1.48 Kirsch-Howland solutions for the distributions of σ_θ in finite-width sheets containing a circular hole

and the cross-width stress distributions are significantly affected by panel width. Panels with narrower widths exhibit higher stresses at the hole wall. For a very narrow panel, the stresses near the panel edge are below the nominal stress level (i.e., $\sigma_\theta/S < 1$). This is caused by the "load balancing" act mentioned earlier. Generally, the high stresses near the hole edge use up a good portion of the input loads. Therefore, the stress levels decay away toward the panel edge. The stresses near the hole in a very narrow panel are excessively high, and so are the loads required to produce these stresses. Consequently, only a small percentage of input load is left for the remaining area.

It is desirable for the Howland solution to be expressed in a close-form format like that for the infinite sheet. In the mid-1980s Saff developed a close-form equation that can be used for any combination of hole shape (circular or elliptical), hole size to sheet width ratio, and material (isotropic or orthotropic). Saff's work is discussed in Appendix 2.

Next, consider an axially loaded member that does not have a constant cross section throughout its length (Fig. 1.49a). The average stress in the large section is P divided by the large cross-sectional area. The average stress in the small section is P divided by the small cross-sectional area. Because there is no shear stress other than P/A in these cross sections, the principal stresses are the same as their respective P/A stresses. As before, some imaginary flow lines near the edge of the panel (of constant width) will be affected by the "squeeze action" in the area where change of width takes place (Fig. 1.49a). The corners that are identified as "A" on each side of the

intersection are totally stress free, which means the materials at these corners are not moving. However, the materials along the flow lines are moving. Consequently, shear stress develops between these two chunks of materials. The added shear stress component increases the magnitude of principal stress in this local area. Figure 1.49(b) shows a sketch of the relative magnitude of principal stresses at various locations on the fillet surface. The severity of stress concentration can be reduced by reducing the sharpness of the intersecting corner. For example, add a radius to smooth out the transition, as shown in Fig. 1.49(b). Some sample numerical values are shown in Fig. 1.50. The data in the figure (Ref 1.19) indicate that the stress concentration factor decreases as the two widths become more nearly equal and that the factor decreases rapidly as the fillet radius increases.

1.7.2 Elastic Stress Concentration Factors for Pin-Loaded Lugs

The previous section discussed the stress concentration at an open hole subjected to uniform loading. Now consider the case of an open hole subjected to a concentrated load (Fig. 1.51a). This type of loading is most relevant to the situation of a lug loaded by a pin. It has long been recognized that some portions of the hole make contact with the pin under load. Determining the stress concentration factor for a lug requires us-

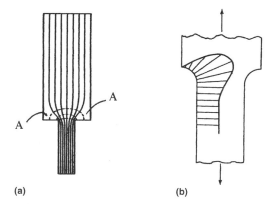

(a)   **(b)**

Fig. 1.49 Stress concentration at a change in cross section. (a) Change in density of the flow lines. (b) Principal stresses at a fillet

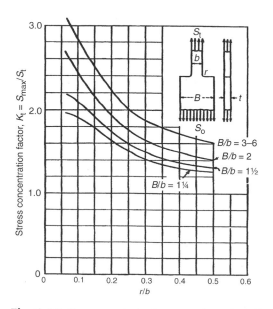

Fig. 1.50 Stress concentration factors at a change in width. Source: Ref 1.19

ing a known contact pressure distribution as input loads. Therefore, determining the contact pressure distribution is a first step in stress analysis of lugs. In the past, stress analysts used an assumed contact pressure distribution. Two examples are shown in Fig. 1.51(b) and (c). Actually, the real pin-and-lug contact area and the pressure distribution over that area are something like those shown in Fig. 1.32. With a finite element code, it is possible to model the pin and the lug together. Briefly, the pin-lug model is constructed with all the nodes along the pin surface connected to the nodes on the hole inner surface. Those nodes with identical nodal coordinates along the contact surface are connected with spring elements. The advantage of using spring elements is that their length is unspecified; the spring can be used as a "zero-length" element to connect two nodes, and can be assigned any level of rigidity. A concentrated force is applied at the center of the pin to simulate pin loading and is reacted at the base of the lug. Under load, some of the spring elements will be in compression, others in tension. The indication is that the compression side of the pin is moving toward the bore of the hole, in the direction of loading. Iteration is then performed by removing those tensionally stressed spring elements, followed by incrementally increasing the applied load to further determine which spring elements are to be removed. The procedure is repeated until the pin is fully loaded to a desired level. The zone where the spring elements remain after the final analysis will be considered the contact surface due to the given loading.

Stress concentration factors for straight and tapered lugs (see Fig. 1.34) were developed at Lockheed (Ref 1.20–1.22). A wide range of pin diameter to lug width ratio was included in the two-dimensional analysis scheme. The concentrated loads were applied at 0°, 45°, and 90° directions. Contact pressure distribution for each lug and loading combination was determined using a procedure similar to that discussed above. Figures 1.52 and 1.53 show typical results for straight and tapered lugs subjected to axial loading. In either case, the pattern of bearing pressure is fairly close to the uniform pressure distribution, with minor variations here and there depending on lug geometry. Also depending on lug geometry, the pin-bearing pressure may or may not precisely cover the entire upper half of the pin hole area (i.e., ±90° respect to the line of loading). For the lugs in Fig. 1.52 and 1.53, the pin and lug contact area lies in between

±85°. The contact area for the lug shown in Fig. 1.32 lies in between approximately ±100°. Also shown in Fig. 1.53 are the contact pressure distributions for lugs that contain cracks of different sizes. The significance of contact pressure on the strength of a cracked lug will be discussed in Chapter 5. For lugs loaded at directions other than 0°, the pin-bearing pressure distributions are all different (Ref 1.20–1.22). Figure 1.54 shows the points of peak stresses in tapered lugs subjected to three different load orientations. Except for the 90° case, the peak stresses are in the cross section approximately perpendicular to the loading direction.

Two finite element models were constructed for showing how the stress in a lug is affected by the input loads. The models used a single-point force and uniform radial forces as input loads. The dimensions of the lug models were chosen to be the same as one of the Lockheed lugs—a round-headed lug with an outer diameter equal to 1.5 times the inner diameter. The NASTRAN code was used to determine the cross-sectional stress distribution, with the results plotted in Fig. 1.55. The Lockheed result for the same lug, which had the predetermined lug-pin contact pressure as input loads, is also plotted in Fig. 1.55 for comparison. In this figure, the X-axis represents the distance between the inner and outer surfaces having the origin set at the hole wall. The Y-axis represents the ratio of the tangential stress (in the axial direction) to the average bearing stress, which is the total input load divided by the bearing area. Evidently, all three stress distributions shown in Fig. 1.55 are not the same due to differences in loading methods. Thus, predetermination of the pin-lug contact pressure is a prerequisite step in stress analysis of lugs.

With the above modeling in mind, the Lockheed data are presented in Fig. 1.56 to 1.59, starting with the stress concentration factors for

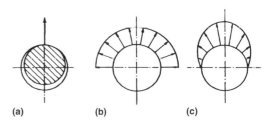

Fig. 1.51 Possible loading conditions for a pin-loaded hole. (a) Loaded by a concentrated force. (b) Uniform contact pressure. (c) Contact pressure having a cosine distribution

the straight lugs (both axially loaded and transversely loaded) in Fig. 1.56. Note that the stress concentration factors for the transversely loaded lugs were not developed by FEA. Instead, they were estimates and their values justified by test data (Ref. 1.9). Stress concentration factors for other loading directions (between 0° and 90°) can be obtained by interpolation.

Note that the symbol K_{tb} is used in Fig. 1.56 for the stress concentration factor. In the present case, it stands for the ratio of the peak tangential stress to the lug bearing stress. Referring to the dimensions shown in Fig. 1.34, the bearing stress, σ_{br} or S_{br}, is defined as $P/(D \cdot t)$. In stress analysis of lugs, stress concentration factor of a straight lug also can be interpreted as K_{tn}, or the regular K_t, with K_{tn} being the ratio of the peak tangential stress to the lug net section stress (which is equal to $P/([W - D] \cdot t)$. The regular K_t is the ratio of the peak tangential stress to the gross area stress (which is equal to $P/W \cdot t$). All three types of stress concentration factors are interrelated. They can be converted from one to the other, as follows:

$$K_{tn} = K_{tb} \cdot (W - D)/D \qquad \text{(Eq 1.75a)}$$

$$K_t = K_{tb} \cdot W/D \qquad \text{(Eq 1.75b)}$$

$$K_t = K_{tn} \cdot D/(W - D) \qquad \text{(Eq 1.75c)}$$

To illustrate, let us now convert the K_{tb} to K_{tn} for the straight lug under axial loading (Fig. 1.57). Take a point from the converted and the original curves; for instance, the K_{tn} value for $D/W = 0.67$ ($W/D = 1.5$) is 2.2, and the K_{tb} value for the same W/D ratio is 4.4. After going through a simple calculation, it can be proved that the peak stress for both cases has the same value. Therefore, stress concentration factors can be presented in many ways. The user must simply be aware of the consequence, and choose one that suits the situation. The additional curves shown in Fig. 1.57 are stress concentration factors for straight lugs obtained from the literature. It shows that the Lockheed solution holds close agreements with the solutions of Frocht and Hill (Ref 1.13) and Heywood (Ref 1.24), up to a $W/$

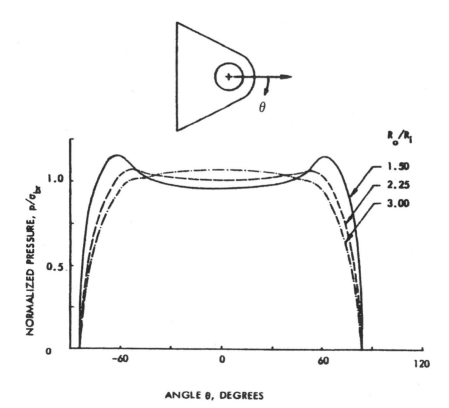

Fig. 1.52 Pin-bearing pressure distributions along the contact surface of an axially loaded tapered lug. R_0 and R_i are the outer and inner radii of the lug, and σ_{br} equals pin load divided by the lug bearing area. Source: Ref 1.20

D ratio of 3 (or $D/W = 0.333$). It is also shown that the stress concentration factors for the round and the square lugs are nearly the same.

The Lockheed finite element data also contain K_{tb} values for eccentric lugs and tapered lugs. An eccentric lug is a straight lug that has a shorter or longer head as compared to the regular half-circular head. A graphical presentation of the eccentric lug solutions is given in Fig. 1.58. Stress concentration factors for the tapered lugs are given in Fig. 1.59.

1.8 Stress Analysis of Cracks

A notch acts as a stress raiser such that the stress at the notch root is actually higher (sometimes much higher) than the remotely applied stress. In the preceding section, the Kirsch-Howland solutions for the redistribution of stress near a circular hole are discussed. Now imagine the circular hole is elongated in one direction to become an ellipse. For an infinite plate with remote uniform stress S, the stress concentration factor

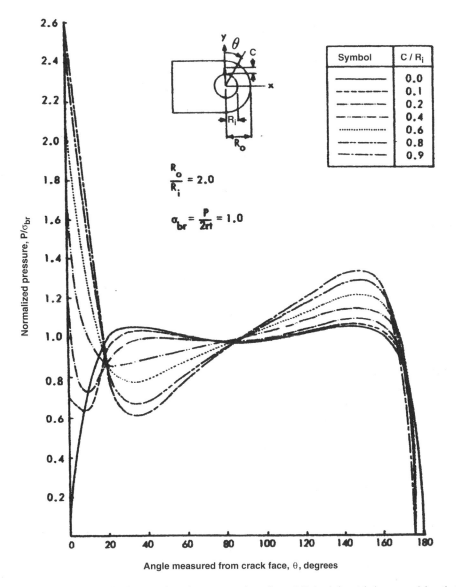

Fig. 1.53 Pin-bearing pressure distributions along the contact surface of an axially loaded straight lug. c, crack length. Source: Ref 1.21

for an elliptical hole of semimajor axis a (on the X-axis) and end radius ρ is equal to (Ref 1.15):

$$K_t = 1 + 2(a/\rho)^{1/2} \qquad \text{(Eq 1.76a)}$$

or

$$K_t = 1 + (2a/b) \qquad \text{(Eq 1.76b)}$$

where b is the semiminor axis (on the Y-axis) of the ellipse, and ρ is the radius of curvature. For $\rho = a$ or $b = a$ (i.e., a circle), $K_t = 3$, the well-known stress concentration factor for a circular hole in an infinite sheet. As ρ or b approaches zero, the ellipse may be regarded as a crack hav-

ing an infinite stress concentration at the crack tip. However, the problem with this result is that K_t would be equal to ∞ for any crack size of $2a$. The quantity $K_t = \infty$ is not useful, either. Therefore, a different approach is required to define a parameter that can relate fracture strength and crack size.

1.8.1 Stresses and Displacements at the Crack Tip

Referring to the coordinate system shown in Fig. 1.60, the local stress distributions near the crack tip are given by (Ref 1.25–1.28):

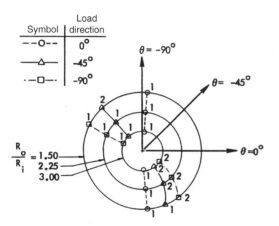

Fig. 1.54 Fatigue critical locations of tapered lugs subjected to various load orientations. Source: Ref 1.20

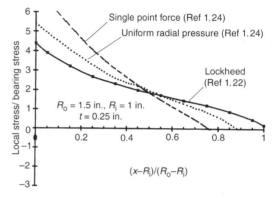

Fig. 1.55 Cross-sectional stress distribution in a round-headed straight lug: Comparison of results among different methods of loading. Note: R_0 and R_i are the outer and inner radii of the lug, x is the distance between R_i and R_0, local stress is the computed elastic tangential stress in the Y-direction, bearing stress equals to pin load divided by the lug bearing area. Source: Ref 1.23

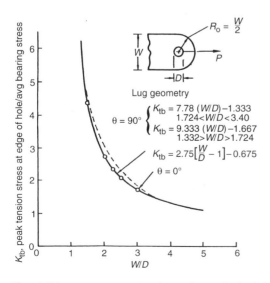

Fig. 1.56 Stress concentration factors for axially loaded straight lugs with $2R_0/W = 1.0$. Source: Ref 1.9

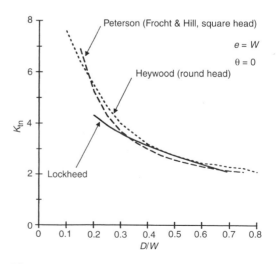

Fig. 1.57 Comparison of stress concentration factor solutions for straight lugs loaded in tension

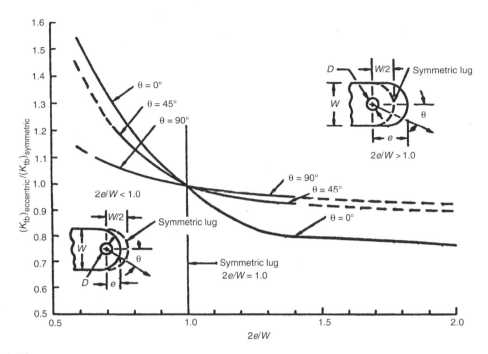

Fig. 1.58 Stress concentration factors for eccentric lugs. Source: Ref 1.9

$$\sigma_y = \frac{K_1}{\sqrt{2\pi r}} \cos\frac{\theta}{2}\left[1 + \sin\frac{\theta}{2}\sin\frac{3\theta}{2}\right] + O(r^{1/2})$$

$$\sigma_x = \frac{K_1}{\sqrt{2\pi r}} \cos\frac{\theta}{2}\left[1 - \sin\frac{\theta}{2}\sin\frac{3\theta}{2}\right] + O(r^{1/2})$$
$$+ \sigma_{xO}$$

$$\tau_{xy} = \frac{K_1}{\sqrt{2\pi r}} \sin\frac{\theta}{2}\cos\frac{\theta}{2}\cos\frac{3\theta}{2} + O(r^{1/2})$$

(Eq 1.77)

where K_1 is the stress intensity factor. The subscript 1 refers to "crack opening mode," which means that the crack is normal to the applied tension load. Here, σ_{xO} is a nonsingular term; $O(r^{1/2})$ stands for higher-order terms. These two terms are generally omitted from the above equations. As r becomes small compared to planar dimensions (in the X-Y plane), the magnitudes of the nonsingular terms become negligible compared to the leading $1/\sqrt{r}$ term. Therefore, under ordinary circumstances inclusion of these terms is unnecessary.

For plane strain (Eq 1.11b) condition (with higher-order terms omitted):

$$\sigma_z = \nu(\sigma_x + \sigma_y)$$ (Eq 1.78)

$$\tau_{xz} = \tau_{yz} = 0$$ (Eq 1.79)

with displacements given by:

$$v = \frac{K_1}{G}\sqrt{\frac{r}{2\pi}}\sin\frac{\theta}{2}\left[2 - 2\nu - \cos^2\frac{\theta}{2}\right]$$

$$u = \frac{K_1}{G}\sqrt{\frac{r}{2\pi}}\cos\frac{\theta}{2}\left[1 - 2\nu + \sin^2\frac{\theta}{2}\right]$$

$$w = 0$$ (Eq 1.80)

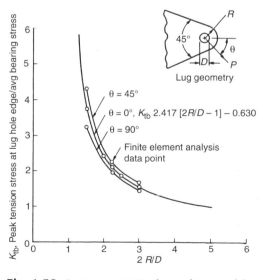

Fig. 1.59 Stress concentration factors for tapered lugs. Source: Ref 1.9

where r is the absolute distance from the crack tip (i.e., for a point ahead or behind the crack tip); v, u, and w are the displacements corresponding to the directions of y, x, and z respectively. The constant G is the shear modulus of elasticity (Eq 1.2).

For plane stress (with higher-order terms omitted):

$$\sigma_z = \tau_{xz} = \tau_{yz} = 0 \qquad \text{(Eq 1.81)}$$

with displacements given by:

$$v = \frac{K_1}{G}\sqrt{\frac{r}{2\pi}}\sin\frac{\theta}{2}\left[2 - 2\left(\frac{v}{1+v}\right) - \cos^2\frac{\theta}{2}\right]$$

$$u = \frac{K_1}{G}\sqrt{\frac{r}{2\pi}}\cos\frac{\theta}{2}\left[1 - 2\left(\frac{v}{1+v}\right) + \sin^2\frac{\theta}{2}\right]$$

$$w = -\frac{K_1}{G}\frac{z}{\sqrt{2\pi r}}\left(\frac{v}{1+v}\right)\cos\frac{\theta}{2} \qquad \text{(Eq 1.82)}$$

It should be emphasized that K is in no way related to plane stress or plane strain, because in either case the crack-tip stress distributions are unchanged. The difference between the hypothetical plane stress and plane strain loci is not due to the stress components in the plane, because they are identical. Referring to Eq 1.77, K is not a function of the coordinates r and θ, but depends on the configuration of the body, including the crack—that is, crack size, overall and local geometry, crack morphology, crack location, and loading condition. As will be discussed in Chapter 4, the difference between plane stress and plane strain hinges on the presence or absence of transverse constraint in material deformation in the vicinity of the crack tip.

Speaking in terms of solid mechanics, the state of stress in a plate—whether it is plane stress or plane strain—is determined by the stress and displacements in the Z-direction (i.e., thickness direction) of the plate, whether or not σ_z, τ_{xz}, τ_{yz}, and w are equal to zero. Translated into plain layman language, this simply means that when a plate is subjected to in-plane loading, there is no stress acting normal to the free surfaces of the plate. When the plate is sufficiently thin, the Z-directional stress across the thickness of the plate can be considered negligible. Therefore, the stress state in a thin sheet is usually plane stress (i.e., stresses are acting only on the X-Y plane). Conversely, the Z-directional stress inside a very thick plate is nonzero, whereas the Z-directional strain is zero. Therefore, it is considered a plane strain condition. In some literature, especially earlier writings, Roman numerals I, II, and III are used to designate the three cracking modes. In this book, Roman numerals I and II are reserved for the plane strain cases, whereas Arabic numerals 1, 2, and 3 are used for an unspecified state of stress (from plane stress to any degree of mixture between plane stress and plane strain). The tearing mode is neither plane stress nor plane strain.

When a cracked plate is subjected to arbitrary loading, the stress field near the crack tip can be divided into three basic types (Fig. 1.61). In addition to K_1, the terminology for the other two crack-tip displacement modes are K_2 for the shear mode and K_3 for the tearing mode. These three crack-tip displacement modes are also known as mode 1, mode 2, and mode 3, respectively. The tensile mode, or opening mode, is associated with local displacement in which the crack surfaces move directly apart. The shear mode, or sliding mode, is characterized by displacements in which the crack surfaces slide

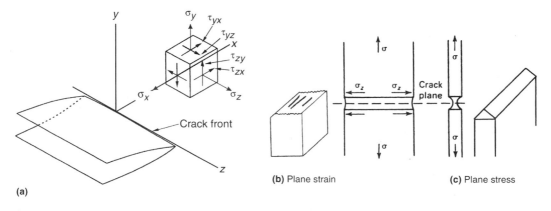

Fig. 1.60 Crack phenomena: (a) Coordinates used to describe stresses near a crack tip. (b) When the section is thick in the Z-direction, deformation is constrained and plane-strain conditions ($\varepsilon_z = 0$) dominate with $\sigma_z = v(\sigma_x + \sigma_y)$. (c) When the section thickness in the Z-direction becomes small, lateral constraint is reduced and deformation occurs in all directions with plane-stress conditions ($\sigma_z = 0$).

over one another in the direction perpendicular to the leading edge of the crack. The tearing mode, or torsion mode, results in the crack surfaces sliding with respect to one another in the direction parallel to the leading edge of the crack. The stress and displacement fields near the crack tip are characterized by the stress intensity factor K_1, K_2, or K_3, or the combination of any two (or all three) of these modes. The linear superposition of these three modes is sufficient to describe the most general case of crack-tip stress fields. In any event, K without a subscript is often used to mean K_1. Discussions about mode 2 and mixed mode 1 and mode 2 are often found in the literature. Reference 1.11 devotes a full chapter to this subject. Stress intensity factor solutions and related cracking mechanisms in mode 3 can only be found in a handful of literature and in handbooks for stress intensity factors (e.g., Ref 1.28).

Stress and displacement field equations for modes 2 and 3, similar to those for mode 1, are available. They are given below.

Mode 2 field equations. Stress equations are:

$$\sigma_y = \frac{K_2}{\sqrt{2\pi r}} \sin\frac{\theta}{2} \cos\frac{\theta}{2} \cos\frac{3\theta}{2} + O(r^{1/2})$$

$$\sigma_x = -\frac{K_2}{\sqrt{2\pi r}} \sin\frac{\theta}{2}\left[2 + \cos\frac{\theta}{2}\cos\frac{3\theta}{2}\right]$$
$$+ O(r^{1/2}) + \sigma_{xo}$$

$$\tau_{xy} = \frac{K_2}{\sqrt{2\pi r}} \cos\frac{\theta}{2}\left[1 - \sin\frac{\theta}{2}\sin\frac{3\theta}{2}\right] + O(r^{1/2})$$
$$\text{(Eq 1.83)}$$

For plane strain, with higher-order terms omitted:

$$\sigma_z = \nu(\sigma_x + \sigma_y) \qquad \text{(Eq 1.84)}$$

$$\tau_{xz} = \tau_{yz} = 0 \qquad \text{(Eq 1.85)}$$

with displacements given as:

$$v = \frac{K_2}{G}\sqrt{\frac{r}{2\pi}}\cos\frac{\theta}{2}\left[-1 + 2\nu + \sin^2\frac{\theta}{2}\right]$$

$$u = \frac{K_2}{G}\sqrt{\frac{r}{2\pi}}\sin\frac{\theta}{2}\left[2 - 2\nu + \cos^2\frac{\theta}{2}\right]$$

$$w = 0 \qquad \text{(Eq 1.86)}$$

For plane stress, with higher-order terms omitted:

$$\sigma_z = \tau_{xz} = \tau_{yz} = 0 \qquad \text{(Eq 1.87)}$$

with displacements given as:

$$v = \frac{K_2}{G}\sqrt{\frac{r}{2\pi}}\cos\frac{\theta}{2}\left[-1 + 2\left(\frac{\nu}{1+\nu}\right) + \sin^2\frac{\theta}{2}\right]$$

$$u = \frac{K_2}{G}\sqrt{\frac{r}{2\pi}}\sin\frac{\theta}{2}\left[2 - 2\left(\frac{\nu}{1+\nu}\right) + \cos^2\frac{\theta}{2}\right]$$

$$w = \frac{K_2}{G}\cdot\frac{z}{\sqrt{2\pi r}}\cdot\left(\frac{\nu}{1-\nu}\right)\cdot\sin\frac{\theta}{2}$$
$$\text{(Eq 1.88)}$$

Mode 3 field equations are:

$$\sigma_x = \sigma_y = \sigma_z = \tau_{xy} = 0 \qquad \text{(Eq 1.89)}$$

with shear stresses of:

$$\tau_{xz} = -\frac{K_3}{\sqrt{2\pi r}}\sin\frac{\theta}{2} + O(r^{1/2}) + \sigma_{xo}$$

$$\tau_{yz} = \frac{K_3}{\sqrt{2\pi r}}\cos\frac{\theta}{2} + O(r^{1/2}) \qquad \text{(Eq 1.90)}$$

and displacements of:

$$v = u = 0 \qquad \text{(Eq 1.91)}$$

$$w = \frac{K_3}{G}\sqrt{\frac{2r}{\pi}}\sin\frac{\theta}{2} \qquad \text{(Eq 1.92)}$$

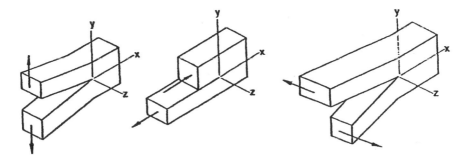

Fig. 1.61 Basic modes of crack surface displacements for isotropic materials

1.8.2 Stress Intensity Factor

The stress intensity factor K is actually a physical quantity, not a factor. By definition, a factor (e.g., the stress concentration factor) is unitless (or dimensionless). However, it is clear in Eq 1.77 that K has a unit of stress times the square root of the crack length (a quantity needed for balancing the stress on the left, and the $1/\sqrt{r}$ on the right, of Eq 1.77). K has been called the stress intensity factor because it appears as a factor in the crack-tip stress and displacement equations. As shown in Chapter 5 and in other sections of this book, K is often expressed as a closed-form solution that includes a conglomeration of many dimensionless factors to account for the unique geometry under consideration. In Chapter 4, K will be introduced as a fracture index, a fracture toughness equivalent. That is certainly a physical quantity. In the remainder of this book, K is referred to as stress intensity or stress intensity factor as the situation demands. In any event, K is mathematically analyzable, and it characterizes the stress and displacement distributions at the crack tip, so it also characterizes the behavior and the criticality of the crack. The solution for K consists of terms representative of stress (or load), crack length, and geometry. It fully accounts for the geometry of a local area in a structure in question, the crack morphology, and how the load is applied. A K-expression is generally written as:

$$K = S\sqrt{\pi a} \cdot \beta \qquad \text{(Eq 1.93)}$$

where S is a stress term, a is an appropriate crack size, and β is the so-called geometric correction factor, all of which are discussed in detail in Chapter 5. For now, remember that K is the driving force for crack propagation and fracture. The stress term in a K-expression is the applied remote gross area stress unless otherwise specified. For K_1, S is the tension stress component normal to the crack; for K_2, S is the shear stress component parallel to the crack, and so on. It also should be noted that in some earlier literature the $\sqrt{\pi}$ term was omitted in the crack-tip stress field equation. For example, Eq 1.77 was written as $\sigma_y = K/\sqrt{2r} \cdot F(\theta)$ instead of $\sigma_y = K/\sqrt{2\pi r} \cdot F(\theta)$. Accordingly, the equation for K was expressed in terms of $\sqrt{a}$ instead of $\sqrt{\pi a}$.

REFERENCES

1.1. C.T. Wang, *Applied Elasticity*, McGraw-Hill, 1953

1.2. A. Nadai, *Theory of Flow and Fracture of Solids*, Vol 1, 2nd ed., McGraw-Hill, 1950

1.3. S. Timoshenko and J.N. Goodier, *Theory of Elasticity*, 3rd ed., McGraw-Hill, 1970

1.4. W.M. Murray and P.K. Stein, *Strain Gage Techniques*, Massachusetts Institute of Technology, 1958

1.5. C.C. Perry and H.R. Lissner, *The Strain Gage Primer*, McGraw-Hill, 1955

1.6. R.J. Roark and W.C. Young, *Formulas for Stress and Strain*, 5th ed., McGraw-Hill, 1975

1.7. S. Timoshenko, *Strength of Materials, Part I: Elementary Theory and Problems*, 2nd ed., D. Van Nostrand, 1941

1.8. S. Timoshenko, *Strength of Materials, Part II: Advanced Theory and Problems*, 2nd ed., D. Van Nostrand, 1941

1.9. J.C. Ekvall, Static Strength Analysis of Pin-Loaded Lugs, *J. Aircraft*, Vol 23, 1986, p 438–443

1.10. V.E. Saouma and I.J. Zatz, An Automated Finite Element Procedure for Fatigue Crack Propagation Analysis, *Eng. Fract. Mech.*, Vol 20, 1984, p 321–333

1.11. A.F. Liu, *Structural Life Assessment Methods*, ASM International, 1998

1.12. A.F. Liu and J.J. Gurbach, Application of a p-Version Finite Element Code to Analysis of Cracks, *AIAA J.*, Vol 32, 1994, p 828–835

1.13. R.E. Peterson, *Stress Concentration Factors*, John Wiley & Sons, 1974

1.14. W.D. Pilkey, *Peterson's Stress Concentration Factors*, 2nd ed., John Wiley & Sons, 1997

1.15. H. Neuber, *Theory of Notch Stresses: Principles for Exact Calculation of Strength with Reference to Structural Form and Material*, 2nd ed., Springer-Verlag, Berlin, 1958 (English transl., Edwards Brothers, Ann Arbor, MI)

1.16. G.N. Savin, *Stress Concentrations Around Holes*, Pergamon Press, 1961

1.17. G.N. Savin, *Stress Distributions Around Holes*, Naukova Dumke Press, Kiev, 1968 (English transl., NASA Technical Translation, NASA TT F607, 1970)

1.18. R.C. Howland, On the Stresses in the Neighborhood of a Circular Hole in a Strip under Tension, *Philos. Trans. Roy. Soc. (London) A*, Vol 119, 1930, p 49–86

1.19. S. Timoshenko and W. Dietz, Stress Concentration Produced at Holes and Fillets, *Trans. Am. Soc. Mech. Eng.*, Vol 47, 1925, p 199

1.20. K. Kathiresan, T.M. Hsu, and J.L. Rudd, Stress and Fracture Analysis of Tapered Attachment Lugs, *Fracture Mechanics: 15th Symposium,* STP 883, ASTM, 1984, p 72–92

1.21. T.M. Hsu, Analysis of Cracks at Attachment Lugs, *J. Aircraft,* Vol 18, 1981, p 755–760

1.22. K. Katherisan, T.M. Hsu, and T.R. Brussat, "Advanced Life Analysis Methods: Crack Growth Analysis Methods for Attachment Lugs," Report AFWAL-TR-84-3080, Vol II, Air Force Flight Dynamics Laboratory, Wright-Patterson Air Force Base, Sept 1984

1.23. A.F. Liu, unpublished data

1.24. R.B. Heywood, *Designing Against Fatigue of Metals,* Reinhold Publishing, 1962

1.25. G.R Irwin, Relation of Stresses Near a Crack to the Crack Extension Force, *Ninth Int. Congr. Appl. Mech.,* Vol 8, 1957, p 245

1.26. G.R Irwin, Analysis of Stresses and Strains Near the End of a Crack Transversing a Plate, *J. Appl. Mech. (Trans. ASME),* Vol 24, 1957, p 361

1.27. G.R Irwin, Fracture, *Hanbuch der Physik,* Vol VI, Springer-Verlag, Berlin, 1958, p 551–590

1.28. H. Tada, P.C. Paris, and G.R. Irwin, *Stress Analysis of Cracks Handbook,* 3rd ed., American Society of Mechanical Engineers, 2000

SELECTED REFERENCES

- Y.C. Fung, *Foundations of Solid Mechanics,* Prentice-Hall, 1965

- H.A. Kuhn, Overview of Mechanical Properties and Testing for Design, *ASM Handbook,* Vol 8, *Mechanical Testing and Evaluation,* ASM International, 2000, p 49–69

- J.D. Landes, W.T. Becker, R.S. Shipley, and J. Raphael, Stress Analysis and Fracture Mechanics, *ASM Handbook,* Vol 11, *Failure Analysis and Prevention,* ASM International, 2002, p 460–483

CHAPTER 2

Deformation and Fracture Mechanisms and Static Strength of Metals

THE THREE TYPES OF IDEAL SUB-STANCES (as discussed in Chapter 1) are the Hooke solid, the St. Venant solid, and the Newtonian liquid. Under load, most engineering materials exhibit some kind of mixed deformation behavior that is either a combination of the Hooke solid and the St. Venant solid, or the Hooke solid and the Newtonian liquid. Chapter 1 also introduced stress analysis, which is a basic step in determining design allowable loads. On the basis of continuum solid mechanics, working solutions of stress/strain and load relationships for some structural elements are derived and presented. In this chapter, the phenomena of deformation and fracture are discussed in more detail for metals and engineering alloys.

Many factors influence deformation and fracture in metals. Of these, the three major areas of concern are mechanical, metallurgical, and environmental factors. Mechanical factors include geometry, loading condition, and loading rate. Structural geometry and loading condition are not just required elements in stress analysis, but also contribute to how a structural part will fail. Loading condition (including loading rate) plays an important role in the mode of structural failure, interacting with structural geometry and ambient environment to govern whether the failure mode is ductile or brittle. On a macroscopic scale, all these are important elements in structural design and fracture prevention.

On a microscopic scale, metallurgical factors such as lattice structure, phase constituents, interstitial alloying or impurities, and grain-boundary behavior can influence the strength and toughness of metals. Interactions between metals and ambient environment (such as tem-perature or corrosive media, etc.) also are crucial to the microprocess of deformation and fracture in metals. In addition, mechanical behavior can be altered by complex thermomechanical effects such as heat treatment, aging, environmentally assisted cracking, and strain aging. Therefore, both the macroscopic and microscopic aspects of metal deformation and fracture need to be understood.

Understanding the deformation mechanisms of metals has been, and still is, a subject of major interest in the study of mechanical metallurgy. There is no shortage of published materials regarding this subject. At a basic level, metals are crystalline with typical lattice structures of one of three types: face-centered cubic (fcc), body-centered cubic (bcc), or hexagonal close-packed (hcp). The deformation and fracture mechanisms for one type of crystal structure differ from the other types. Thus, under a set of known loading conditions, the mechanical behaviors in different types of metals may not be the same owing to their metallurgical details. Explanations of the crystal structures and their slip systems can be found in Appendix 1, which also discusses alternative mechanisms such as twinning and cleavage fracture. While the discussions throughout this book are filled with macroscale aspects of failure, efforts are made to balance them with discussions of microscopic mechanisms whenever applicable.

2.1 Elastic and Plastic Behavior

Before examining the macro- and microscale aspects of deformation and fracture, it is useful to study the load-carrying characteristics of an engineering material through its stress-strain curve obtained from a tension test. The information provided by such a curve is particularly useful to those involved with the plastic working of metals.

2.1.1 Stress-Strain Behavior under Tension Load

Many basic mechanical properties of a material, and how the loading variables can affect mechanical behavior, can be learned from a tensile curve. First, the basic data that can be reduced from a tensile curve are presented; then other parameters and load response phenomena are described.

The tension test is widely used to provide basic design information on the strength of materials and as an acceptance test for the specification of materials. In the tension test, a specimen is subjected to a continually increasing uniaxial tensile force while simultaneous observations are made of specimen elongation. An engineering stress-strain curve is constructed from the load-elongation measurements. A typical engineering stress-strain curve is shown in Fig. 2.1. The stress used in this stress-strain curve is the average longitudinal stress in the tensile specimen. It is obtained by dividing the load (P) by the original cross-sectional area of the specimen (A_0):

$$S = P/A_0 \qquad \text{(Eq 2.1)}$$

The strain used for the engineering stress-strain curve is the average linear strain, which is obtained by dividing the elongation (the amount of extension) by the original gage length that was marked on the specimen:

$$\varepsilon = (l - l_0)/l_0 \qquad \text{(Eq 2.2)}$$

where ε is the average linear strain (also known as the engineering strain, or conventional strain), l_0 is the original gage length, and l is the gage length corresponding to a given point on the stress-strain curve. The engineering unit for strain is dimensionless (mm/mm, or in./in.) for fractional change in length per initial length.

The shape and magnitude of the stress-strain curve of a metal depend on its composition, heat treatment, state of stress, and prior history of plastic deformation; they are also influenced by the loading rate and temperature during testing. The parameters used to describe the stress-strain curve of a material are tensile strength (ultimate strength), yield strength (or yield point), percent elongation, and reduction of area. The first two are strength parameters; the last two indicate ductility. The stress-strain curve also is used to measure the modulus of elasticity (Young's modulus) and the modulus of resilience. These parameters are defined in more detail in the following.

Modulus of Elasticity (E). In Fig. 2.1, point A is called the proportional limit and point B is called the elastic limit. Within the proportional limit the strain will follow the original path going back to the origin of the stress-strain curve during unloading. Elastic limit is the maximum stress that a material is capable of sustaining without any measurable permanent strain (deformation) remaining after complete release of load. The elastic limit is very difficult to measure, but does exist. It is derived from linearity and is permissible in the theory of elasticity as long as there is zero deformation upon complete release of load.

The slope of the initial linear portion of the stress-strain curve is the modulus of elasticity, that is, the ratio of stress (up to the proportional limit) to the corresponding strain. This linear load-deformation behavior is known as Hooke's law:

$$E = S/\varepsilon \qquad \text{(Eq 2.3)}$$

Here E is a measure of the stiffness of the material; the greater the modulus, the smaller the

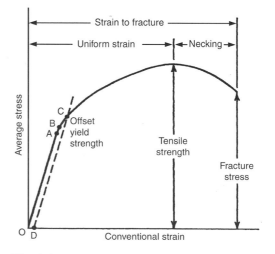

Fig. 2.1 Typical engineering stress-strain curve for a ductile tension-test specimen

elastic strain resulting from the application of a given stress. A material that has a higher Young's modulus is said to be stiffer than one with a lower Young's modulus. Modulus of elasticity has the same dimension as stress because it results from dividing the stress by strain.

Tensile strength (ultimate strength) is defined as:

$$F_{tu} = P_{max}/A_0 \qquad \text{(Eq 2.4)}$$

where P_{max} is the maximum load reached during the test.

Yield strength is defined as:

$$F_{ty} = P_y/A_0 \qquad \text{(Eq 2.5)}$$

where P_y is the load at a selected point on the stress-strain curve (e.g., point C in Fig. 2.1). Stress-strain curves for many materials have the same shape as shown in Fig. 2.1. However, stress-strain curves for mild steels and some other materials do not have a smooth transition between the elastic and plastic regions. As shown in Fig. 2.2, it has a rather sharply defined yield point (point A), followed by a discontinuous region (B, D, and C), and then takes off again at point C. Points A and B (or C) are commonly referred to as the upper and lower yield point, respectively.

For materials that do not have an upper (and/or lower) yield point, the yield stress is customarily determined by drawing a line parallel to the straight portion of the stress-strain curve starting from a selected point on the abscissa of the diagram (point D in Fig. 2.1). The yield stress is taken as the stress at the intersection of this straight line with the stress-strain curve (point C in Fig. 2.1). The magnitude of strain at point D is an arbitrary amount of permanent strain that is considered admissible for general purposes. The value of this strain is 0.002, which is universally adopted by material testing engineers, and is known as the 0.2% offset. For materials that do not have a linear relationship between stress and strain, even at very low stresses, the offset yield stress has to be determined by drawing the line (from the 0.002 strain) parallel to the line that determines the tangent modulus, or the chord modulus (Ref 2.1). Determination of either the tangent modulus or the chord modulus should be taken in the area below the elastic limit.

Elongation. Percent elongation is defined as:

$$\% \text{ elongation} = (l_f - l_0)/l_0 \qquad \text{(Eq 2.6)}$$

This quantity is the same as the conventional strain at fracture:

$$\varepsilon_f = (l_f - l_0)/l_0 \qquad \text{(Eq 2.7)}$$

Here l_f is the final gage length determined by putting the broken pieces back together after specimen failure and then measuring the distance between the gage marks (Fig. 2.3). However, some modern computer-controlled testing systems can obtain data from an extensometer that is left on the testpiece through fracture. In this case, the computer may be programmed to report the elongation as the last strain value obtained prior to some event, perhaps the point at which the applied force drops to 90% of the maximum value recorded. There has been no general agreement about what event should be the trigger. Users and manufacturers of such testing machines agree that different guidelines should be applied to different materials.

Elongation in an unnotched tension-test bar of a ductile material is uniform until the onset of necking. During the stage of uniform elongation, the amount of strain is proportional to the length of the test section. Once necking begins, plastic deformation becomes concentrated in the necked region. Thus, the elongation to fracture (ε_f) depends on the gage length over which the measurement is taken. The smaller the gage length, the greater the contribution to the overall elongation from the necked region and the higher the value of ε_f.

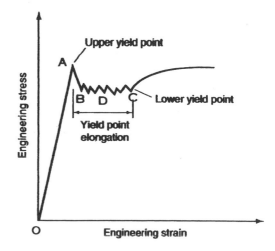

Fig. 2.2 Typical yield behavior

A set of test data that demonstrates the effect of gage length is shown in Fig. 2.3. The values of five final gage lengths are labeled in Fig. 2.3(b), along with a sketch of the broken pieces. The original gage lengths for these gage marks were 25, 75, 125, 175, and 225 (all in mm), respectively. The computed percentage elongation is plotted in Fig. 2.3(a), which shows that a large percentage elongation comes from the shortest original gage length.

The elongation tends to be independent of gage length when a very long gage length is used. However, it is not practical because a very long specimen would have to be used. In practice, the gage length is equal to 4D for round specimens, or 50 mm (2 in.) for rectangular specimens. For economic reasons a subsize specimen (with 25 mm, or 1 in., gage length or shorter) is often used by the vendor for material certification testing and by the procurer for acceptance checking. Nevertheless, when reporting percentage elongation, the gage length should be always noted.

While other issues relevant to selection of testpiece dimensions, shape, and the required

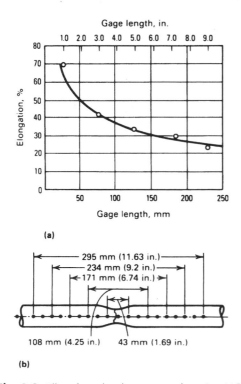

(a)

(b)

295 mm (11.63 in.)
234 mm (9.2 in.)
171 mm (6.74 in.)
108 mm (4.25 in.)   43 mm (1.69 in.)

Fig. 2.3 Effect of gage length on percent elongation. (a) Percent elongation as a function of gage length for a fractured tension testpiece. (b) Distribution of elongation along a fractured tension testpiece. Original spacing between gage marks, 12.5 mm (0.5 in.) Source: Ref 2.1

gage length are discussed in Holt's article (Ref 2.1), there is a way to avoid the complications resulting from necking. That is to base the percentage elongation on the uniform strain out to the point at which necking begins. However, because the engineering stress-strain curve often is quite flat in the vicinity of necking, it may be difficult to establish the strain at maximum load without ambiguity.

Typically, necking begins at the point of maximum load on a stress-strain curve (Fig. 2.1). This is the point when the increase in the load-carrying ability of a ductile metal (due to strain hardening) is overcome by the increase in stress due to the decrease in the cross-sectional area of the specimen. Mathematically, the point of necking at maximum load ($dP = 0$) can be described in calculus as:

$$dP = (\sigma)(dA) + (A)(d\sigma) = 0 \qquad \text{(Eq 2.8)}$$

If the material deforms with a constancy of volume, then:

$$\frac{dl}{l} = -\frac{dA}{A} = d\varepsilon \qquad \text{(Eq 2.9)}$$

With these two relations, then the point of necking is defined in terms of true stress (σ)* and true strain (ε) as follows:

$$\frac{d\sigma}{d\varepsilon} = \sigma \qquad \text{(Eq 2.10)}$$

Substituting Eq 2.10 into the general relation for strain hardening ($\sigma = K\varepsilon^n$) also provides a simple relation for uniform strain (ε_u):

$$\varepsilon_u = n \qquad \text{(Eq 2.11)}$$

where n is the strain-hardening exponent.

Reduction of area is defined as:

$$A_d = (A_0 - A_f)/A_0 \qquad \text{(Eq 2.12)}$$

where A_f is the final dimension of the cross-sectional area.

Modulus of Resilience. Resilience is the ability of a material to absorb energy when deformed elastically and then return to its original shape upon release of load. In other words, it is the amount of energy stored in a material when

---

*True stress and true strain are discussed in section 2.1.2.

loaded to its elastic limit. It is represented by the area under the stress-strain curve up to the elastic limit (see the shaded areas in Fig. 2.4), that is:

$$W_R = \frac{1}{2} \sigma_0 \varepsilon_0 = \frac{1}{2E} \sigma_0^2 \qquad \text{(Eq 2.13)}$$

Of materials A and B in Fig. 2.4, material A has greater resilience because of its higher yield strength and higher modulus of elasticity.

2.1.2 True-Stress/True-Strain Curve

The engineering stress-strain curve does not give a true indication of the deformation characteristics of a metal because it is based entirely on the original dimensions of the specimen. Actually, these dimensions change continuously during the test. Also, ductile metal that is pulled in tension becomes unstable and necks down during the course of the test, because the cross-sectional area of the specimen decreases rapidly during necking. The P/A stress based on original area likewise decreases, producing a falloff in the stress-strain curve beyond the point of maximum load. Actually, the metal continues to strain-harden all the way up to fracture so that the stress required to produce further deformation should also increase. If the true stress, based on the actual cross-sectional area of the specimen, is used, it is found that the stress-strain curve increases continuously up to fracture. If the strains are also based on instantaneous measurements, the curve obtained is known as a true-stress/true-strain curve.

The true stress is the load at any instant divided by the cross-sectional area of the specimen at that instant:

$$S' = P/A \qquad \text{(Eq 2.14)}$$

To compute the true strain (also known as the natural strain, or logarithmic strain), the change in length is referred to the instantaneous gage length, rather than the original gage length. The strain at a certain time during the test is equal to the sum of all the instantaneous strains accumulated up to that instant:

$$\varepsilon' = \sum \frac{l_1 - l_0}{l_0} + \frac{l_2 - l_1}{l_1} + \frac{l_3 - l_2}{l_2} + \text{ etc.}$$
$$\text{(Eq 2.15a)}$$

or

$$\varepsilon' = \int_{l_0}^{l} \frac{dl}{l} = \ln\left(\frac{l}{l_0}\right) \qquad \text{(Eq 2.15b)}$$

Because the volume remains essentially constant during plastic deformation, Eq 2.15(b) can be written in terms of either length or area:

$$\varepsilon' = \ln\left(\frac{A_0}{A}\right) \qquad \text{(Eq 2.15c)}$$

The concept of constant volume also dictates that the sum of three principal strains is equal to zero:

$$\varepsilon_1 + \varepsilon_2 + \varepsilon_3 = 0 \qquad \text{(Eq 2.16)}$$

Equation 2.16 is valid within the elastic limit where the conventional and natural strains are the same. Beyond the elastic limit, the strains in Eq 2.16 should be the natural strains.

By manipulating the above equations, we can arrive at a shortcut of calculating true stress and true strain from engineering stress and strain, respectively. The results are:

$$S' = S (\varepsilon + 1) \qquad \text{(Eq 2.17a)}$$

$$\varepsilon' = \ln (\varepsilon + 1) \qquad \text{(Eq 2.17b)}$$

Note that Eq 2.17(a) and (b) are accurate up to maximum load (beginning of necking). Beyond this point, the major portion of the strain is localized at the neck,* and these equations are not applicable. The true stress and true strain should

---

*The phenomenon of necking and its effect on the values of true stresses (after neckdown) is discussed later in section 2.5.1.

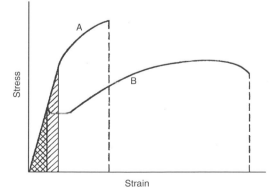

Fig. 2.4 Schematic comparison of stress-strain curves for two structural steels. Source: Ref 2.2

be calculated using Eq 2.14 and 2.15(c). Actually, the strain in the uniformly extended regions outside the neck remains constant (at a value that corresponds to the maximum load).

There is another good reason for using the natural strain instead of the conventional strain. Incompatibility exists between conventional strains in tension and compression tests. For example, when a tensile bar is stretched to a length that is twice the original length, the conventional linear strain is then:

$$\varepsilon_t = (2L_0 - L_0)/L_0 = 1.0 \quad \text{(Eq 2.18a)}$$

where the subscript t stands for tension. It seems logical that a compression test would produce the same amount of strain (although opposite in sign) when the cylinder is compressed to one-half its original height. However, using conventional strain:

$$\varepsilon_c = (L_0/2 - L_0)/L_0 = -1/2 \quad \text{(Eq 2.18b)}$$

where the subscript c stands for compression. It is seen that the computed conventional strains (Eq 2.18a and b) are far from similar. However, using Eq 2.17(b) we obtain:

$$\varepsilon_t' = \ln(\varepsilon + 1) = \ln 2 \quad \text{(Eq 2.18c)}$$

$$\varepsilon_c' = \ln(\varepsilon + 1) = \ln(1/2) = -\ln 2 \quad \text{(Eq 2.18d)}$$

Thus, the natural strains are compatible in both cases.

Many mathematical equations have been proposed to represent the monotonic stress-strain curve. As far back as 1946, Osgood had counted 17 such equations; there are many more today, including two presented here. The Ramberg-Osgood equation (Ref 2.3) has two terms:

$$\frac{\varepsilon}{\varepsilon_0} = \varepsilon^e + \varepsilon^p = \frac{\sigma}{\sigma_0} + \alpha\left(\frac{\sigma}{\sigma_0}\right)^{1/n} \quad \text{(Eq 2.19a)}$$

where $\varepsilon_0 = \sigma_0/E$, σ_0 is the yield stress, and E is the Young's modulus of the material. The coefficient α and the strain-hardening exponent n are determined from the experimental data by a best-fit procedure. Alternatively, Eq 2.19(a) can be written as:

$$\varepsilon = \frac{\sigma}{E} + \left(\frac{\sigma}{\zeta}\right)^{1/n} \quad \text{(Eq 2.19b)}$$

where ζ is a coefficient with dimensions of stress. In a similar manner, the material constants ζ and n are obtained by curve-fitting of the test data. The first term in Eq 2.19(a) and (b) represents the elastic part of the curve, and the second term represents the plastic part.

The stress-strain relationship of materials in the work-hardening range (i.e., the second term of Eq 2.19a and b) is of fundamental importance in engineering applications. In the engineering community, it is common practice to use a very simple equation to express the true-stress/true-strain relationship. It has been assumed that the true-stress/true-strain curve takes the form of a parabola:

$$\sigma = K \cdot \varepsilon^n \quad \text{(Eq 2.20)}$$

Here K is called the strength coefficient (the true stress at $\varepsilon = 1$), n is the linear slope of a log-log plot (double-natural-log scale), implying that the data will lie along a straight line. The slope n is also known as the strain-hardening exponent, because the true stress is increasing (keeping up) with strain. This equation is valid only from the beginning of plastic flow to the maximum load at which the specimen begins to neck down. There are several variations of Eq 2.20, depending on strain rate and temperature; for details, see Dieter (Ref 2.2). The strain-hardening exponent n is commonly used as a convenient indicator of a material's mechanical behavior—alloy formability, in particular. Like modulus of elasticity and Poisson's ratio, it appears in many analytical and semiempirical solutions whenever a strength component is involved. It is frequently used to make correlations with many other types of mechanical testing parameters in an attempt to show that such material behavior can be empirically estimated from the result of a simple tensile strength test.

2.1.3 Yield Phenomenon

Upper and Lower Yield Points. The tensile stress-strain curve of a mild steel shows a localized transition region in the early part of the curve. Rather than exhibiting a gradual transition from elastic to plastic behavior as shown in Fig. 2.1, it goes from elastic to plastic deformation with a yield point (see Fig. 2.2). The load increases steadily with elastic strain, drops suddenly, fluctuates about some approximately constant value of load, and then rises with further strain. The load at which the sudden drop occurs is called the upper yield point. The constant load is called the lower yield point, and the elongation that occurs at constant load is called the yield point elongation (or discontinuous yielding).

At the upper yield point, a discrete band of deformed metal, often readily visible with the eye, appears at a stress concentration such as a fillet. Coincident with the formation of the band, the load drops to the lower yield point (point *B* in Fig. 2.2). The band then propagates along the length of the specimen, causing discontinuous yielding. Usually, several bands will form at several points of stress concentration. These bands, generally inclined at an angle of approximately 45° with respect to the direction of tension, are called Lüders bands or Lüders lines (Fig. 2.5). When the test is in progress, more Lüders bands will form and propagate toward the middle of the specimen. The flow curve during the yield point elongation will be irregular, each jog corresponding to the formation of a new Lüders band. The sketches identified as *B*, *D*, and *C* in Fig. 2.5 correspond to the region *BDC* in Fig.

2.2. After the Lüders bands have propagated to cover the entire length of the specimen test section, the flow will increase with strain in the usual manner. This marks the end of the yield point elongation.

The upper yield point is usually associated with low-carbon steels (more accurately, non-deoxidized low-carbon steels). This type of yield point is also found in nonferrous alloys, depending on the experimental conditions and the composition. Often, the sharp upper yield point may be suppressed due to slow testing speed, less than perfect axial alignment of the specimen, a hard (rigid) testing machine, load frequency, subambient temperatures, and so on. When this happens, a curve of the type shown in Fig. 2.6 results. The flat portion of the curve, which shows a constant yield point (YP), represents continuous yielding.

Work Hardening. The true-stress/true-strain curve is also known as a flow curve since it represents the basic plastic flow characteristics of the material. That is, it shows the stress required to cause the metal to flow plastically to any given strain. Any point on the flow curve can be considered as the yield stress for a metal strained in tension by the amount shown on the curve. Thus, if the load is removed at some point and then reloaded, the material will behave elastically throughout the entire range of reloading. As shown in Fig. 2.7, when a metal is stressed

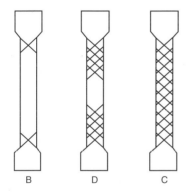

Fig. 2.5 Schematic Lüders bands

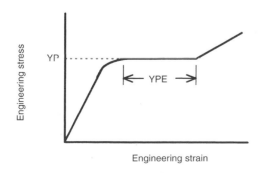

Fig. 2.6 Engineering stress-strain curve without a sharp yield point

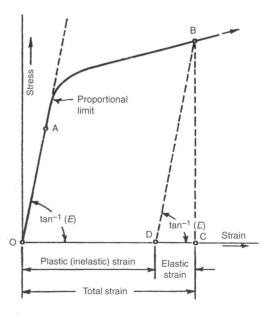

Fig. 2.7 Schematic elastic and inelastic strain. Source: Ref 2.4

to point A below the proportional limit, it will return to the origin upon unloading. If it is stressed to point B, the total strain is OC, of which DC is elastic and OD is plastic (or inelastic). Upon relief of stress, the relationship between stress and strain is expressed by the line BD. The elastic strain is recoverable. However, the plastic strain is nonrecoverable.

If the metal is reloaded, the new stress-strain curve will follow DB and continue along the original path as shown in Fig. 2.7. Following this scenario, one can see that the elastic limit of the material was lifted to point B after the material had been plastically deformed. This behavior is called "work hardening" because the material now has greater resistance to plastic deformation. Generally, the stress-strain curve on unloading from a plastic strain will not be exactly linear and parallel to the elastic portion of the curve. Upon reloading, the curve will generally bend over as the stress approaches the original value of stress from which it was unloaded. With a little additional plastic strain, the stress-strain curve becomes a continuation of the main curve as if the unloading had not occurred (Fig. 2.8).

To look at the sharp yield point phenomenon from another angle, consider the schematic representation of a flow curve of a low-carbon steel shown in Fig. 2.9. If a testpiece initially loaded in tension beyond the elastic limit is then unloaded, the unload path is parallel to the initial load path but offset by the set; on reloading in tension, the unloading path will be followed. Region A of Fig. 2.9 shows that the low-carbon steel was strained plastically (past the yield point elongation) to a strain corresponding to point X. The specimen was then unloaded and reloaded

to point Y without appreciable delay or heat treatment prior to reloading (region B). Upon reloading, the yield point did not occur because the dislocations had been torn away from the atmosphere of carbon and nitrogen atoms. Then the test was disrupted again and unloaded from point Y. The test resumed after the specimen was aged at room temperature for a few days (or at a moderate temperature, say 150 °C, or 300 °F, for several hours). The yield point reappeared as shown. Furthermore, the yield point was higher than before because of the diffusion of carbon and nitrogen atoms to the dislocations to form new atmospheres of interstitials anchoring the dislocations. This yield point recurrence phenomenon is called "strain aging."

Keep in mind that the total amount of elongation (or total strain) remains the same before and after the entire cold-working process no matter how many times unloading and reloading occur. That is, the metal's capability to deform has been reduced after each unloading. In this case, ε_3 is considered the total strain for the virgin material. In other words, if the test was continued from start to finish without any unloading and reloading, the total strain would be 0 to ε_3. If we consider the first reloading (from ε_1) as a new test, and again assume the test would continue through the finish, the total strain this time would be $\varepsilon_3-\varepsilon_1$ because the portion from 0 to ε_1 was lost. The two parts of the tests would join together and seem like a single test without any interruption. Similarly, the total amount of strain available to the second reloading would be $\varepsilon_3-\varepsilon_2$. Again, the entire stress-strain curve would act like a continuous one, as if the little lump

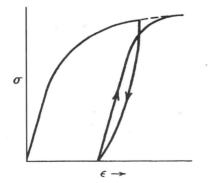

Fig. 2.8 Typical unload and reload behavior in a true-stress/true-strain curve

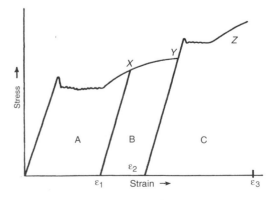

Fig. 2.9 Stress-strain curves for low-carbon steel showing strain aging. Region A, original material strained through yield point. Region B, immediately retested after reaching point X. Region C, reappearance and increase in yield point after aging at 150 °C (300 °F). Source: Ref 2.2

that indicates the increase in yield stress was not there.

This scenario illustrates what happens if a test is stopped, unloaded, and restarted. It also illustrates one of the problems that can occur when testing pieces from a material that has been formed into a part (or otherwise plastically strained before testing). An example is a testpiece machined from a failed structure to determine the tensile properties. If the testpiece is from a location that was subjected to tensile deformation during the failure, the properties obtained are probably not representative of the original properties of the material. Lastly, the ductility of the testpiece would be fully recovered if it were fully annealed each time before reloading. This phenomenon is true whether or not the material has a sharp yield point.

2.1.4 Stress-Strain Behavior under Compression Load

Compression loads occur in a wide variety of material applications, such as upper wing skins, steel building structures, concrete bridge supports, and rolled and forged billets. Characterizing the material response to these loads requires tests that measure the compressive behavior of the materials. In addition, characterizing the mechanical behavior of anisotropic materials often requires compression testing. For isotropic polycrystalline materials, compressive behavior is correctly assumed identical to tensile behavior in terms of elastic and plastic deformation. Usually, the absolute value of compression yield strength and the value of tensile yield strength are the same. However, their ultimate strength may not be the same because their fracture behaviors are not the same, as we shall see later. Highly textured alloys and unidirectionally reinforced composites have different tension and compression stress-strain curves.

A compression test is a convenient method for determining the stress-strain response of materials at large strains ($\varepsilon > 0.5$) because the test is not subject to the necking that occurs in a tension specimen. Compression testing is particularly well suited for brittle materials because compressive stress alone creates a favorable condition for deformation by slip. When a compression load is applied to a cylindrical specimen, however, two problems arise: buckling and barreling. Figure 2.10 illustrates these phenomena, along with some other typical deformation modes in compression.

Buckling is a failure mode characterized by an unstable lateral material deflection caused by compressive stresses. Buckling is controlled by selecting a specimen geometry with a low cylinder height to diameter ratio; $h_0/d_0 < 2$ is preferred, and a specimen having a height to diameter ratio of 1 is often used.

Barreling is the generation of a convex surface on the exterior of a cylinder that is deformed in compression. The cross section of the specimen is barrel shaped. Barreling is caused by the friction between the end faces of the stud and the anvils that apply the load. As the stud decreases in height its diameter increases, because the volume of an incompressible material must remain constant. As the material spreads outward over the anvils, it is restrained by friction at this interface. Thus, the material at the end of the stud deforms less compared to that at the midlength of the stud, where the material is free to deform. Because barreling increases with the d_0/h_0 ratio, the force to deform a compression cylinder increases with d_0/h_0, as shown in Fig. 2.11. The problem of barreling can be minimized by lubricating the anvils and the end surfaces of the stud.

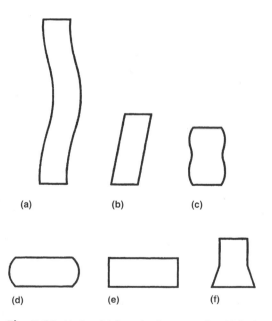

Fig. 2.10 Modes of deformation in compression. (a) Buckling, when $h_0/d_0 > 5$. (b) Shearing, when $h_0/d_0 > 2.5$. (c) Double barreling, when $h_0/d_0 > 2$ and friction is present at the contact surfaces. (d) Barreling, when $h_0/d_0 < 2$ and friction is present at the contact surfaces. (e) Homogenous compression, when $h_0/d_0 < 2$ and no friction is present at the contact surfaces. (f) Compressive instability due to work-softening material. The subscript 0 stands for original dimension. Source: Ref 2.5

Another problem associated with compression tests is the application of compression loads to a thin plate having a rectangular cross section. Although some special jigs are available for this purpose, it is almost impossible to obtain a compression stress-strain curve for a thin plate by taking a simple compression test. However, it is possible to construct a compression stress-strain curve by combining the stress-strain curves that are obtained from tension and bending tests of the same material and the same cross-sectional area. A simple procedure using the Herbert equation (for plastic bending) is demonstrated below.

From Fig. 1.26, we find that the radius of curvature for the outmost fibers is:

$$\rho_t = \rho + \eta_t, \qquad \rho_c = \rho - \eta_c \qquad \text{(Eq 2.21a)}$$

Since $1/\rho = \varepsilon_t/\eta_t$ and $1/\rho = \varepsilon_c/\eta_c$, we have:

$$\frac{1}{\rho} = \frac{1 + \varepsilon_t}{\rho_t} \qquad \text{(Eq 2.21b)}$$

and

$$\frac{1}{\rho} = \frac{1 - \varepsilon_c}{\rho_c} \qquad \text{(Eq 2.21c)}$$

Therefore, from Eq 1.51,

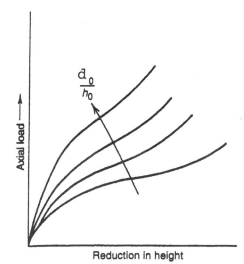

Fig. 2.11 Load-displacement curves for compression tests with specimens having different initial values of d_0/h_0. The subscript 0 stands for original dimension.

$$\theta = \frac{l}{\rho} = \frac{l(1 + \varepsilon_t)}{\rho_t} = \frac{l(1 - \varepsilon_c)}{\rho_c} \qquad \text{(Eq 2.22)}$$

Therefore,

$$\varepsilon_t = \frac{1}{l}(\theta \cdot \rho_t - l) \qquad \text{(Eq 2.23)}$$

and

$$\varepsilon_c = \frac{1}{l}(l - \theta \cdot \rho_c) \qquad \text{(Eq 2.24a)}$$

or

$$\varepsilon_c = \frac{1}{l}(l - \theta \cdot (\rho_t - t)) \qquad \text{(Eq 2.24b)}$$

because $\rho_c = \rho_t - t$. From Eq 1.57 we have:

$$\sigma_c = \frac{\sigma_t \dfrac{d}{d\theta}(M\theta^2)}{bt^2\sigma_t\theta - \dfrac{d}{d\theta}(M\theta^2)} \qquad \text{(Eq 2.25a)}$$

or

$$\sigma_e = \frac{\sigma_t\left(2M + \theta\dfrac{dM}{d\theta}\right)}{bt^2\sigma_t - \left(2M + \theta\dfrac{dM}{d\theta}\right)} \qquad \text{(Eq 2.25b)}$$

By conducting a bending test and measuring θ and ρ_t (as a function of M), we can determine ε_t from Eq 2.23. Separately, we can construct a tensile stress-strain curve from the result of a tensile test. We can then obtain all the σ_t values corresponding to each ε_t on the tensile stress-strain curve. Finally, we can obtain the compression stress σ_c by substituting σ_t into Eq 2.25(a) or (b). A compression stress-strain curve can be drawn by pairing σ_c with ε_c, which are obtained from Eq 2.24(b).

Alternatively, we can construct an M versus θ curve, similar to Fig. 1.27, and transform Eq 2.25(b) into graphical form:

$$\sigma_c = \frac{\sigma_t(2\overline{PC} + \overline{PB})}{bt^2\sigma_t - (2\overline{PC} + \overline{PB})} \qquad \text{(Eq 2.25c)}$$

Therefore, a compression stress-strain curve can be constructed using Eq 2.24(b) paired with Eq 2.25(b) or (c).

2.2 Yield Criteria

The main reason for determining the magnitudes and directions of the principal stresses and strains and of the maximum shearing stresses in a structural member is to use them in evaluating whether the design (including the dimensions and the selected material) can withstand the intended service load. There are two general types of failure criteria:

- Yielding criteria
- Fracture criteria

Failure from yielding occurs when the monotonic load on a part results in stresses that exceed the yield strength of the material. If the piece is subjected to a set of combined loads (biaxial or triaxial), yielding occurs when the level of the effective stress (which accounts for the effect of each stress component) is equal to the material tensile yield strength. In certain situations, effective strain is used as the criterion for failure so that a part is designed not to exceed a certain amount of permanent deformation in its operational life. When the von Mises equivalent stress $\overline{\sigma}$ is adopted as the effective stress (Eq 1.27a to 1.28b), the material is considered yielded when $\overline{\sigma}$ is equal to or exceeds the material tensile yield strength F_{ty}.

In a similar manner, Nadai (Ref 2.6) introduced the concept of octahedral stresses, which is based on the stresses acting on the octahedral plane (Fig. 2.12). The octahedral shearing stress whose normal makes equal angles with σ_1, σ_2, and σ_3 is given by:

$$\tau_{oct} = \frac{1}{3} \cdot \sqrt{(\sigma_1 - \sigma_2)^2 + (\sigma_2 - \sigma_3)^2 + (\sigma_3 - \sigma_1)^2}$$

(Eq 2.26a)

or

$$\tau_{oct} = \frac{1}{3} \cdot \sqrt{(\sigma_x - \sigma_y)^2 + (\sigma_y - \sigma_z)^2 + (\sigma_z - \sigma_x)^2 + 6(\tau_{xy}^2 + \tau_{yz}^2 + \tau_{zx}^2)}$$

(Eq 2.26b)

The octahedral strain is given by:

$$\gamma_{oct} = \frac{2}{3} \cdot \sqrt{(\varepsilon_1 - \varepsilon_2)^2 + (\varepsilon_2 - \varepsilon_3)^2 + (\varepsilon_3 - \varepsilon_1)^2}$$

(Eq 2.26c)

or

$$\gamma_{oct} = \frac{2}{3} \cdot \sqrt{(\varepsilon_x - \varepsilon_y)^2 + (\varepsilon_y - \varepsilon_z)^2 + (\varepsilon_z - \varepsilon_x)^2 + \frac{3}{2}(\gamma_{xy}^2 + \gamma_{yz}^2 + \gamma_{zx}^2)}$$

(Eq 2.26d)

Comparing these equations with the equations for $\overline{\sigma}$ and $\overline{\varepsilon}$ in Chapter 1, we see that:

$$\tau_{oct} = \frac{\sqrt{2}}{3} \overline{\sigma}$$

(Eq 2.27a)

and

$$\gamma_{oct} = \sqrt{2} \cdot \varepsilon$$

(Eq 2.27b)

When we use the octahedral shear stress (or the von Mises stress) and the octahedral shear strain (or the von Mises strain) as yield criteria, we consider the effective stress as a measure of the state of stress with respect to its propensity for causing a plastic distortion. The effective strain is regarded as a measure of the magnitude of the plastic distortion or change of shape.

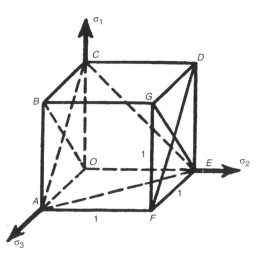

Fig. 2.12 Octahedral plane (*ACE*) and planes of maximum shear (*ACDF* and *BGEO*)

Another known criterion of failure is called the maximum shear stress theory, or the Tresca criterion of failure (also known as the Guest-Mohr criterion). It simply states that yielding will occur when the maximum shear stress reaches a critical value equal to the shearing yield stress in a uniaxial tension test. As evidenced by the Mohr circle, the maximum shear stress has a value of half the difference between the highest and the lowest principal stresses. It is customarily assumed that σ_1 is the highest and σ_3 is the lowest. Then:

$$\tau_{max} = (\sigma_1 - \sigma_3)/2 \qquad \text{(Eq 2.28a)}$$

which acts on the planes 45° to σ_1 and σ_3 and parallel to σ_2 (see Fig. 2.12), and consequently the corresponding maximum strain will be:

$$\gamma_{max} = \varepsilon_1 - \varepsilon_3 \qquad \text{(Eq 2.28b)}$$

In the case of uniaxial tension, from which the basic mechanical properties of materials are developed,

$$\tau_{max} = \sigma_1/2 = \sigma_{max}/2 \qquad \text{(Eq 2.29)}$$

Although the maximum shearing stress is one-half the maximum normal stress, this stress is sometimes the controlling factor when considering the strength of materials that are much weaker in shear than in tension. Comparing Eq 2.29 with Eq 2.27(a), interestingly, we discover that the yield criterion for a multiaxial stress system is very much the same as for simple tension. This is because by analogy τ_{oct} is equivalent to τ_{max}, $\overline{\sigma}$ is equivalent to σ_{max}, and the factor $\sqrt{2}/3$ in Eq 2.27(a) is approximately equal to ½ (0.471 to be exact). Rewriting Eq 1.30(a) for pure shear stress yields:

$$\tau = \overline{\sigma}/\sqrt{3} \qquad \text{(Eq 2.30)}$$

Thus, using the von Mises stress criterion (also known as the von Mises-Hencky criterion), the yield strength in shear as determined from a torsion test should be only 57.7% of the tensile yield strength. It is higher than 50% per the maximum shear stress theory and 47.1% per the octahedral stress theory.

In the case of combined tension and torsion, the yield criteria derived from the Tresca and the von Mises theories would be, respectively (Ref 2.6):

$$\sigma^2 + 4\tau^2 = \sigma_0^2 \qquad \text{(Eq 2.31a)}$$

and

$$\sigma^2 + 3\tau^2 = \sigma_0^2 \qquad \text{(Eq 2.31b)}$$

where σ is the axial tensile stress, τ is the shear stress, and σ_0 is the material tensile yield strength (a constant, same as F_{ty}) for pure tension. These equations can be transformed into an elliptical format:

$$\left(\frac{\sigma}{\sigma_0}\right)^2 + 4\left(\frac{\tau}{\sigma_0}\right)^2 = 1 \qquad \text{(Eq 2.32a)}$$

and

$$\left(\frac{\sigma}{\sigma_0}\right)^2 + 3\left(\frac{\tau}{\sigma_0}\right)^2 = 1 \qquad \text{(Eq 2.32b)}$$

Checking the applicability of these equations, Taylor and Quinney (Ref 2.7) conducted tests for aluminum, copper, and steels; their results are shown in Fig. 2.13 and 2.14. Comparing the correlations made with the Tresca and the von Mises theories, the latter is better.

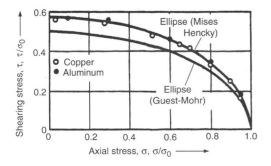

Fig. 2.13 Comparison between failure criteria: correlation with tension-torsion test data of aluminum and copper. Source: Ref 2.7 (graphs adopted from Ref 2.6)

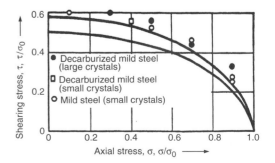

Fig. 2.14 Comparison between failure criteria: correlation with tension-torsion test data of steels. Source: Ref 2.7 (graphs adopted from Ref 2.6)

2.3   Fracture Criteria

The prevention of yielding is an important factor in structural design, and the von Mises, Nadai, and Tresca equations are customarily used to determine conditions for yielding. Of course, the other important criterion of failure is fracture from either brittle or ductile cracking. Brittle and ductile fractures represent different processes of cracking, and so different fracture criteria are needed for ductile and brittle fracture under static (monotonic) loads. From the standpoint of continuum mechanics, brittle cracking is largely an elastic process with a rapid release of stored mechanical energy. In this case, the work (or expended energy) of the brittle cracking process is the energy needed to create new surfaces from material separation. In contrast, the continuum mechanics of ductile fracture involves plastic deformation, which thus results in the expenditure of mechanical energy prior to fracture.

The occurrence of a ductile or brittle fracture depends on a variety of factors, including material condition, strain rate, temperature, and geometric features (e.g., notches, preexisting cracks, and/or discontinuities). However, ductility and brittleness of a material (in a given geometry/service condition) also depends on the state of stress. Therefore, it is necessary to understand not only why some materials are inherently ductile or brittle, but also how the state of stress influences the fracture process and the determination of fracture criteria. This section briefly reviews how fracture behavior is influenced by the state of stress (i.e., tension, shear, torsion). The next section describes the fracture mechanisms and appearances of ductile and brittle materials under the controlled conditions of a standard tension test.

Effect of Stress State on Fracture. From the standpoint of mechanics, fracture from a monotonic load may occur from material separation in a plane approximately normal to a tensile stress or by a shear stress that causes glide (slip) and deformation. Fracture by separation normally leads to brittle fracture, as indicated by the appearance of a flat fracture surface and little or no macroscopic yielding prior to the fracture. Separation from a tensile stress can occur due to brittle fracture mechanisms (i.e., cleavage or intergranular fracture) or from geometric conditions (notches, cracks, section size) that constrain deformation and thereby induce plane-strain crack growth under tension. These various mechanisms are described in section 2.4.

Fracture by glide is a ductile-type fracture from deformation along a shear plane, whereby one part of the body is sheared against the other. This can result in a fracture surface with shear lips that correspond to fracture along the shear plane. From the standpoint of mechanics, failure could be caused by shear alone, without a tension stress component in a stress system. Figure 2.15 demonstrates this concept via a simple compression test of a soil cylinder. Since there is no tension stress in a simple compression test (as depicted by a Mohr circle), the piece must fail by shear, the only available stress component that could cause failure.

To be clear, Fig. 2.15 is intended only to demonstrate a failure that is attributed to shear stress alone, not to demonstrate a compression failure. Actually, compression failure is much more complicated than this, because a variety of deformation mechanisms may cause brittle or ductile fracture. For example, barreling (or bulging) during compression testing of metals occurs due to friction at the contact surfaces between the anvils and the end surfaces of the stud. For a given axial compressive strain, the bulge profile would create a tensile strain in the circumferential direction, which is maximum at midheight of the stud. The tensile stress that associates with this circumferential strain could cause crack initiation on the hoop plane. The final fracture plane would be vertical, instead of on a slanted angle. On the other hand, the stud might be compressed to a very thin piece without fracture if friction is eliminated.

Fig. 2.15   Shear fracture of a soil specimen. Source: Ref 2.8

In light of the foregoing discussion, the maximum normal stress and the maximum shear stress are important stress quantities. The maximum normal stress corresponds to the principal stress σ_1. Thus:

$$\sigma_{max} = \sigma_1 \qquad \text{(Eq 2.33)}$$

The maximum shear stresses for the three basic loading cases (Fig. 2.16) are given as:

For simple tension, $\tau_{max} = \sigma_1/2 = \sigma_{max}/2$
$$\text{(Eq 2.34a)}$$

For simple compression, $\tau_{max} = -\sigma_3/2; \sigma_{max} = 0$
$$\text{(Eq 2.34b)}$$

For torsion, $\tau_{max} = \sigma_1 = \sigma_{max}$ \qquad (Eq 2.34c)

The values of σ_{max} and τ_{max}, and sometimes the planes on which they act, are the stress factors of greatest interest.

Fracture from a tension-stress state promotes brittle-like fracture profiles, because it causes separation of the planes normal to it. In contrast, shear stress causes slip, and hence is associated with ductile fracture. It is believed that the rate

of plastic flow is closely related to τ_{max}. The process of flow is fundamentally a shearing operation, in which one layer of the body slides past an adjacent layer. Thus, on a macroscopic scale, where there is no shear stress there will be no plastic deformation or flow. This is also the case for hydrostatic tension (or so-called triaxial stresses), where all three principal stresses are in tension of equal magnitude. There is no shear stress in this stress state. Thus, it is a fully brittle condition. As for the case of hydrostatic compression, it is just the same as for hydrostatic tension; that is, it produces no shear. However, it also has no tension stress. Therefore, a material subjected to hydrostatic compression would carry the pressure up to its elastic limit, then crumble. This is not the same kind of brittle fracture (i.e., by separation) as generally expected. Slippage can occur if we superimpose some compression load (in one direction) to the hydrostatic pressure.

Torsional failure can occur in one of two modes—shear or tension—because the magnitudes of τ_{max} and σ_{max} are equal. The yield strength in shear is only about 47 to 58% of the tensile yield strength for a given material, depending on the yield criterion. The applied load

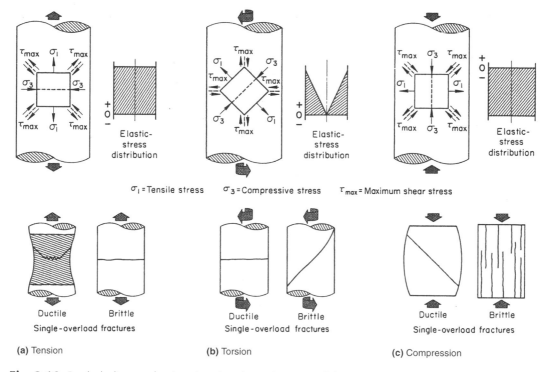

$\sigma_1 = $ Tensile stress $\sigma_3 = $ Compressive stress $\tau_{max} = $ Maximum shear stress

(a) Tension   (b) Torsion   (c) Compression

Fig. 2.16 Free-body diagrams showing orientation of normal stresses and shear stresses in a shaft under simple (a) tension, (b) torsion, and (c) compression loading, and the single-overload fracture behavior of ductile and brittle materials

will reach the yield strength for shear before it reaches the tensile yield strength of the material, and thus shear mode failure will prevail.

As shown in Fig. 2.16(b), there are two planes for maximum shear. That means the fracture path may lie on a plane that is either 90° or 0° to the longitudinal axis. The likely outcome is that fracture initiates on the surface (of the circular bar) and then progresses inward along the 90° plane. Most the time the fracture path stays on the 90° plane until complete fracture occurs.

This type of fracture mode is shown in Fig. 2.17. Depending on the situation, the initial fracture path may turn from the 90° plane into the 0° plane. A brittle material might exhibit a failure mode that is lying on the 45° plane, which is perpendicular to maximum tension, possibly due to an insufficient amount of (or lack of) plastic strains in the stress-strain curve. Since this plane bisects the angle between the two planes of maximum shear and makes a 45° angle with the longitudinal and transverse directions, it results in a helical fracture.

For a structural member of which the geometry and metallurgical factors and service condition are given, the material can exhibit varying degrees of ductility depending on the states of stress. Compression stress component(s) promote a shear mode of failure. Therefore, adding compression stresses to the stress system always results in a beneficial mode of failure. An important state of stress is pure shear. Other than

torsion, a pure shear condition can be reached by applying equal amounts of tensile and compressive stresses ($\sigma_3 = -\sigma_1$, $\sigma_2 = 0$) to a unit cube (Ref 2.2). Compression stress components are often found in plastic working of metals such as rolling, in which σ_2 (along the plate width direction) is nearly zero, and σ_3 (acting on the plate surface by the rollers) is compression.* In the case of extrusion, both σ_2 and σ_3 are in compression.

According to the theory of plasticity and assuming that $\sigma_1 > \sigma_2 > \sigma_3$, we can express each shear stress in a three-dimensional slip system as follows (Ref 2.6):

$$\tau_1 = |\sigma_2 - \sigma_3|/2$$
$$\tau_2 = |\sigma_3 - \sigma_1|/2 \qquad \text{(Eq 2.35)}$$
$$\tau_3 = |\sigma_1 - \sigma_2|/2$$

By going through several simple exercises (Table 2.1), we can demonstrate how tension and compression stress components affect the magnitude of maximum shear stress. For those data listed in Table 2.1, it has been assumed that σ_1 is the maximum normal stress and that the magnitude of σ_1 in each case has been kept constant.

---

*In practice, the strain is positive in the rolling direction but negative in the thickness direction. The sheet width is assumed unchanged after rolling (i.e., $\varepsilon_2 = 0$), and thus a plane strain condition is assumed.

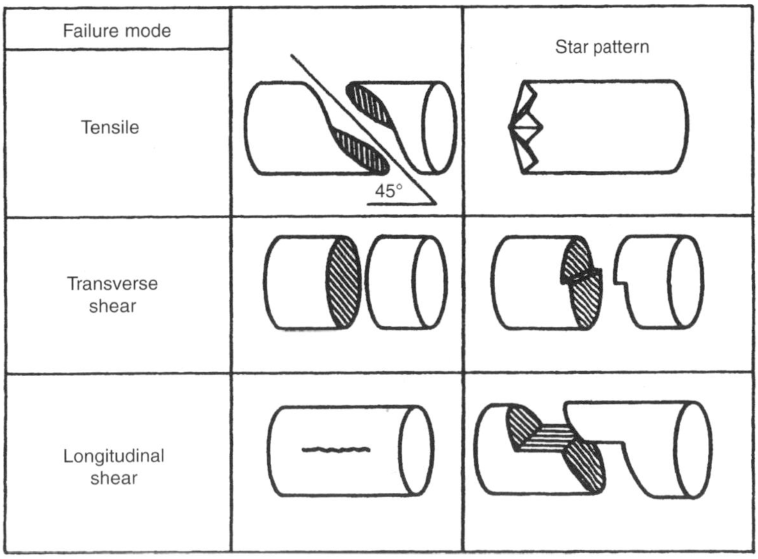

Failure mode		Star pattern
Tensile		
Transverse shear		
Longitudinal shear		

Fig. 2.17 Typical appearances of torsional fractures. Source: Ref 2.9

The magnitude of the normal stresses in the other two directions had fractions of the magnitude of σ_1 (in either positive or negative direction). Table 2.1 also shows that τ_2 is the maximum shear stress in all cases. With reference to the baseline value, which is the τ_{max} value for uniaxial tension, the τ_{max} value for the other cases changes along with the states of stress. With a higher τ_{max} value, the substance in question is more ductile. Otherwise, it would be less ductile or even brittle.

2.4 Fracture Mechanisms and Appearances

In very general terms, when crack size (a) in a stressed part reaches a critical size (a_{cr}), the fracture process occurs almost instantaneously with complete and sudden separation of the part. This stage of final fracture is referred to as overload fracture, as discussed in more detail in this section. However, it must be recognized that many overload failures occur after subcritical ($a < a_{cr}$) crack growth from various progressive damage mechanisms such as fatigue or environmentally assisted cracking. This section focuses mainly on overload fractures under static loads, while fatigue fracture is discussed in more detail in Chapter 3. Environmental effects on cracking under static load are introduced in section 2.8.

Overload cracking can be categorized into three general types of mechanisms:

- *Brittle overload* from cleavage (i.e., transgranular brittle cracking)
- *Ductile overload* failures that involve the fracture mechanisms of ductile tearing and/ or microvoid formation caused by transgranular slip
- *Stress rupture* (sometimes referred to as decohesive rupture), when a pair of free surfaces are created from a preexisting grain boundary or second-phase boundary

Ductile and brittle cracking are the two main types of overload fracture, while stress rupture includes various types of mechanisms that may be either brittle or ductile. For example, a brittle stress-rupture failure may occur from grain-boundary embrittlement, while a ductile stress-rupture failure may occur from grain-boundary slip (i.e., time-dependent creep deformation in polycrystalline metals). These underlying microscopic mechanisms are discussed in more detail in section 2.4.2.

When examining fracture surfaces, it is important to obtain an overall perspective from both macroscopic and microscopic study. Examination beyond the fracture surface also provides information. For example, visual inspection of a fractured component may indicate events prior to fracture initiation, such as a shape change indicating prior deformation. Metallographic examination of material removed far from the fracture surface also can provide information regarding the penultimate microstructure, including the presence of cold work (bent annealing twins, deformation bands, and/or grain shape change), evidence of rapid loading and/or low-temperature service (deformation twins), and so forth. These types of investigative methods are also important in the analysis of fractures.

In general, identifying the cause and corrective action of a fracture benefits by the careful documentation of various macroscopic and microscopic observations. Typically, observation begins with a visual examination of a fracture surface, where general features and surface roughness can be revealed under favorable lighting. Examination at low magnification (about $15\times$ or less) can also reveal features regarding the nature of the fracture path. Metallographic and fractographic techniques then can be used to reveal microscopic features. The failed piece may be properly sectioned for preparation of metallographic samples and examination under an optical microscope. Alternatively or in addition, an electron microscope—typically a scanning electron microscope (SEM), or a transmission electron microscope (TEM)—with higher magnifications and depth of field may be used to make direct examination of the raw fracture surface.

Table 2.1 Maximum shear stress as a function of states of stress

Case	Applied σ components	Computed τ components	Maximum τ (τ_{max})	Comment
1	$\sigma_1 = \sigma_{max}$	$\tau_1 = 0$	$\sigma_1/2$	Baseline
	$\sigma_2 = 0$	$\tau_2 = \sigma_1/2$		
	$\sigma_3 = 0$	$\tau_3 = \sigma_1/2$		
2	$\sigma_1 = \sigma_{max}$	$\tau_1 = \sigma_1/4$	$\sigma_1/2$	Unchanged
	$\sigma_2 = \sigma_1/2$	$\tau_2 = \sigma_1/2$		
	$\sigma_3 = 0$	$\tau_3 = \sigma_1/4$		
3	$\sigma_1 = \sigma_{max}$	$\tau_1 = 0$	$\sigma_1/4$	Less ductile
	$\sigma_2 = \sigma_1/2$	$\tau_2 = \sigma_1/4$		
	$\sigma_3 = \sigma_1/2$	$\tau_3 = \sigma_1/4$		
4	$\sigma_1 = \sigma_{max}$	$\tau_1 = 0$	$\frac{3}{4}\sigma_1$	More ductile
	$\sigma_2 = -\sigma_1/2$	$\tau_2 = \frac{3}{4}\sigma_1$		
	$\sigma_3 = -\sigma_1/2$	$\tau_3 = \frac{3}{4}\sigma_1$		
5	$\sigma_1 = \sigma_{max}$	$\tau_1 = \sigma_1/2$	σ_1	More ductile
	$\sigma_2 = 0$	$\tau_2 = \sigma_1$		
	$\sigma_3 = -\sigma_1$	$\tau_3 = \sigma_1/2$		

Both the macro- and microscale appearances of fracture-surface features can tell a story of how and sometimes why fracture occurred in terms of the following information:

- Crack initiation site and crack propagation direction
- Mechanism of cracking and the path of fracture
- Load conditions (tension, bending, shear, monotonic, or cyclic)
- Environment
- Geometric constraints that influenced crack initiation and/or crack propagation
- Fabrication imperfections that influenced crack initiation and/or crack propagation

This section does not discuss how to perform an analysis, nor does it present exhaustive coverage on the subject of fractography. Instead, some typical appearances are described for certain types of fractures to illustrate general fracture characteristics and how classifications of fracture are made on the basis of the appearance of their fracture surfaces, the nature of the fracture path, and the amount of plastic deformation prior to fracture. Selected references on fractography are listed at the end of this chapter.

2.4.1 Ductile Fracture

Ductile Tension-Test Fracture. The mechanisms and appearances of ductile fracture are best introduced with the simple example of an unnotched bar subjected to conventional tension testing at quasi-static strain rates (<0.1 s^{-1}). In general, tension test specimens of a ductile ma-

terial have a visible region of necking (Fig. 2.18), while a brittle specimen results in fracture with little or no visible evidence of any necking (Fig. 2.19). Necking is a region of strain localization that forms when the increase in stress due to decrease in the cross-sectional area of the specimen becomes greater than the increase in the load-carrying ability of the metal due to strain hardening (see section 2.1.1). Necking generally occurs at the point of maximum load on the engineering stress-strain curve (Fig. 2.1). Prior to the onset of necking, strain is uniform along the gage length, while plastic deformation becomes concentrated in the necked region of the tension specimen. The size of the neck and the extent to which it is visible depends primarily on strain hardening and strain-rate hardening (when temperatures are below 0.4 T_m of the metallic material, where T_m is the material melting point on the Kelvin scale).

The resulting necked region is, in effect, a mild notch, which introduces a complex state of stress that has a large tensile-hydrostatic component (see also "Necking of an Unnotched Tensile-Test Specimen" in section 2.5.1). This tensile-hydrostatic (or triaxial) stress is highest in the center of the specimen, where microvoids occur from the tensile separation of the ductile matrix from harder second-phase particles or inclusions (which are present in most commercial alloys). This central region of the fracture surface also is typically flat (at the macroscale), which indicates separation from a tension stress state. Thus, even though the ductile fracture involves deformation, the microscopic mechanism of crack initiation involves the brittlelike effect

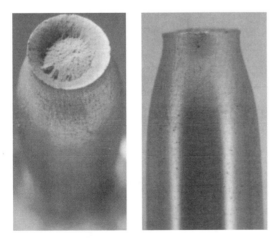

Fig. 2.18 Cup-and-cone fracture of a low-carbon steel bar under tension. Source: Ref 2.10

Fig. 2.19 Brittle fracture of a smooth (unnotched) tensile test specimen. Source: Ref 2.10

of tensile separation in the region of a deformation-induced notch.

The microvoids also grow and connect to ultimately form and fracture near the central region of the specimen. However, as the central crack grows, the material in the outer annulus deforms by stresses along the shear plane (45° to the direction of the tensile load). This results in distinctive shear lips of ductile fracture and the classic cup-and-cone profile of the fracture surface. One piece has a conelike profile, while the other piece (shown in Fig. 2.18) has the surface profile of a cup. The ratio of the area of the flat-face region to the area of the shear lip usually increases with section thickness. The area of the shear lip also depends on the extent of necking.

The amount of necking depends on the extent of strain and strain-rate hardening, which in turn are influenced by factors such as temperature and material condition. Lowering the temperature below room temperature generally increases the strain-hardening exponent, n. This increases the strain to neck formation. In contrast, an ideal plastic material (in which no strain hardening occurs) becomes unstable in tension and begins to neck as soon as yielding occurs.

Most metals and alloys (not heavily cold worked) undergo strain hardening, which tends to increase the load-carrying capacity of the specimen as deformation increases. Commercial engineering materials also typically contain inclusions, second-phase particles, and other constituents. These microscopic constituents can influence the process of fracture nucleation and crack growth. In an ideal material containing neither inclusions nor second phases, ductile fracture would be expected to occur by slip and possibly twinning, resulting in complete reduction in area. Alternately, cleavage across a grain on a single plane would be expected to result in a smooth fracture surface. Such results are sometimes observed in high-purity single-crystal specimens, but are seldom seen in commercial engineering materials.

Fracture by uninterrupted plastic deformation is a special type of plastic fracture. For metals that do not work harden much, the metal under tension would be drawn down almost to a chisel edge or a point before breaking apart (see Fig. 2.20a). This special circumstance can hardly be termed "fracture" in a normal sense, as there is no fracture surface at all. It is usually referred to as "rupture" because the process arises from prolonged shear on slip planes within the worked region of the crystals, which finally at one point shear apart. It should also be noted that strain rate and adiabatic deformation also play an important role in this type of behavior. That is, high temperature and very slow strain rates can result in extensive uninterrupted deformation.

Another form of uninterrupted plastic deformation, which may or may not result in a "chisel point" type fracture, is the shearing of a single crystal. As described in Appendix 1, deformation of a single crystal is governed by the critical resolved stress and slip occurs in an active slip plane and slides in a specific direction. Figure 2.20(b) shows a copper-aluminum single crystal that has gone through such a prolonged extension. Final separation of the specimen eventually occurs by "shearing off" at one of the slip planes. Whether a single crystal will fracture by shearing off or by drawing down to a chisel point depends on the slip system of a particular crystal (Ref 2.17).

General Macroscopic Appearance of Ductile Fractures. For ductile fracture, macroscopic features include:

- A relatively large amount of plastic deformation precedes the fracture.
- Shear lips are usually observed at the fracture termination areas.
- The fracture surface may appear to be fibrous or may have a matte or silky texture, depending on the material.
- The cross section at the fracture is usually reduced by necking.
- Crack growth is slow.

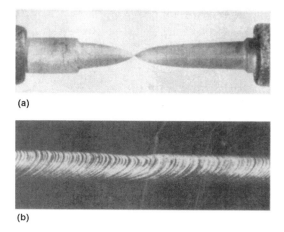

(a)

(b)

Fig. 2.20 Examples of uninterrupted shear failure. (a) Polycrystalline aluminum bars pulled at 600 °C (1110 °F). Source: Ref 2.6. (b) Extended copper-aluminum single crystal. Source: Ref 2.6

Ductile fractures often progress as single cracks, without many separated pieces or substantial crack branching at the fracture location. The region of crack initiation typically has a dull fibrous appearance that is indicative of cracking by microvoid coalescence. The crack profiles adjacent to the fracture are consistent with tearing. The fracture surface may have radial markings, chevrons, and/or shear lips, depending on the specimen geometry and material condition.

Depending on the state of stress and geometric constraints on macroscopic ductility, the fracture of a ductile material may occur by plane strain, plane stress, or mixed mode. In general, these variations in fracture profiles are related to fracture toughness, which depends on section thickness (B) and the crack size (a) of a preexisting discontinuity such as a crack or notch. Figure 2.21 is a schematic illustration of this for an inherently ductile material with varying section thickness (see also section 4.4.1 for a more detailed discussion of this subject).

Plane-strain fracture is characterized by a flat surface perpendicular to the applied load. Plane-stress fracture occurs when shear strain becomes the operative mode of deformation and fracture (as maximum stresses occur along the shear plane from the basic principles of continuum mechanics). In plane-stress cracking, the fracture profile is characterized by shear lips, which are at about a 45° oblique angle to the maximum stress direction (although this angle may vary depending on material condition and loading condition).

The classic cup-and-cone appearance that results from ductile fractures of unnotched cylindrical tension test specimens is a good example of mixed-mode fracture. In this case, crack initiation near the specimen center occurs from fracture under triaxial tension and thus has a flat fracture surface normal to the applied load. When fracture reaches the region near the outer surface, deformation by slip (shear) becomes dominant, and the stress state of fracture changes to plane stress. However, shear lips are not necessarily the definitive characteristic of a ductile fracture. The macroscale appearance of a fracture is also influenced by geometric constraints and stress-state conditions. For example, even though a flat center region of crack initiation is characteristic of tensile (brittlelike) separation, the flat fracture region of a ductile fracture also has a dull fibrous appearance with small microscopic dimples (see section 2.4.3). Microscopic dimples are indicative of separation

from displacement within a ductile matrix (or localized regions) of a material.

2.4.2 Brittle Fracture

Materials that do not develop a neck before fracture are generally considered brittle. When the specimen lacks ductility (due to low temperature, environment, strain rate, or the material itself), the fracture is brittle and occurs by separation on a plane that is normal to the direction of the applied load. Due to absence of necking, deformation is approximately uniform until fracture occurs from complete separation under tension. However, lack of ductility depends on a number of other factors, such as environment, strain rate, and the internal state of stress created (influenced by part geometry and discontinuities in the material). Therefore, lack of ductility is not just due to the material itself, but is also influenced by complex relationships of stress state, part geometry, localized deformation, and internal discontinuities.

Conversely, a fracture surface normal to the applied load also is not necessarily a definitive indication of an inherently brittle material or a brittle mechanism (such as cleavage or intergranular fracture, discussed in section 2.4.3). This important distinction is indicated in the central region of the classic ductile fracture in Fig. 2.18. Initial deformation causes a region of localized strain (i.e., necking), which is essentially a mild notch that results in the development of hydrostatic tensile stresses in the interior of a tension test specimen. This "triaxial" state of tension then causes crack initiation (void formation) by tensile separation around small inclusions, second-phase particles, or discontinuities in the center region of the tension bar. In effect, ductile mechanisms lead to a stress state that causes tensile separation around less ductile

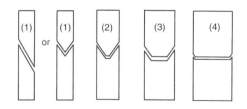

Fig. 2.21 Schematic of variation in fracture behavior and macroscale features of fracture surfaces for an inherently ductile material. As section thickness increases, plane-strain conditions develop first along the centerline and result in a flat fracture surface. With further increases in section thickness, the flat region spreads to the outside of the specimen, decreasing the widths of the shear lips.

inclusions or second phases in the specimen material.

In general, brittle fracture can be distinguished by these characteristics:

- Little or no visible plastic deformation precedes the fracture.
- The fracture surface is generally flat and perpendicular to the loading direction and to the component surface.
- The fracture may appear granular or crystalline and is often highly light reflective. Facets may also be observed, particularly in coarse-grain steels.
- Chevron patterns may be present.
- Rapid crack growth results in catastrophic failure, sometimes accompanied by a loud noise.

Brittle overload failures, in contrast to ductile overload failures, are characterized by little or no macroscopic plastic deformation. Brittle fracture initiates and propagates more readily than ductile fracture or for so-called "subcritical" crack propagation processes such as fatigue or stress-corrosion cracking (SCC). Because brittle fractures are characterized by relatively rapid crack growth, the cracking process is sometimes referred to as being "unstable" or "critical" because the crack propagation leads quickly to final fracture.

The macroscopic behavior of brittle fracture is essentially elastic up to the point of failure. The energy of the failure is principally absorbed by the creation of new surfaces—that is, cracks. For this reason, brittle failures often contain multiple cracks and separated pieces, which are less common in ductile overload failures. All brittle fracture mechanisms can exhibit chevron or herringbone patterns that indicate the fracture origin and direction of rapid fracture progression. Herringbone patterns are unique microscopic features of brittle fractures. Ductile cracking, which occurs by microvoid coalescence, does not result in a herringbone pattern. On a microscopic scale, the features and mechanisms of fracture may have components of ductile or brittle crack propagation, but the macroscopic process of fracture is characterized by little or no work expended from deformation.

2.4.3 Microscopic Aspects of Fracture

Although the examination of a fracture surface begins logically with macroscopic observations, the microscopic aspects of fracture are also essential in understanding the causes of fracture. In general, fracture on a microscale can be distinctively defined in terms of transgranular brittle fracture (cleavage), transgranular slip (ductile fracture), or intergranular fractures. These three types of fracture paths result in distinctly different fracture surfaces, as seen in Fig. 2.22. These distinct microscopic appearances provide important information on the underlying mechanisms of fracture.

On a microscopic level, most engineering materials (alloys) are polycrystalline solids that consist of many grains (crystals) with grain-boundary regions between the crystals. Typically, the grain boundaries are stronger than individual grains in properly processing polycrystalline plastic/elastic solids. The reason for this can be understood in simple terms. The grain boundaries are disruptions between the crystal lattice of individual grains, and this disruption provides a source of strengthening by pinning the movement of dislocations. Thus, a finer-grain alloy imparts more grain-boundary regions for improved strength. Moreover, because a greater number of arbitrarily aligned grains are achieved when grain size is reduced, the stressed material has more opportunity to allow slip and thus improve ductility.

However, the grain boundaries are also a region with many faults, dislocations, and voids. This relative atomic disarray of the grain boundaries, as compared to the more regular atomic arrangement of the grain interiors, provides an easy path for diffusion-related (thermally activated) alterations. For this reason, grain boundaries can be a preferential region for congregation and segregation of impurities, preferential phase precipitation, and/or absorption of environmental species. These thermally activated alterations are potential mechanisms for weakening or embrittlement along the grain boundaries. In addition, grain-boundary regions are weakened when the temperature is high enough to activate diffusion-induced flow deformation along grain boundaries at stresses below the yield strength. This onset of time-dependent flow (i.e., creep deformation) is roughly representative of the viscoelastic deformation that occurs when the temperature is higher than 0.4 T_m.

This basic overview of the relative strength of grains and grain boundaries provides a general framework to describe the basic mechanisms of transgranular and intergranular fracture. Transgranular fracture of a crystalline material can occur by either a brittle process of cleavage or by the ductile process of microvoid formation.

These two mechanisms of transgranular fracture are very distinct in terms of appearance and have clear causes. Brittle transgranular fracture of a crystalline material takes place by cleavage along low-index crystallographic planes within grains, while ductile transgranular fracture occurs when small voids (microvoids) form and coalesce in the region of fracture. These microvoids leave distinctive concave depressions called dimples on both surfaces of the fracture, as described in more detail later in the next section.

Intergranular fractures are also very distinctive in appearance (Fig. 2.22a), but the underlying causes or mechanisms can be complex and varied. Grain boundaries are weakened in various ways, and the surface of an intergranular fracture does not necessarily reveal the evidence of the underlying mechanism that leads to fracture along a grain-boundary region. This illustrates the need for careful analysis of the overall circumstances that lead to fracture. In many instances, SEM fractography provides a means to identify the fracture path as intergranular, but it

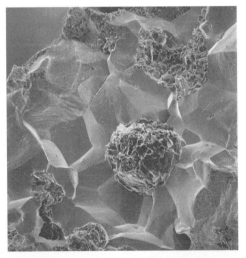

(a)

(b)

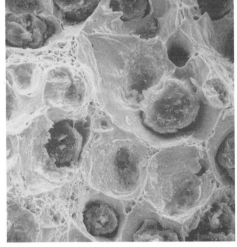

(c)

Fig. 2.22 SEM images of (a) intergranular fracture in ion-nitrided layer of ductile iron (ASTM 80-55-06), (b) transgranular fracture by cleavage in ductile iron (ASTM 80-55-06), and (c) ductile fracture with equiaxed dimples from microvoid coalescence around graphite nodules in a ductile iron (ASTM 65-40-10). Picture widths are approximately 0.2 mm (0.008 in.) from original magnifications of 500×. Courtesy of Mohan Chaudhari, Columbus Metallurgical Services

may yield little other information. Additional important information can be obtained by chemical analysis in the SEM and by microstructural examination. Use of the SEM for microstructural examination in addition to optical light examination may be appropriate depending on the scale of microconstituents present. Important clues on the underlying cause of intergranular separation also may be revealed by fractography at assorted magnifications.

Transgranular Ductile Fracture (Transgranular Slip and Microvoid Formation). In terms of inherent material structure of crystalline materials, the deformation processes of slip and twinning compete with the brittle fracture process of cleavage (see Appendix 1). Cleavage is a brittle process that occurs on the plane of maximum normal stress, while slip mechanisms are associated with plastic deformation and ductile fracture. At temperatures lower than 0.4 T_m, plastic deformation occurs by transgranular slip and/or twinning in the crystalline lattice. If other events do not intervene, this deformation culminates in fracture first by strain localization (necking or shear band formation), and then final fracture occurs in the volume or region of strain concentration. At temperatures of ~0.4 T_m or higher, however, deformation can occur by slip and viscous grain-boundary flow, as the grain-boundary regions become weakened at high temperature. Thus, the predominant fracture path becomes intergranular in the region of creep (time-dependent) deformation at temperatures of ~0.4 T_m or higher.

The overall process of ductile fracture is illustrated by the preceding example of a ductile fracture in an unnotched tension test specimen (Fig. 2.18). Ductile fractures are uniquely characterized by microvoids that form in the region of high stress. In an unnotched tension test bar, microvoids nucleate and grow in the central region, where the diffuse notch (created by necking) causes separation due to triaxial (hydrostatic) tension. These voids coalesce and join together to form a microscopic crack. At the same time, more small cavities are formed and distributed over the remaining section of the testpiece. A typical example is presented in Fig. 2.23, which shows a cross section of a tensile specimen containing numerous voids at a stage between necking and final fracture. The fracture process consists of these voids joining on the plane perpendicular to the loading direction and coalescing into a central crack. This crack grows until it approaches the outer annulus, where a change in fracture path occurs as the process approaches final fracture. At some point (depending on ductility), final fracture occurs along the shear plane and results in shear lip. The amount of shear lip varies, depending on ductility, strain rate, and temperature.

The surfaces of plastic fractures are characterized by microscopic "dimples," also called "cupules." These dimples represent the numerous concave depressions left on the opposite fracture faces of the broken specimen. Dimples on fracture surfaces are observed in many materials, including carbon and alloy steels, austenitic steels, alloys of aluminum, titanium, and copper, and plastics. It has been suggested that dimples represent the coalesced voids, and the voids initiate from inclusions or intermetallic particles (Ref 2.11, 2.12).

Dimples also can take different shapes, depending on loading condition (Fig. 2.24). Round dimples occur from separation under tension, while the dimples have an elongated parabolic shape when ductile fracture occurs from shear, torsion, and tearing (or bending). In general, the concavity of the parabola is oriented toward the direction of relative displacement of the other half of the specimen, and the axis of symmetry

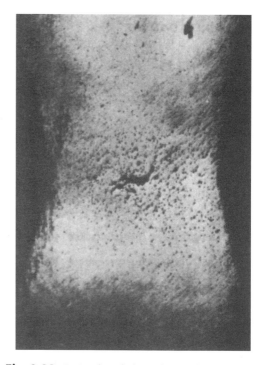

Fig. 2.23 Section through the neck area of a tensile specimen of copper showing cavities and crack formed at the center of the specimen as the result of void coalescence. Source: Ref 2.11

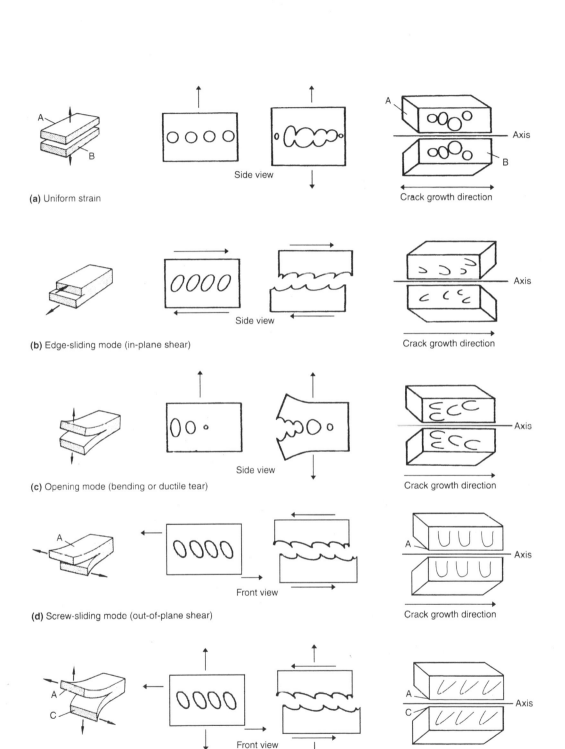

(a) Uniform strain

(b) Edge-sliding mode (in-plane shear)

(c) Opening mode (bending or ductile tear)

(d) Screw-sliding mode (out-of-plane shear)

(e) Opening mode with screw sliding (c + d)

Fig. 2.24 Schematic of plastic fracture. (a) Normal plastic (formation of round dimples). (b) Shear plastic (formation of elongated dimples pointing in the direction of shear on each fracture surface). (c) Tear plastic (formation of elongated dimples pointing in the direction opposite to the direction of each propagation. (d) Dimple elongation from out-of-plane shear. (e) Dimple elongation from mixed mode of screw sliding with ductile tear (c + d). See Ref 2.13 for other mixed modes.

is parallel to the direction of propagation of the rupture front (Fig. 2.24). Thus, when the two opposite surfaces of rupture due to in-plane shear (Fig. 2.24b) are examined, the concavity of these dimples is turned in opposite directions on the opposite faces. Parabolic dimples from torsional loading are shown in Fig. 2.25. An example of matching surfaces is shown in Fig. 2.26 for dimple shape and orientation for bending/ductile tearing (i.e., opening mode I in Fig. 22.4c). In the case of bending, there may be some question as to how often elongated dimples are seen in bending loading (or seen at all), because dimples on typical compact tension specimens (axial + bending) appear mostly equiaxed.

Transgranular Brittle Fracture (Cleavage). Cleavage is a low-energy fracture that propagates along well-defined low-index crystallographic planes known as cleavage planes. Theoretically, a cleavage fracture should have perfectly matching faces and should be completely flat and featureless. However, engineering alloys are polycrystalline and contain grain and subgrain boundaries, inclusions, dislocations, and other imperfections that affect a propagating cleavage fracture so that true, featureless cleavage is seldom observed. These imperfections and changes in crystal lattice orientation, such as possible mismatch of the low-index planes across grain or subgrain boundaries, pro-

duce distinct cleavage fracture surface features, such as cleavage steps, river patterns, feather markings, herringbone patterns, and tongues (Ref 2.15–2.17). Cleavage fracture also occurs in ceramics, inorganic glasses, and polymeric materials (see Chapter 7).

The cleavage mode of fracture is controlled by tensile stresses acting normal to a cleavage plane and is brittle in nature (see Appendix 1). Its fracture surface, which is caused by cleavage, appears at low magnification to be bright or granular, owing to reflection of light from the flat cleavage surfaces. It exhibits a river pattern when examined under an electron microscope. It occurs in bcc and hcp metals, particularly in irons and steels, below the ductile-to-brittle transition temperature (DBTT). Cleavage fracture is very seldom found in fcc metals. The fcc metals (e.g., copper, aluminum, nickel, and austenitic steels) have a large number of slip systems (12), which is one reason why they exhibit high ductility. The fcc metals also are more closely packed (i.e., a shorter distance exists between atoms in the crystal cell) than are bcc metals. This partly explains why cleavage fracture does not normally occur in the matrix of fcc metals.

However, it is not just a matter of the multiplicity of ways for slip or cleavage to occur. Two other factors also control the inherent ductile-brittle behavior of crystalline materials:

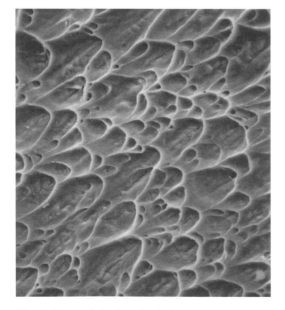

Fig. 2.25 Parabolic shear dimples from torsional fracture of cast Ti-6Al-4V alloy. Original magnification, 2000×

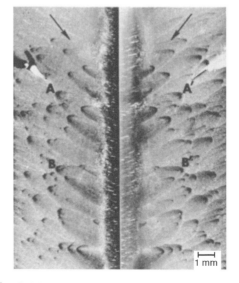

Fig. 2.26 Fracture markings on plexiglass. TEM, matching fracture surfaces. Note the matching features A, A' and B, B' on the two fracture faces. The parabola markings are similar to the plastic dimples observed in a tear ductile fracture of a metallic material. Source: Ref 2.14

- Critical shear stress required to initiate slip
- Critical normal stress required to propagate a cleavage crack

As long as there is only one type of atom in the lattice, the shear stress for slip is low. The presence of foreign atoms raises this stress and, depending on the location of the foreign atoms (random or periodic), may cause a severe loss in ductility, as is the case for the ordered intermetallic alloys. Thus, the critical stresses required for slip and cleavage also determine whether or not cleavage occurs.

The cleavage process in bcc and hcp metals occurs by separation normal to crystallographic planes of high atomic density. Microscopic examination of a fracture surface from cleavage typically reveals distinctive "river lines" indicative of propagation by fracture along nearly parallel sets of cleavage planes. The direction of crack propagation is indicated by the "flow" of the river lines as marked by an arrow in Fig. 2.27. Other distinctive microscopic surface features of cleavage fractures include:

- *Feather markings* (Fig. 2.28 and 2.29) are fan-shaped arrays of very fine cleavage steps on a large cleavage facet. The apex of the fan points back to the fracture origin.

- *Herringbone patterns* consist of a series of nested Vs. This pattern is a microscopic version of chevron marking, but is created by a different mechanism. The central spine of a herringbone is created by cleavage on a {100} plane and continued intermittent lateral crack expansion of the crack on {1,1,2} twinning planes.
- *Tongues* (Fig. 2.30 and 2.31) are microscale fractographic features that can form during cleavage in materials that mechanically twin. The high speed of cracking during cleavage produces a local strain rate too high for slip processes to provide all the accommodation required; thus, many twins are normally formed just ahead of the moving crack tip.
- *Wallner lines* (Fig. 2.32) are a distinct pattern of intersecting sets of parallel lines, usually producing a set of V-shape lines, sometimes observed when viewing brittle fracture surfaces at high magnification in an electron microscope. Wallner lines are attributed to interaction between a shock wave and a brittle crack front propagating at high velocity. Sometimes Wallner lines are misinterpreted as fatigue striations.

It bears repeating that complete examination of a fracture surface should be done at various levels of magnification by light and electron microscopy. For example, the fracture surface

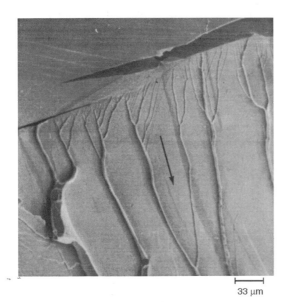

33 μm

Fig. 2.27 Fracture surface from a ferritic steel (Fe-0.01C-0.24Mn-0.02Si, heat treated at 950 °C for 1/2 h, air cooled). The fracture was generated by impact at −196 °C (321 °F). Cleavage steps beginning at the twin at top form a sharply defined river pattern. The arrow indicates crack propagation direction. Source: Ref 2.18

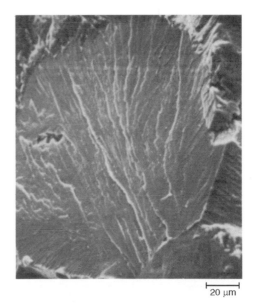

20 μm

Fig. 2.28 Feather pattern on a single grain of a chromium steel weld metal that failed by cleavage

shown in Fig. 2.33 represents a cleavage fracture in low-carbon martensitic steel. Both the low-magnification light microscope and the SEM and TEM electron microscopes were used for making the direct examinations. It can be seen that much different images are obtained for the same local spot on a fracture surface.

Quasi-Cleavage. Totally brittle fracture in metals at the microscopic level ("ideal cleavage" or "pure cleavage") occurs only under certain well-defined conditions (primarily when the component is in single-crystal form and has a limited number of slip systems) and is correctly described as "cleavage fracture." More com-monly in metals, the fracture surface contains varying fractions of transgranular cleavage and evidence of plastic deformation by slip. Grains oriented favorably with respect to the axis of loading may slip and exhibit ductile behavior, whereas those oriented unfavorably cannot slip and will exhibit transgranular brittle behavior.

When both transgranular fracture processes operate intimately together, the fracture process is termed "quasi-cleavage." The dividing line between cleavage and quasi-cleavage is some-what arbitrary. The term quasi-cleavage applies when significant dimple rupture and/or tear ridges accompany the cleavage morphology.

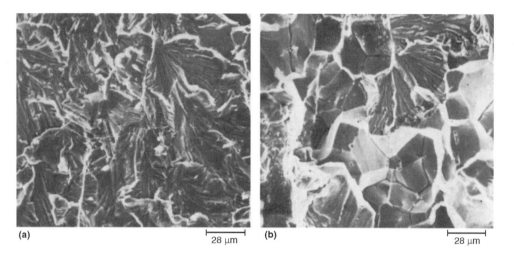

(a) 28 μm   (b) 28 μm

Fig. 2.29 Surface of a fracture in type 316 stainless steel resulting from SCC by exposure to a boiling solution of 42 wt% $MgCl_2$. The fracture in general exhibited the fan-shaped or feather-shaped transgranular cleavage features shown in (a). In a hasty scrutiny, the presence of local areas of "rock candy" intergranular fracture, such as those in (b), might be missed. This raises the question of whether corrosion-generated hydrogen caused local embrittlement. The separated-grain facets in (b) show no trace of corrosion. Both at 350×

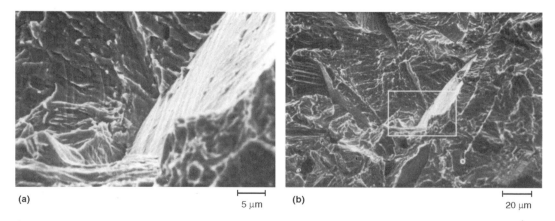

(a) 5 μm   (b) 20 μm

Fig. 2.30 Cleavage fracture from bend testing of 201 nickel in hydrogen atmosphere. (a) Ledgelike character of cleavage facets with small tongues on the bright facet (SEM, original magnification at 2000×). (b) Lower magnification view (original magnification at 500×) with higher-magnification area in (a) indicated by the rectangle.

The fracture surface is typically dominated by cleavage, but there are usually small patches of microvoid coalescence present or thin ribbons of microvoid coalescence contained in the fracture surface. As the patches increase, the fracture surface is more accurately described as (microscale) mixed cleavage and microvoid coalescence. Another term is "cleavage with ductile tear ridges."

Quasi-cleavage should not be confused with the decohesion along certain crystallographic planes that can occur by shear, by sliding off (plastic shear), or by separation along weak, still poorly defined interfaces. This type of decohesion has been referred to as glide-plane decohesion. Quasi-cleavage fractures also should not be confused with those in which cleavage appears in brittle second phases with the characteristic dimples of microvoid coalescence appearing in the more ductile matrix. In quasi-cleavage, there is no apparent boundary between a cleavage facet and a dimpled area bordering the cleavage facet (Fig. 2.34).

Figure 2.35 is a schematic representation of quasi-cleavage. The occurrence of quasi-cleavage is usually distinguished by:

- Initiation within facet boundaries—in contrast to fracture by cleavage, which usually initiates from one edge of the region being cleaved (Fig. 2.36)
- Cleavage steps appearing to blend directly into tear ridges of the adjacent dimpled areas

Many high-strength engineering metals fracture by quasi-cleavage, which is a mixed mechanism involving both microvoid coalescence and cleavage. When tested under embrittling conditions, such as those imposed by corrosive mediums or triaxial stress states, quasi-cleavage can occur in metals that normally are not known to have active cleavage planes (e.g., austenitic stainless steels, and nickel and aluminum alloys). One explanation is that facets that exhibit quasi-cleavage features fracture ahead of the moving crack front; then, as the stress increases, the cleavage facet extends by tearing into the matrix around it by microvoid coalescence.

Quasi-cleavage fracture surfaces appear in steels from (a) sudden or impact loading, (b) low temperature, (c) high levels of constraint (ambient temperature), or (d) in heavily cold-worked parts (ambient temperature). Quasi-cleavage, or cleavage in complex microstructures, is more difficult to identify than the cleavage found in low-carbon steel made up of ferrite and pearlite. When identification is uncertain, it is essential to relate the fracture features to the microstructure, including the prior austenite grain size, the martensite plate size, and the distribution, size, spacing, and volume fraction of fine carbide particles precipitated during tempering.

In steels that have been quenched to form martensite and then tempered to precipitate a fine network of carbide particles, the size and orientation of the available cleavage planes within a grain of prior austenite may be small and poorly defined. Small cleavage facets usu-

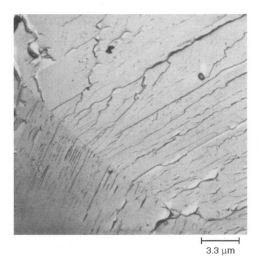

3.3 µm

Fig. 2.31 Cleavage fracture in Armco iron broken at −196 °C (−321 °F), showing river patterns, tongues, and (from bottom right to top left) a grain boundary. TEM p-c replica, 3000×

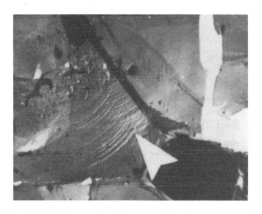

Fig. 2.32 TEM image of fracture surface from a cemented carbide (94WC-6Co) after four-point bending test. The trapezoidal WC grain at center (transgranular fracture) exhibits Wallner lines (indicated by arrow), which result from the interaction between the advancing crack front and a simultaneously propagating elastic wave. Fracture surface was etched in 5% HCl to remove the cobalt. TEM Formvar replica, 12,000×

ally can initiate at precipitated carbide particles with the ill-defined cleavage facets connected by tear ridges and shallow dimples. Quasi-cleavage facets on a fracture surface of a quenched-and-tempered 4340 steel specimen broken by impact at −196 °C (−321 °F) are shown in Fig. 2.37. The poorly defined cleavage facets are connected by tear ridges and shallow dimples. Until the work of Inoue et al. (Ref 2.19) and Beachem (Ref 2.20), these facets were not considered true cleavage planes (hence the term "quasi-cleavage"). It is now evident, however, that the quasi-cleavage facets in quenched-and-tempered steels do conform to cleavage on {100} planes.

Intergranular fracture occurs in polycrystalline materials when crystals (grains) separate

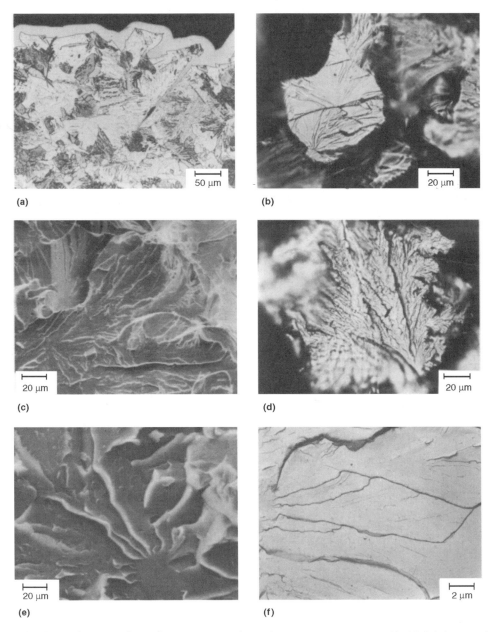

(a)

(b)

(c)

(d)

(e)

(f)

Fig. 2.33 Cleavage fracture in a low-carbon martensitic steel. (a) Light microscope cross section with nickel plating at top showing the fracture profile. (b) Direct light photograph. (c) Direct SEM fractograph. (d) Light fractograph of replica. (e) SEM fractograph of replica. (f) TEM fractograph of replica. Source: Ref 2.10

from each other along the grain boundaries. With a few exceptions, intergranular fractures are macroscopically brittle with little or no mechanical work expended as part of the fracture process. However, the micromechanisms of intergranular fracture may be brittle or ductile, depending on how the grain-boundary regions are weakened or embrittled. For example, grain boundaries may become embrittled by a film of a brittle phase or by the segregation of an impurity in the boundary region. In this case, the mechanism of intergranular cracking may be brittle by cleavage in the brittle phases that congregate in the grain boundaries. Conversely, intergranular fracture may also occur from ductile micromechanisms (slip) that involve localized formation of ductile microvoids in the region near the grain boundaries. For example, voids along the grain boundaries may form at a particle in the grain boundary or in a precipitate-free zone (PFZ) adjacent to the grain boundary. This feature is sometimes referred to as dimpled intergranular fracture.

The interpretation of intergranular fracture is more complex than the distinct mechanisms of transgranular fracture (i.e., fatigue, cleavage, or ductile with microvoid coalescence). However, the appearance of intergranular fracture is relatively easy to recognize, and the causes are fairly limited. The presence of intergranular fracture (especially in the region of crack initiation) also is often helpful in narrowing the potential cause for failure. Some common circumstances that have been known to induce intergranular cracking have been classified into four general categories (Ref 2.21):

- Presence of grain-boundary precipitates
- Thermal treatment or exposure that causes segregation of certain impurities to the grain boundaries without an observable second phase
- Stresses applied at elevated (creep-regime) temperatures

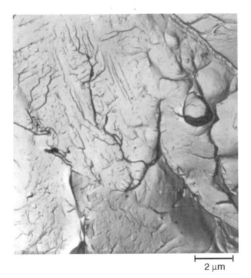

Fig. 2.34 Effect of quasi-cleavage—mixed cleavage and microvoid coalescence—on the fracture surface appearance of 17-PH stainless steel. TEM p-c replica, 4900×

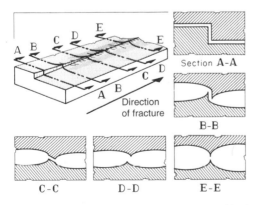

Fig. 2.35 Fracture model showing a cleavage step blending with a tear ridge in a quasi-cleavage fracture surface. At top left is the lower surface of a fracture, showing a step at the lower left and a ridge at the upper right. At right and at bottom are sections through the fractured member, showing profiles of both the upper and the lower fracture surfaces.

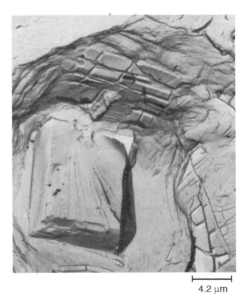

Fig. 2.36 Cleavage in a large second-phase particle on a fracture surface of A-286 steel

- Environmentally assisted alteration or weakening of the grain boundaries by various mechanisms such as hydrogen embrittlement, liquid-metal embrittlement, solid-metal embrittlement, oxidation or reduction potentially in the grain boundaries, radiation embrittlement, and SCC

Large grain size also plays a role in causing a change from transgranular to intergranular cracking and can enhance any of the above mechanisms.

Mechanisms of Intergranular Fracture. On an atomic scale, crack growth occurs by any one or a combination of the following:

- Tensile separation of atoms (decohesion)
- Shear movement of atoms (dislocation egress or insertion)
- Removal or addition of atoms by dissolution or diffusion

All of these processes can occur preferentially along the grain boundary by various phenomena, such as:

- Segregation of embrittling elements to the grain boundaries
- More rapid diffusion of elements along grain boundaries than along grain interiors
- More rapid nucleation and growth of precipitates in grain boundaries than in grain interiors
- Greater adsorption of environmental species in the grain-boundary regions

These basic mechanisms of intergranular cracking are more varied than those of transgranular fracture. However, except for conditions of creep stress rupture at elevated temperatures, intergranular fracture is not common in properly processed materials in a benign environment. There also are a fairly limited number of circumstances of intergranular fracture associated with improper processing of a material and/or some aggressive service environment. Some specific situations include (Ref 2.22, 2.23):

- High-carbon steels with a pearlitic microstructure
- Segregated phosphorus and cementite at prior-austenite grain boundaries in the high-carbon-case microstructures of carburized steels
- Stress-relief cracking
- Grain-boundary carbide films due to eutectoid divorcement in low-carbon steels
- Grain-boundary hypereutecoid cementite in carburized or hypereutectoid steels
- Iron nitride grain-boundary films in nitrided steels
- Temper embrittlement in heat-treated steels due to segregation of phosphorus, antimony, arsenic, or tin
- Embrittlement of copper due to the precipitation of a high density of cuprous oxide particles at the grain boundaries
- Embrittlement of steel due to the precipitation of MnS particles at the grain boundaries as a result of overheating

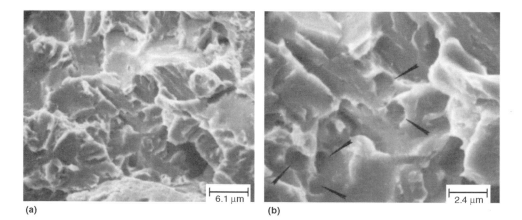

(a) 6.1 μm   (b) 2.4 μm

Fig. 2.37 Quasi-cleavage in the surface of an impact fracture in a specimen of 4340 steel. The same area is shown in both SEM fractographs, but at different magnifications. The small cleavage facets in martensite platelets contain river patterns and are separated by tear ridges. Shallow dimples, marked by arrowheads, are also visible. Direction of crack propagation is from bottom to top in each fractograph. The specimen was heat treated at 845 °C (1550 °F) for 1 h, oil quenched, and tempered at 425 °C (800 °F) for 1 h. Fracture was by Charpy impact at −196 °C (−321 °F). (a) 1650×. (b) 4140×

- Grain-boundary carbide precipitation in stainless steels (sensitization)
- Improperly precipitation-hardened alloys, resulting in coarse grain-boundary precipitates and a denuded region (PFZ)
- Embrittlement of molybdenum by interstitials (carbon, nitrogen, oxygen)
- Embrittlement of copper by antimony
- Reduction of Cu_2O in tough-pitch copper by hydrogen
- Hydrogen embrittlement by grain-boundary absorption of hydrogen
- Stress-corrosion cracking (sometimes intergranular, but also transgranular)
- Liquid metal induced embrittlement (LMIE), for example, mercury in brass, lithium in type 304 stainless steel
- Solid metal induced embrittlement (SMIE)

In all these cases, SEM fractography can provide the means to identify the fracture path. However, it cannot yield sufficient information on the underlying causes (or mechanisms) of intergranular fracture. Thus, additional important information may be needed in terms of chemical analysis or fractographic examination at assorted magnifications. The following sections briefly describe appearances for three general categories of intergranular fracture:

- Intergranular brittle cracking
- Dimpled intergranular fracture
- Intergranular fracture surfaces with corrosion products

Intergranular brittle cracking typically has a relatively "clean" fracture surface with the faceted appearance of cracking along grain contours. The general appearance of intergranular brittle fracture may include:

- Brittle second-phase particles and/or films in grain boundaries
- Fracture where no film is visible and, due to impurities, atom segregation at the grain boundary
- Environmentally induced embrittlement where there is neither a grain-boundary precipitate nor solute segregation

Grain-boundary segregation of elements (such as oxygen, sulfur, phosphorus, selenium, arsenic, tin, antimony, and tellurium) is known to produce intergranular brittle fractures. Studies of the effects of such impurities in pure iron have been greatly aided by the development of Auger electron spectroscopy. In the case of brittle grain-boundary films, it is not necessary for the film to cover the grain boundaries completely; discontinuous films are sufficient. Some common examples of intergranular embrittlement by films or segregants include:

- Grain-boundary carbide films in steels
- Iron nitride grain-boundary films in nitrided steels
- Temper embrittlement of alloy steels by segregation of phosphorus, antimony, arsenic, or tin
- Grain-boundary carbide precipitation in austenitic stainless steels (sensitization)
- Embrittlement of molybdenum by oxygen, nitrogen, or carbon
- Embrittlement of copper by antimony

Grain-boundary strengthening is characteristic of intergranular fractures caused by embrittlement. Intergranular brittle fracture can usually be easily recognized, but determining the primary cause of the fracture may be difficult. Fractographic examinations can readily identify the presence of large fractions of second-phase particles on grain boundaries. Unfortunately, the segregation of a layer a few atoms thick of some element or compound that produces intergranular fracture often cannot be detected by fractography.

Dimpled intergranular fractures result in low macroscopic ductility (i.e., the grains separate rather than deform) and a fracture surface that reveals microscopic dimples at higher magnifications (typically on the order of 1000 to $5000\times$). For example, a fracture surface from stress rupture of a nickel-base alloy is shown in Fig. 2.38 at two levels of magnification. The higher-magnification image reveals a dimpled topology on the grain facets.

Another example is shown in Fig. 2.39 for a high-purity aluminium-copper precipitation-hardened alloy with a coarse grain structure. In this example (with the coarse grain size), ductility is limited, and the yield strength in the local region of the grain boundary becomes lower than the matrix; thus, fracture tends to develop first within the grain-boundary zone by microvoid coalescence. The two levels of magnification provide a useful combination of images, one demonstrating the intergranular fracture path and the other revealing the microscopic mechanism of microvoid coalescence.

Other circumstances of dimpled intergranular fracture include:

- Uniform void nucleation aided by the formation of methane bubbles at the grain boundaries (Ref 2.24)
- Void nucleation at precipitates in the grain boundaries of precipitation-hardening alloys (such as Al-Mg-Zn alloys) with large precipitates on grain boundaries (Ref 2.25) and wide PFZs

- Voids aided by impurities that adsorb strongly on the grain-boundary surface
- Stress-relief cracking of chromium-molybdenum steels (Ref 2.24)

Intergranular fracture surfaces with corrosion products can provide some evidence of cause. For example, corrosion products are fre-

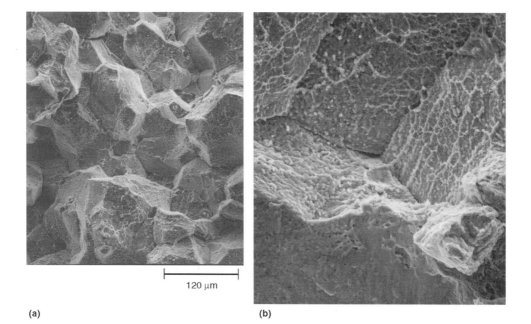

(a)   120 μm   (b)

Fig. 2.38 SEM image of the fracture surface of a nickel-base alloy (Inconel 751, annealed and aged) after stress rupture (730 °C, or 1350 °F; 380 MPa, or 55 ksi; 125 h). (a) Low-magnification view, with picture width shown at approximately 0.35 mm (0.0138 in.) from original magnification of 250×. (b) High-magnification view, with picture width shown at approximately 0.1 mm (0.004 in.) from original magnification of 1000×. Courtesy of Mohan Chaudhari, Columbus Metallurgical Services

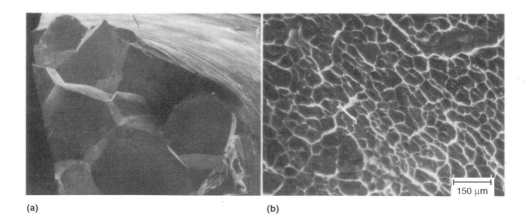

(a)   150 μm   (b)

Fig. 2.39 SEM fractographs of the tensile test fracture surface of a high-purity, coarse-grained Al-4.2Cu alloy with (a) intergranular facets at low magnification (10×) and (b) uniform dimples on one facet at higher magnification (67×). The microstructure indicated alloy depletion at the grain boundaries. Source: Ref 2.23

quently observed on the separated grain facets of fractures of intergranular stress-corrosion cracking (IGSCC) (Fig. 2.40). Intergranular fractures from stress rupture may also have oxidation products. For example, Fig. 2.41 is an example of turbine blade failure from a combustion gas environment. The original reference reported it as an IGSCC failure, when in all likelihood the fracture is one of creep rupture with oxidation products on the surface. Oxides are commonly observed on the fracture surface of creep fractures.

This illustrates the importance of using all available information sources, such as stress analysis and surface chemical analysis. Depending on the environment, cracks from hydrogen embrittlement may also reveal corrosion products on a fracture surface. Analysis of the cause of fracture in metal parts and components that have been exposed to corrosive environments is often difficult because of interactions of fracture mechanisms or because fractures generated by different mechanisms have similar appearances.

2.4.4 Macroscopic Aspects of Overload Failures

In terms of macroscopic behavior, overload cracking is either ductile or brittle, but the entire fracture may occur from different combination:

1. Totally ductile
2. Totally brittle

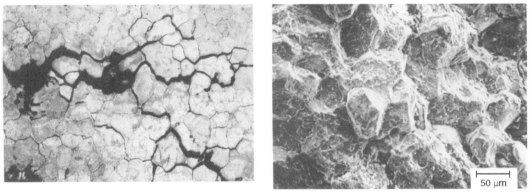

(a)　　　　　　　　　　　　(b)

Fig. 2.40 Intergranular fracture of 201-T6 cast aluminum after SCC testing. (a) Optical micrograph, Keller's etch, approximately 75×. (b) SEM image of fracture surface. Source: Ref 2.26

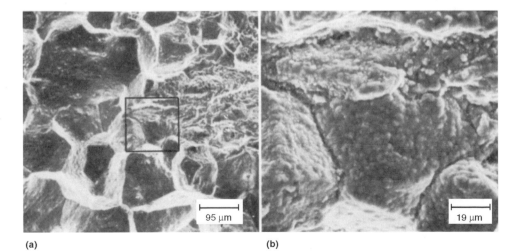

(a)　　　　　　　　　　　　(b)

Fig. 2.41 Surface from fractured U-700 turbine blade. (a) Region with transgranular and intergranular fracture feature. (b) Debris on intergranular facets, may be indicative of oxidation at high temperature after creep cracking. Source: Ref 2.23

3. Initially brittle, then ductile
4. Initially ductile, then brittle
5. Mixed mode (ductile and brittle)

In the last two cases (4 and 5), the ductile appearance may not be directly visible at the macroscale. Initially ductile fractures (case 4) are usually associated with rising-load ductile tearing, or the initial ductility may be inferred by transverse strain at the crack tip. The size of the plastic zone may be microscale in this case. Mixed-mode ductile and brittle cracking (case 5) would be inferred due to the presence of an intimate mixture of cleavage and microvoid coalescence at the microscale (quasi-cleavage) or by the presence of shear lips at the macroscale.

On a macroscopic scale, ductile and brittle fractures often are simply determined by visual examination for evidence of whether the part is deformed excessively, not at all, or somewhere in between. In the case of a tensile bar, for example, one might judge the degree of brittleness of that material by its stress-strain curve (i.e., by the amount of elongation or reduction in area before final fracture). More often, however, the macroscale appearance is insufficient to convey the full story about a fracture. The microscopic mechanisms and appearances of fracture also are needed for thorough understanding and testing of hypotheses in a case study of a fracture. Uncovering both the macro- and microscale mechanisms allows the source and cause of a given failure to be identified, and thus a course of corrective actions can be reached more reliably.

Surface roughness and optical reflectivity also provide qualitative clues to events associated with crack propagation. For example, a dull/matte surface indicates microscale ductile fracture, while a shiny, highly reflective surface indicates brittle cracking by cleavage or intergranular fracture. In addition, when intergranular fracture occurs in coarse-grain materials, individual equiaxed grains have a distinctive rock-candy appearance that may be visible with a hand lens. In terms of documenting surface conditions, one major problem with optical (light) macroscopic or microscopic examination of fracture surfaces is its inability to obtain favorable focus over the entire surface if the magnification exceeds 5 to 10×. Therefore, SEM also has become a standard metallographic tool in failure analysis.

Surface roughness provides clues as to whether the material is high strength (smoother) or low strength (rougher) and whether fracture occurred as a result of cyclic loading. The surfaces from fatigue crack growth are typically smoother than monotonic overload fracture areas. The monotonic overload fracture of a high-strength quenched and tempered steel is significantly smoother overall than is the overload fracture of a pearlitic steel or annealed copper. Also, fracture surface roughness increases as a crack propagates, so the roughest area on the fracture surface is usually the last to fail. Fracture surface roughness and the likelihood of crack bifurcation also increase with magnitude of the applied load and depend on the toughness of the material. Brittle failures often contain multiple cracks and separated pieces, while ductile overload failures often progress as single cracks, without many separated pieces or substantial crack branching at the fracture location.

Macroscopic features typically help identify the fracture initiation site and crack propagation direction. For example, crack branching and T-junctions (Fig. 2.42) can indicate the direction of crack propagation and location of crack initiation. Similar techniques also apply when brittle materials fracture into multiple pieces (Fig. 2.43). The orientation of the fracture surface, the location of crack initiation site(s), and the crack propagation direction should correlate with the internal state of stress created by the external loads and component geometry. In general, frac-

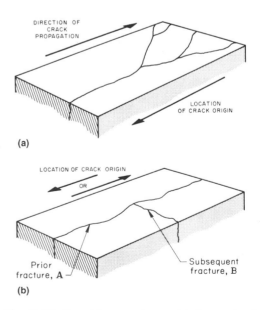

(a)

(b)

Fig. 2.42 General features to locate origin from crack path branching (a) and sequencing of cracking (b) by the T-junction procedure, where fracture A precedes and arrests fracture B

ture initiates in the region where local stress (as determined by the external loading conditions, part geometry, and/or macroscopic and microscopic regions of stress concentration) exceeds the local strength of the material. This includes microscale discontinuities (such as an inclusion or forging seam) and macroscopic stress concentrations (such as a geometric notch or other change in cross section).

The fracture surface orientation relative to the component geometry may also exclude some loading conditions (axial, bending, torsion, monotonic versus cyclic) as causative factors. For example, crack initiation is not expected along the centerline of a component loaded in bending or torsion, even if a significant material imperfection is present at that location, because

no normal stress acts at the centerline. (There is a shear stress at this location in bending, but in a homogeneous material, it is too small to initiate fracture. That might not be the case for a laminated structure loaded in bending.) Likewise, the profile of a fracture surface can indicate the direction of crack growth. For example, the region of plane-strain fracture indicates the direction of fracture in a shear overload fracture of annealed iron sheet (Fig. 2.44).

Under the right conditions, fracture surfaces may also have radial marks and chevrons, which are macroscopic surface features that indicate the region of crack initiation and propagation direction. They are common and dominant macroscopic features of the fracture of wrought metallic materials, but are often absent or poorly defined in castings. The "V" of a chevron points back to the initiation site, and a sequence of "V"s across the fracture surface indicates the crack propagation direction. The appearance of chevrons or radial marks near the crack origin depends in part on whether the crack-growth velocity at the surface is greater or less than that below the surface. If crack-growth velocity is at a maximum at the surface, radial marks have a fan-shape appearance (Fig. 2.45). If crack-growth rate is greatest below the surface, the result is chevron patterns (Fig. 2.46).

In rectangular sections, specimen dimensions can affect the appearance of radial markings and chevron patterns. For example, the macroscale fracture appearances of unnotched sections are shown schematically in Fig. 2.47 for sections with various width-to-thickness (w/t) ratios. The w/t ratio influences the ability of the sample to maintain a unidirectional state of stress during tension. In a thick section (top), strain in the width direction is constrained and thus tends to a condition of plane-strain (mode I) fracture. In this case, a large portion of the fracture surface

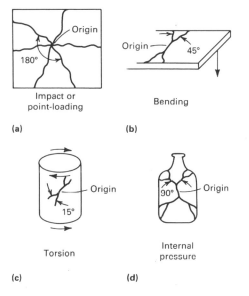

Fig. 2.43 Characteristics of crack direction and branching in fractures of brittle materials from (a) impact, (b) bending, (c) torsion, and (d) internal pressure

Fig. 2.44 Fracture surface showing a localized zone of plane-strain fracture (left) from shear overload failure of annealed Armco iron sheet at −196 °C (−321 °F). The configuration indicates that the fracture propagated from left to right in this view. Light fractograph, 5×

is comprised of radial markings or chevron patterns indicative of rapid, unstable cracking. At higher *w/t* ratios, the radial zone is suppressed in favor of a larger shear-lip zone. In very thin sections (bottom), plane-stress conditions apply, and the fracture surface is comprised almost entirely of a shear lip outside the fibrous zone of crack initiation. Figure 2.48 is a schematic of

radial marks and chevrons when fracture initiates from surface notches.

If conditions are right, radial markings associated with rapid or unstable crack propagation may also appear on the fracture surface of bar sections. The extent of radial markings depends on ductility, as seen in Fig. 2.49 from tension testing of unnotched bars after at different temperatures. As temperature decreases, ductility

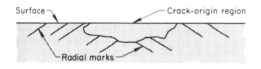

Fig. 2.45 Radial marks typical of crack propagation that is fastest at the surface (if propagation is uninfluenced by part or specimen configuration)

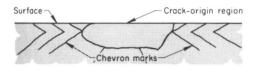

Fig. 2.46 Chevron patterns typical when crack propagation is fastest below the surface. It is also observed in fracture of parts having a thickness much smaller than the length or width (see middle illustration in Fig. 2.47).

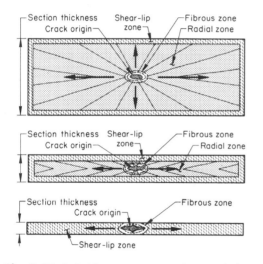

Fig. 2.47 Typical fracture appearances for unnotched prismatic tension test sections

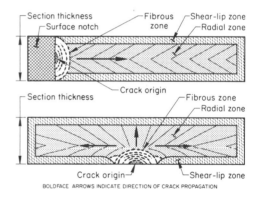

BOLDFACE ARROWS INDICATE DIRECTION OF CRACK PROPAGATION

Fig. 2.48 Schematic of typical fracture appearances for edge- and side-notched rectangular tension test sections

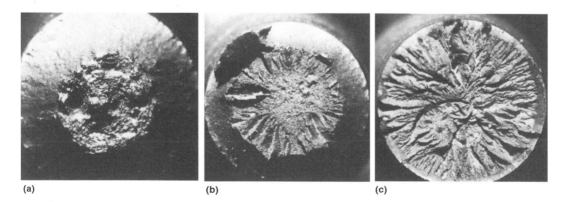

(a) (b) (c)

Fig. 2.49 Fracture surfaces of unnotched AISI 4340 steel specimens (heat treated to a hardness of 35 HRC) after tension testing at three different temperatures. (a) A shear lip surrounding a fibrous region is visible in the specimen tested at 160 °C (320 °F). (b) At a lower test temperature (90 °C, or 195 °F), a radial fracture zone formed around the fibrous region, which formed first; also, the shear lip here is smaller. (c) No fibrous region formed in the specimen tested at −80 °C (−110 °F). Instead, fracture formed a radial zone that extends nearly to the specimen surface and terminates in a narrow shear lip. Original magnifications all at 15×

and the extent of shear-lip formation are reduced. The fracture surface of radial marks is visually distinct from the fibrous region near the

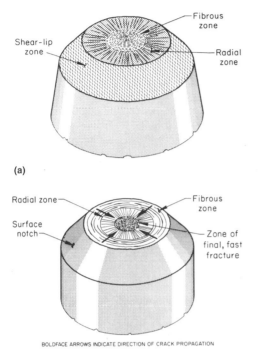

(a)

(b)

BOLDFACE ARROWS INDICATE DIRECTION OF CRACK PROPAGATION

Fig. 2.50 General fracture-surface regions from ductile fracture of an unnotched (a) and notched (b) tension test bar. (a) Radial zones on an unnotched point to the region of crack initiation near the center of the specimen. (b) In the notched tensile specimen, the fibrous zone surrounds the radial zone, because fracture initiates near the root of the notch (and completely around the specimens in this idealized case without additional stress raisers). Fast final fracture occurs in the center.

center on an unnotched bar (Fig. 2.49b). The radial markings (sometimes called a radial shear, star, or rosette) are perpendicular to the crack front. With an unnotched tension-test bar, radial marks point to crack initiation in the central region of the bar (Fig. 2.50a). This is not true with notched bars (Fig. 2.50b), where crack initiation occurs near the root of the notch, and where the radial zone points to the region of fast final fracture near the central region of the notched bar. The extent of radial marking depends on the degree of ductility, as shown in Fig. 2.51 from tension testing of notched 4340 bar at various temperatures. As an illustration of environmental effects, tension-overload fractures of a notched quenched and tempered 4340 steel at room temperature are presented in Fig. 2.52, which shows fracture surfaces from sustained-load cracking after hydrogen charging. Note that the progressive decrease in fracture stress from Fig. 2.52(a) to (d) is related to a progressive increase in the size of the fibrous zone in the outer region by the notch.

Cracks propagating from a preexisting stress raiser or notch (adapted from Ref 2.15) may propagate totally in plane stress with net-section yielding, totally in plane strain, or there may be a fracture transition as the crack propagates. In some instances, buckling may also occur. In terms of crack initiation, the likelihood of brittle crack initiation at the free surface is not high, unless the material is inherently brittle. Initial cracking is typically by a ductile mechanism (with three types of appearances: a tear zone, microvoid coalescence, or a tear zone followed by microvoid coalescence), but further cracking

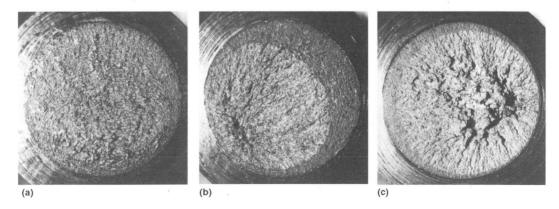

(a)   **(b)**   **(c)**

Fig. 2.51 Overload fracture in notched AISI 4340 steel specimens (35 HRC) from tension testing at three different temperatures. (a) The surface of the specimen tested at −40 °C (−40 °F) shows only fibrous marks. (b) The specimen tested at 80 °C (−110 °F) has a fibrous zone that surrounds a radial zone, which is off-center because of nonsymmetrical crack propagation. (c) The specimen tested at −155 °C (−245 °F) has a small annular zone of fibrous fracture, with prominent radial marks in the central region of final fast fracture. All at ~17×

may change to cleavage or quasi-cleavage as the crack reaches some (small) critical length, depending on the temperature, loading rate, and grain size. This is common in bcc ferrous materials. In steels, three types of crack initiation are found, depending, in part, on the temperature. In the first case, crack-tip blunting followed by ductile crack propagation by tearing initiates at the free surface. In the second, quasi-cleavage fracture and/or ductile microvoid coalescence, fracture initiates at the location of maximum constraint. In the less common third case, the material has sufficient toughness and sufficient crack blunting occurs that ductile fracture occurs on a shear plane at the crack tip. If blunting occurs, the peak stress is reduced, and the stress falls off more gradually behind the notch.

Figure 2.53 shows schematically the differences in cracking behavior in ductile alloys that exhibit a ductile-brittle transition with temperature. The presence of the stretch zone can be used to quantitatively estimate the magnitude of the nominal stress and the fracture toughness.

2.5 Fracture Strengths

Mechanical behavior of metals is influenced by many factors, including grain size, grain structure, temperature, the geometry of a structural piece, condition of the applied load (including loading speed and type of loading), and so on. Generally speaking, it is known that a fast loading rate will raise the yield point to a higher

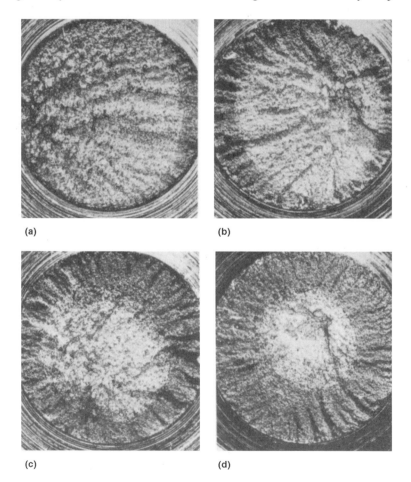

(a)

(b)

(c)

(d)

Fig. 2.52 Effect of sustained loading with hydrogen charging on the fracture-surface characteristics of notched specimens of quenched and tempered AISI 4340 steel tension tested at room temperature. (a) The specimen with a relatively small fibrous zone at the right edge was broken by tension overload with notched tensile strength of 2005 MPa (291 ksi). The specimen in (a) was not charged with hydrogen, while the three other specimens were charged with hydrogen and then subjected to sustained loading as follows: (b) Broke in 1.65 h under a stress of 1380 MPa (200 ksi). (c) Broke after 5.35 h under a stress of 1035 MPa (150 ksi). (d) Broke after 5.5 h under a stress of 690 MPa (100 ksi). Notch radius was 0.025 mm (0.001 in.). Microstructure of all four specimens was tempered martensite. All at ~8×

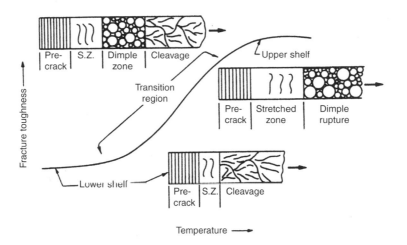

Fig. 2.53 Schematic of the brittle-to-ductile fracture transition. The relative area on the fracture surface of the three microscale fracture mechanisms (stretch zone, dimple zone, and cleavage zone) are indicated. Source: Ref 2.27

value (Fig. 2.54). It is also known that the grain size effect is a reverse function of fracture stress, yield stress, reduction in area, and transition temperature (Ref 2.28–2.30). A finer grain will result in higher values of all these parameters.

As for fracture and yield stresses, the difficulty of crack propagation in finer grains is due to the boundaries between differently oriented grains. A smaller grain means a larger grain-boundary area that wraps around the individual grain. At room temperature as well as at lower temperatures the grain-boundary material is strong, acting as a barrier for dislocation pileup. Thus, slip is difficult within each individual grain. The result is that fracture is more difficult with smaller grains, leading to a higher fracture stress. Theoretical and experimental formulas for the grain-size effect on yield point, fracture stress, and ductile-brittle transition are given in the literature (Ref 2.29–2.31). The famous Petch equation relates yield stress to grain size (Ref 2.30).

Strength and ductility of an alloy are significantly affected by alloying elements, which contribute to a variety of grain structures. Consider two series of binary alloy systems—one ranging in composition between two pure metals A and B, the other with pure metals C and D. The first set of alloys, over the entire composition range shown in Fig. 2.55, is solid solutions with a homogeneous grain structure. The flow stresses of such so-called solid solutions are generally distinctly higher and the ductility correspondingly lower than those of the adjacent pure metal. A maximum in flow stress and minimum in ductility occur at some intermediate alloy. The magnitude of these effects is very different for different alloy systems.

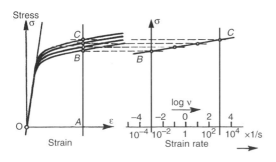

Fig. 2.54 Schematic influence of strain rate on yield stress. Source: Ref 2.6

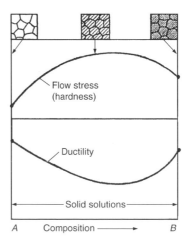

Fig. 2.55 Structure and forming properties of solid solutions of two metals. Source: Ref 2.32

The second set of alloys consists entirely of heterogeneous grain structures. The flow stresses and ductility of such alloys are frequently represented by straight lines between the properties of the pure metals (Fig. 2.56). Of course, many other factors can contribute to variations in grain structures, such as heat treatment, cold work, and so on, and lead to different results.

A metal may be ductile under one set of conditions and brittle under another. Ductility and brittleness, then, are properties that must be considered in terms of a particular set of testing or service conditions. A test may not always reliably indicate the relative ductility of two metals in service. These differences can always be traced to either (a) different states of stress in the two tests or service conditions, (b) differences in the speed with which the metal is deformed, and how it is deformed, or (c) the temperature during deformation. Relative ductility refers to the amount of deformation that can be accomplished before the metal breaks.

Stress State. As noted, triaxial tension increases brittleness, because the maximum shear stress is reduced by the presence of the other two stress components. Hydrostatic tension is a limiting case where the Mohr's circle has been reduced to a single point and all the shear stress has vanished. On the other hand, adding compression component(s) to a primary tension system will raise the magnitude of the shear component, thereby increasing ductility (see Table 2.1).

Type of Applied Stress. Tension, torsion, and bending significantly influence ductility. Compared to tension or bending, torsion retains ductility in most circumstances. Two metals may have approximately the same degree of ductility in tension as measured by reduction of area and elongation in an ordinary tension test. However, they may exhibit quite different degrees of ductility in a notch bend test as measured by the energy absorbed in a Charpy or Izod test. Thus, speed (i.e., loading rate) and/or geometry will cause some ductile materials to become macroscale brittle. However, it has been pointed out by Orowan (Ref 2.33) and Gensamer (Ref 2.4) that high loading speed is not the main cause of increased tendency toward brittleness, but notch does play an important role.

Temperature. Strength is generally lower at higher temperatures, and vice versa. Under monotonic loading, a metal is also more ductile at higher temperatures, and vice versa. However,

it will be brittle under the creep rupture condition.

A bcc metal will change from ductile to brittle at a certain temperature (or within a narrow range of temperatures). This is called the ductile-to-brittle transition temperature (DBTT); the transition usually occurs at low temperatures, and in steels depends on grain size and strain rate. An increase in strain rate (except for high-strength steels) and an increase in grain size (Ref 2.17) will raise the DBTT. For steels with reduced ductility after strain aging, the DBTT may be raised. Hexagonal close-packed metals also cleave, some even at temperatures near room temperature.

However, the stresses necessary to cause cleavage in hcp metals vary with purity, and there is some question as to whether cleavage can occur with sufficient purity. For example, the critical normal stress for zinc single crystals (containing 0.03 to 0.53% Cd) varies from 1.9 to 11.8 MPa (0.19 to 1.2 kg/mm^2) (Ref 2.2).

Generally speaking, there is no ductile-to-brittle transition in fcc metals; the increase in yield stress can go from room temperature all the way down to liquid-nitrogen temperature (-196 °C, or -321 °F). However, it has been reported that brittle fractures were found in iridium and rhodium. Additionally, nitrogenated austenitic stainless steels and austenitic stainless steels in SCC conditions may possibly cleave (Ref 2.15). For a given metal, the temperature (or temperature range) that causes the transition may shift as it is affected by the other factors mentioned earlier. The right combination of all

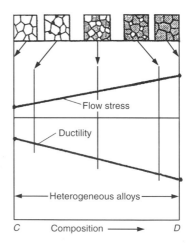

Fig. 2.56 Structure and forming properties of heterogeneous mixtures of two metals. Source: Ref 2.32

these factors is necessary to make a ductile metal become brittle, or vice versa.

Example. Figure 2.57 (Ref 2.4) qualitatively shows that the type of loading and temperature influence whether a structural part fails in a ductile or brittle manner. Ductility or brittleness depends on whether the part first reaches the maximum shear stress required for flow or the maximum normal stress required for fracture. Equations 2.34(a) and (c) show that σ_{max}/τ_{max} equals 2 in tension, but equals 1 in torsion. We'll now see that σ_{max}/τ_{max} is greater than 2 in a notched member, depending on the sharpness of the notch.

By σ_{max}/τ_{max} equals 1 in torsion we mean that in a torsion test, $\tau_{max} = \sigma_{max}$ is maintained in the stress system throughout the test. Recalling the criteria for yield discussed earlier, the shear yield strength is only a fraction of the tensile yield strength. That is, the yield strength in shear as determined from a torsion test should be only 57.7% of the tensile yield strength, based on the von Mises stress criterion. The $\tau_{yield}/\sigma_{yield}$ ratio will be even lower if the Tresca or the octahedral stress theory is used. All of this means that in a torsion test the yield strength for shear is reached at a stress level that is still lower than the yield strength for tension. Thus, a specimen will fail by shear, and hence is ductile. By comparison, a tension specimen will fail by normal stress and thus is brittle.

Figure 2.57 shows that critical shear stress is significantly affected by temperature, whereas critical normal stress is not. For illustration purposes, we locate the temperatures corresponding to the transition (from ductile to brittle) for each type of loading condition. With decreasing temperature, brittleness will occur in tension when the critical normal stress for fracture becomes less than twice the critical shear stress for flow. In torsion, this does not occur until the ratio becomes less than unity. Thus, ductile failure is maintained to a much lower temperature in torsion than in tension. Similarly, the DBTT would be much higher for a notched member. This also means that at a sufficiently high temperature, where the critical shear stress is so low and the corresponding σ_{max}/τ_{max} ratio is always greater than the theoretical value, ductile failure will occur under almost any loading conditions.

2.5.1 Notch Strengths

Notches appear in many practical structures or machine components. Well-known examples include a groove in a shaft, a shoulder or abrupt change of cross section, and a circular hole. Extreme notch forms are blunt cracks and sharp cracks. Stresses (tensile or otherwise) in structural components can cause failure, especially when a geometric discontinuity, or notch, causes stress concentration and raises the local stress. Thus, the susceptibility of an engineering part to failure can increase significantly at the presence of a notch.

Engineering predictions of fracture from stress concentrations require the development of methods that realistically take into account the

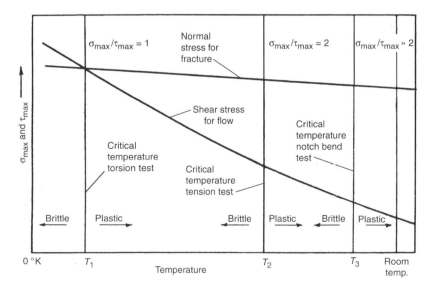

Fig. 2.57 Variation of critical shear stress for flow and of critical normal stress for fracture with temperature. Source: Ref 2.4

complex interplay between stress distribution and material behavior. Figure 2.58 shows the geometric details of a notched tensile specimen. The introduction of the notch produces a condition of biaxial stress at the root of the notch and triaxial stress at the interior of the specimen. The presence of transverse stress at the notch increases the resistance to flow and decreases the ratio of shear stress to tensile stress. Notch strength is defined as the maximum load divided by the original cross-sectional area at the notch. Comparing the notch strength to the ultimate tensile strength of a smooth specimen (unnotched) gives a measure of the notch sensitivity of a given material. The material is notch brittle when the notch strength is lower than the ultimate tensile strength of a smooth specimen. Otherwise, it is notch ductile. Normally, high-strength steels are prone to notch brittleness, whereas steels of lower strengths are somewhat notch ductile.

Notch sensitivity also depends on three other factors: the depth and sharpness of the notch (which sets up various degrees of stress triaxiality), high strain rate, and low temperature. All three factors promote brittle fracture, causing a normally ductile material to fail in a brittle manner. Stress concentration not only induces higher

(than the applied) stress at the notch root, but in some cases also causes a triaxial stress state, thus altering the load response behavior of the structural member (or the test coupon).

Consider two extreme cases: a sharp notch and a mild notch. The notch geometry shown in Fig. 2.58 can be considered a sharp notch. The stress distribution in a round bar with a sharp notch is schematically shown in Fig. 2.59. The geometry and the stress distribution of a mild notch are shown in Fig. 2.60. The stress profiles of these two configurations obviously are far from similar. Among other differences, the maximum longitudinal stress is at the notch root of a sharp notch but at the interior of the specimen with a mild notch.

The effects of strain rate and temperature are discussed later in this chapter, and in Chapter 4 together with discussion of dynamic fracture. In the following paragraphs, the mechanical behavior from a notch is examined more thoroughly.

Necking of an Unnotched Tensile Test Specimen. As noted, necking usually starts right when the tension test reaches maximum load based on the following rationale: At a certain point during the occurrence of strain hardening, the rate of change of area at the weakest section of the test specimen may become so great that the load required to maintain the flow drops off, despite the strain-hardening effect. This point on the true-stress/true-strain curve is

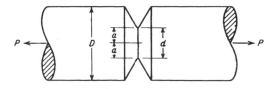

Fig. 2.58 Drawing of a notched round bar

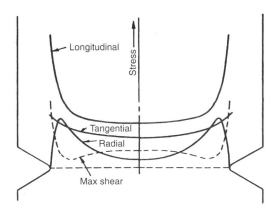

Fig. 2.59 Local stress distributions at the notch of a round notched bar in tension

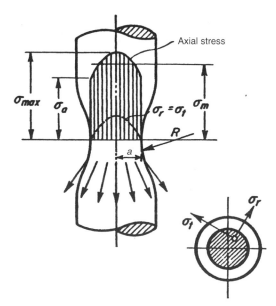

Fig. 2.60 Stress distribution at the neck of a round tensile bar

the ultimate load that corresponds to the ultimate stress. At that point, material distant from the weak section would have been strain-hardened sufficiently to support the reduced load without further strain. All of the strain is then concentrated in the vicinity of the weak section, and so a neck develops.

The necked region in the tensile specimen is in effect a mild notch, causing a complex triaxial state of stress in that area. The material adjacent to the neck restrains its development. Radial and tangential stresses are thereby induced in addition to the axial stress. A model simulating the neck (as a notched round bar with a radius of curvature R for the notch) is shown in Fig. 2.60. The stresses across the cross section of the neck consist of a uniform tension component and a hydrostatic tension component. In this figure, the axial stress is not just on the surface of the neck (σ_a), but is also distributed across the cross section with a maximum (σ_{max}) near the center. The hydrostatic tension stress is the composite of the radial stress (σ_r), the tangential stress (σ_t), and a portion of the axial stress (all of the same magnitude and distribution profile). The axial component of the hydrostatic tension is represented by the difference between σ_{max} and axial stress at the surface of the neck (σ_a); that is, $\sigma_{max} - \sigma_a$, the parabolic area depicted in Fig. 2.60. The mean axial stress (σ_m) at the core is equal to P (the applied load) divided by πa^2 (the area of the minimum circular cross section). This quantity, labeled as σ_m in the figure, is actually the instantaneous average true stress (Eq 2.14). However, for the stress profile shown in Fig. 2.60, the "truly" true-stress component must be the uniform axial component right at the neck (i.e., σ_a), because a hydrostatic component of stress does not contribute to plastic flow. An adjustment to the average true stress (at points after necking) is required. Correction formulas (as functions of a and R) have been developed by Davidenkov and Spirdonova (Ref 2.34):

$$\sigma_a = \frac{\sigma_m}{1 + (a/4R)} \qquad \text{(Eq 2.36a)}$$

and by Bridgman (Ref 2.35):

$$\sigma_a = \frac{\sigma_m}{(1 + 2R/a) \cdot \ln(1 + a/2R)} \qquad \text{(Eq 2.36b)}$$

To determine σ_a, one needs to measure a and R during a tensile test. Then, the portion of the true-stress/true-strain curve after necking can be

plotted by replacing the σ_m values with the values of σ_a. Experimental data correlation with the Bridgman correction was later conducted by Marshall and Shaw (Ref 2.36). Their results indicate that the Bridgman correction correlates very well with steel (hot-rolled SAE 4140 steel, 256 HB), but not so well with copper (electrolytic copper annealed 1 h at 500 °C, or 930 °F, 64 HB). However, Trozera (Ref 2.37) has found that the Bridgman correction correlated very well with extruded type 1100 aluminum. Clearly this type of correction is highly empirical and requires measuring the notch contour R frequently during the test. Because R is not easy to measure, it is not a practical way of collecting data for the last few steps of a tensile test.

For a test specimen made of a plate with a rectangular cross section (Fig. 2.61), the problem of necking is more complicated. Not only is there a redistribution of stresses at the neck, but the magnitude and the profile of these stresses over the neck cross section may depend on the width-to-thickness ratio (b_0/h_0) of the test specimen (where b is width, h is thickness, and the subscripts 0 stand for original width and original thickness, respectively). Aronofsky (Ref 2.38) worked out two solutions, one for $b_0/h_0 = 6$ and another for $b_0/h_0 = 10$. He also worked out a solution for round bar for making comparisons with flat plates as well as with the round bar solutions of Bridgman (Ref 2.35) and Davidenkov and Spirdonova (Ref 2.34).

The stress distributions for the round bar confirm those previously shown in Fig. 2.60; that

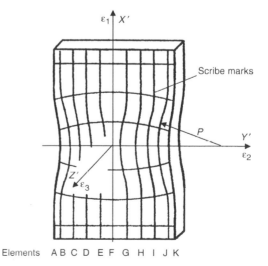

Fig. 2.61 Schematic view of the neck of a flat tensile bar

is, there is a uniform axial stress component plus a hydrostatic tension component at the neck. Thus, the existence of a uniform axial stress component in the round bar is confirmed. As for the flat plates, the resulting stress profiles right at the onset of specimen fracture are shown in Fig. 2.62 and 2.63. There are only two principal stresses in the plate, σ_1 and σ_2, which are the counterparts of σ_{max} and σ_r, respectively. Likewise, $(\sigma_1 - \sigma_2)$ is the counterpart of σ_a. Because the octahedral shear stresses in these plates are not uniform across the plate width, $(\sigma_1 - \sigma_2)$ is not uniform across the plate width. On the contrary, Aronofsky (Ref 2.38) claimed that the octahedral shear stress was uniform across the neck of the round bar; therefore, σ_a was uniform. As shown in Fig. 2.62 and 2.63, the stress distribution for $(\sigma_1 - \sigma_2)$ starts from 620 MPa (90 ksi) on the plate edge and increases to 690 MPa (100 ksi) at the midwidth of the plate. The stress profiles (and magnitude) for both b_0/h_0 ratios are the same. The maximum stress level at the midwidth of the plate (i.e., σ_1) was also found to be approximately the same ($\sim$830 MPa, or $\sim$120 ksi) in both cases. Therefore, plate width-to-thickness ratio does not seem to be the problem, at least within the b_0/h_0 range in the investigation.

Yielding of a Notch. Structural details often play an important role in contributing to fatigue and fracture behavior. In Chapter 1, it was demonstrated that a nonuniform distribution of stresses occurs in the local area in the vicinity of a geometric stress raiser, such as a notch or a cutout. As presented in the following, this idea can be applied to other situations as long as there is a stress concentration in the structure.

Taking a circular hole, for example, the local tangential stress at the hole edge is at least three times the applied far-field uniform tension stress. The local stresses gradually decay away and eventually converge to the nominal stress field at a distance away from the hole. In the theory of elasticity, this local stress distribution is represented by the commonly known Kirsch/Howland solutions (presented in Chapter 1). Because the material cannot forever follow Hooke's law at all levels of the applied load, in reality it follows the stress-strain relationship of the tensile stress-strain curve. Therefore, the material at the hole edge will undergo plastic deformation whenever the applied gross area stress exceeds one-third of the material tensile yield strength. Precisely, nonlinear behavior starts at the proportional limit, a point that is slightly below the yield stress on the tensile stress-strain

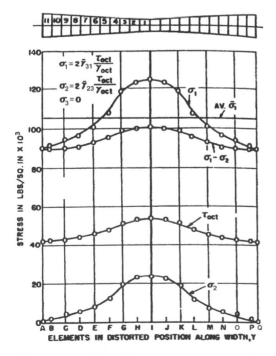

Fig. 2.62 Distribution of stresses in minimum section at fracture, flat specimen $b_0/h_0 = 6$. Source: Ref 2.38

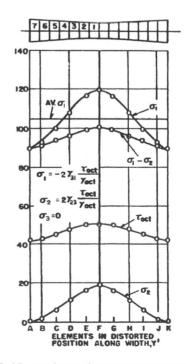

Fig. 2.63 Distribution of stresses in minimum section at fracture, flat specimen $b_0/h_0 = 10$. Source: Ref 2.38

curve. However, for most purposes of engineering application, this distinction can be regarded as unimportant.

Once the material starts to yield, the stress-concentration factor as well as the entire local stress distribution over the plate width, which is normally represented by the Kirsch/Howland solutions, will be altered. Depending on the applied stress level, the stress-concentration factor (the local stress to far-field stress ratio at the hole edge) is gradually reduced as the applied load increases. The cross-the-width stress distribution also changes accordingly. The result of an elastic-plastic analysis has shown that the local stress distributions vary depending on the applied stress level.

An example of such variation is shown in Fig. 2.64. In this figure, the local tangential stresses (corresponding to applied stress levels ranging from 83 to 248 MPa, or 12 to 36 ksi) were computed by using the NASTRAN structural analysis computer code. The stepwise linear option of the NASTRAN code was used to determine the plastic stresses. The tensile stress-strain curve that was obtained for the same material used in this example is shown in Fig. 2.65. Since the proportional limit and the tensile yield strength for the 2024-T351 alloy were 345 MPa (50 ksi) and 372 MPa (54 ksi), respectively, the stress distribution for an 83 MPa (12 ksi) applied stress level is considered pure elastic. Figure 2.64 also shows that the local stresses deviate from the elastic stress curve as soon as the applied stress level exceeds one-third the proportional limit of the

material. Should the applied stress continue to increase beyond 248 MPa (36 ksi), the plastic stress curve would eventually become flat, meaning that net section yielding across the entire plate width had been reached. The effect of stress concentration (K_t) disappears ($K_t = 1$). In other words, when a notched member (e.g., a testpiece containing an open hole) is subjected to monotonic increasing load, the K_t at the hole edge will decrease continuously as the loads continue to increase.

What this means is that a stress concentration does not necessarily cause a structural member to fail prematurely in monotonic loading. The notion that a notched member will fail as soon as the stress level reaches a value that is equivalent to the material tensile ultimate strength divided by K_t is not always valid. This is because the hole edge will yield before it reaches that stress level, and thus the testpiece can continue to take a higher load. The chance is that this process will continue and allow the applied load to go higher and higher. Then, final failure might occur with little or no dependence on the initial K_t value.

The foregoing discussion is only a rationale of an anticipated phenomenon; there is no analytical or empirical solution to assess fracture strength of this nature. Though the quantity is limited, the data in Table 2.2 may help to shed some light on the magnitude of this problem. Four specimens were made of 2024-T351 aluminum, each with a circular hole with a diameter of 12.7 or 19 mm (0.5 or 0.75 in.). Each specimen also contained a very tiny crack at the edge of the hole. However, for a moment, ignore the crack with the idea that we can use these data to prove a point. The calculated net section fracture stresses of these specimens are just slightly

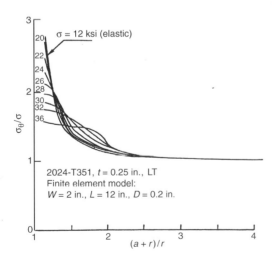

Fig. 2.64 Uncracked stress distribution in the vicinity of an open hole as a function of applied uniform far-field stress levels. Source: Ref 2.39

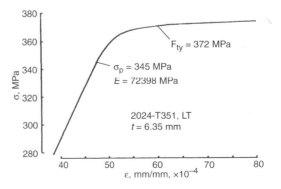

Fig. 2.65 Tensile stress-strain curve for a 2024-T351 aluminum alloy plate. Source: Ref 2.39

below the material tensile yield strength. Thus, the initial K_t did not cause these specimens to fail prematurely, and their net section fracture stresses were very high, close to the material tensile yield strength.

In conclusion, it thus seems necessary that notch strength should be determined by testing on a case by case basis. Nonetheless, the importance of K_t should still be recognized, even though K_t may not contribute detrimental effects to monotonic loading. Stress-concentration factors certainly play an important role in fatigue crack initiation and fatigue life (see Chapter 3).

2.5.2 Effect of Specimen Size

Necking of a tensile bar is very common in mechanical testing of ductile materials, and formation of microvoids during necking is conceivably the first step in the fracture process. The second step is the coalescing and joining of these microvoids. Inhomogeneities such as point defects, inclusions, and second-face particles increase the probability of void formation. Inhomogeneities actually are crack initiation sites even without necking.

According to Weibull's statistical theory of fracture (Ref 2.40, 2.41), the number of imperfections in a specimen increases with increasing specimen size. The probability of having a larger

initial cavity, or flaw (a weakest link), present in the material also increases as specimen size increases. Consequently, the fracture strength will decrease with increasing specimen size. This theory has been expressed as:

$$\frac{F_2}{F_1} = \left(\frac{V_2}{V_1}\right)^{-1/m}$$

(Eq 2.37)

where F is the fracture strength of a specimen having a test volume V, and m is an experimentally determined factor in Weibull's function. Thus, a large size effect is predicted for materials having a low m value, while no effect will be predicted for homogeneous materials with an m value equal to infinity.

The size effect is also applied to materials that do not form voids during fracture. For some materials, such as glass, dangerous cracks or flaws are usually formed on the specimen surface, and its fracture strength will depend on the specimen surface area instead of volume. For special material forms, such as wire and fiber stocks and glass tubing, the tensile strength may depend on specimen length.

2.5.3 Interaction between Temperature and Strain Rate

Tensile strength of an alloy usually decreases continuously with increasing temperature, except when unusual microstructural changes occur. At the same time, ductility increases as the strength decreases at a change in temperature. The variation of the tensile properties of steel with temperature is schematically shown in Fig. 2.66. The change with temperature in the shape of the engineering stress-strain curve in mild steel is shown schematically in Fig. 2.67.

Static-strength test data for five metals of the bcc or fcc crystal structure are shown in Fig. 2.68 and 2.69 to illustrate the temperature effect on yield strength and ductility. Figure 2.68 shows the variation of yield strength with temperature for four bcc metals (iron, molybdenum, tungsten, and tantalum) and an fcc nickel. Figure

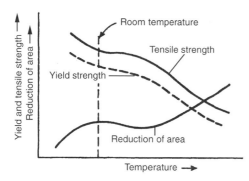

Fig. 2.66 Variation of tensile properties of steel with temperature. Source: Ref 2.2

Table 2.2 Fracture stresses of open hole specimens

Specimen ID	Hole diam		Thickness		Net area		Fracture load, 10^3 N	Net stress	
	mm	in.	mm	in.	mm²	in.²		MPa	ksi
A2-22	19	0.75	6.528	0.257	870.51	13.56	310.8	357.03	51.74
A2-26	12.7	0.5	6.553	0.256	915.45	14.19	332.7	363.43	52.67
A2-35	12.7	0.5	6.553	0.256	915.45	14.19	330.5	361.02	52.32
A2-36	12.7	0.5	6.528	0.257	911.96	14.14	324.7	356.05	51.60

Material: 2024-T351 aluminum plate, F_{ty} = 372 MPa (54 ksi); specimen width: 152.4 mm (6 in.). Source: Ref 2.39

2.69 shows the variation of the reduction in area with temperature for the same metals. These test results agree with the general trend depicted in Fig. 2.66, and their characteristics are discussed below.

The tensile yield strengths of the bcc metals seem to divide themselves into three regions: a high-strength region at low temperatures, a low-strength region at higher temperatures, and a region between these two. Except for tantalum, the ductility (reduction in area) of the bcc metals inversely corresponds to yield strength. That is, when the strength for a metal is high, its ductility generally is low, and vice versa. We may think that a metal of high strength and low ductility is most likely brittle, and that a metal of low strength and high ductility is ductile. For instance, tungsten is completely brittle at (and below) 100 °C (210 °F) and totally ductile at temperatures above 500 °C (930 °F). The temperatures in between these two regions are called the ductile-to-brittle transition temperature(s).

To the contrary, the yield strength and reduction in area for nickel (fcc) do not vary much over a wide temperature range. The lack of a ductile-to-brittle transition in nickel is a general characteristic of fcc metals. Tantalum is an exceptional case; it shows no ductility transition, although the yield stress increases rapidly at low temperature.

Similar to room-temperature behavior, the magnitude of fracture stress in high temperatures also becomes increasingly dependent upon the rate of straining: The higher the stain rate, the higher the tensile strength. The effects attributed to temperature and strain rate are illustrated by the test data in Fig. 2.70. To explore the strain

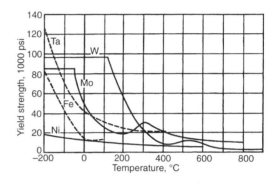

Fig. 2.68 Effect of temperature on the yield strength of bcc tantalum, tungsten, molybdenum, and iron and fcc nickel. Source: J.H. Bechtold (graphs adopted from Ref 2.2)

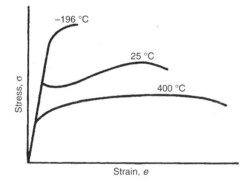

Fig. 2.67 Changes in engineering stress-strain curves of mild steel with temperature. Source: Ref 2.2

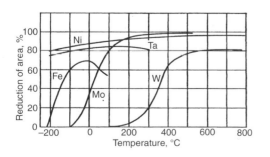

Fig. 2.69 Effect of temperature on the reduction of area of tantalum, tungsten, molybdenum, iron, and nickel. Source: J.H. Bechtold (graphs adopted from Ref 2.2)

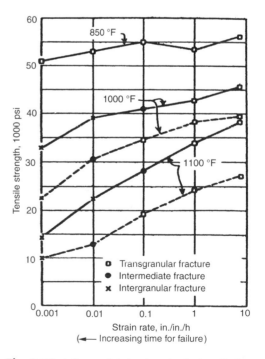

Fig. 2.70 Influence of strain rate on tensile strength of pearlitic (solid lines) and spheroidized (dashed lines) carbon-molybdenum steel at 455, 540, and 595 °C (850, 1000, and 1100 °F); controlled strain-rate tests. Source: Ref 2.42

rate effect further, the investigators conducted two more experiments on a 0.5% Mo steel and switched the loading rate in the middle of a test. As shown in Fig. 2.71, one test started at a higher strain rate, and then switched to a lower strain rate at maximum load. The second test did just the opposite. When we try to connect the first part of the stress-strain curve for one test to the second part of the stress-strain curve for the other test, the final curve seems to come from a single test. This phenomenon is true for both tests. Note that the slight drop in nominal stress at the point of changing strain rates was a result of the time interval consumed in changing gears to effect the change in loading rate (Ref 2.42). Nonetheless, these experiments showed that prior strain rate had a minor influence on the subsequent stress-strain behavior, which depended primarily on the strain rate from the maximum load to failure.

Another quite interesting and important effect is that fracture occurs differently at high temperatures than at low temperatures. Instead of necking down (in a ductile metal), the specimen fractures rather abruptly, although considerable plastic strain may have occurred. Microexamination reveals that the path of fracture often is intergranular at high temperatures. It also reveals that intergranular fracture that shows little or no necking apparently starts at the surface and progresses inward. A hypothetical mechanism for the intergranular fracture is that above some temperature of equicohesion, the grain boundary is weaker than the material within the grain and fracture occurs at the grain boundaries. However, the grain boundary is stronger at temperatures below this; consequently, fracture occurs through the grains. The grain-boundary failure mechanism is revisited later in section 2.8.1.

Let's reexamine Fig. 2.70 and 2.71 for a moment. They show that for tests conducted at higher temperatures (540 and 595 °C, or 1000 and 1100 °F), the specimen tested at a faster rate exhibits a transgranular fracture, while the specimen tested at a slower rate exhibits an intergranular fracture. At a lower temperature of 455 °C (850 °F), the fracture paths are transgranular at all strain rates. A fracture path of an intermediate type (probably a mixture of transgranular and intergranular fracture) is also observed. This type of fracture occurs at 540 or 595 °C (1000 or 1100 °F), with intermediate to high strain rates. A comparison of fracture types in the "single-rate" and "two-rate" tests is shown in Table 2.3. The fracture behavior of a given two-rate specimen corresponds with the fracture type that normally occurs at the temperature and rate for the second part of the test.

In summary, this type of fracture depends not only on temperature but also on strain rate. The two effects are of opposite sign; that is, intergranular fracture is more likely to occur with increasing temperature but with decreasing strain rate. Thus, at constant temperature, transgranular or intergranular fracture may occur, depending on the strain rate, while at constant strain rate the type of fracture depends on the temperature.

2.5.4 Transition Temperature

This behavior can be studied by conducting one or all of the following tests: ordinary tension test, notched bar tension test, notched bar impact test, and residual strength test of a precracked specimen. This section and section 2.5.5 are devoted to tension tests, while the other two types of tests are discussed in Chapter 4. Discussions in this section are based on investigations by Hahn et al. (Ref 2.43). The results of those investigations are presented in Fig. 2.72 for the temperature dependence of fracture stress, yield stress, and microcracking of ferrite (bcc iron).

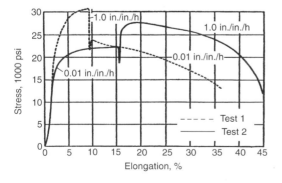

Fig. 2.71 Effect of change of strain rate on tensile properties of carbon-molybdenum steel at 595 °C (1100 °F). Source: Ref 2.42

Table 2.3 Comparison of single-rate and two-rate tests on a pearlitic 0.5% Mo steel at 595 °C (1100 °F)

Type of test	Strain rate, in./in./h	Elongation, %	Reduction in area, %	Type of fracture
Two-rate	0.01–1.0	41	75	Transgranular
Two-rate	1.0–0.01	43	43	Intergranular
Single-rate	1.0	56.5	80.5	Transgranular
Single-rate	0.01	35	42.5	Intergranular

Source: Ref 2.42

The temperature scale is divided into six regions. In region *A*, the specimen undergoes necking and breaks with a characteristic cup-and-cone fracture. Cleavage is not observed. In region *B*, necking still occurs, but fracture starts in the necked region with a fibrous crack at the center of the specimen and changes to cleavage in an annular rim before rupture is complete. The fracture appearance, which is expressed in terms of the percentage of fibrous area, thus indicates the size of the initiating fibrous crack prior to cleavage. The critical size of fibrous crack to initiate the cleavage decreases as temperature is lowered in this region. The fracture transition is analogous to 50% fibrous appearance convention in Charpy specimens.

At the dividing line between regions *B* and *C*, fracture stress and reduction in area fall abruptly. This behavior defines the ductility transition temperature T_d. Microcracks are observed at the lower end of region *B* and in region *C*. Below T_d, fracture is completely cleavage and is initiated by cleavage microcracks. However, it should be noted that despite these observations, microcracks are not the only requirement for brittle fracture.

In region *D*, the yield and fracture stresses are almost identical, but yielding is always observed prior to fracture. Fracture occurs at the lower yield stress. The fracture stress increases because the yield stress is increasing with decreasing temperature. Microcracks are found in this temperature range (most of them about the length of a ferrite grain), but only in areas of discontinuous strain.

In region *E*, cleavage fracture occurs abruptly. The cup-and-cone characteristic of yielding is not observed. Fracture stress values scatter about the extrapolated upper yield stress curve. It appears that the behavior in region *E* is an extension of region *D*, and that cleavage fracture takes place in the very first location to undergo discontinuous yielding. Microcracks are not found in this region; presumably the first one propagates to failure. The borderline between regions *D* and *E* is defined as the microcrack transition temperature (T_m). Below T_m, fracture occurs at upper yield stress rather than at the lower yield stress, because the testing machine cannot unload rapidly enough.

In region *F*, samples break abruptly, with the fracture stress values falling below extrapolations of the upper and lower yield stress curves. This region has often been identified with complete brittle fracture—that is, cleavage fracture

without prior plastic deformation. However, results often show that cleavage fracture in this region is associated with mechanical twinning, although the exact role of twinning is not clear. Also note that mechanical twins are observed at temperatures as high as T_d, but it is only in region *F* that they appear to be the source of initiating fracture.

2.5.5 Effect of Notch on Transition Temperature

Despite its limited applicability, the concept of fracture occurring on attainment of a critical value of the normal stress has been useful in explaining the processes of flow and fracture. Figure 2.57 has been widely employed to show qualitatively the variation with temperature of the critical normal stress to produce fracture and of the critical shear stress to produce plastic flow. The variation of normal stress for fracture is much less than the variation of the shear stress for flow. With increasing stress, either flow or fracture will occur, depending on whether the critical shear stress or the critical normal stress is reached first. The ratio of the greatest normal stress to greatest shear stress varies with the type of test, being 1 for torsion, 2 for tension, and greater than 2 for notch tests, depending on the sharpness of the notch.

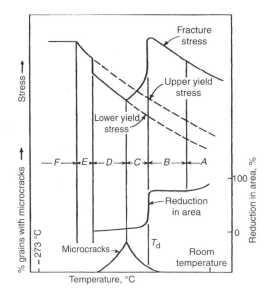

Fig. 2.72 Temperature dependence of fracture stress, yield stress, and microcrack frequency for mild steel. Source: Ref 2.43

With decreasing temperature, brittleness will occur in tension when the critical normal stress for fracture becomes less than twice the critical shear stress for flow by the Tresca criterion; in torsion this does not occur until the ratio becomes less than 1. Thus, toughness is retained to a lower temperature in torsion than in tension, while similarly, toughness is retained to a lower temperature in tension than in a notch test. The reason for metals remaining tough to a lower temperature in simple tension than in notch tension is because the normal stress for fracture depends on the state of stress. The triaxial stress at the notch promotes brittleness, and thus brittle fracture can occur at higher temperatures.

According to a similar concept that explains the relationship between notch sensitivity and temperature, the ratio of notch yield stress to simple tension yield stress cannot be greater than 3 (Ref 2.33). The reasoning is that the existence of a transition temperature is due to the difference in the way resistance to shear and cleavage changes with temperature. The relative values of these parameters, the multiplicity of the slip and cleavage systems, and the specific grain orientation determine whether the fracture will be ductile or brittle. Factors that increase the critical shear stress for slip, without at the same time raising the fracture stress, will favor brittle fracture. Decreasing the temperature and/or increasing the strain rate will have this effect.

In Fig. 2.73, the curve marked S_y is the yield stress in simple tension, S_f is the fracture stress, T_1 is the transition temperature in simple tension, and T_2 is the transition temperature in notch tension. Using Fig. 2.53, Orowan (Ref 2.33) has concluded the following:

- Full brittleness is attained at temperatures below T_1 where $S_f < S_y$.
- Between T_1 and T_2 the material is ductile in the tensile test but notch brittle ($S_y < S_f < 3S_y$).
- For temperatures above T_2 the material is fully ductile ($S_f > 3S_y$).

2.6 Residual Stresses

Residual stresses are internal stresses that can exist in a body when it is free from external forces. Residual stresses are produced and retained in a body whenever the body undergoes nonuniform plastic deformation. Fundamentally, residual stresses arise from two sources: thermal gradients and mechanical strain gradients.

A large variety of residual stresses are derived indirectly from deformations accompanying processes characterized by nonuniform cooling. The main types of such cooling stresses are:

- Casting stresses
- Quenching stresses
- Transformation stresses
- Weld stresses

All of these stresses result from a nonhomogeneous thermal contraction or a volume change during a phase transformation, at temperatures low enough that stresses cannot be relaxed by plastic flow. As long as the temperature of some parts of the metal remains within the range of hot deformation, only very small stresses are retained, as the hotter sections are able to yield and thus relieve the stresses.

For quenching stresses, residual stresses are produced during quenching by deformation associated with shrinkage during cooling. Generally, the first portion of the part to cool is left in residual compression, while the last portion to cool is left in residual tension. Therefore, one can expect that compressive residual stress would be on the surface and tensile residual stress would be in the center, in all three dimensions. The amount of compression stress should be balanced by the same amount of tension stress in order to maintain equilibrium. For high-hardenability steels, the residual stress distribution may be different if there is a phase transformation during cooling and if the new phase is brittle. It may have tensile residual stresses in the hoop and axial directions at the surface (if not relieved by quench cracking).

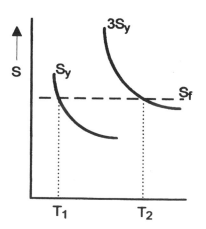

Fig. 2.73 Explanation of the transition temperature: regions of full brittleness (up to T_1), notch brittleness (T_1 to T_2), and full ductility (above T_2). Source: Ref 2.33

Residual stresses can occur in any multiple-piece tool or structure formed by shrinking a heated tubular part on a core or pressing an insert into an undersize hole, or as the result of misfit or mismatch in an assembly. Internal stresses are also present in any bimetal composed of two metals with different coefficients of thermal expansion.

Residual stresses may cause a beneficial as well as a deleterious effect. Superposition of the applied external stresses and the residual stresses is a valid procedure for obtaining the resultant stress distribution. Depending on the usage of the part, residual stresses that are induced by mechanical prestressing can be either harmful or beneficial. When a part is subjected to external loads in tension, compressive residual stresses counteract the applied tensile stresses and are thus beneficial. In contrast, a part that contains tensile residual stresses will yield plastically at a lower value of applied stress than a part without residual stress. With time, tensile residual stress will cause SCC even in the absence of external forces. Residual stresses (tension or compression) also significantly affect fatigue life (see Chapter 3).

During machining, warping can occur, depending on the magnitude of the residual stresses, their distribution, and the quantity and location of the removed metal. When part of a body containing residual stresses is machined away, the residual stresses in the material removed are also eliminated. This upsets the static equilibrium, and the body distorts to establish a new equilibrium condition. From a researcher's standpoint, warping provides a means for measuring the magnitude of residual stress.

Residual stresses may originate directly from a nonuniform cold working process. Tensile yielding under an applied load will result in compressive residual stresses when the load is released, and vice versa. For instance, in the case of sheet rolling, plastic flow occurs only near the surfaces of the metal sheet (Fig. 2.74a). In other words, the surface fibers of the sheet are cold worked and tend to elongate, while the center fibers tend to restrain the surface fibers from elongating. The layer on the sheet surface is in tension during cold working; it is left in a state of compressive residual stress when the external load is removed. Because the residual stress system existing in a body must be in static equilibrium, the total force acting on any plane through the body must be zero. Therefore, the compressive residual stress on the sheet surface must be balanced by some residual stress in the center of the sheet, which has to be in tension (Fig. 2.74b).

The magnitude of the residual stress at the surface of a rolled strip depends on the roll diameter, friction, sheet thickness, and reduction in thickness. Actually, a residual stress system should be considered as three dimensional. However, in most cases, one of the directions is of little importance and can be ignored. In some cases, because of symmetry, only the residual stress in one direction need be considered. The residual stress distribution given in the example of Fig. 2.74(b) comes from rolling of a sheet to which the reduction in sheet thickness is small. That is, the plastically deformed region does not extend much below the outer surface. When the reduction in thickness is large, and the plastic flow penetrates through the sheet thickness, the residual stress pattern will be the complete reverse of that shown in Fig. 2.74(b).

Residual stresses can be regarded as the difference between the acting stresses during the deformation under load and the stresses of an elastic condition corresponding to the same load. A general rule is that only the final deformation process determines the resulting residual stress pattern. Then the residual stresses are superimposed to the applied stresses upon reloading. We will demonstrate this concept with several examples.

Autofrettage. One of the better known examples is "autofrettage," a process that is used to improve the strength of gun barrels. This is done by expanding a tube to the plastic range and setting up a residual stress distribution across the thickness of the tube upon unloading. In the process of autofrettage, residual compressive stresses are formed in the gun barrel in such

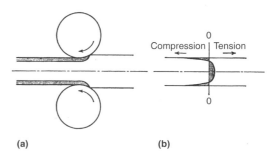

(a)   **(b)**

Fig. 2.74 Rolling of a metal sheet. (a) Elongation on surface of the sheet. (b) Resulting distribution of longitudinal residual stress over thickness of sheet (schematic). Note: The residual stress pattern would be the reverse of those shown here for large reduction of thickness. Source: Ref 2.2

a way that they will partially counterbalance the tension stresses created by the explosion.

Figure 2.75 illustrates the autofrettage process. The across-the-thickness plastic stress distribution in part 2 is obtained from plasticity analysis (Eq 1.69e), where the plastic stress distribution is caused by an internal pressure p_2. This pressure level is high enough to cause yielding of the entire cylinder wall. Part 1 is the elastic stress distribution of the same applied internal pressure, but computed on the basis of elasticity analysis (Eq 1.69a).

The residual stress distribution in part 3 of the figure is obtained from superposition of 1 and 2—that is, subtracting the elastic stress by the plastic stress. Owing to plastic deformation, considerable compressive tangential stresses are produced in the portion of the cylinder wall. Generally, residual stresses are considered as elastic stress only. The maximum value that the residual stress can reach is the current value of the elastic limit of the material. A stress in excess of this value, with no external force to oppose it, will relieve itself by plastic deformation until it is lowered to the value of the yield stress. Therefore, it has been assumed that such a compressive stress distribution is less than the yield point stress and that no yielding occurs during unloading.

Reloading will just reverse the process: The resultant stress distribution will be that represented by the curve shown in part 2 of Fig. 2.75. The maximum resultant stress is the yield strength of the material, and no yielding occurs during reloading. In any event, the maximum benefit of the procedure can be obtained by an additional stress relief treatment at low temperatures. After stress relieving, the resulting stress distribution is more uniform than the pure elastic strain and thus will resist a considerably higher pressure without yielding. The resultant stress distribution during reloading will be that given by part 5 of Fig. 2.75.

Residual Stresses in the Vicinity of a Stress Raiser. In a preloaded part with a stress raiser (e.g., a notch or a circular hole), a residual stress distribution can result from geometric-induced yielding. This can be explained by continuing where we left off with the example of a circular hole in a sheet (Fig. 2.64). Previously, we saw that the Kirsch/Howland solutions cannot always obtain the local stresses in the vicinity of a stress raiser. The local stress distribution for a given geometry is not a single curve and is not always elastic. Rather, it is a group of curves

whose magnitudes and shapes depend on the applied stress levels.

Upon unloading from a given point in the plastic range of the tensile stress-strain curve, the plastically deformed material in the vicinity of the hole will be subjected to restoring forces by the surrounding elastic material. In other words, the sheet material follows Hooke's law during unloading. The result is the creation of a residual stress field. The residual stress at a given local point in the sheet (i.e., at a distance from the hole edge) is simply computed by subtracting the elastic stress (i.e., Eq 1.72 or equivalent) from the plastic stress. An example of the results, for which the residual stress corresponds to a local stress distribution that in turn corresponds to a 248 MPa (36 ksi) applied stress, is depicted in Fig. 2.76. This depiction fits with a

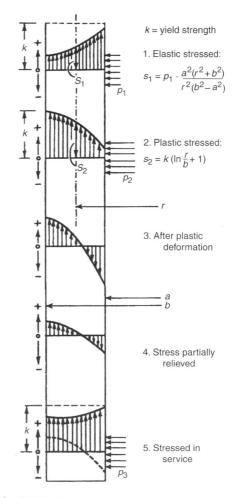

k = yield strength

1. Elastic stressed:

$$s_1 = p_1 \cdot \frac{a^2(r^2 + b^2)}{r^2(b^2 - a^2)}$$

2. Plastic stressed:

$$s_2 = k \left(\ln \frac{r}{b} + 1 \right)$$

3. After plastic deformation

4. Stress partially relieved

5. Stressed in service

Fig. 2.75 Schematic stress distributions in a thick-wall cylinder produced during different stages of autofrettage and during the service of the gun barrel

generalized concept that is graphically presented below.

General Graphical Representation of Geometric-Induced Yielding at a Stress Raiser. Although it might be somewhat oversimplified, the following graphical procedure serves to illustrate the scenario of geometric-induced yielding. To begin, let us just pick any applied stress level that is higher than one-third of the proportional limit of the material. Using Eq 1.72, the elastic tangential stress distribution (corresponding to a given applied load) is computed and plotted as the curve ABC on the upper left of Fig. 2.77. Here we only use three hypothetical data points for the demonstration; the points A, B, and C are the computed elastic stresses at distances A, B, and C from the edge of the notch. Next, we place a hypothetical tensile stress-strain curve to the right of the curve ABC. The stress levels of A, B, and C on the left are projected to the right as points A_1, B_1, and C_1. We place these points on a line that is the imaginary extension of the linear elastic stress-strain relationship while knowing, of course, that these stresses cannot exist because they are significantly higher than the elastic limit of the mate-

rial. In reality, the local stresses in the vicinity of the hole are limited by the material tensile properties as explained earlier with Fig. 2.64. In other words, these points should correspond to points A_2, B_2, and C_2 on the material tensile curve instead. These points of plastic stress actually translate to the plastic stress distribution A', B', and C'. This is the kind of plastic stress distribution shown in Fig. 2.64.

Upon unloading, the stresses after plastic deformation (points A_2, B_2, and C_2) will drop in accordance with Hooke's law. Because the sheet tends to return to a strain-free condition after all the loads are removed, these locations (that have been plastically stretched) become regions of compressive stress when the local strain returns to zero (i.e., points A_3, B_3, and C_3 in Fig. 2.77). The residual stress distribution is finally represented by A'', B'', and C''. It is clear in Fig. 2.77 that preyielding produces a significant amount of compressive residual stress at the hole edge and in its vicinity. The starting stress level will be A_3 when the part is reloaded. If the part is reloaded in tension, the part will not yield until it reaches the stress level of A_2. Referring back to the example given in Fig. 2.64 and the applied

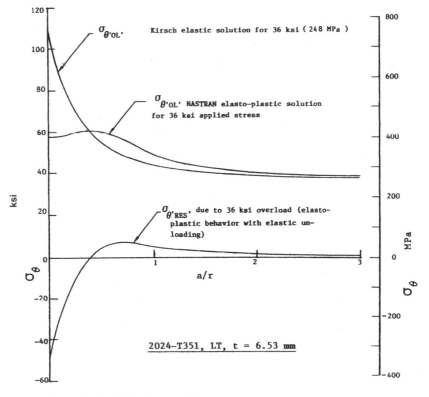

Fig. 2.76 Tangential stresses in the vicinity of an open hole

stress level of 248 MPa (36 ksi), the sheet with a circular hole yielded at the hole and produced a residual stress pattern like that shown in Fig. 2.76. Consequently, when the sheet is reloaded in tension, it will not yield until the applied stress level is above 248 MPa (36 ksi).

2.7 Material Toughness

Structural members may fail prematurely at a stress level much below the ultimate strength, or even below tensile yield strength. Yet fracture stresses have been inconsistent among materials with different combinations of static strength and ductility. Conventional wisdom has suggested that other than the commonly known mechanical properties (such as tensile strength, etc.) there must be an inherent material property that provides resistance to premature failure of a structure. This is called material toughness, or simply toughness. The toughness of a material is its ability to absorb energy primarily in the plastic range under monotonic loading since the area under the elastic portion of the curve is small relative to the plastic area. Toughness is a

commonly used concept, but toughness as a quantity is difficult to define from a simple conventional tensile test.

Toughness Expressed as Area under the Stress-Strain Curve. One simple way of describing toughness is to consider it as the total area under the stress-strain curve. This area is indicative of the amount of work done per unit volume, a quantity that comprises both strength and ductility. Figure 2.4 shows two stress-strain curves representing high- and low-toughness steels. Steel A has a higher yield strength and tensile strength than steel B. However, steel B is more ductile and has a greater total elongation. The total area under the stress-strain curve is greater for steel B, which therefore is a tougher material.

Several mathematical representations for the area under the stress-strain curve have been suggested (Ref 2.2). For ductile metals that have a stress-strain curve like that of steel B, the area under the curve (which implicates the total energy) can be approximated by either of the following equations:

$$W_T = \frac{1}{2}\, \varepsilon_f (F_{ty} + F_{tu})$$

(Eq 2.38a)

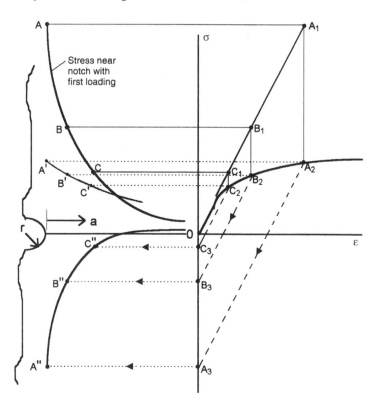

Fig. 2.77 Schematic residual stress distribution resulting from geometric-induced yielding at a stress raiser

or

$$W_T = \varepsilon_f \cdot F_{tu} \qquad \text{(Eq 2.38b)}$$

For relatively brittle materials, the stress-strain curve is sometimes assumed to be a parabola, and the area under the curve is approximated by:

$$W_T = \frac{2}{3} \varepsilon_f \cdot F_{tu} \qquad \text{(Eq 2.39)}$$

Apart from the fact that the above equations are only approximations, the engineering stress-strain curves shown in Fig. 2.4 do not represent the true behavior in the plastic range. The inaccuracy that is injected by these equations increases for more ductile materials. When a material becomes more and more ductile, the plastic range occupies a much larger area as compared to the elastic range (compare the stress-strain curves for materials A and B in Fig. 2.4). It is conceivable that the area under a true-stress/true-strain curve might provide a better indication of material toughness. However, remember that the result of a tensile test only provides a qualitative indication of material toughness. It is meaningful in the sense of being a quick and economical way to identify the fracture-resistance capability of a material (in the first order of magnitude) during the course of material development or selection of material for a design.

Cockcroft-Latham Ductile Fracture Criterion (Workability). Prevention of ductile cracking is an important concern in metal-forming processes such as forging, extrusion, and rolling. These operations must be properly designed so that the required deformation does not induce cracking in the manufactured piece. This determination of suitable conditions to prevent cracking during metalworking is generally referred to as workability, as discussed in more detail in Ref 2.44 and Ref 2.45.

A common fracture criterion with bulk deformation is that proposed by Cockcroft and Latham (Ref 2.46), whereby fracture occurs when the plastic strain energy per unit volume (W) reaches a critical value:

$$\int_0^{\varepsilon_f} \overline{\sigma} d\overline{\varepsilon} = W \qquad \text{(Eq 2.40)}$$

where W is the plastic strain energy density, ε_f is the strain at fracture, and $\overline{\sigma} = f(\overline{\varepsilon})$ is the cur-

rent value of flow stress. The key to this concept is that fracture is based on the largest existing tensile stress (not on an average stress at fracture) (Ref 2.44). Thus, the effective stress is multiplied by a dimensionless stress-concentration factor to give the Cockcroft-Latham criterion:

$$\int_0^{\varepsilon_f} \overline{\sigma}\left(\frac{\sigma^*}{\overline{\sigma}}\right) d\overline{\varepsilon} = \int_0^{\varepsilon_f} \sigma^* d\overline{\varepsilon} = C \qquad \text{(Eq 2.41)}$$

where σ^* is the maximum tensile stress in the workpiece and C is the Cockcroft-Latham constant. The Cockcroft-Latham method is sometimes referred to as a "ductile fracture criterion," because it can be used to evaluate fracture conditions with significant plastic deformation. However, it should be recognized that tensile separation is the underlying fracture concept, rather than fracture by shear glide.

The Cockcroft-Latham method has been used successfully to predict fracture in edge cracking in rolling and free-surface cracking in upset forging under conditions of cold working. Examples are given in Ref 2.47. In the past, one difficulty has been the requirement to determine the maximum tensile stress in the workpiece (Ref 2.44); however, this is no longer a difficulty with computer modeling methods (Ref 2.48). In fact, most commercial finite-element models for large plastic deformation include the ability to map-out the Cockcroft criteria over the deforming body. More of an impediment is the lack of a well-proven method for experimentally determining C under the experimentally difficult temperature and strain rate conditions found in hot working (Ref 2.44).

Griffith's Criterion of Brittle (Elastic) Fracture. Toughness can also be understood in terms of Griffith's criterion of brittle (elastic) fracture. This seminal contribution of A.A. Griffith (Ref 2.49) is based on the recognition that the driving force for crack extension is a balance of the release of stored (elastic) energy from crack extension and the energy needed to create new surfaces. This fundamental concept can be understood by considering the idealized example of a through-thickness crack (of length $2a$) in a infinitely wide sheet (Fig. 2.78) that is stressed with elastic strain. The elastic energy (U_e) stored in the system (per unit thickness) is:

$$U_e = -\frac{\pi a^2 S^2}{E} \qquad \text{(Eq 2.42)}$$

where S is the applied stress and E is the modulus of elasticity. The minus sign indicates that this quantity would be released from the system as the crack length (a) increases. The system also has energy associated with the total area of the crack surface, and this surface energy of the crack (per unit thickness) is calculated as follows:

$$U_s = 2(2a)\gamma \qquad \text{(Eq 2.43)}$$

where γ is the specific surface energy (surface tension), and the $2 \times 2a$ denotes that the crack length ($2a$) has two surfaces, top and bottom.

Griffith's criterion of fracture states that the crack will propagate under constant applied stress, S, if an incremental increases in crack length produces no net change in the total energy of the system; that is, the derivative with respect to a is zero for the quantity $U_e + U_s$ as follows:

$$\frac{\partial U}{\partial a} = 0 = \frac{\partial}{\partial a}\left[4a\gamma - \frac{\pi a^2 S^2}{E}\right] = 4\gamma - \frac{2\pi a S^2}{E}$$
$$\text{(Eq 2.44)}$$

which gives:

$$S_{cr} = \left(\frac{2E\gamma}{\pi a}\right)^{1/2} \text{ or } 2\gamma = \frac{\pi a S_{cr}^2}{E} \qquad \text{(Eq 2.45)}$$

where S_{cr} is the critical stress that results in unimpeded growth of the crack. This represents brittle fracture with the sudden and rapid release of stored elastic energy in the stressed component. The resulting fracture is macroscopically brittle with a fracture surface that is normal to the direction of tensile separation. This criterion of brittle (elastic) fracture is the underlying foundation of fracture mechanics, as discussed in more detail in Chapter 4.

Alternatively, crack growth also can occur when stresses are below S_{cr}. This condition is referred to as subcritical crack growth. During subcritical crack growth, a small crack slowly grows over time, often due to fatigue or environmental factors. In this situation, the crack grows until cataclysmic fracture occurs when the crack reaches a critical length (a_{cr}) such that:

$$a_{cr} = \frac{2\gamma E}{\pi S^2} \qquad \text{(Eq 2.46)}$$

Although the concepts of the Griffith criterion apply to the specific case of fracture under elas-

tic conditions, this fracture criterion has been instrumental in defining the terms "material toughness" or "fracture toughness" in a practical engineering context. Recognizing that a structural part often contains a cracklike damage prior to catastrophic failure, fracture toughness is used to measure the material's ability to resist rapid crack propagation. Energy absorption is also used to characterize this quantity. So, we may literally take material toughness as an inherent quality of a material and fracture toughness as the quality that provides a means to measure the residual strength of a cracked body. The concepts of residual strength and fracture toughness, and the techniques for determining structural residual strength, are discussed in detail in Chapters 4 to 6.

2.8 Deformation and Fracture under Sustained Loads

Prior to the onset of rapid fracture, many types of subcritical ($a < a_c$) crack growth can occur from progressive (time-dependent) damage mechanisms. Fatigue crack growth, for example, is one of the most prevalent progressive mechanisms of subcritical crack growth, as described in more detail in Chapters 3 and 5. However, several types of progressive crack growth occur under monotonic (or sustained) loads. This section briefly describes the following three mechanisms of deformation and fracture under sustained loads:

- Creep and stress rupture
- Stress corrosion
- Hydrogen embrittlement

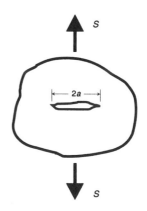

Fig. 2.78 Cracked body subjected to uniaxial tension

2.8.1 Creep and Stress Rupture

Creep is a high-temperature phenomenon. In crystalline solids, such as metals and ceramics, creep is a concern when the service temperature is somewhere above one-half the absolute melting temperature (T_m). However, it should be noted that measurable creep strains can occur in some crystalline materials at temperatures as low as one-fourth of the absolute melting temperature. As for noncrystalline materials, such as polymers and glasses, creep occurs at temperatures above the glass transition temperature of the material.

Creep, or creep deformation, is any permanent inelastic strain that occurs when a material is subjected to a sustained stress. The rate of this deformation depends not only on the magnitude of the applied stress, but also on temperature and time. Thus, it is appropriate to consider creep as a kinetic process. When designing a structural component operating at high temperature for a prolonged period, it is essential to obtain information about the creep and stress-rupture properties of a material for the intended usage. This is normally accomplished by conducting creep and stress-rupture tests. Creep tests measure the amount of creep strain as a function of time (at a given temperature and stress level). Stress-rupture tests provide information regarding design allowable as functions of time and temperature.

In a creep test, a constant load is applied to a tensile specimen and maintained at a constant temperature. The strain of the specimen is determined as a function of time, resulting in a typical creep curve such as that shown in Fig. 2.79. Curve A illustrates the idealized shape of a creep curve. The slope at each point of this curve ($d\varepsilon/dt$ or $\dot{\varepsilon}$) is referred to as the creep rate. At the beginning of a creep test, a certain amount of elastic and plastic deformation takes place immediately after the load is applied. This initial deformation is indicated in Fig. 2.79 as the quantity ε_0. This is followed by a period of decreasing strain rate, identified as first-stage creep (or primary creep, or transient creep), which can be compared to the onset of work hardening that would occur at lower temperatures. The material then enters a stage in which the strain increases linearly with time, identified as second-stage creep (or secondary creep, or steady-state creep). The creep rate changes little with time in this stage, and finally increases rapidly with time until fracture. The last period is called third-stage creep (or tertiary creep). By not counting the initial deformation ε_0, it is customarily considered that the entire creep curve consists of three parts: a period of decreasing creep rate, a minimum creep rate, and finally a period of accelerating creep rate.

The extent of the initial deformation, ε_0, which occurs immediately after the load is applied, depends on the nature of the material under test and the chosen conditions of temperature and load. The properties of a material in this respect can be fully determined by the normal tensile test carried out at the required temperature. Although the applied stress is below the yield stress, not all the instantaneous strain is elastic. Most of this strain is recoverable upon release of the load; some is recoverable with time. These are called elastic and anelastic strains, respectively. Some remaining plastic strain may be nonrecoverable.

Although the instantaneous strain is not really creep, it is important because it may constitute a considerable fraction of the allowable total strain in machine parts. Sometimes the instantaneous strain is subtracted from the total strain in the creep specimen to give the strain due only to creep. This type of creep curve starts at the origin of the coordinates. Families of creep curves obtained at different stresses and a constant temperature, or vice versa, for most polycrystalline materials tend to look like those shown in Fig. 2.80. It is clear in this figure that the lower the stress (and the lower the temperature), the longer the time before the minimum creep rate is attained, and the longer the period over which this rate remains substantially constant. It is also suggested that this type of curve is typical of the creep behavior within the range of 0.4 to 0.6 T_m (Ref 2.51).

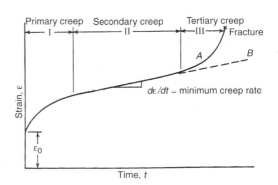

Fig. 2.79 Tyical creep curve showing the three stages of creep. Curve A, constant-load test; curve B, constant-stress test. Source: Ref 2.2

The transient creep stage is generally associated with the appearance of slip bands in the grains of the metal. In the initial stage of decreasing creep rate, the effect of strain hardening is believed to outweigh the softening. In addition, it has been shown that during this stage of creep the metal crystals tend to rotate slightly to reorientate themselves in the direction of the applied stress. It is believed that the decreasing creep rate is attributed to strain hardening and to this reorientation behavior. However, the primary creep of polycrystalline metals is partly or wholly recovered from removal of the load at the test temperature. Sample test data of this kind (reported by H.J. Tapsell and L.E. Prosser in 1934) is shown in Fig. 2.81(a). The recovery of creep strain (in a Ni-Ci-Mo steel at 450 °C,

or 840 °F, after partial removal of the applied stress) is evident in this figure. A generalization of this phenomenon is depicted schematically in Fig. 2.81(b). In a recent book by Webster and Ainsworth (Ref 2.51), the authors claim that recovery after unloading (by a small amount) is attainable in the period of second-stage creep.

During the steady-state creep period, the opposing forces (strain hardening and thermal softening and damage processes) balance one another. Similarly, the final period of accelerating creep rate is considered to be from a predominance of softening and the increase in net-section stress. An accelerating creep rate from an increase in net-section stress is evidenced from constant-load test data (see curve A in Fig. 2.79). The initial applied stress is usually the reported value of stress. This is the normal way of developing an engineering creep curve, because engineering parts are subjected to applied loads. However, it has been pointed out that the applicability of constant-load data should be restricted at 1% creep strain. Above that, constant-stress conditions must be strictly imposed in order to obtain valid data (Ref 2.52). In contrast, it has also been suggested that the distinction between a constant-stress test and a constant-load test is unimportant up to 5% of creep strains (Ref 2.51).

Usually, at small strains, the two methods give essentially the same results. Under a constant-stress condition, the loads are required to adjust to lower levels time after time. The second stage may last throughout, taking much longer to finish the test, and the third stage of creep may never be reached. The creep curve would appear like curve B in Fig. 2.79. However, accelerated

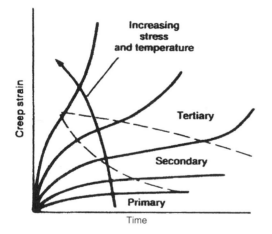

Fig. 2.80 Schematic illustration of creep curve shapes as functions of stress and temperature. Source: Ref 2.50

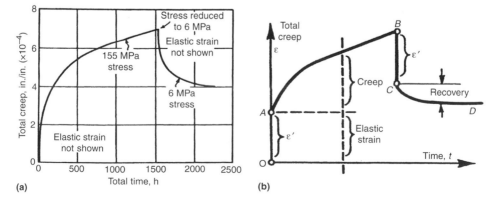

Fig. 2.81 Creep recovery. (a) Test data for Ni-Cr-Mo steel at 450 °C (840 °F). Source: Ref 2.42. (b) Schematic representation of the phenomenon. Loading produces an immediate elastic strain followed by viscous flow. Unloading produces an immediate elastic recovery followed by additional recovery over a period of time. Source: Ref 2.6

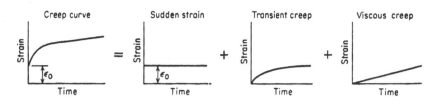

Fig. 2.82 Andrade's model of a constant stress creep curve. Source: Ref 2.53

creep might still be found in constant-stress data when metallurgical changes occur in the metal (Ref 2.2).

Creep Mechanisms. In explaining and characterizing the creep curve of Fig. 2.79, Andrade in 1914 (Ref 2.53) created a model that represents the initial and the linear regions of the creep curve. His model, shown in Fig. 2.82, led us to believe (for a long time) that creep is a viscous phenomenon. However, engineers later found that many creep mechanisms exist; depending on the combination of stress and temperature, viscous behavior does not prevail in all the identifiable mechanisms.

From a metallurgist's point of view, the creep phenomenon largely hinges on the grain boundary. The atomic processes of flow in an incoherent boundary between grains are more like those in a viscous fluid than like plastic glide in a crystal. The activation energy to overcome barriers between various atomic configurations in the boundary is supplied almost entirely from thermal fluctuations. There is no elastic limit; at sufficiently low stresses, Newton's law of viscosity is obeyed. At high temperature where the viscosity is low, we must deal with fluid grain boundaries constrained by relatively rigid grains. This constraint restricts the amount of grain-boundary deformation so that the total deformation of the sample is small at high temperature and low stresses. Plastically weak surfaces are embedded in a relatively unyielding matrix, a condition that is conductive to fracture.

Two types of grain-boundary failure have been observed, both of which produce intergranular cracking. In the first, which occurs particularly at triple points where three grain boundaries meet, stress concentration is sufficient to start cracks there (Ref 2.54). Such processes are geometrically very similar to the plastic-shear process. This type of grain-boundary failure is prevalent for high stresses, where the total life is short. Figure 2.83 presents models that depict formation of a crack at a triple point as the result of grain-boundary sliding. A micrograph of

grain-boundary sliding is shown in Fig. 2.84. When a low level of stress is applied for a long time, the second type of grain-boundary failure appears. Small holes form on grain boundaries, especially those perpendicular to the tensile axis, where they grow and coalesce. The growth of holes is due to the movement of lattice vacancies. The micrograph in Fig. 2.85 shows voids coalesced on grain boundaries.

Creep strength is the opposite of ordinary tensile strength at room and lower temperatures. The grain-boundary material is stronger at room and lower temperatures. Materials that have small grains exhibit higher strength because the grain-boundary area that surrounds the grain is larger. The grain boundary also acts as a barrier for dislocation pileup. While at high temperature, the grain boundaries prove disadvantageous to creep strength. For a material with

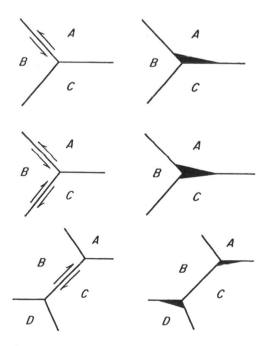

Fig. 2.83 Models illustrating how intergranular cracks form due to grain-boundary sliding. Source: Ref 2.55

small grains, the larger areas of grain boundaries provide good sources and sinks of vacancies and more triple-point sites for the same volume of material. The grain boundaries are weaker at high temperatures. Therefore, creep resistance can be improved by increasing the grain size (to reduce the grain-boundary area) or developing an elongated grain structure through directional solidification.

From an engineer's point of view, the second-stage minimum creep rate ($\dot{\varepsilon}_{min}$) and the time to rupture (t_r) are the important parameters in engineering applications. Specifically, their dependence on stress and temperature are of utmost interest to the designer. This dependence varies with the applicable creep mechanism, which can be grouped into one of two general categories: diffusion creep and dislocation creep. Diffusion creep is favored at high temperatures and low stresses, while dislocation is more dominant at low temperatures and high stresses.

At high temperatures (above 0.6 T_m) and low stresses, the steady-state creep rates in this regime are postulated to vary linearly with stress.

Thus, we can borrow an equation for viscous flow from Chapter 1, where the rate of shear strain is proportional to the shear stress:

$$\tau = \eta \cdot d\gamma/dt \qquad \text{(Eq 2.47)}$$

In this equation, τ is the shear stress, γ is the shear strain, and η is the coefficient of viscosity. At higher stress levels, and at temperatures below 0.4 T_m, deformation is primarily controlled by dislocation glide. In this low-temperature regime, diffusion is considered negligible and does not contribute to deformation process. The material essentially undergoes plastic deformation in this temperature regime.

In the intermediate-temperature regime (in between 0.4 and 0.6 T_m), the creep deformation mechanism can be a mix of those in the high- and the low-temperature regimes. The creep rate of second-stage creep varies nonlinearly with stress. Mathematically, the stress can be plotted as an exponential function (semilog) of the minimum creep rate:

$$\sigma = C_2 \log(\dot{\varepsilon}_{min}/C_1) \qquad \text{(Eq 2.48)}$$

where C_1 and C_2 are constants. Or, one may plot the minimum creep rate as a power function (log-log) of stress:

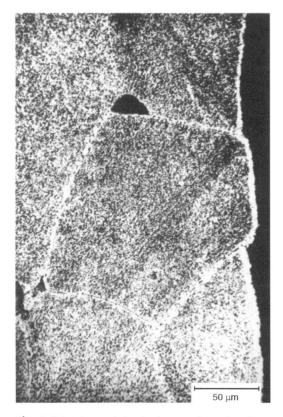

Fig. 2.84 Micrograph showing intergranular cracking due to grain-boundary sliding. Source: Ref 2.51

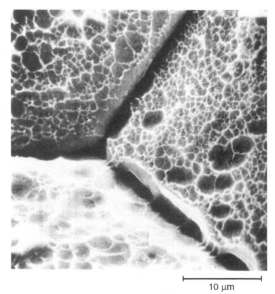

Fig. 2.85 Fracture surface of a tensile bar showing linking up of cavities on grain boundaries perpendicular to the maximum principal stress axis. Source: Ref 2.51

$$\dot{\varepsilon}_{min} = A\sigma^n \qquad \text{(Eq 2.49)}$$

where A and n are constants derived from test data. Both mathematical relations exhibit good correlation with test data. However, these equations present a drawback: Neither relation has any real physical significance because neither can represent the zero rate corresponding to zero stress. Despite this, the power law representation (Eq 2.49) has been accepted and received wide use. In addition, because one contribution to creep is a thermally activated diffusion process, its temperature sensitivity would be expected to obey an Arrhenius-type expression, with a characteristic activation energy (Q) for the rate-controlling mechanism. Equation 2.49 can therefore be rewritten as (Ref 2.56):

$$\dot{\varepsilon}_{min} = A\sigma^n \exp\!\left(\frac{-Q_C}{RT}\right) \qquad \text{(Eq 2.50)}$$

where A and n are constants for a given condition, Q_C is the activation energy for creep, T is the absolute temperature (in degrees Kelvin), and R is the universal gas constant; A, n, and Q_C must be derived from test data.

Creep test results indicate that the values for n and Q are both variable with respect to stress and temperature. An example showing the change in the value of n is given in Fig. 2.86. A distinct break in the curve is evident, with $n = 4$ at lower stresses and $n - 10$ at higher stresses. The breaks in the curves occur at stresses at which the fracture mode changes from intergranular to transgranular. More examples for the variations of n and Q in other steels are given in Viswanathan's book (Ref 2.50). The same data are also available in Volume 8 of the *ASM Handbook* (Ref 2.57).

Perhaps Ashby's fracture mechanism map (Ref 2.58–2.60) is best for identifying fracture modes at any combination of stress and temperature. As shown in Fig. 2.87, this map plots the normalized stress τ/G, or σ/E, against the normalized temperature T/T_m. Boundaries that divide the regions for possible fracture modes are shown.

There are numerous theories and suggestions as to the nature and the reason for the variations of n and Q, and ways to modify the power law equations. Meanwhile, however, industrial practice has continued to ignore these controversies and to use Eq 2.50 with discretely chosen values of n and Q. Because variations in n and Q are generally interrelated and self-compensating, no major discrepancies in the end results are noted.

A question now arises as to exactly when to use what equation to calculate the creep rate. Frost and Ashby (Ref 2.61) have come up with a graphical tool called the "deformation mechanism map." Similar to the fracture mechanism map, this map also plots the normalized stress, τ/G, or σ/E against normalized temperature T/T_m. The map basically presents the conditions under which the various creep mechanisms predominate. As shown in an example of such a map for pure nickel (Fig. 2.88), each divided field represents the range of stress and temperature over which a particular mechanism is expected to be the principal creep process. These maps are created using test data to determine the necessary material properties and constants in equations that describe each mechanism. Field boundaries are drawn where two mechanisms contribute equally to the overall creep rate. In addition, the map includes contours of constant strain rate and can be used to establish the process controlling creep at a given stress and temperature. A large number of maps of this kind, for a variety of materials, are compiled in Ref 2.61.

In the high-stress and low-temperature regime (the area marked as "plasticity" in the upper left corner of the map in Fig. 2.88) deformation is primarily controlled by dislocation glide. The

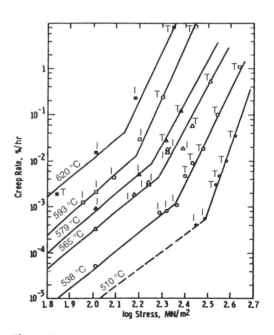

Fig. 2.86 Variation of minimum creep rate with stress for a nomalized-and-tempered 1.25Cr-0.5Mo steel. The letters T and I denote transgranular and intergranular failure, respectively. Source: Ref 2.50

bottom half of the map is diagonally (sort of) divided into two major fields, identified as "power law creep" and "diffusional flow." For a given material, the area of power law creep may be expanded or contracted by changing the grain size or by other metallurgical and processing means. In other words, one can move the mechanism boundary to higher τ/G and T/T_m ratios by means of alloying, grain control, and so on. Further, each of the two main creep fields can be subdivided, as indicated by the dashed lines in Fig. 2.88. In the power law creep regime, the area for high-temperature (HT) creep and the area for low-temperature (LT) creep may possess different sets of n and Q values. The creep rates in the HT and the LT regimes are controlled by lattice diffusion and core diffusion, respectively. It has been shown that in the lattice self-diffusion controlled regime, the activation energy for creep is the same as the activation energy for lattice self-diffusion. This is indicated in Fig. 2.89, where a large quantity of data for

metals and ceramics is compiled. Therefore, in the lattice diffusion regime, we can substitute Q_C with Q_L. In the diffusion flow regime, we can use Eq 2.50 instead of Eq 2.47 and let $n = 1$, because the behavior of a metal may be compared with that of a viscous material and creep rates are in proportion to stress. The appropriate value of activation energy depends on whether grain boundary or lattice diffusion controls.

Stress Rupture. The stress-rupture, or creep-rupture, test has been widely used for assessing the relative merits of various alloys for high-temperature service. The test is similar to the ordinary creep test, but concentrates entirely on the time required for fracture under a range of stresses. The ordinary high-temperature tensile test is a form of stress-rupture test in which the ultimate tensile stress causes fracture in a very short time. When lower initial steady stresses are used, the time to fracture becomes progressively longer. From the test results, a diagram of the type shown in Fig. 2.90 can be plotted. Straight

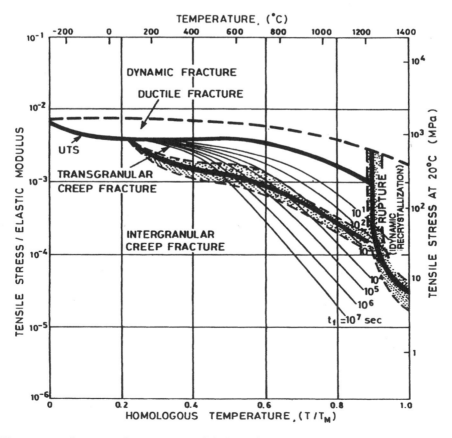

Fig. 2.87 Fracture mechanism map for an 80Ni-20Cr solid solution showing regions of fracture modes and lines of equal rupture life. Source: Ref 2.58

lines are normally obtained in a log-log plot, but a break (or double slope with a knee point) in a log-log plot are also observed in some materials. The break is typically associated with a metallurgical change.

It must be pointed out that the measure of strength obtained by stress-rupture testing is not necessarily related directly to measurements of the stress required to produce specified creep rates. In designing for high-temperature service, however, the information obtained from both tests is supplementary. In fact, creep data can be presented in the same format as Fig. 2.90. In this case, the time to rupture (i.e., t_r) is replaced by minimum creep rate, or the time to a specified creep strain (say 1%). In any event (i.e., σ versus t_r or σ versus $\dot{\varepsilon}_{min}$), the supposedly straight lines in a log-log plot may not always be straight when attempts are made to include a wide spread of test data. A break (sometimes more than one break) in the slope of the line may be observed. Similar to Fig. 2.86, a change in fracture mode that coincides with this break has also been re-

ported. This is because different n and Q are needed to describe different creep mechanisms.

Determination of creep-rupture behavior under the conditions of intended service requires extrapolation and/or interpolation of raw data. Time-temperature creep parameters can be devised for superimposing all the data onto one master curve for a given material. The techniques used include graphical methods, time-temperature parameters, and methods used for estimations when data are sparse or hard to obtain. Though this may be oversimplified, the basis of the creep parameters is that time and temperature have similar effects. In other words, the same creep behavior is obtained at the same stress in a short-time test at a high temperature as is obtained in a long-time test at a lower temperature. Basically, parametric techniques incorporate time and temperature test data into a single expression. When test data recorded over adequate times and at temperatures above the service temperature are incorporated into a master curve, the stress for the service-temperature

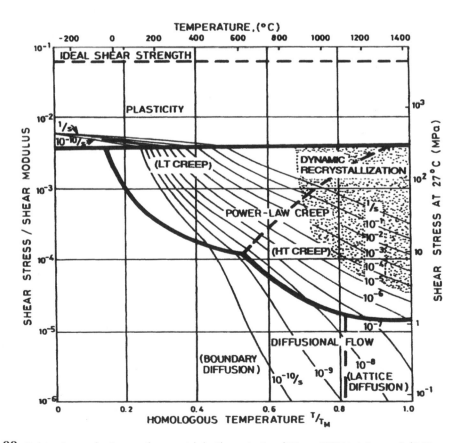

Fig. 2.88 Deformation mechanism map for pure nickel with a grain size of 0.1 mm (0.004 in.). Source: Ref 2.61

conditions can be read directly from the master curve. Although numerous studies have considered the relative merits of the many proposed parameters and techniques, no one parameter has emerged as universally superior to all others. A frequent finding is that different parameters provide best fit to different portions of the same data.

Extensive reviews of the parametric techniques are available in the literature (e.g., Ref 2.63). Descriptions of several commonly used methods are frequently found in books on mechanical metallurgy and high-temperature materials (e.g., Ref 2.2, 2.50, 2.51, and 2.64). This book provides descriptions for only two of these methods. The Larson-Miller parameter (Ref 2.65) is the first of this kind of grouping methods. It remains popular among engineers, because it directly deals with the three main variables in a stress-rupture test. It treats stress as a function of a lumped parameter that includes temperature and rupture time. In other words, it converts those data directly from Fig. 2.90 to the plot that uses the new format. No physical constants are required. The second method to be discussed is the Monkman-Grant relationship (Ref 2.66), which provides a way to estimate the rupture time from minimum creep rates.

The Larson-Miller parameter is defined as $T_A \cdot (C + \log t)$. Here T_A is the absolute temperature in Rankine or Kelvin units. The unit of Rankine is customarily used so that the Larson-Miller parameter is expressed as: $(460 + °F) \cdot$ $(C + \log t)$. The C is an experimentally determined constant. When limited test data points are available, C is assumed to have a value of 20. In a stress-rupture test, the time t (in hours) is time to rupture t_r. An example of the Larson-Miller plot is shown in Fig. 2.91.

In the following, we will demonstrate how to determine the parameter C and create a master curve for a test data set by going through an example that was originated by Voorhees and Prager (Ref 2.64). In Fig. 2.92(a), a set of stress-rupture test data is shown for the nickel-base alloy Inconel 718 as log-stress versus log-time to rupture. The data are then replotted as constant-time curves on coordinates of stress versus temperature in Fig. 2.92(b). To that graph, dashed horizontal lines have been added for stress levels of 550, 620, 760, and 830 MPa (80, 90, 110, and 120 ksi). Values for T at the intercepts of these dashed lines and the constant-time curves have been read off and plotted in Fig. 2.92(c) on coordinates of $\log(t)$ versus $10^4/T_A$. By extending the data in Fig. 2.92(c), a plausible set of converging isostress lines merges to the ordinate. Thus, $\log(t) = C$ at that point; it defines the optimum value of C for the data involved. The Larson-Miller equation for this data set becomes:

$$\sigma = T_A \cdot (25 + \log t_r) \qquad (Eq\ 2.51)$$

For each data point in the original set of stress, time, and temperature data, the proper value can be substituted in Eq 2.51 and plotted as shown in Fig. 2.92(d).

The Monkman-Grant relation is effectively a critical strain criterion (Ref 2.66). It states that

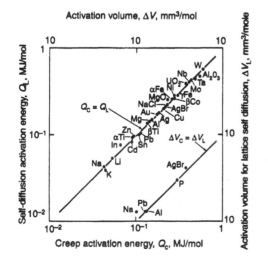

Fig. 2.89 Comparison of activation energies and activation volumes for steady-state creep and lattice self-diffusion for various materials above 0.5 T_m. Source: Ref 2.62

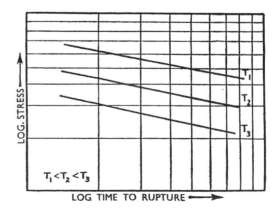

Fig. 2.90 Typical plot of stress rupture results at several temperatures

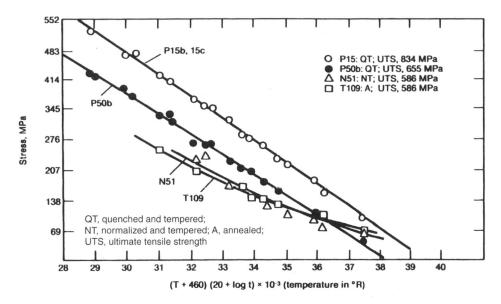

Fig. 2.91 Larson-Miller plot for 2.25Cr-1Mo steels under different heat treatment conditions. Ref 2.50

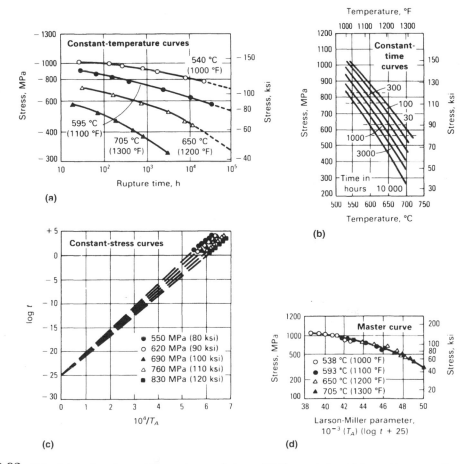

Fig. 2.92 Method of creating a Larson-Miller master curve for Inconel 718 from experimental stress-rupture curves. Source: Ref 2.64

the strain accumulated during secondary creep is a constant at failure so that the product of the minimum creep rate and the time to rupture, t_r, is a constant:

$$\dot{\varepsilon}_{min} \cdot t_r = C_{MG} \qquad \text{(Eq 2.52)}$$

where C_{MG} is the Monkman-Grant constant. An example of a plot of the Monkman-Grant relationship for a 2.25Cr-1Mo steel is given in Fig. 2.93. A C_{MG} value of 4.4 (± 1.2) is derived from the figure. For a given service condition—for example, for a known temperature and stress level combination—we can calculate $\dot{\varepsilon}_{min}$ from Eq 2.50. Finally, we can calculate t_r using Eq 2.52.

2.8.2 Stress Corrosion

Another important mechanism of sustained-load fracture is SCC. The conventional definition for stress corrosion (or stress-corrosion cracking) applies to the chemical interaction between the corrosive media and the metal, whether it is a test coupon or a structural member. It simply means that the part is simultaneously subjected to stress and corrosion, where the stress may be sustained from a constant load over a substantial time period, and where the stress has to be in tension (whether it is an applied stress or residual stress). It is recognized that in order for stress corrosion to occur, the presence of both tensile stress and a corrosive environment is required. Environments that induce stress-corrosion failure are specific to particular metals, and only a limited range of environments can cause cracking in any one metal.

Many of these environments are likely to be encountered in everyday usage. The presence of oxygen is important in most of them. The amount of interstitial in an alloy contributes to stress-corrosion susceptibility of that alloy. The so-called extra-low interstitial (ELI) that improves SCC resistance in titanium alloys is one example. Although a list such as that shown in Table 2.4 can be used as a general guideline for materials selection, SCC depends on many factors other than the bulk environment. Environments that cause SCC are usually not necessarily aqueous, and specific environment parameters must be in specific ranges for cracking to occur. These include, but are not limited to:

- Temperature
- pH
- Electrochemical potential
- Solute species
- Solute concentration
- Oxygen concentration

Changing any of these environmental parameters may significantly affect the crack nucleation process or the rate of crack propagation. Extensive reviews on the theories and mechanisms of SCC are given in Ref 2.67 to 2.69.

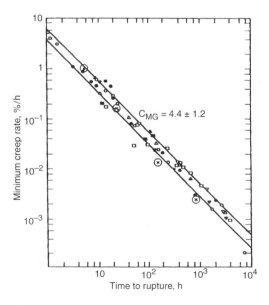

Fig. 2.93 Monkman-Grant relationships between minimum creep rate and time to rupture for 2.25Cr-1Mo steel. Source: Ref 2.50

Table 2.4 Alloy/environment systems exhibiting SCC

Alloy	Environment
Carbon steel	Hot nitrate, hydroxide, anhydrous ammonia, and carbonate/bicarbonate solutions
High-strength steels	Aqueous electrolytes, particularly when containing H_2S
Austenitic stainless steels	Hot, concentrated chloride solutions; caustics, saline solution, and chloride-contaminated steam
High-nickel alloys	High-purity steam, hot caustics
Aluminum alloys	Aqueous Cl^-, Br^-, and I^- solutions, including contaminated water vapor
Titanium alloys	Aqueous Cl^-, Br^-, and I^- solutions; methanol organic liquids; N_2O_4; hydrochloric acid
Magnesium alloys	Aqueous Cl^- solutions
Zirconium alloys	Aqueous Cl^- solutions; organic liquids; I_2 at 350 °C (660 °F)
Copper alloys	Ammonia and amines for high-zinc brasses; ammoniacal solutions for α brass; range of solutions for other specific alloys
Gold alloys(a)	Chlorides, particularly ferric chloride; ammonium hydroxide; nitric acid

(a) Alloys containing less than 67% Au

To determine the susceptibility of metals to stress corrosion, smooth specimens are used. Test setups are typically like those shown in Fig. 2.94. This type of test setup is called the constant-extension test. Stressing of the sheet specimen is accomplished by bending. Using either one of these devices, one can bend the flat sheet specimen to a desired stress level by adjusting the y dimension—that is, the deflection of the beam. Then the stressed specimen is left inside a tank of corrosive solution. Several specimens can be tested at the same time, with each specimen bent to a different stress level, producing a stress versus time-to-failure curve. Eventually, there may be a threshold stress level below which no failure will occur (Fig. 2.95). The following equations are used to calculate the maximum stress at the midpoint of the specimen:

For a four-point loaded specimen:

$$\sigma = \frac{12 \cdot t \cdot y \cdot E}{3H^2 - 4A^2} \qquad \text{(Eq 2.53a)}$$

For a three-point loaded specimen:

$$\sigma = \frac{6 \cdot t \cdot y \cdot E}{H^2} \qquad \text{(Eq 2.53b)}$$

where E is the modulus of elasticity; the dimensions t, y, H, and A are defined in Fig. 2.94. Constant-extension tests are widely used because of the ease of specimen preparation and the ability to test a large number of specimens at one time. However, there is one major drawback: Once a stress-corrosion crack is formed, the gross section stress decreases, which will eventually cause the crack to stop. There are many other types of stress-corrosion tests. Among them, either constant load or constant strain rate is used (Ref 2.67, 2.68). ASTM standards for stress-corrosion testing are listed in Table 2.5. Since the mid-1960s, emphasis has shifted from testing of smooth specimens to testing of precracked specimens. The term "stress-corrosion cracking" is almost exclusively used in the field of fracture mechanics. Fracture-mechanics-based studies of SCC occupy the main body of today's literature. The fracture mechanics aspects of SCC are discussed in more detail in Chapter 4.

From metallurgical failure analysis, engineers become aware of stress corrosion as a failure mechanism. Such knowledge is helpful in developing new alloys, improving an existing alloy, or selecting an alloy for a certain application. When a part is submerged in a corrosive agent (e.g., salt water) and under a sustained load, intergranular corrosion pits often form on the part's external surfaces. Multiple corrosion pits are always present, and these pits can eventually act as crack initiation sites. The resulting fracture is often (but not always) intergranular.

The general perception has been that stress-corrosion fracture is often intergranular, because grain boundaries are a logical path and sink for chemical diffusion. In fact, most engineering alloys do exhibit intergranular cracking in the case of stress corrosion. However, cracks can propagate along other forms of dislocations inside the grain (e.g., a twin plane, Fig. 2.96). Therefore, transgranular cracks are observed in aus-

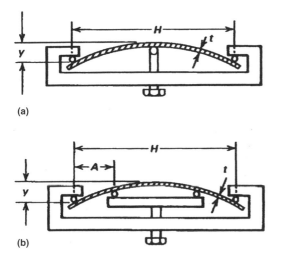

(a)

(b)

Fig. 2.94 Schematic specimen and holder configurations for bent-beam specimens. (a) Four-point loaded specimen. (b) Three-point loaded specimen

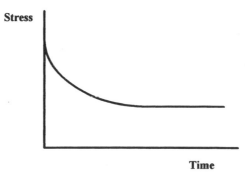

Fig. 2.95 Schematic stress corrosion delayed fracture curve (applied constant stress versus time to fracture)

tenitic stainless steels where mechanical or annealing twins are present. Sometimes both transgranular and intergranular cracks are found adjacent to each other (Fig. 2.97). Through the years, more and more transgranular stress-corrosion fractures have been observed in various alloy systems, leading engineers to accept transgranular as an alternate fracture mode in SCC.

These stress-corrosion fractures are functions of the combinations of many factors, including chemistry, heat treatment, and product form. For example, a magnesium alloy may exhibit intergranular fracture after furnace cooling, but its fracture path would be transgranular if it were water quenched (Fig. 2.98).

Branching also is generally thought to be a positive indication of SCC, whether the fracture path is intergranular or transgranular. A typical branched crack in an aluminum forging is shown in Fig. 2.99. However, lack of branching is not evidence that SCC is not involved.

2.8.3 Hydrogen Embrittlement

Hydrogen embrittlement can occur in most alloy systems. Body-centered cubic and hexagonal close-packed metals are most susceptible to hydrogen embrittlement; high-strength steels and titanium alloys have drawn most attention. Severe embrittlement can be produced in many metals by very small amounts of hydrogen; as little as 0.0001 wt% H_2 can cause cracking in steel. Hydrogen may be introduced during melting and entrapped during solidification, or it may be picked up during heat treatment, electroplating, acid pickling, or welding. Hydrogen-embrittled alloys exhibit reduced ductility in short-time tensile tests.

Titanium alloys show a strong microstructural dependence on hydrogen embrittlement. For ex-

Table 2.5 ASTM standards for stress-corrosion testing

ASTM No.	Title
G 30	Standard Practice for Making and Using U-Bend Stress-Corrosion Test Specimens
G 35	Standard Practice for Determining the Susceptibility of Stainless Steels and Related Nickel-Chromium-Iron Alloys to Stress-Corrosion Cracking in Polythionic Acids
G 36	Standard Practice for Evaluating Stress-Corrosion-Cracking Resistance of Metals and Alloys in a Boiling Magnesium Chloride Solution
G 37	Standard Practice for Use of Mattsson's Solution of pH 7.2 to Evaluate the Stress-Corrosion Cracking Susceptibility of Copper-Zinc Alloys
G 38	Standard Practice for Making and Using C-Ring Stress-Corrosion Test Specimens
G 39	Standard Practice for Preparation and Use of Bent-Beam Stress-Corrosion Test Specimens
G 41	Standard Practice for Determining Cracking Susceptibility of Metals Exposed under Stress to a Hot Salt Environment
G 44	Standard Practice for Exposure of Metals and Alloys by Alternate Immersion in Neutral 3.5% Sodium Chloride Solution
G 47	Standard Test Method for Determining Susceptibility to Stress-Corrosion Cracking of 2xxx and 7xxx Aluminum Alloy Products
G 58	Standard Practice for Preparation of Stress-Corrosion Test Specimens for Weldments
G 64	Standard Classification of Resistance to Stress-Corrosion Cracking of Heat Treatable Aluminum Alloys
G 103	Standard Test Method for Performing a Stress-Corrosion Cracking Resistance of Low Copper 7xxx Series Al-Zn-Mg-Cu Alloys in Boiling 6% Sodium Chloride Solution

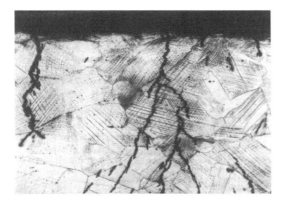

Fig. 2.96 Transgranular SCC in a drillcollar made of cold-worked Cr-Mn-N austenitic stainless steel (P530HS). Note the branch cracks passing through grains and along slip bands and twin boundaries. Source: Ref 2.70

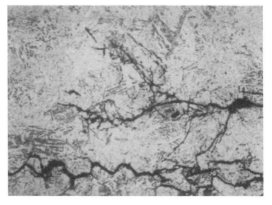

Fig. 2.97 Photomicrograph of SCC in a sample coming from a steam turbine disk made of A470 class 4 alloy steel forging. Note that the two immediately adjacent cracks exhibit opposite types of fracture: transgranular for the upper crack, intergranular for the lower crack. Source: Ref 2.71

ample, Ti-6Al-4V is a two-phase (α + β) alloy at room temperature. In this and other alpha-beta titanium alloys, the morphology of the α-phase depends on heat treatment and processing. For example, if a titanium alloy is heated to the region of 100% β and then cooled back to the α + β region, the newly formed α nucleates and grows into lamellar morphology (from slow cooling after the beta heat treatment) or a needlelike acicular structure if the cooling rate is faster. Subsequent deformation in the α + β re-

gion then can be done to break up the lamellar or acicular α-phase morphologies into a more equiaxed microstructure.

The occurrence of hydrogen embrittlement is influenced by the morphology of the α-phase in α-β titanium alloys. When two different microstructures were tested in gaseous hydrogen, it was found that acicular structure is more susceptible to embrittlement than equiaxed structure (Ref 2.74). The microscopic fracture path in the equiaxed structure appears transgranular through the α-phase grains. In a continuous β-matrix (acicular α), the fracture path is pressure dependent: At higher hydrogen pressures, the fracture is intergranular along prior-β and transformed α-grain boundaries, while at lower pressures, cracking is transgranular through the β-grains. It has been suggested that when a continuous network of β-phase is present, it provides a short-circuit for rapid hydrogen diffusion into the titanium. Intergranular fracture takes place due to increased hydrogen pressure or lowered surface energy at the grain boundaries. When a continuous α-phase is present, no such rapid transport path exists, and hydrogen must interact directly with the α-phase, forming a thin hydride layer; therefore, the fracture path will be transgranular.

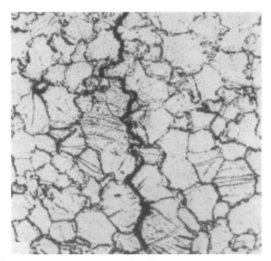

(a)

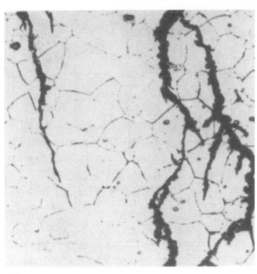

(b)

Fig. 2.98 Stress-corrosion cracking in an extruded Mg-6Al-1Zn alloy tested in a salt-chromate solution. (a) Intergranular crack propagation in the face-cooled alloy. (b) Transgranular crack propagation in the water-quenched material. Source: Ref 2.72

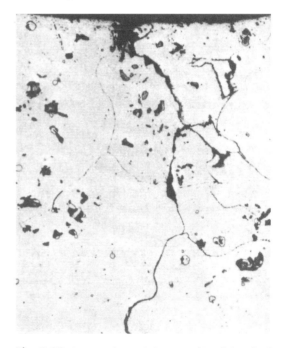

Fig. 2.99 Intergranular cracks in a gas turbine disk made of 2014-6 aluminum. Note crack initiation at a corrosion pit (or pits) and branching along grain boundaries, typical of stress-corrosion failure. Source: Ref 2.73

Whether or not an alloy is already contaminated with hydrogen, exposure to hydrogen-containing compounds can result in absorption of hydrogen atoms and cause embrittlement. In fact, hydrogen gas that is generated during stress-corrosion testing might well diffuse into the test coupon. The final failure could be identical to hydrogen embrittlement. In addition to hydrogen gas, hydrogen sulfide (or hydroxide, etc.) also contributes to SCC. Therefore, the mechanisms between stress corrosion and hydrogen embrittlement are sometimes difficult to separate. Because embrittlement through hydrogen absorption involves diffusion, hydrogen embrittlement tests are usually performed the same as SCC tests. Stresses versus time-to-failure data similar to stress corrosion are usually obtained. A common method of studying hydrogen embrittlement is to charge notched tensile specimens with known amounts of hydrogen, load them to different stress levels in a dead-weight machine, and observe the time to failure. Again, like SCC, the use of precracked specimens is popular.

Hydrogen embrittlement and stress corrosion differ by the interactions of the specimens with applied currents. Stress-corrosion cracking can be influenced by a combination of anodic and/or cathodic reactions that influence crack growth mechanisms. Cases where the applied current makes the specimen more anodic and accelerates cracking are considered to be SCC; that is, the anodic-dissolution process contributes to SCC. On the other hand, when cracking is accentuated by current in the opposite direction, the hydrogen-evolution reactions are accelerated. It is conceivable that the process of cracking is dominated by hydrogen embrittlement. While temperature can be a factor affecting SCC in metals, hydrogen embrittlement is negligible in high or low temperatures; it happens in room or slightly above room temperature. External or internal stresses are prerequisites for stress corrosion or SCC. However, hydrogen embrittlement caused by diffusion of atomic hydrogen into steel can result in strength degradation with or without applied stresses.

Stress-corrosion cracking in some material/environment combinations is the result of hydrogen-induced subcritical crack growth. Hydrogen-induced cracking (HIC), also known as hydrogen blistering, can cause internal SCC initiation. This type of SCC is commonly observed in lower-strength steels in aqueous hydrogen sulfide or other hydrogen charging environ-

ments. For HIC, initiation is generally at non-metallic inclusions and results from corrosion generated hydrogen diffusing into and accumulating in the steel (Ref 2.69, 2.75).

A final note on hydrogen embrittlement and stress corrosion is that although their mechanisms share many similarities, and the test methods are similar, the method of prevention based on initiation can be different. Theories and models concerning various types of hydrogen embrittlement are thoroughly documented in Ref 2.68 and 2.75 to 2.77. Discussion of scientific research results of hydrogen embrittlement is beyond the scope of this book. The fundamentals of hydrogen embrittlement are well covered in the references just cited.

REFERENCES

2.1. J.M. Holt, Uniaxial Tension Testing, *ASM Handbook,* Vol 8, *Mechanical Testing and Evaluation,* ASM International, 2000, p 124–142

2.2. G.E. Dieter, Jr., *Mechanical Metallurgy,* McGraw-Hill, 3rd ed., 1986

2.3. W. Ramberg and W.R. Osgood, "Description of Stress-Strain Curves by Three Parameters," Report NACA TN 402, National Advisory Committee for Aeronautics, 1943

2.4. M. Gensamer, *Strength of Metals Under Combined Stresses,* American Society for Metals, 1941

2.5. H.A. Kuhn, Uniaxial Compression Testing, *ASM Handbook,* Vol 8, *Mechanical Testing and Evaluation,* ASM International, 2000, p 143–151

2.6. A. Nadai, *Theory of Flow and Fracture of Solids,* Vol 1, 2nd ed., McGraw-Hill, 1950

2.7. G.I. Taylor and H. Quinney, The Plastic Distortion of Metals, *Trans. R. Soc. (London) A,* Vol 230, 1931 p 323–362

2.8. M. Reiner, *Deformation, Strain and Flow: An Elementary Introduction to Rheology,* H.K. Lewis & Co., 1960, p 97

2.9. C. Lipson, Why Machine Parts Fail, Part 6—Torsional Fractures, *Mach. Des.,* 1950

2.10. G.F. Vander Voort, Visual Examination and Light Microscopy, *ASM Handbook,* Vol 12, *Fractography,* ASM International, 1987, p 91–165

2.11. K.E. Puttick, Ductile Fracture in Metals, *Philos. Mag.,* Vol 4, 1959, p 964–969

2.12. H.C. Rogers, The Tensile Fracture of Ductile Metals, *Trans. AIME,* Vol 218, 1960, p 498–506

2.13. D.J. Wulpi, *Understanding How Components Fail,* 2nd ed., ASM International, 1999

2.14. R.M.N. Pelloux, "An Analysis of Fracture Surfaces by Electron Microscopy," Report D1-82-0169-R1, Boeing Scientific Research Laboratories, Dec 1963; *Met. Q.,* Nov 1965, p 34

2.15. W.T. Becker and D. McGarry, Mechanisms and Appearances of Ductile and Brittle Fracture in Metals, *ASM Handbook,* Vol 11, *Failure Analysis and Prevention,* ASM International, 2002, p 587–626

2.16. C.R. Brooks and A. Choudhury, *Failure Analysis of Engineering Materials,* McGraw-Hill, 2002

2.17. W.T. Becker and S. Lampman, Fracture Appearance and Mechanisms of Deformation and Fracture, *ASM Handbook,* Vol 11, *Failure Analysis and Prevention,* ASM International, 2002, p 559–586

2.18. Atlas of Fractographs, *ASM Handbook,* Vol 12, *Fractography,* ASM International, 1987, p 223

2.19. T. Inoue, S. Matsuda, Y. Okamura, and K. Aoki, The Fracture of a Low Carbon Tempered Martensite, *Trans. Jpn. Inst. Met.,* Vol 11, 1970, p 36–43

2.20. C.D. Beachem, Orientation of Cleavage Facets in Tempered Martensite (Quasi-Cleavage) by Single Surface Trace Analysis, *Metall. Trans.,* Vol 4, 1973, p 1999–2000

2.21. C.L. Briant and S.K. Banerji, Intergranular Fracture in Steel: The Role of Grain-Boundary Composition, *Int. Met. Rev.,* No 4, 1978, p 164–196

2.22. C.J. McMahon, Jr., Mechanism of Intergranular Fracture in Alloy Steels, *Mater. Charact.,* Vol 26 (No. 4) June 1991, p 269–287

2.23. S. Lampman, Intergranular Fracture, *ASM Handbook,* Vol 11, *Failure Analysis and Prevention,* ASM International, 2002, p 641–649

2.24. P.G. Shewmon, Grain Boundary Cracking, *Metall. Mater. Trans. B,* Vol 29, June 1998, p 509–518

2.25. S.P. Lynch, Mechanisms of Intergranular Fracture, *Mater. Sci. Forum,* Vol 46, 1989, p 1–24

2.26. *Met. Eng. Q.,* May 1976, p 41

2.27. R.C. Bates, Modeling of Ductile Fracture by Microvoid Coalescence, *Fracture,* J.M. Wells and J.D. Landes, Ed., TMS-AIME, 1984

2.28. M. Gensamer, The Structure of Metals and the Strength of Structures, *Trans. AIME,* Vol 215, 1959, p 2

2.29. A.H. Cottrell, Theory of Brittle Fracture in Steel and Similar Metals, *Trans. AIME,* Vol 212, 1958, p 192

2.30. N.J. Petch, The Cleavage Strength of Polycrystals, *J. Iron Steel Inst.,* Vol 174, 1953, p 25

2.31. N. Louat and H.L. Wain, Brittle Fracture and the Yield Point Phenomenon, *Fracture,* B.L. Averbach, D.K. Felbeck, G.T. Hahn, and D.A. Thomas, Ed., Technology Press and John Wiley & Sons, 1959

2.32. G. Sachs, *Fundamentals of the Working of Metals,* Pergamon Press, 1954

2.33. E. Orowan, Fracture and Strength of Solids, *Phys. Soc. Prog. Rep.,* Vol 12, 1949, p 185–232

2.34. N.N. Davidenkov and N.I. Spirdonova, Analysis of Tensile Stress in the Neck of an Elongated Test Specimen, *Proc. ASTM,* Vol 46, 1946, p 1147–1158

2.35. P.W. Bridgman, The Stress Distribution at the Neck of a Tension Specimen, *Trans. ASM,* Vol 42, 1944, p 553–572

2.36. E.R. Marshall and M.C. Shaw, "The Determination of Flow Stress from a Tensile Specimen," Massachusetts Institute of Technology Liaison Office Publ. No. 51-102, May 1951

2.37. T.A. Trozera, On the Use of the Bridgman Technique for Correcting Stresses Beyond Necking, *Trans. ASM,* Vol 56, 1963, p 780–782

2.38. J. Aronofsky, Evaluation of Stress Distribution in the Symmetrical Neck of Flat Tensile Bars, *J. Appl. Mech.,* March 1951, p 75–84

2.39. A.F. Liu, *Structural Life Assessment Methods,* ASM International, 1998

2.40. W. Weibull, A Statistical Theory of the Strength of Metals, *Proc. R. Swed. Inst. Eng. Res.,* Vol 193, 1939, p 151

2.41. W. Weibull, A Statistical Distribution Function of Wide Applicability, *J. Appl. Mech.,* Vol 18, 1951, p 185–232

2.42. G.V. Smith, *Properties of Metals at Elevated Temperatures,* McGraw-Hill, 1950

2.43. G.T. Hahn, B.L. Averbach, W.S. Owen, and M. Cohen, Initiation of Cleavage Microcracks in Polycrystalline Iron and Steel, *Fracture,* B.L. Averbach, D.K. Felbeck, G.T. Hahn, and D.A. Thomas, Ed., Technology Press and John Wiley & Sons, 1959

2.44. G. Dieter, Evaluation of Workability for Bulk Forming Process, *ASM Handbook,* Vol 14A, *Metalworking: Bulk Forming,* ASM International, to be published 2005

2.45. G.E. Dieter, H.A. Kuhn, and S.L. Semiatin, Ed., *Handbook of Workability and Process Design,* ASM International, 2003

2.46. M.G. Cockcroft and D.J. Latham, Ductility and the Workability of Metals, *J. Inst. Met.,* Vol 96, 1968, p 33–39

2.47. H.A. Kuhn, Workability Theory and Application in Bulk Forming Processes, *Handbook of Workability and Process Design,* G.E. Dieter et al., Ed., ASM International, 2003, p 172–187

2.48. H.-S. Kim, Y.-T. Im, and M. Geiger, Prediction of Ductile Fracture in Cold Forging of Aluminum Alloys, *J. Manuf. Sci. Eng (Trans. ASME),* Vol 121, 1999, p 336–344

2.49. A.A. Griffith, The Theory of Rupture, *Proc. First International Congress of Applied Mechanics,* 1924, p 55–63

2.50. R. Viswanathan, *Damage Mechanisms and Life Assessment of High Temperature Components,* ASM International, 1989

2.51. G.A. Webster and R.A. Ainsworth, *High Temperature Component Life Assessment,* Chapman & Hall, 1994

2.52. J.C. Earthman, Introduction to Creep and Stress-Relaxation Testing, *ASM Handbook,* Vol 8, *Mechanical Testing and Evaluation,* ASM International, 2000, p 361–362

2.53. E.N. da C. Andrade, *Proc. R. Soc. (London),* Vol 90A, 1914, p 329–342; *Creep and Recovery,* American Society for Metals, 1957, p 176–198

2.54. C. Zener, *Elasticity and Anelasticity of Metals,* Univerity of Chicago Press, 1948

2.55. H.C. Chang and N.J. Grant, *Trans. AIME,* Vol 206, 1956, p 544–550

2.56. O.D. Sherby and P.M. Burke, Mechanical Behavior of Crystalline Solids at Elevated Temperature, *Prog. Mater. Sci.,* Vol 13, 1968, p 325–390

2.57. Creep and Creep Rupture Testing, *ASM Handbook,* Vol 8, *Mechanical Testing and Evaluation,* ASM International, 2000, p 369–382

2.58. M.F. Ashby, C. Gandhi, and D.M.R. Taplin, Fracture-Mechanism Maps and Their Construction for FCC Metals and Alloys, *Acta Metall.,* Vol 27, 1979, p 699–729

2.59. C. Gandhi and M.F. Ashby, Fracture-Mechanism Maps for Materials Which Cleave: FCC, BCC and HCP Metals and Alloys, *Acta Metall.,* Vol 27, 1979, p 1565–1602

2.60. R.J. Fields, T. Weerasurya, and M.F. Ashby, Fracture Mechanisms in Pure Iron, Two Austenitic Steels and One Ferritic Steel, *Metall. Trans. A,* Vol 11, 1980, p 333–347

2.61. H.J. Frost and M.F. Ashby, *Deformation Mechanism Maps,* Pergamon Press, 1982

2.62. W.D. Nix and J.C. Gibeling, Mechanisms of Time-Dependent Flow and Fracture of Metals, *Flow and Fracture at Elevated Temperatures,* ASM International, 1985, p 1–63

2.63. S.S. Manson and C.R. Ensign, A Quarter Century of Progress in the Development of Correlation and Extrapolation Methods for Creep Rupture Data, *J. Eng. Mater. Technol., (Trans. ASME) Ser. H,* Vol 101, 1979, p 317–325

2.64. H.R. Voorhees and M. Prager, Assessment and Use of Creep-Rupture Properties, *ASM Handbook,* Vol 8, *Mechanical Testing and Evaluation,* ASM International, 2000, p 383–397

2.65. F.R. Larson and J. Miller, Time-Temperature Relationship for Rupture and Creep Stress, *Trans. ASME,* Vol 74, 1952, p 765–771

2.66. F.C. Monkman and N.J. Grant, An Empirical Relationship Between Rupture Life and Minimum Creep Rate in Creep-Rupture Tests, *Proc. ASTM,* Vol 56, 1956, p 593–620

2.67. G.H. Koch, Stress-Corrosion Cracking and Hydrogen Embrittlement, *ASM Handbook,* Vol 19, *Fatigue and Fracture,* ASM International, 1996, p 483–506

2.68. Y. Katz, N. Tymiak, and W.W. Gerberich, Evaluation of Environmentally Assisted Crack Growth, *ASM Handbook,* Vol 8, *Mechanical Testing and Evaluation,* ASM International, 2000, p 612–648

2.69. W.R. Warke, Stress Corrosion Cracking, *ASM Handbook,* Vol 11, *Failure Analysis and Prevention,* ASM International, 2002, p 823–867

2.70. K.A. MacDonald and H. Aigner, Some Case Studies of Failed Austenitic Drillcollars, *Engineering Failure Analysis,* Vol 3, Pergamon Press, 1996, p 281–298

2.71. P.F. Timmins, *Solutions to Equipment Failures,* ASM International, 1999

2.72. W.K. Miller, Stress-Corrosion Cracking of Magnesium Alloys, *Stress-Corrosion Cracking,* R.H. Jones, Ed., ASM International, 1992, p 251–264

2.73. A.F. Liu, unpublished data

2.74. H.G. Nelson, *Hydrogen in Metals,* American Society for Metals, 1974, p 445–464

2.75. P.F. Timmins, Failure Control in Process Operations, *ASM Handbook,* Vol 19, *Fatigue and Fracture,* ASM International, 1996, p 468–482

2.76. P.F. Timmins, *Solutions to Hydrogen Attack in Steels,* ASM International, 1997

2.77. R. Gibala and R.F. Hehemann, Ed., *Hydrogen Embrittlement and Stress-Corrosion Cracking,* American Society for Metals, 1984

SELECTED REFERENCES

• L.E. Alban, *Systematic Analysis of Gear Failures,* ASM International, 1985

• D.A. Aliya and R.J. Shipley, Ed., *Failure Prevention through Education: Getting to the Root Cause,* ASM International, 2000

• K.A. Esaklul, Ed., *Handbook of Case Histories in Failure Analysis,* ASM International, Vol 1, 1992, Vol 2, 1993

• *Failure Analysis Library on CD-ROM, Disk 1: Principles and Practices of Failure Analysis and Fractography; Disk 2: Failure Analysis Applications and Case Histories,* ASM International, 1999

• A.J. McEvily, *Metal Failures: Mechanisms, Analysis, Prevention,* John Wiley & Sons, 2002

• I. Le May, Examination of Damage and Material Evaluation, *ASM Handbook,* Vol 11, *Failure Analysis and Prevention,* ASM International, 2002, p 351–370

• R.E. Peterson, Interpretation of Service Fractures, *Handbook of Experimental Stress Analysis,* M. Hetenyi, Ed., John Wiley & Sons, 1957

CHAPTER 3

Fatigue Strength of Metals

THE DEFINITION OF "FATIGUE" according to ASTM Standard E 1150 reads as follows: "The process of progressive localized permanent structural damage occurring in a material subjected to conditions that produce fluctuating stresses and strains at some point or points and that may culminate in cracks or complete fracture after a sufficient number of fluctuations." In terms of the stress-strain conditions leading to fatigue damage or fracture, Fine and Chung (Ref 3.1) further define fatigue as "a progressive, localized, and permanent structural change that occurs in a material subjected to repeated or fluctuating strains at nominal stresses that have maximum values less than (and often much less than) the static yield strength of the material." They also state: "Fatigue damage is caused by the simultaneous action of cyclic stress, tensile stress, and plastic strain. If any one of these three is not present, a fatigue crack will not initiate and propagate. The plastic strain resulting from cyclic stress initiates the crack; the tensile stress promotes crack growth (propagation)." Moreover, "although compressive stresses will not cause fatigue, compressive loads may result in local tensile stresses. Microscopic plastic strains also can be present at low levels of stress where the strain might otherwise appear to be totally elastic."

What this amounts to in everyday engineering problems is that a part may fail prematurely after being subjected to a large number of fluctuating stresses, even when the nominal stress level is below the yield strength. The same is true for an engineering part subjected to repeated straining, particularly when the strains are beyond the material's elastic limit. Therefore, either stress or strain is a variable in fatigue life analysis. Consequently, fatigue methods are divided into two categories: stress based and strain based. Both types of fatigue analysis are discussed in this chapter, primarily for metals. Fatigue phenomena for nonmetals are discussed in Chapters 7 and 8, although the terminologies and the analysis and data presentation methods are applicable to both.

3.1 Mechanical Behavior under Cyclic Loads

3.1.1 Significance of Load Reversal

Back in 1886, experiments by J. Bauschinger indicated that the yield strength in tension or compression was reduced after applying a load of the opposite sign that caused inelastic deformation. Thus, one single reversal of inelastic strain can change the stress-strain behavior of metals. This basic phenomenon is illustrated graphically by the σ-ε plot in Fig. 3.1. Consider a segment of a tensile stress-strain curve OAB in Fig. 3.1(a), where point A is the yield stress (or elastic limit, whichever the case may be) in tension. Similarly, there is a segment of a compression stress-strain curve $OA'B'$ with its compression yield stress (or elastic limit in compression) at point A'. These two curves are similar and symmetrically positioned. In other words, the degrees of work hardening and the absolute yield stresses are the same for both tension and compression.

Now, consider further loading along the tension curve with stretching (past the elastic limit) to point B, followed by unloading to point C. As described in Chapter 2, we can reload it from point C to point E (in tension). Or, we can continue the unloading and push it into the compression zone as shown by the segment $CD'E'$. Here, points D and D' are the new yield stresses for tension and compression, respectively. For comparison purposes, flip over the compression part and group both curves together as shown in Fig. 3.1(b). We find that point D' is lower than point D and that the segment $D'E'$ is lower than DE. While the yield stress in tension was increased by strain hardening from A to B, the yield stress in compression was decreased. This behavior is equally true whether the material is initially subjected to tension or compression and followed by a reverse loading. Thus, the yield stress of the material has been lowered after reversing the load direction. This is called the "Bauschinger effect." The significance of this phenomenon is that the material deformation behavior and strength might have been altered after a part had been put into service, because service history usually includes cyclic loads.

Further, in conducting failure analysis, if we machine a test coupon from a location that has been subjected to plastic deformation, the stress-strain properties will possibly differ from those had the material not been so strained. Remember that the Bauschinger effect is a bulk phenomenon, whereas fatigue is a local process. Therefore, although small inhomogeneities in the material may not measurably affect the Bauschinger effect, fatigue life may be drastically reduced.

3.1.2 Plastic Hysteresis

Although the Bauschinger effect was recognized prior to the 20th century, further studies on how metals respond to cyclic loading were not extensively undertaken until much later. The most comprehensive summaries of research results were published in the mid-1960s by Morrow (Ref 3.2) and Manson et al. (Ref 3.3, 3.4). Highlights of these documents are presented below.

Cyclic Strain Hardening and Softening. Consider the stress-strain history shown in Fig. 3.2. First, tension stress was applied and the test went beyond point A_0 (the elastic limit) to point B_0, an arbitrary preselected strain level of $+0.048$. It was then unloaded from point B_0 and continued to point B_1 at a strain of -0.048 (another arbitrary selected strain level). Then ten-

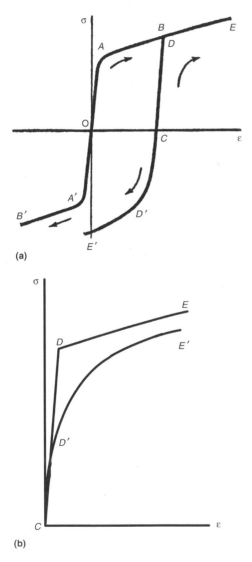

(a)

(b)

Fig. 3.1 Schematic of load reversal. (a) Comparison of initial loadings (*OAB* and *OA'B'*) and reversed loading in the post-plasticity range (*DCD'E'*). (b) Comparison of reloaded tension (*CDE*) and reversed compression (*CD'E'*) behavior in the post-plasticity range

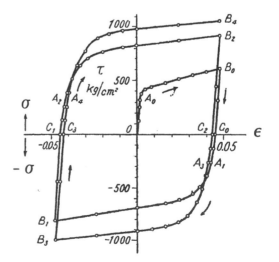

Fig. 3.2 Schematic of plastic hysteresis phenomenon

sion load was applied again, passing point A_2 and stopped at point B_2. The strain level was again $+0.048$. A loop that is completed with points $B_0C_0A_1B_1C_1A_2B_2$ is called the plastic hysteresis (or simply the stress-strain hysteresis), an imitation of the magnetic hysteresis. The terminology that is used to characterize a hysteresis loop is given in Fig. 3.3. This type of alternating tension and compression operation can be repeated many times, each time passing through (or stopping at) the preselected strain level: $+0.048$ or 0.048, depending on the loading direction. As shown in Fig. 3.2, the second loop is completed with points $B_2C_2A_3B_3C_3A_4B_4$. We observed that the magnitude of the highest stresses corresponding to a selected strain level had been increased gradually after each loop, that is, $B_0 < B_2 < B_4$. This is known as "cyclic hardening," a phenomenon that is true for other loading systems such as bending and torsion. According to Morrow (Ref 3.2), thin-wall tubular specimens are ideal for this purpose. The stress may be assumed uniform across the wall when it is tested in torsion. In an axial tension-compression test, the tube helps reduce the chances of buckling.

Depending on the initial condition of an alloy, the maximum stress at each loop may decrease, instead of increase, with repeated straining. Thus, it is called "cyclic softening." If a metal is initially softened, it will cyclically harden, and vice versa. Examples of oxygen-free high-conductivity (OFHC) copper are given in Fig. 3.4, showing that fully annealed metals harden, heavily cold-worked metals soften, and partial annealing results in an intermediate behavior. Figure 3.5 schematically compares the hardening and softening behavior.

After studying a number of metals, including several aluminum alloys, titanium, and steels, Manson et al. (Ref 3.3, 3.4) proposed a general rule that indicates whether an alloy undergoes cyclic hardening or softening. This rule, by which one can predict from monotonic stress-strain properties alone, states that if the ratio of the ultimate tensile strength to the 0.2% offset yield strength is greater than 1.4, hardening will occur. If this ratio is smaller than 1.2, softening will occur. Prediction cannot be made for ratios between 1.2 and 1.4, but the material should be relatively stable. Therefore, from the material selection point of view, selecting materials that have an ultimate to yield strength ratio greater than 1.2 is desirable in order to avoid cyclic softening during service life. To check the applica-

bility of this rule, Feltner and Landgraf (Ref 3.6) examined test results for 35 materials with available monotonic properties and cyclic stress-strain curves. The materials included 17 steels, 10 pure metals and alloys, four nickel-base superalloys, two aluminum alloys, a titanium alloy, and a cast iron. Their results show that 27 predictions (out of the 35 materials) worked. Of the remaining items, six had no predictions, and only two predictions did not work.

In these discussions, plastic hysteresis behavior has been described in terms of strain-controlled tests. The same type of phenomenon can be shown using stress-controlled experiments. Figure 3.6 shows typical hysteresis loops between fixed limits of stress. However, the strain-controlled method is considered most suitable for studying the cyclic stress-strain behavior (Ref 3.6).

Cyclic Stress-Strain Curve. Although Manson's rule reliably predicts how a material will behave under cyclic loading, determining the amount by which the monotonic properties will change is another matter. Figure 3.2 shows that in a strain-range-controlled test the hysteresis loops will continually change. The magnitude of stress increases from the tip of one loop to the tip of the next loop. However, notice that the rate

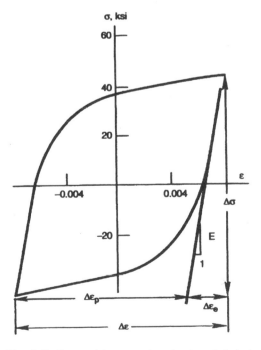

Fig. 3.3 Characterizing parameters of a stress-strain hysteresis loop based on strain-controlled, complete reverse loading with mean strain equal to zero

of increase actually decreases gradually. That means that the hysteresis loops will stop changing after a certain number of reversals. Let's say, for the case of Fig. 3.2, the cyclic hardening process had stabilized at point B_4; that is, the stress will not increase further even if more reversals are made after point B_4. We can repeat this type of test on other specimens with different preselected cyclic strain levels. As shown in Fig. 3.7, each test will produce a stable loop, and the tip of that stable loop gives a pair of values representing the stabilized stress and strain of that test condition. By connecting the tips of the stable loops from several companion tests at different strain ranges, a smooth curve is formed; this is called the cyclic stress-strain curve.

An example of a cyclic stress-strain curve for 4340 steel developed using this procedure is shown in Fig. 3.8, which illustrates both the tension and compression part of the curve. Note that each hysteresis loop actually represents the stable loop resulting from an individual test.

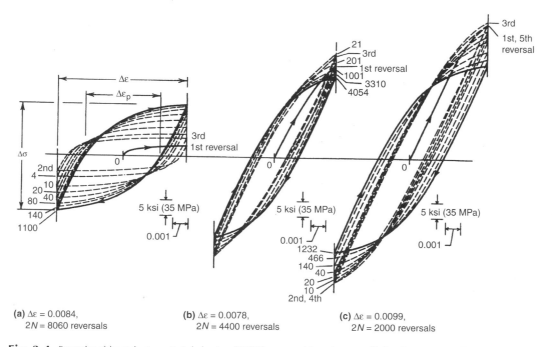

(a) $\Delta\varepsilon = 0.0084$, $2N = 8060$ reversals

(b) $\Delta\varepsilon = 0.0078$, $2N = 4400$ reversals

(c) $\Delta\varepsilon = 0.0099$, $2N = 2000$ reversals

Fig. 3.4 Examples of the early stress-strain behavior of OFHC copper subjected to controlled cyclic strain. (a) Fully annealed showing cyclic hardening. (b) Partially annealed. (c) Severely cold worked showing cyclic softening. Source: Ref 3.5

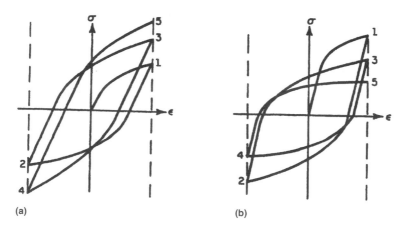

(a)

(b)

Fig. 3.5 Schematic comparison of strain hardening (a) and strain softening (b) phenomena

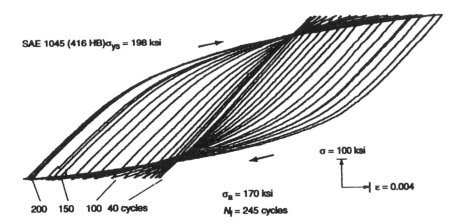

Fig. 3.6 Schematic of cyclic softening by stress-controlled cyclic loading

They are put together this way simply for demonstration purposes. The data points are the tips of those loops.

For metals that cyclically harden, the cyclic stress-strain curve will be above the monotonic stress-strain curve (Fig. 3.9a); for metals that soften, the cyclic stress-strain curve will be below the monotonic stress-strain curve (Fig. 3.9c). The behavior shown in Fig. 3.9(b) is referred to as cyclically stable. Other cyclically stable materials include 7075-T73 aluminum alloy (Ref 3.7). One might ask how soon the hysteresis would reach its stable condition. It could vary from 10 to 50% of the material's fatigue life, depending on who provides the answer. The variation amounts to 20% (Ref 3.2), or 50% (Ref 3.3, 3.4), or 10 to 20% (Ref 3.6), or 20 to 40% (Ref 3.7), regardless of the material's initial condition.

Morrow (Ref 3.2) has pointed out that the cyclic stress-strain curve always has a character-istically smooth shape regardless of the original shape of the monotonic curve. For example, the upper yield point in the monotonic stress-strain curve of the mild steel would be completely removed after a number of repeated cycles. He also showed that the curve formed by connecting the tips of the stable hysteresis loops could be fitted by a parabolic equation similar to that used to fit the monotonic stress-strain curve. That is:

$$\sigma_a = \sigma_f' \left(\frac{\Delta \varepsilon_p}{2\varepsilon_f'}\right)^{n'} \qquad \text{(Eq 3.1a)}$$

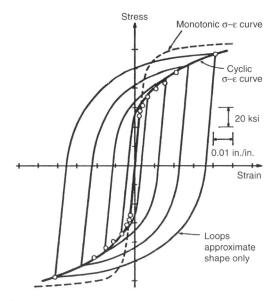

Fig. 3.8 Cyclic stress-strain curve for SAE 4340 steel obtained by connecting the tips of several stable loops of separate tests. Cyclic softening is apparent by comparing the cyclic stress-strain curve with the material's original monotonic stress-strain curve. Source: Ref 3.3

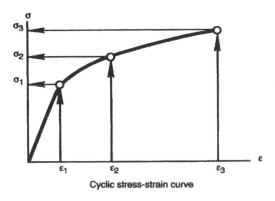

Fig. 3.7 Construction of a cyclic stress-strain curve by joining tips of stabilized hysteresis loops. Source: Ref 3.7

where σ_a is the stable stress amplitude ($= \Delta\sigma/2$), $\Delta\varepsilon_p/2$ is the plastic strain amplitude, and n' is the cyclic strain-hardening exponent. According to Morrow,

$$\sigma_f' = \sigma_f \qquad \text{(Eq 3.1b)}$$

and

$$\varepsilon_f' = \varepsilon_f \qquad \text{(Eq 3.1c)}$$

where σ_f and ε_f are the monotonic true stress and true strain at fracture, respectively. Later, Feltner and Landgraf (Ref 3.6) showed that the value of ε_f' is better correlated with the cyclic stress-strain curve. That is:

$$\varepsilon_f' = 0.002\left(\frac{\sigma_f}{\sigma_y'}\right)^{1/n'} \qquad \text{(Eq 3.1d)}$$

where σ_y' is the 0.2% offset of the cyclic stress-strain curve.

In general, the entire cyclic stress-strain curve can be expressed in the form of the Ramberg-Osgood equation:

$$\frac{\Delta\varepsilon}{2} = \frac{\Delta\varepsilon_e}{2} + \frac{\Delta\varepsilon_p}{2} \qquad \text{(Eq 3.2a)}$$

and

$$\frac{\Delta\varepsilon}{2} = \frac{\Delta\sigma}{2E} + \left(\frac{\Delta\sigma}{2\zeta'}\right)^{1/n'} \qquad \text{(Eq 3.2b)}$$

where $\Delta\varepsilon_e$ is the elastic strain range, and $\Delta\varepsilon_e/2$ is the elastic strain amplitude ($= \sigma_a/E$). Rearranging Eq 3.1(a) and 3.2(b) yields:

$$\frac{\Delta\varepsilon}{2} = \frac{\sigma_a}{E} + \varepsilon_f'\left(\frac{\sigma_a}{\sigma_f'}\right)^{1/n'} \qquad \text{(Eq 3.2c)}$$

Note that Eq 3.1(a) is a straight line when plotted to log-log coordinates. An average value for n' is equal to 0.15 for most metals; some materials may be as low as 0.05 or as high as 0.25 (Ref 3.7, 3.8). By comparing the value of n' to the monotonic strain-hardening exponent n for the same metal, Landgraf et al. (Ref 3.9) concluded that metals with low monotonic strain-hardening exponents ($n < 0.1$) will cyclically soften, while those with high monotonic strain-hardening exponents ($n > 0.2$) will cyclically harden.

Shortcut Method of Generating the Cyclic Stress-Strain Curve. By definition, we know

that the cyclic stress-strain curve is obtained by connecting the tips of the stable hysteresis loops from several companion specimens tested at different completely reversed strain ranges. Because this method requires many specimens and is time consuming, Landgraf et al. (Ref 3.9) explored into several alternative methods. They concluded that the incremental step test method, which uses only one specimen, is most effective for obtaining a cyclic stress-strain curve. In this test procedure, explained in detail in Ref 3.9, a specimen is subjected to blocks of gradually in-

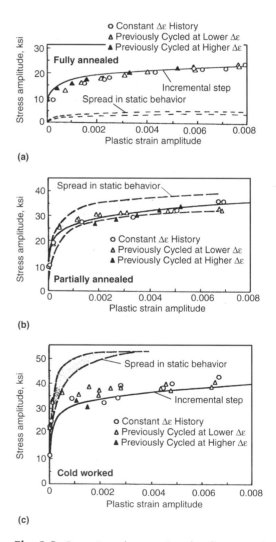

(a)

(b)

(c)

Fig. 3.9 Comparison of monotonic and cyclic stress-strain curves for three initial conditions of OFHC copper. Comparisons are also made between test methods for obtaining the cyclic stress-strain curve. The solid line came from a single incremental step test. The open circles came from companion tests of different fixed strain ranges. The triangles came from a procedure called the multiple step test, which is not covered in this book. Source: Ref 3.5

creasing and then decreasing stain amplitudes (Fig. 3.10). In short, the variables involved in programming the test with a closed-loop servo-control machine are the maximum strain amplitude attained and the number of cycles in a given block. A maximum strain-amplitude of $\pm 1.5\%$ was used in Landgraf's tests. The advantage of this test method is that the hysteresis loops will quickly stabilize after only a few blocks of cyclic applications. Figure 3.11 shows results for cold-worked OFHC copper. Here, the hysteresis loops corresponding to each step are plotted for each block of testing. The progress toward material stabilization is visualized based on the appearance of the loops in the consecutive blocks. A fully stabilized block is recognized by judging that no further change of the connected loci is apparent.

The cyclic stress-strain curves for OFHC copper in three initial conditions are shown in Fig. 3.9. The cyclic stress-strain curve developed using the incremental step method compared favorably with the other test methods. Note that if a test is started at the maximum strain amplitude,

it is possible to obtain the initial portion of the monotonic tension curve for comparison with the cyclic curve. Also, if the specimen is pulled to failure after the incremental step test, the resulting stress-strain curve will be nearly identical to the one obtained from the stabilized condition.

3.2 Microscopic and Macroscopic Aspects of Fatigue and Crack Propagation

By observation, "fatigue" is a phenomenon leading to fracture under repeated or fluctuating stresses having a maximum value less than the ultimate tensile strength of the material. A material will fail prematurely under repeated loading condition whether it behaves in the manner of cyclic softening or cyclic hardening. While Bauschinger was the first to investigate material resistance to fatigue loading, A. Wohler's work in 1871 is credited as the first to recognize fatigue failure in an engineering structure (railway carriage axles, to be exact). Since then, fatigue has been a major consideration in design of engineering structures. Researchers have extensively studied the nature and mechanisms of fatigue, and engineering analysis procedures have been established and periodically updated. The engineering approach has been to experimentally determine the fatigue life (in terms of number of load cycles to failure) at a selected stress level, or to determine the stress level at which the material will never fail. Figure 3.12 schematically represents this type of engineering data. Simply plotting the applied stress level against the number of cycles to failure, it is commonly known as the *S-N* curve.

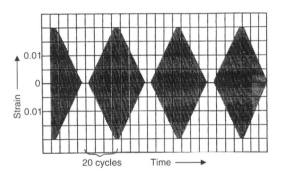

Fig. 3.10 Schematic strain-time record of an incremental step test

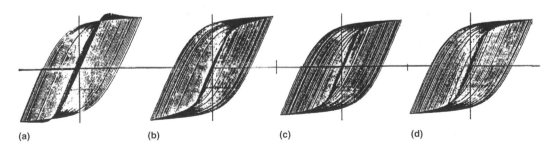

(a)   (b)   (c)   (d)

Fig. 3.11 Incremental cyclic strain test result for cold-worked OFHC copper. (a) Starting with a virgin specimen, the cyclic strain is incrementally increased, producing larger and larger hysteresis loops. (b) When the maximum strain range is reached, the strain range is incrementally decreased until the behavior is nominally elastic. (c) The strain is gradually increased to the same maximum strain range. Note the change in the cyclic stress-strain curve. (d) The strain is cyclically decreased again. Note that the cyclic stress-strain curve is nearly stabilized (as before). The maximum strain amplitude used in this test was ±1%. Source: Ref 3.5

The general features of a typical fatigue fracture consist of three stages, as illustrated in Fig. 3.13:

- Initial fatigue damage leading to crack nucleation and crack initiation
- Progressive cyclic growth of a crack (crack propagation) until the remaining uncracked cross section of a part becomes too weak to sustain the loads imposed
- Final, sudden fracture of the remaining cross section

Final fracture occurs during the last stress cycle when the cross section cannot sustain the applied load. The final fracture—which is the result of a single overload—can be brittle, ductile, or a combination of the two. For some materials, such as relatively brittle cast iron, there may be no distinct difference between overload and fatigue appearances.

These three regions are shown in Fig. 3.14(a) and (b), photographs of a fractured piston rod with a distinctive region of fatigue crack propagation and a fibrous surface and radial markings in the region of final fast fracture. The relative portions of the fracture surface from fatigue crack propagation and the final overload region depend on the level of applied stress relative to the nominal strength of the material and the part surface geometry. The overload region is usually visibly rougher than the fatigue region. The general variations of fast-fracture zones on round, square, and rectangular sections are shown schematically for various types of load conditions and stress concentration in Fig. 3.15.

3.2.1 Fatigue Crack Initiation

Crack initiation frequently is the major component of fatigue life, and so identification of the location and nature of origin sites is important in failure analysis. Typically fatigue cracks initiate near the surface, where nominal stresses are often higher (e.g., in bending), and where geometric variations at surfaces (such as machining marks, surface flaws, notches, etc.) cause stress concentration. Material heterogeneities (such as inclusions, second-phase particles, voids, microcracks) also can act as stress concentrators. These areas of stress concentration permit local permanent plastic deformation at nominal stresses below yield. If crack initiation occurs at locations on two closely spaced parallel planes, then the cracks eventually join to form a steplike feature or ratchet mark (Fig. 3.13).

The mechanisms of fatigue crack initiation have been studied with highly homogeneous, single-phase metals (without the heterogeneities described above). In these studies of "unflawed" material, the processes of damage accumulation are also favored at surfaces, and it is generally agreed that fatigue cracks are formed by the gradual development of shear bands in the material and the initiation of cracks within the shear bands. These shear bands are called persistent slip bands (PSBs), which develop during the very early stage of fatigue. During reversal of stress cycles, some of these bands get pushed in,

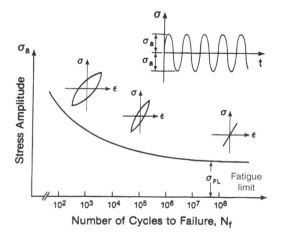

Fig. 3.12 Schematic representation of engineering fatigue data, commonly known as the *S-N* curve. Also shown are schematics of stable hysteresis loops at various fatigue lives.

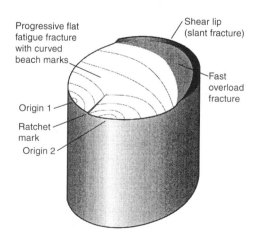

Fig. 3.13 General features of fatigue fractures

or out, with respect to the surface of the part (or testpiece), as shown in Fig. 3.16. These dislocated PSBs are called the intrusion and extrusion, respectively. In this early stage under repetitive stressing, continued surface roughening and damage accumulation ultimately results in crack nucleation at the surface. Cyclic damage can also accumulate at grain boundaries, particularly at elevated temperature or under conditions that cause grain-boundary embrittlement, and at twin boundaries.

The exact mechanism as to how a crack develops from the extrusions/intrusions of the PSBs is not completely certain. Several theories have been suggested (Ref 3.10–3.12), but none is completely satisfactory. However, it is clear that most fatigue cracks originate near the surface. This is confirmed by numerous test results and failure analyses. There are exceptions. For example, components that have been case hardened, surface hardened, or shot peened can exhibit subsurface initiation. Such surface treatments introduce compressive surface stresses (favorable for improving fatigue resistance by reducing the driving force for initiation) and/or increase surface fatigue strength. Under some conditions, these gradients in driving force and resistance result in subsurface fatigue crack initiation.

Fatigue caused by contact stress, such as rolling-contact loading, also frequently initiates subsurface cracking. Stresses generated by contact between two surfaces characteristically display a maximum component of shear stress beneath the surface. For sufficiently high levels of contact stress, this can result in subsurface initiation and growth of fatigue cracks that result in pitting and spalling. Finally, significant subsurface discontinuities or material defects can foster subsurface fatigue crack initiation.

3.2.2 Fatigue Crack Propagation

While fatigue crack initiation may occupy a significant portion or majority of the overall time (duration) of the fatigue process, the area of fatigue propagation on the fracture surface usually occupies a much larger area than the origin

(a) (b)

Fig. 3.14 Two views of the surface of a fatigue fracture in a 145 mm (5¾ in.) diam threaded piston rod of heat-treated AISI 4340 steel (341 HB hardness). Beyond the zone of fatigue crack growth, failure was by radial fibrous fast fracture. (a) The full-face view [unreadable] clearly shows beach marks in the region of the primary fatigue crack, at top. Secondary fatigue cracks are visible on either side of it, as well as the bottom. ~0.5×. (b) The oblique view reveals that the fatigue cracks began at the roots of the threads, the origin of the primary crack being two threads to the right of the abutting secondary cracks. The thread roots were apparently too sharp, causing stress concentrations locally. 0.67×

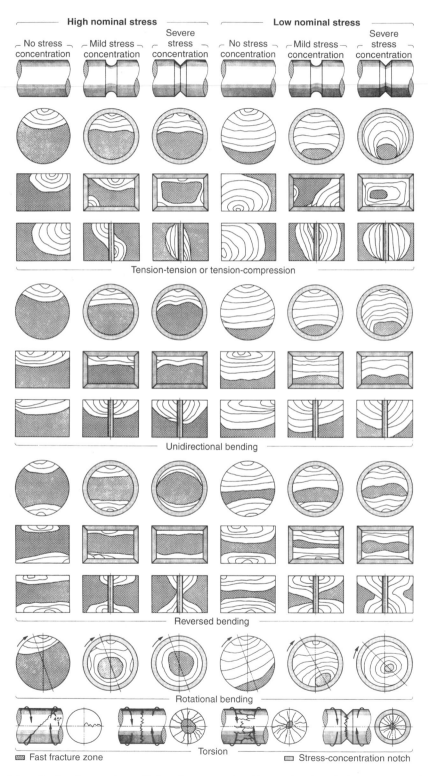

Fig. 3.15 Schematic of marks on surfaces of fatigue fractures produced in smooth and notched components with round, square, and rectangular cross sections and in thick plates under various loading conditions at high and low nominal stress

site(s). This can be important from a practical standpoint in failure analysis. It is also important to recognize the different stages of fatigue crack growth, and the mechanical relation between crack growth rates and the cyclic stress intensity range (ΔK) at the crack tip (see Chapter 5). This is shown in Fig. 3.17 with the classic log-log plot of crack growth rate per cycle (da/dN) versus ΔK.

After a fatigue microcrack develops, the first stage of crack growth may involve propagation on specific crystallographic planes with greatest resolved fluctuating shear stress. This initial region of fatigue crack growth historically has been referred to as stage I. In this region, fatigue

cracks nucleate and coalesce by slip-plane fracture, extending inward from the surface at approximately 45° to the stress axis (Fig. 3.18). Then at some point, a transition occurs from stage I crack growth to stage II crack growth on planes normal to the fluctuating tensile stress. This transition occurs when the fracture path changes from one or two shear planes in stage I to many parallel plateaus separated by longitudinal ridges in stage II (Fig. 3.18). The plateaus are usually normal to the direction of maximum tensile stress, and further propagation is perpendicular to the tensile stress axis.

This division between stages of propagation is a natural one considered by early researchers investigating fatigue processes. However, stage I propagation occurs over a limited region (characteristically spanning no more than a few near-surface grains) adjacent to origin sites. In addition, many commercial alloys (notably most steels) typically do not exhibit any detectable stage I propagation. After initiation in such alloys, growth immediately occurs by so-called stage II propagation. Notches, sharp corners, or preexisting cracks can also eliminate detectable stage I propagation in many metals.

Stage II propagation is favored in materials with easy cross-slip (wavy slip). The stage I

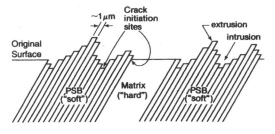

Fig. 3.16 Schematic of slip profile during cyclic loading. Source: Ref 3.9

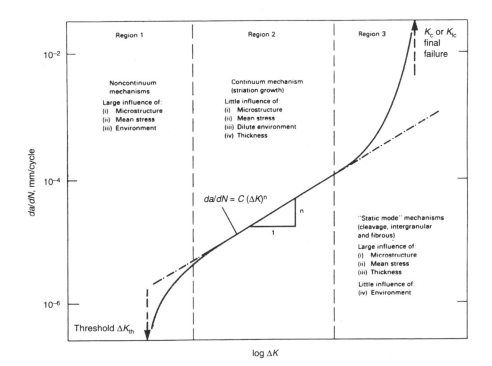

Fig. 3.17 Schematic illustration of variation of fatigue crack growth rate, da/dN, with alternating stress intensity, ΔK, in steels, showing regions of primary crack growth mechanisms

propagation mechanism is favored in materials that exhibit planar slip and in some alloys strengthened by coherent precipitates (such as age-hardened aluminum) and under fluctuating loads low enough that the cyclic fracture process zone is small relative to a characteristic microstructural scale parameter. Hence, stage I propagation is often favored by large grain size. Some alloys (such as certain nickel-base superalloys and cobalt-base alloys) can display very extensive regions of propagation on specific crystallographic planes (so-called stage I propagation) with little or no discernible stage II type of growth prior to final fracture. Depending on loading and geometry, extensive propagation in these materials occurs on alternate crystallographic planes such that growth follows a zigzag or serrated faceted pattern, on average normal to the direction of fluctuating tension.

In terms of crack growth rates, the behavior of metals generally can be divided into three regions in a log-log plot of da/dN versus ΔK (Fig. 3.17). In region 1, crack growth is negligible for an almost unlimited number of cycles near the fatigue crack propagation threshold (ΔK_{th}), where the crack growth rate becomes diminishingly small. As the crack grows, the stress intensity increases at the crack tip, and crack growth rates accelerate until they reach region 2, where crack growth rates are mainly governed by ΔK. Microstructure and environmental factors have little influence on crack growth rates in region 2. In contrast, microstructure and the environment can influence rates at low values of ΔK. Environmental factors and microstructural variations become more dominant at low values of ΔK, because small cracks are more prone to effects from material heterogeneities or the extended times associated with regions of slow crack growth.

The mechanisms of fatigue can also vary as function of stress intensity. For example, face-centered cubic (fcc) metals (such as austenitic stainless steel and aluminum) can fracture crystallographically under conditions of very low-stress high-cycle fatigue (i.e., propagation at low ΔK near the fatigue threshold). Flat facets can also appear at near-threshold fatigue crack growth rates in materials such as carbon steel (Fig. 3.19). Variations in fatigue features are also shown in Fig. 3.20 for different crack growth rates of a nickel alloy. Region 2 is characterized by the occurrence of fatigue striations common to many (but not all) fatigue fractures. Region 3 has features of fatigue (striations) and overload (in this case, dimples of ductile overload).

3.2.3 Fatigue Fracture Appearance

Fatigue cracks are typically transgranular, but intergranular fatigue does occur in some circumstances such as high-strain (low-cycle) fatigue or at low ΔK values when environmental factors become more dominant or when microstructural variations influence the crack path. Fatigue fractures also are typically macroscale brittle, even though the microscopic mechanisms involve slip and are thus microscale ductile.

On a macroscopic scale, long-life fatigue failures are brittle. There is little or no net-section

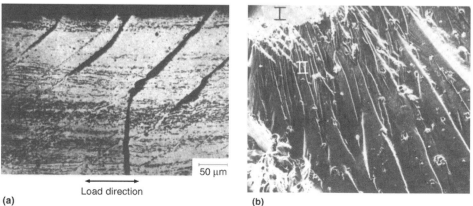

(a) Load direction

50 μm

(b)

Fig. 3.18 Transition from stage I to stage II fatigue. (a) Change in fracture path from stage I (top of photo) to stage II (bottom). (b) Transition from stage I to stage II of a fatigue fracture in a coarse-grain specimen of aluminum alloy 2024-T3. Source: Ref 3.13

distortion; the fracture is typically macroscopically flat and generally transverse to the direction of cyclic or fluctuating tensile stress that caused fatigue. In contrast with tensile overload fracture, which generally has more or less shear lip (slant 45° fracture) along free surfaces, propagating fatigue fractures typically intersect free surfaces at right angles. This provides a tool for helping to identify fatigue locations. However, at high loads/high fatigue crack growth rates in relatively thin components made of relatively tough material, a transition to propagation on a slant plane can occur (Fig. 3.21).

In common with other progressive fracture modes such as stress corrosion cracking (scc), field fatigue fractures are frequently decorated by more or less curved marking referred to as beach marks or clamshell marks and arrest marks (Fig. 3.14 and 3.22). When examined by SEM at higher magnifications, beach marks also can be resolved into hundreds or thousands of fatigue striations (Fig. 3.23) that occur from the alternating blunting and resharpening of the crack during the fatigue process. These marking can help delineate the area of crack initiation, because beach marks and fatigue striations radiate away from the origin as a series of concentric arcs. The crack initiation site(s) can be identified by drawing an imaginary radius perpendicular to their direction and centered at the origin.

Beach marks and striations are special features commonly associated with fatigue. However, beach marks and striations are not always present on the surface of a fatigue fracture. Beach marks are produced by a change in crack growth conditions, such as a change in environment or stress level or a pause in stress cycling (interruption in service). Therefore, beach marks are not found in most laboratory tests conducted under uniform loading and environmental conditions. Moreover, the presence of beach marks is not necessarily conclusive evidence of fatigue fracture, because beach marks may also occur from other types of subcritical crack growth such as stress corrosion cracking. For example, Fig. 3.24 shows beach marks from SCC.

Crack propagation variations in anisotropic or inhomogeneous materials may also occasionally produce beach marks that are difficult to interpret. In highly anisotropic material such as spring wire, for example, beach marks are not always present. Beach marks are more likely to be present if the spring material is comparatively soft (40 to 45 HRC), but they are seldom present when higher-strength spring materials fail by fatigue. Beach marks may also be removed by surface rubbing during the compression cycle. For example, Fig. 3.25 shows multiple points of crack initiation from cyclic reversed bending. Cracks initiated at several locations on two sides near the three o'clock and nine o'clock positions, where maximum tensile stresses developed during cyclic reversed bending. Rubbing has removed evidence on the fracture surface of crack initiation, but the initiation locations can still be identified by the presence of ratchet marks. In this instance, the overload region (visibly rougher along the north-south diameter of the section) also indicates completely reversed loading ($S_{max}/S_{min} = -1$), because the regions of fatigue crack growth are the same size, causing the overload region to be centered on the original neutral axis.

Striations are characteristically mutually parallel and at right angles to the local direction of

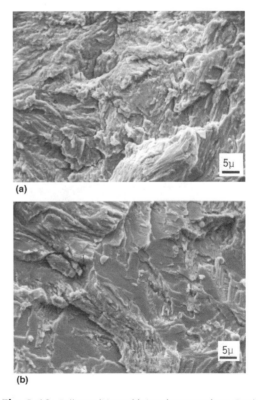

(a)

(b)

Fig. 3.19 Fully pearlitic steel fatigue fracture surfaces. Crack growth direction is from left to right in both images. (a) Intermediate crack growth rate (~0.1 µm/cycle), and (b) low crack growth rate (~0.001 µm/cycle). No fatigue striations were resolved by scanning electron microscopy (SEM) at any crack growth rate. Unmarked facets are more prevalent at low (near-threshold) growth rates. Source: Ref 3.14

crack propagation. They vary in striation-to-striation spacing with cyclic stress intensity (ΔK), and they are equal in number to the number of load cycles (under cyclic stress-loading conditions). Fatigue striations do not cross one another, but they may join and form a new zone of local crack propagation. If the component has been subjected to uniformly applied loads of sufficient magnitude, the distance between two adjacent striations (i.e., a single advance of the crack front) is a measure of the rate of propagation per stress cycle. However, if the loading

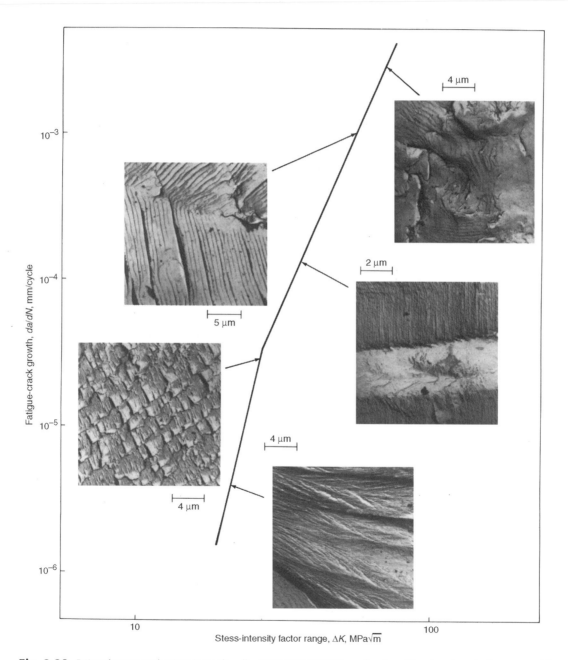

Fig. 3.20 Fatigue fracture mechanisms in Incoloy alloy X750 (UNS N07750) as a function of the stress-intensity factor range, ΔK. Test conditions: 24 °C (75 °F), 300 cycles/min, $R = 0.05$. The plot of fatigue crack growth rate, da/dN, versus ΔK shows that at high ΔK, the fatigue fracture surface exhibited well-defined striations and dimples. At progressively lower values of ΔK, a combination of fatigue fissures (associated with small secondary cracks) and striations was observed. At low ΔK, a highly faceted fracture surface resulted due to crystallographic fracture along intense slip bands.

is nonuniform, there are wide variations between a given stress cycle series and the spacing of the striations; each stress cycle does not necessarily produce a striation. Further, under nonuniform loading conditions, the lower-amplitude stress cycles may not be of sufficient magnitude to produce resolvable striations.

In addition, not all fatigue fractures exhibit striations. Although the presence of striations establishes fatigue as the mode of failure, their ab-

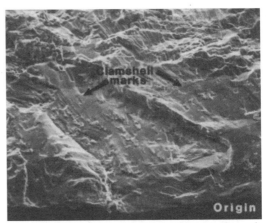

(a)
100 μm

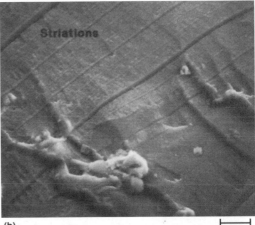

(b)
10 μm

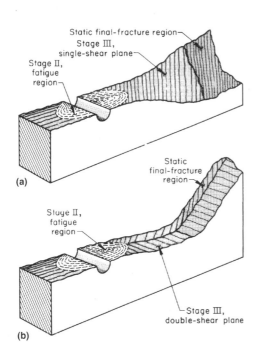

Fig. 3.21 (a) Single-shear and (b) double-shear fracture planes that are 45° to the direction of loading

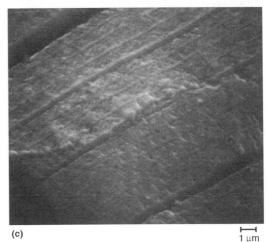

(c)
1 μm

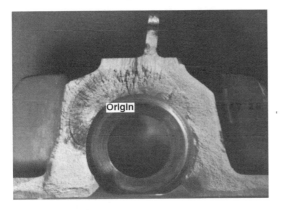

Fig. 3.22 Forged aluminum alloy 2014-T6 aircraft component that failed by fatigue. Characteristic beach marks are evident. See also Fig. 3.23

Fig. 3.23 A series of low- to high-magnification micrographs of the specimen shown in Fig. 3.22. Note that as magnification increases, progressively finer striations are resolved. (a) 80× (b) 800× (c) 4000×

sence does not eliminate fatigue as a possibility. For example, striations are usually well defined in aluminum alloys fatigued in air, but do not form if the component is tested under vacuum (Ref 3.16); the same holds true for titanium alloys (Ref 3.17). Further, the fidelity of the striations changes with composition. For example, striations are often prominent in aluminum alloys, but are often poorly defined in ferrous alloys. Oxidation, corrosion, or mechanical damage can obliterate striations.

Figure 3.26 shows an example of fatigue striations. The amount of crack extension due to a stress cycle can be determined by measuring the width of the striation spacing. It has been reported (Ref 3.19) that there are two types of fatigue striations: brittle and ductile (Fig. 3.27). Brittle striations are very rare, whereas ductile striations are more common. Ductile striations have also been observed in polymers, in which their formation is not necessarily related to the crystallographic nature of metals. The ductile striations in the 7178 aluminum alloy shown in Fig. 3.26 exhibit a typical appearance. Brittle striations found in 2014 aluminum alloy are

shown in Fig. 3.28. As noted, fatigue striations in other metallics and alloys (including steels) may not be as sharply defined as those in high-strength aluminum alloys.

3.3 Fatigue Life, Crack Initiation, Crack Growth, and Total Life

For the purposes of standardization, the term "fatigue life" must be defined. For stress-life data such as those shown in Fig. 3.12, tests usually are carried out until the specimen breaks into two parts. Before the advent of fracture mechanics technology, there was no need to distinguish between crack initiation and propagation lives. Laboratory fatigue test specimens usually are very small, and complete failure (fracture) of a specimen generally occurs shortly after the crack has initiated. Therefore, the two stages of failure (initiation and propagation) are not distinguished. Conventional fatigue data that represent the total fatigue life of a part may well be primarily the crack initiation life plus an insignificant amount of crack growth life. With fracture mechanics, it is now recognized that both phases require attention if the total life of an initially uncracked element is to be determined. As this book will show, the mechanisms of fa-

Fig. 3.24 Beach marks on a 4340 steel part caused by SCC. Tensile strength of the steel was approximately 1780 to 1900 MPa (260 to 280 ksi). The beach marks are a result of differences in the rate of corrosion penetration on the surface. They are in no way related to fatigue marks. 4×

Fig. 3.25 Fatigue fracture from reversed bending load. In this example, rubbing has obliterated the early stages of fatigue cracking, but ratchet marks are present to indicate locations of crack initiation. The material is 1046 steel with a hardness of approximately 30 HRC. Source: Ref 3.15

Fig. 3.26 Transmission electron micrography (TEM) of ductile fatigue striations in 7178 aluminum alloy. Arrow indicates the cracking direction. Source. Ref 3.18

tigue and crack propagation are fundamentally different. It is traditionally believed that each reversal causes fatigue damage. There is no distinction as to which way the load changes direction. That is fine for the crack initiation phase. Regarding fatigue crack growth, however, crack extension takes place only in the upstroke (tension) portion of a load cycle. Therefore, life assessments for these two phases of the stress cycle should be treated separately.

What is the definition of "crack initiation"? A convenient answer would be that crack initiation life is the time that an originally uncracked piece takes to develop a very small fatigue crack of a certain size. Beyond that, the process can be considered "crack propagation." This, however, requires definition of the "assumed" crack size that forms the boundary between these two methodologies. Depending on experimental testing and/or the analytical prediction methods involved, investigators have defined crack initiation in various ways. Barsom and McNicol (Ref

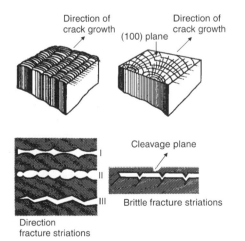

Fig. 3.27 Schematic of different types of ductile and brittle striations. Source: Ref 3.19

Fig. 3.28 TEM of brittle striations in a 2014 aluminum alloy that failed in service. Note the cleavage facets running parallel to the direction of crack propagation and normal to the fatigue striations. Arrow indicates the cracking direction. Source: Ref 3.18

3.20) used a 0.25 mm (0.01 in.) crack as an indication of crack initiation. Kim et al. (Ref 3.21) defined the number of cycles required to develop a crack to 0.3 mm (~0.01 in.) as crack initiation life. Forman (Ref 3.22) considered a crack of 50 to 75 μm (0.002 to 0.003 in.) as the minimum size that can be detected under an optical microscope; however, he also stated that a 0.25 mm (0.01 in.) crack is approximately the minimum length that can be distinguished with the naked eye. He used both definitions in his report. During the F18-A aircraft development period, Northrop used a 0.25 mm (0.01 in.) crack as the dividing line for crack initiation and propagation. Numerous fatigue tests were conducted to justify adopting this approach and use the data in the development of an analysis methodology. The test program included both constant amplitude strain-life data and flight-by-flight spectrum tests representing nine different F18-A components made of aluminum (7050 or 7049), titanium (6Al-4V), or 9-4-0.20 steel (Ref 3.23). Because most fatigue cracks develop at a stress raiser in a structural or machine part, some researchers relate the crack initiation size to structural geometry. Dowling (Ref 3.24) defines this boundary as a crack length equal to one-tenth of the notch radius. This corresponds to a crack size of approximately 0.25 mm (0.01 in.) for hole sizes typical of aircraft structures (i.e., 5 to 6 mm or $^3/_{16}$ to $^1/_4$ in. diameters).

The detection of crack initiation in Jack and Price's experiments (Ref 3.25) was based on a selected value of potential drop in electric current. Clark (Ref 3.26) used a change in ultrasonic signal as a tool to detect crack initiation in his experiments. For the magnitude of change in the ultrasonic signal that he selected, crack initiation corresponds to the development of a crack less than 0.8 mm (0.03 in.) long and less than 0.1 mm (0.005 in.) deep on the notch root of test coupons.

In strain-controlled (low cycle) fatigue testing, crack initiation detection is based on the indication of changes in the elastic or plastic "stiffness" of the specimen. The stress-strain or load-deflection record is readily available during the test and the instrumentation is a part of the test setup. For a complete reverse loading, a chart that shows the load range and/or the tensile load to compression load ratio as a function of time (number of cycles) can be obtained while the test is in progress. The chart would show these values dropping when a fatigue crack initiates (Ref 3.27).

The current approach to assessing structural life tends to divide the estimation of total life into two parts: fatigue crack initiation and propagation of the fatigue crack to final failure. The first part involves using fatigue technology to predict crack initiation life. The second part uses fracture mechanics technology to predict crack propagation life. Again, the problem hinges on determination of a crack size that can be called the fatigue crack. The aforementioned experimental methods seem to do a proper job in determining the crack initiation event. However, two problems remain. First, it is still up to the engineer to establish a criterion for deciding when a crack actually develops. Second, too much experimental work is involved, because crack initiation appears to be material specific. It may even be influenced by loading conditions, or the local geometry of the structural part. Use of a measurable crack size (e.g., 0.25 mm, or 0.01 in.) is rather arbitrary in nature. However, it is a practical and easy way to connect the two methodologies to make a total life prediction. A more precise approach is the one developed by Socie et al. (Ref 3.28). This method can determine crack initiation life based on a "nonarbitrary crack size," which is the size of that crack corresponding to that life (at initiation). The method is largely based on fracture mechanics analysis procedures and some other elements of fatigue analysis not yet mentioned here. Socie's concept of crack initiation is described in Appendix 3.

3.4 Infinite-Life (Stress-Based) Fatigue Strength

Fatigue loading belongs to one of two categories: constant amplitude or variable amplitude (spectrum loading). Schematic representations of these loading profiles, including definitions of the events, are given in Fig. 3.29. In Fig. 3.29(b), the terms (+) range and (−) range signify that the load of that cycle is applied in an increasing or decreasing order. Actually, in most cases only the absolute magnitude of the range is of concern. The figure also shows that the peak or the valley at which the load changes direction is called a reversal. Therefore, one load cycle is made up of two reversals. In fatigue testing, when hold time is not involved, load cycles are usually applied in the form of a sawtooth (Fig. 3.29) or sinusoidal (Fig. 3.12). In either case, the shape of the cycle is symmetrical; that is, it takes the same amount of time to raise the load from the valley to the peak as it does to go down to the valley. It is believed that in room-temperature air, waveform is unimportant. In other words, it is assumed that the fatigue life for a given material will be (more or less) the same for either type of stress wave. (At least, there is insufficient test data that report the differences.) However, it should be noted that if an alloy's absolute melting temperature is low enough (so that the homologous temperature is high enough), it may be susceptible to waveform in room-temperature air.

At this point, it is necessary to define some basic parameters that describe the conditions of fatigue loading. The terminology presented below is also applicable to the strain-based fatigue analysis discussed in Section 3.5. The characteristics of a cycle, whether it is constant amplitude or spectrum loading, is defined by the following terms, in addition to those already shown in Fig. 3.29.

Stress ratio (or load ratio):

$$R = \sigma_{min}/\sigma_{max} \qquad \text{(Eq 3.3a)}$$

or

$$R = P_{min}/P_{max} \qquad \text{(Eq 3.3b)}$$

For instance, $R = -1$ for the complete reverse loading (such as shown in Fig. 3.12).

Stress range:

$$\Delta\sigma = \sigma_{max} - \sigma_{min} \qquad \text{(Eq 3.3c)}$$

Mean stress, equal to the average of maximum and minimum stresses (primarily used in connection with constant-amplitude loading):

$$\sigma_m = (\sigma_{max} + \sigma_{min})/2 \qquad \text{(Eq 3.3d)}$$

Alternating stress (or stress amplitude, also associated with constant-amplitude loading):

$$\sigma_a = (\sigma_{max} - \sigma_{min})/2 \qquad \text{(Eq 3.3e)}$$

Amplitude ratio:

$$A = \sigma_a/\sigma_m \qquad \text{(Eq 3.3f)}$$

Constant-amplitude fatigue tests are conducted by stress controlled either in tension-tension or tension-compression. Compression-

compression loading is also used, although less frequently. The specimens are subjected to either axial or bending loads, or even in torsion. The conventional way to gather and present constant-amplitude fatigue data is in the form of an S-N curve (see Fig. 3.12). For convenience, the values of stress (ordinate) and the number of cycles (abscissa) are plotted in linear and logarithmic scales, respectively. In most cases, the graph is presented with maximum stress versus number of cycles to failure instead of alternating stress versus cycles. Sometimes the main curve is extrapolated back to the ordinate, which limits its maximum to the material's ultimate strength at time zero. Also, as shown in Fig. 3.12, there is a stress level below which no failure will occur. This stress level is called the endurance limit, or fatigue endurance. Some textbooks call it the fatigue limit, or fatigue strength, which seems to be a properly chosen term because fcc metals (such as aluminum alloys) do not exhibit an endurance limit. In other words, the S-N curve will not stop decaying. It is quite common to see S-N curves without an endurance limit. For instance, such behavior is found in some polymers and polymer-matrix composites, and in steels tested at high temperature. On the basis of a number of papers and test reports, Dieter has suggested that strain aging is responsible for

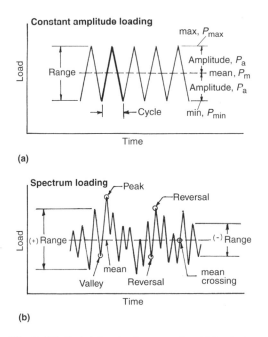

Fig. 3.29 Basic fatigue loading parameters. (a) Constant amplitude loading. (b) Spectrum loading

the knee in the *S-N* curve (Ref 3.29). The tests he reviewed covered a broad base of metals, including carbon steels, titanium, molybdenum, and aluminum with 7% Mg. It seems that when a material shows strain aging in a tension test, it also will exhibit an endurance limit in a fatigue test.

For practical reasons, fatigue tests usually terminate at 10^8 or 5×10^8 cycles, whether or not a runout has occurred. In design practice, a fatigue limit is defined as a selected number of cycles to failure, and is an engineering judgment or design policy. A fatigue limit that is based on 10^7 or 10^8 cycles to failure (whether or not there is a runout in the *S-N* curve) is quite common.

At low strain amplitudes (long-life fatigue), a material with higher tensile strength normally has better resistance to slip. Because failure due to monotonic loading or fatigue begins with slip of crystallographic planes, the end result is that

both tensile and fatigue strengths are higher. However, in most cases this means that the material simply had a longer crack initiation life. An analogy to this is that cold work, or heat treatment, can increase resistance to fatigue. Thus, in general, fatigue limit is proportional to material tensile strength. Figures 3.30 and 3.31 are sample data for steels from a textbook by Fuchs and Stephens (Ref 3.30). These data, which originated from three other sources (Ref 3.31–3.33), provide evidence in support of this claim. Note that the proportionality between tensile and fatigue strengths is not necessarily linear throughout the full range of strength level. The correlations of Fig. 3.30 and 3.31 show that fatigue strength approaches a limiting value at some higher ultimate tensile strengths.

When an *S-N* diagram is used for design, a design stress level is usually selected to be below the endurance limit, the design stress level

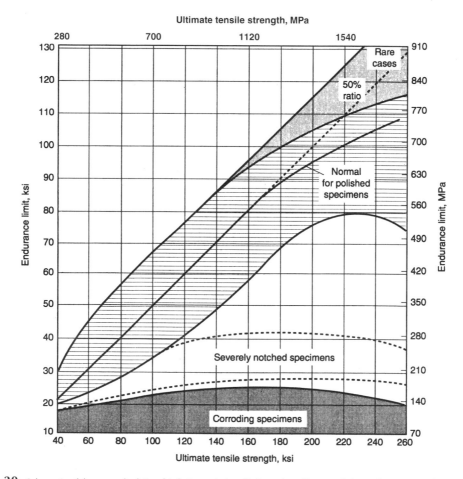

Fig. 3.30 Schematic of the general relationship between fatigue limits and tensile strength for steels. Source: Ref 3.30

is set at 10^8 cycles (or some other preselected number of cycles, as mentioned earlier). In other words, the candidate material should possess an endurance limit (or its equivalent, such as the value of stress at $N \geq 10^8$) that is higher than the design stress level. This design approach is called the infinite-life approach, or design for infinite life. The *S-N* curves for several materials are used for comparison purposes so that the most suitable material can be selected for a design. In reality, most structural or machine parts are not subjected to constant-amplitude loading. The actual life of that design would be determined by analysis, using some kind of damage cumulation method to assess the sum of damages attributed to all the different stress levels. In this regard, the primary function of a material's *S-N* curve is to provide quick reference in the course of material selection. Change of design stress level or material, or both, may be required in order to meet the life requirement.

Another feature of Fig. 3.12 is that some regions of the graph are marked with a hysteresis loop. These loops are not part of the graph, but are there simply to point out that at low stress levels, the material behavior is purely elastic. The material exhibits plastic hysteresis behavior at the region where the stress levels are high. The higher the stress level, the larger the plastic hysteresis loop. The illustration is simple; however, the implication is quite significant, as we shall see later in Section 3.5. For now, simply remember that researchers have divided fatigue phenomena into "high-cycle fatigue" and "low-cycle fatigue." Low-cycle fatigue, or high-strain fatigue, is tentatively defined as the fatigue

mechanism that controls failures occurring at $N < 10^4$ cycles and typically is of concern in cases of significant cyclic plasticity. Actually, the term low-cycle fatigue is not precisely defined. Some literature gives allowance in a range between 10^2 and 10^6 cycles. Most test reports show data that are tested up to 10^5 cycles.

3.4.1 Mean Stress Effects and Constant-Life Diagram

The interrelated terms σ_a, σ_m, $\Delta\sigma$, A, and R in Eq 3.3(a) to 3.3(f) are all functions of σ_{max} and σ_{min}. Thus, an R value is sufficient to identify an *S-N* curve. Figure 3.12 schematically shows a set of fatigue data plotted as an *S-N* curve, and that the stresses were in complete reverse around a zero mean. In this case, $R = -1$. In fact, the mean stress of a cycle does not have be zero. The stress ratio R would have a different value when $\sigma_m \neq 0$. Experimental data show that for a given σ_{max}, the resulting fatigue lives vary with different values of σ_m. Figure 3.32(a) is a plot of such data, where each *S-N* curve is identified with its loading condition by an R value. Note that as R becomes more positive, which is equivalent to increasing the mean stress, the measured fatigue endurance increases. This also means that tensile mean stresses are detrimental and compressive mean stresses are beneficial. Figure 3.32(b) shows another way of plotting the same data set using alternating stress versus cycles at a constant mean stress. Note that as the mean stress increases, the allowable alternating stress decreases. There is yet another way to plot the data,

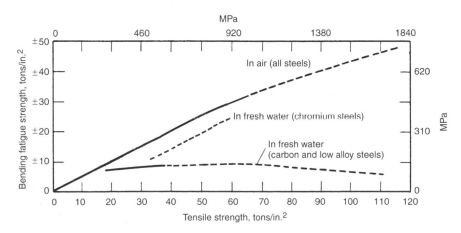

Fig. 3.31 Fatigue limit as a function of tensile strength for steels tested in room-temperature air or subjected to water spray. Source: Ref 3.30

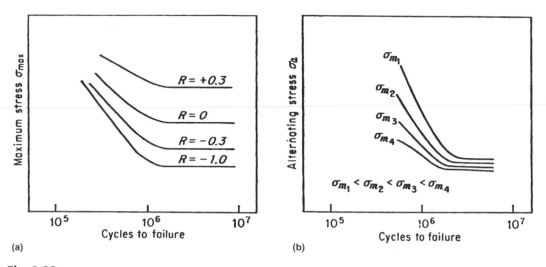

Fig. 3.32 Two methods of plotting constant-amplitude fatigue data

where maximum stresses are plotted against cycles to failure at a constant mean stress. Though not shown in these figures, note that the fatigue strength is limited by the tensile strength of the material; therefore, all the S-N curves in Fig. 3.32(a) and (b) are merged to a stress level at $\sigma_{max} = F_{TU}$ at the ordinate (cycle = one).

For design purposes, data often must be presented by methods other than S-N curves. Certain values of N may require determination of safe combinations of σ_a and σ_m, or σ_{max} and σ_m, during the design development of a structural member. This is accomplished by constructing a graph that shows the relationship between the stress components in question at a constant number of cycles. Such graphs are called constant-life diagrams, R-M diagrams, or S-S diagrams. The latter designations are British and European, respectively. A constant-life diagram is constructed using actual test data originally presented as a family of S-N curves. The conversion procedure is illustrated in Fig. 3.33. Figure 3.34 is an example of the finished constant-life diagram where both the unnotched and notched specimen data are plotted on the same graph. These data may well be plotted in separate graphs because there will be data for other K_t values as well.

3.4.2 Cumulation of Damage

The goal of fatigue analysis is to determine the fatigue life of a designed component that is subjected to various stress levels during its service life. A vehicle, machine, or device may be designed to carry out several functions. For example, the same aircraft may be required to fly several different types of missions with different

Typical S–N curves

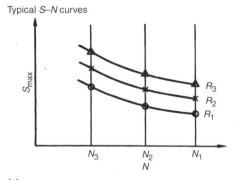

(a)

Typical constant-life diagram

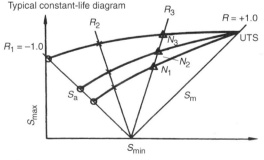

(b)

Fig. 3.33 Constructing a constant-life diagram from a set of S-N curves containing three R ratios, where R = −1 is the fully reversed cyclic loading condition and R = +1 is the monotonic ultimate strength of the material (no cyclic load). (a) Typical S-N curves. (b) Typical constant-life diagram

mixtures of stress levels and number of cycles. The input loads that an aircraft component experiences during long- and short-distance flights are quite different. Each occurrence of stress causes some amount of damage to the component material. It is necessary to estimate the total damage caused by all the stress cycles, as well as the damage caused by each individual stress cycle.

Figure 3.35 illustrates the analysis procedures for predicting the fatigue life for a given design

stress or the design stress for a given design life. It also shows the relationships among five major areas in fatigue design and analysis:

- Service-life requirements and planned operational usage
- Fatigue load input spectra representative of the required life and expected operational usage
- Methods to calculate structural element fatigue load response spectra

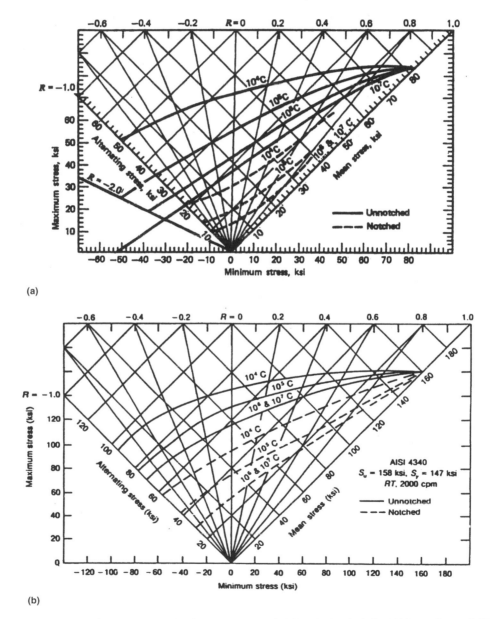

Fig. 3.34 Constant-life diagrams. (a) 7075-T6 aluminum: F_{TU} = 82 ksi, K_t = 3.4 (notched), f = 2000 cpm. Source: Ref 3.34. (b) AISI 4340 steel: K_t = 3.3, f = 2000 cpm. Source: Ref 3.35

- Methods to calculate fatigue damage for expected life and to predict the safe fatigue life
- Structural element fatigue strength properties and the factors that affect fatigue performance

The first three areas involve defining the service usage of a part and converting the anticipated input load magnitudes to stress. For example, different parts in the same aircraft experience different loading condition during their service lives; the load magnitude and the pattern of occurrence in the wing, the fuselage, and the landing gear are all different. The same is true of the service load histories of various parts in a motor vehicle. Development of fatigue stress spectra is not really within the scope of this book.

The method presented below assumes that a finalized stress spectrum is given. However, editing of a given spectrum may be required due to irregularity of stress cycles presenting in the real-time spectrum. Appendix 4 briefly describes a spectrum editing technique. The subject of fatigue performance improvement (listed in the upper right of Fig. 3.35) will be discussed later in this chapter.

The fatigue stress spectra and the material fatigue allowable (in the form of *S-N* curves or constant-life diagrams) are the input data required for fatigue analysis. The fatigue calculations are performed using some damage rule that defines the relationship between the applied stresses and number of cycles to the allowable stresses and number of cycles required causing failure. Of the many damage rules available today, Palmgren-Miner's rule (or simply Miner's rule, as it is often called) is used throughout the engineering community. It is simply a linear summation of fractions of fatigue damage toward crack initiation, or total failure, whichever the case may be. It does not account for load sequence effects. However, the advantage of this method is its simplicity, and its ability to work with available *S-N* data. Its basic equation is expressed as:

$$D = \sum_{i=1}^{k} \frac{n_i}{N_i} \qquad \text{(Eq 3.4a)}$$

where n_i is number of load cycles at the *i*th stress level, N_i is number of cycles to failure for the

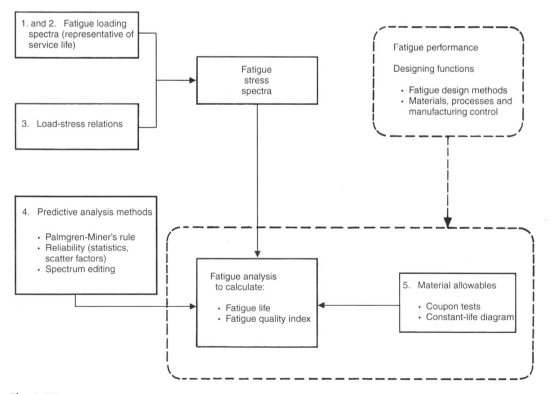

Fig. 3.35 Elements of fatigue design and analysis

*i*th stress level based on constant-amplitude *S-N* data for the applicable K_t value, and *k* is number of stress levels considered in the analysis.

Equation 3.4(a) implies that the total fatigue damage is the sum of the damage for each increment of the loading spectrum. The ratio of the number of applied load cycles to allowable load cycles is the fatigue damage for a given increment of the loading spectrum. Since the loading spectrum represents a specified period of service life (e.g., 1000 missions, or 10,000 hours, etc.,) ideally the calculated total damage factor, *D*, should equal unity, or 0.25 if a safety factor of 4 is desired. The predicted fatigue life for a given fatigue stress spectrum (based on a selected limit design stress level) would be:

$$L_f = \frac{L}{F \cdot D} \qquad \text{(Eq 3.4b)}$$

where L_f is the predicted fatigue life divided by a safety factor, *F* is factor of safety, and *L* is hours or number of missions represented by the spectrum used in the analysis. Each mission, which represents certain hours of operational usage, may contain a number of stress levels and a different number of load cycles at a given stress level.

3.4.3 Other Issues

In addition to the loading condition (i.e., *R*-ratio), a set of constant-amplitude fatigue data must be furnished with the following information:

- *Material:* Anything that can be used to identify the material, such as product form, thickness, grain direction, heat-treat condition, ultimate/yield strength, surface condition (surface finishing and surface treatment), and so forth, in as much detail as possible
- *Frequency:* The number of cycles per second or per minute. The effect of frequency on fatigue life may be insignificant when the test is conducted in room-temperature air (with symmetrical load cycles). A workable frequency range is quoted in the literature as 5 to 100 Hz (Ref 3.30). It is also believed that frequency may cause negligible impact in subzero temperatures. Otherwise, both frequency and cyclic waveform (with or without holdtime) are important factors. For metals, slow frequencies provide time for the corrosive agent to act, or for the dislocations

to mobilize at high temperature, resulting in a shorter fatigue life. The situation for plastics is just the opposite; high frequencies generate heat and cause softening, thereby leading to shorter fatigue lives.
- *Geometry:* That is, with or without a stress raiser, which is commonly identified by the value of a stress concentration factor. Specimen or part sizes and method of loading are other important factors.
- *Environment (including corrosive media and temperature):* Room temperature air if not otherwise specified

The first two items are self-explanatory. The other two issues are described further.

Effect of Specimen Size and Method of Loading. Similar to tensile strength, fatigue strength is influenced by the size (volume) of the part/specimen. As discussed in Chapter 2, a larger specimen has a statistically greater chance of exhibiting more cavities and/or precipitates, resulting in a lower tensile strength. In the case of fatigue, the chance of crack initiation is greater in a larger specimen, because it has a larger surface area. Specimen size effect in fatigue is compounded by the geometry of the test coupon and the way it is loaded. Both Dieter (Ref 3.29) and Bannantine et al. (Ref 3.36) point out that the stress gradient in a specimen affects fatigue strength. Stress gradient can be found in a bending specimen or can be induced by a stress concentration. In the absence of a stress concentration, size effect is less pronounced in axial loading than in bending. A larger specimen under bending will have a less steep stress gradient, whereas the outer fiber stress (of the same magnitude) in a smaller specimen decreases (toward the center) more rapidly. Hence, a larger volume of material in a larger specimen is subjected to the high outer fiber stress, resulting in a greater probability of initiating a fatigue crack. This means crack initiation may come sooner, thus shortening fatigue life.

Now we can compare the fatigue strengths of axially loaded and rotating bending specimens. Consider that both types of specimens are of the same size and subjected to the same stress level; that is, the magnitude of the extreme fiber bending stress in the rotating bending specimen and the uniform tensile stress in the axial specimen are the same. The uniform stress across the entire cross section of the axial specimen creates a worse situation than the stress in a rotating bending specimen, which decreases toward the center of the specimen. On the basis of this idea, Bannantine et al. (Ref 3.36) made estimates relating the fatigue data for rotating bending, axial, and torsion specimens. On the basis of available test data, they estimated that the endurance limit for an axial specimen is approximately equal to 70% of that for a rotating bending specimen of the same size. Furthermore, they estimated the endurance limit for torsion to be between 50 to 60% of that for the rotating bending specimen. Incidentally, this range is right in the proximity of 57.7%, which is the torsion yield strength to tensile yield strength ratio determined by the von Mises criterion. Although the effect of specimen size and loading method on fatigue strength has been recognized for some time, quantitative assessment is far from conclusive. Test data for 4340 steel reported in Ref 3.7 show that the difference in fatigue life between axial fatigue and rotating bending is in the low (to extremely low) cycle fatigue region. The reduction in endurance limit, if any, is insignificant.

To further complicate matters, test reports as well as handbooks often present fatigue data (S-N curves) using different stress parameters. That is, the stress values on a particular graph could represent either the gross area stress or the net section stress of the test coupon. Because test specimens are usually small in size whereas an actual component may be large, fatigue data should be carefully chosen for design analysis. Similarly, the loading condition for a set of fatigue data should match the service condition of the component. If these conditions are not met, some adjustment to the fatigue data, or to the analysis result, is required.

Effect of Local Geometry and Load Transfer. In the previous two chapters, it was shown that holes, notches, and cutouts all act as stress raisers. The local stress at the edge of a stress raiser may be significantly higher than the nominal stress of the part. The ratio of local stress to nominal stress at that location is called the stress concentration factor. The higher the stress concentration factor, the lower the fatigue endurance. Fatigue endurance for a severely notched specimen can be drastically reduced, as schematically shown in Fig. 3.30.

Stress concentration normally plays an important role in fatigue. At low stress levels where the endurance limit is a primary concern, the local stresses (far-field stress times K_t) may still be within the elastic limit of the material. As long as the resulting stress is still elastic, fatigue life at these stress levels will be lowered (com-

pared to the same applied stress level without a stress raiser). The severity of reduction depends on the K_t value. At higher stress levels (or even at low stress levels with a very high K_t), fatigue lives may or may not be affected by yielding of the notch. Just using a K_t factor will not account for the inelastic behavior at the notch. It is true that the local stress at the notch could have been reduced as compared to the original (unyielded) notch. Depending on the original K_t value and the magnitude of the far-field stress, the stress concentration might be moderately or fully reduced. Whatever the case, it is not known how the yielded material (whether in a localized region around the notch or spread across the specimen width) affects the total life of a given test. However, this type of S-N data may handicap the conventional technique for estimating the fatigue life of a structural part subjected to spectrum loading. Perhaps the strain-based approach, presented later in this chapter, is a better way of handling low-cycle fatigue analysis, particularly where notch root plasticity and/or high temperature is involved.

When the geometry and loading condition of a structural part is quite complex, it cannot be represented by a single K_t. In the aircraft industry, a design is judged by its fatigue quality. This is called the fatigue quality index, or effective K_t. Generally, the fatigue quality of the structure is determined from results of fatigue tests of components or the complete structure. The fatigue quality index can be computed at each crack initiation site that is developed during fatigue testing. Fatigue analyses are conducted with a set of constant-amplitude S-N curves that go with various values of K_t obtained from various notched coupons. The stress spectrum that sustained at each critical point to fatigue crack initiation in the test is determined from the spectra of applied loads. Results of the analyses are interpolated to determine the specific S-N curve, which makes the D value in Eq 3.4(a) equal to unity for the test life. The value of K_t associated with the S-N curve is a measure of the fatigue quality index. The better structure is the structure with a lower effective K_t.

In the design stage, aircraft engineers often use the stress severity factor, K_{SSF}, which indicates the severity of stress at the potential crack initiation site. It is an empirical factor that accounts for the geometrical stress concentration effect (K_t) and additional effects such as variations in material properties, product quality, and other analytical uncertainties. These two factors

(K_t and K_{SSF}) both represent the ratio of the peak elastic notch stress at a critical area to the maximum nominal stress S_{max}. Actually, a K_{SSF} includes a baseline K_t in its mathematical expression. Other supplemental factors are added, depending on the situation. Thus, K_{SSF} is often used in fatigue analysis to replace the geometrical stress concentration factor K_t. In any situation that involves K_t, K_{SSF} and K_t are interchangeable (in stress analysis). A stress analyst may choose to compute the stress using $\sigma_{local} = K_t \cdot \sigma_{far\text{-}field}$ in one calculation, or $\sigma_{local} = K_{SSF} \cdot \sigma_{far\text{-}field}$ in another calculation, depending on the situation. This approach is equally applicable to both stress-based and strain-based analysis methods. A brief description of this concept is provided in Appendix 5.

Fatigue analysis of splices, joints, and lugs falls into a category of problems involving load transfer through fasteners from one part to another, which will tend to change the basic fatigue spectrum in the local load transfer area. An extreme example of this is the lug problem. The axially loaded lug hole does not see the compressive loading contained in the fatigue spectrum. Therefore, all the compression loads should be truncated from the applied loads.

The situation is different when the lug is subjected to a transverse load, as shown in Fig. 3.36. Consider location A in the lug. Assuming a load is applied moving to the right, the lug material at location A experiences tension stress from the first half-cycle of the applied load. When the magnitude of the load is reduced from the peak, (i.e., the load moves to the left after it reaches the peak), the material at location A will experience decreasing tension load until it reaches zero. Afterward, the applied load continues to decrease, and is supposed to venture into compressive territory. However, the material at location A will see the load increase again as a tension load, instead of a compression load. The resulting spectrum will be as shown in Fig. 3.36.

A generalized expression can be developed for the axially loaded sheet structure involving attachment that experiences both local stress and bypass stress (Ref 3.23). For the tension case, the notch stress is:

$$\sigma' = [(1 - \eta)K_t^H + \eta K_t^L]\sigma \qquad \text{(Eq 3.5a)}$$

where σ is the gross stress, and η is the load transfer ratio. For the compression case:

$$\sigma' = \sigma(1 - \eta)K_t^H \qquad \text{(Eq 3.5b)}$$

The spectrum can be normalized to a tension load condition, if so desired. In this case, the compression loads are factored by:

$$F_\eta = \frac{(1 - \eta)K_t^H}{(1 - \eta)K_t^H + \eta K_t^L} \qquad \text{(Eq 3.5c)}$$

This rationale satisfies the extreme conditions (open hole and lug problems) and allows for a linear interpolation by superposition for those problems that fall in between. Examples of the effect of loaded holes on notch spectrum modification are shown in Fig. 3.37.

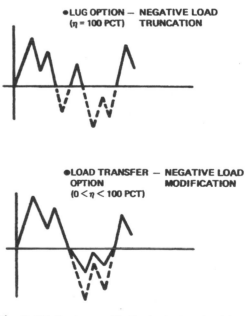

●LUG OPTION — NEGATIVE LOAD
(η = 100 PCT)   TRUNCATION

●LOAD TRANSFER — NEGATIVE LOAD
OPTION   MODIFICATION
(0 < η < 100 PCT)

Fig. 3.37 Spectrum modification for structures involving attachment and bypass loads. The dotted lines belonged to the original spectrum but were subjected to truncation or modification. Courtesy of P.G. Porter

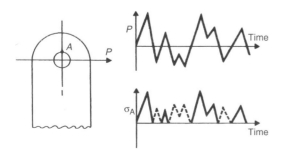

Fig. 3.36 Analytic load manipulation for a transversely loaded lug. Courtesy of P.G. Porter

Corrosion Fatigue. When fatigue loading takes place in a corrosive environment, the fatigue strength of the material will be drastically reduced. Do not confuse stress corrosion cracking and corrosion fatigue cracking. Both types of cracks originate at a corrosion pit. Whereas stress corrosion involves a structural piece subjected to a static sustained load in a corrosive environment, corrosion fatigue is simply fatigue in a corrosive environment. There is a general perception that stress corrosion cracks are usually intergranular, though with exceptions. However, the fracture surface of a corrosion fatigue crack is similar to ordinary fatigue cracks—that is, transgranular, except that the cracking initiates at a corrosion pit, often at the intersection of grain boundaries. Figures 3.38(a) and (b) show the origin and the fracture surface of a corrosion fatigue crack in a rotating bending specimen that failed in a commercial salt solution. The crack initiated from a single origin (a corrosion pit) and developed to a transgranular fracture.

The effect on fatigue strength attributed to corrosion fatigue also depends on how the corrosive environment is applied. Spray, or drip, is more problematic than submersion in fresh or salt water. The mechanism of corrosion fatigue is quite complex. To quote from Bannantine's book (Ref 3.36): "The basic mechanism of corrosion fatigue during the initiation stage can be explained this way. A corrosive environment attacks the surface of a metal and produces an oxide film. Usually, this oxide film would act as a protective layer and prevent further corrosion of the metal. However, cyclic loading causes localized cracking of this layer, which exposes fresh metal surfaces to the corrosive environment. At the same time, corrosion causes localized pitting of the surface, and these pits serve as stress concentrations. The mechanism of corrosion fatigue during the crack propagation stage is very complicated and not well understood."

As seen earlier (Fig. 3.30, 3.31), fatigue strength for a given alloy will be higher when the alloy is heat treated to a higher tensile strength. These figures also show that fatigue strength is decreased by a corrosive environment. Figure 3.31 shows that chromium steels exhibit much better resistance to corrosion fatigue than plain carbon and low-alloy steels. This means that materials that are resistant to corrosion alone will also have good corrosion fatigue properties. The carbon and low-alloy

steels not only had poor corrosion fatigue properties, they also could not take advantage of the higher strength levels that they enjoy in room-temperature air. In other words, the corrosion fatigue strengths for these steels are independent of tensile strengths.

Some general trends are observed in corrosion fatigue. Figure 3.39 shows schematic S-N curves for four different environments in room-tem-

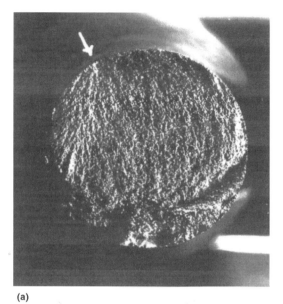

(a)

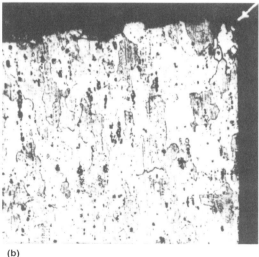

(b)

Fig. 3.38 Fracture surface of a corrosion fatigue crack in a rotating bending specimen of 2014-T6 aluminum alloy. (a) Optical photograph showing the origin and beach marks typical of fatigue fracture. (b) Microphotograph of a section through the fatigue origin (arrow). The fracture surface is predominantly transgranular. Intergranular fracture at (and in the vicinity of) the corrosion pit is apparent. Source: Ref 3.37

perature air. The curves generated from vacuum and air show that even the humidity and oxygen in room-temperature air can slightly reduce fatigue strength. However, room-temperature air is normally regarded as a reference point when evaluating the effect of certain environments on fatigue strength. The "presoak" curve involves rough surfaces caused by corrosion pitting, which act as stress raisers. As shown, the combination of simultaneous environment and re-

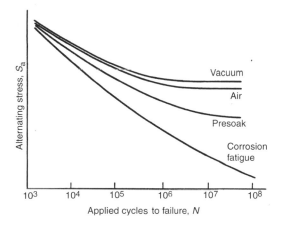

Fig. 3.39 Schematic of relative fatigue behavior under various environmental conditions. Source: Ref 3.30

peated loading certainly is the most severe condition in fatigue testing. The relatively flat long-life *S-N* behavior normally observed in steels is eliminated under corrosion fatigue conditions. Actual sample data for 2014-T6 aluminum alloy tested in room-temperature air and in a synthetic salt solution are shown in Fig. 3.40. The magnitude of fatigue life reduction due to corrosion fatigue for this alloy is considered realistic. The validity of the corrosion fatigue data is substantiated by the fact that the data points obtained from room-temperature air match official data published by Alcoa, indicating that the tests were performed correctly.

Effect of Temperature. Because fatigue strength is somewhat proportional to tensile strength, which is higher at low temperatures, the *S-N* fatigue strength of a material can be expected to be higher at low temperatures. However, strain-life data indicate that fatigue resistance may decrease in the low-cycle fatigue range due to low ductility and low fracture toughness at low temperatures. At short lives, ductility is a principal parameter in strain-controlled behavior, while at longer lives strength is more the controlling parameter. The opposite is true, meaning that fatigue strength could be lower at high temperatures. Meanwhile, the endurance limit for steels disappears due to mo-

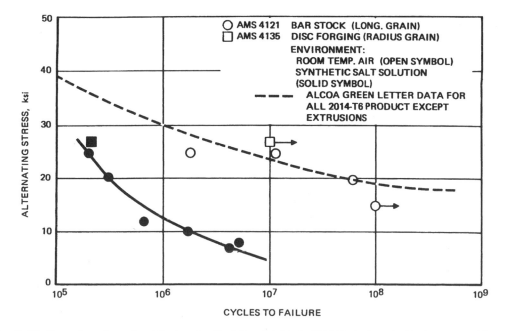

Fig. 3.40 Comparison of smooth-rotating/pure-bending fatigue test data for 2014-T6 aluminum in dripping commercial synthetic solution and in room-temperature air. A flow of liquid around the center section of the specimen was supplied by capillary action during the test. Source: Ref 3.37

bilizing of dislocations. A comprehensive compilation of high-temperature fatigue data is given in Forrest's book (Ref 3.32). Figure 3.41 shows fully reversed fatigue strengths for many types of metals (within their applicable temperature ranges). Fracture surfaces in high-temperature air are usually intergranular due to oxygen attack of the grain boundaries. When a high-temperature fatigue test is conducted in vacuum, the fracture surfaces may be transgranular due to absence of oxygen attack.

Creep-fatigue interaction may occur at temperatures beyond approximately one-half of the melting point of the material. From a metallurgical standpoint, a T_H of 0.3 to 0.4 is considered the temperature range that will show thermal effects. The T_H of 0.5 is the value for initiation of high-temperature creep behavior, at which temperature the crack initiation mechanism often changes from wedge cracking to void formation.

Oxidation plays a key role in high-temperature fatigue and creep. Tests at high temperature in a vacuum or inert atmosphere have shown substantial increases in fatigue-creep resistance compared to high-temperature air tests. In the past, creep-fatigue testing was conducted simply to ascertain the extent of the damaging effect of creep on cyclic fatigue life. Such testing is conducted at an isothermal temperature in conjunction with a regular fatigue test. The addition of creep to a cycle of conventional fatigue loading will invariably reduce the cyclic life, although the clock time to failure may actually be in-

creased. In recent years, strain-controlled tests have been used to conduct creep-fatigue testing. These tests are also designed to evaluate and calibrate the empirical constants that go with a life prediction model. The isothermal hysteresis loops shown in Fig. 3.42 are commonly encountered in creep-fatigue testing. Figure 3.43 is a sample data set (taken from Ref 3.38) developed for AISI type 304 stainless steel tested in air at 650 °C (1200 °F). The tests were conducted under the conditions of (a), (c), (f), and (i) in Fig. 3.42. Only the cycles held in tension (for 1 min or longer) were detrimental. On the basis of a few data points, "compression hold" seems to have a negligible effect. In some cases, it can even be beneficial in that it cancels out the damaging effect caused by the "tension hold" portion of the same loop.

Other effects on fatigue life resulting from high temperatures include thermal fatigue and thermomechanical fatigue. Thermal fatigue is caused by unequal heating of component parts, producing thermal stresses; no mechanical loading is necessarily involved. In other words, thermal fatigue results from the presence of temperature gradients that vary with time to produce cyclic stresses in a structure. Thermomechanical fatigue is the result of thermal (temperature) cycles and mechanical load cycles. These two types of fluctuating cycles are superposed either in phase or out of phase. However, the mechanical loads are not necessarily the externally applied loads. Thermomechanical fatigue may re-

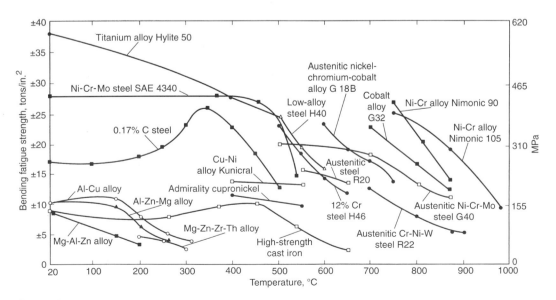

Fig. 3.41 Temperature influence on fully reversed fatigue strengths of metals. Source: Ref 3.32

sult from external constraint of the part while it is subjected to temperature fluctuations. This chapter limits the scope of discussion to the isothermal condition. For more in-depth information, see Ref 3.27 and 3.39.

3.5 Finite-Life (Strain-Based) Fatigue Strength

An *S-N* curve straightforwardly shows the one-to-one relationship between a stress level and the corresponding fatigue life. However, hidden information as to the residual stresses imposed due to material yielding, which may be the result of stress concentration, is not accounted for in the curve's low-cycle fatigue re-

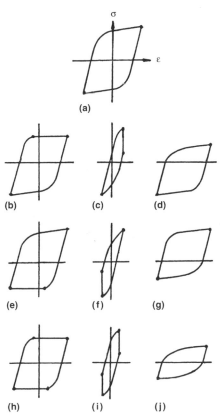

Fig. 3.42 Schematic hysteresis loops encountered in isothermal creep-fatigue testing. (a) Pure fatigue, no creep. (b) Tensile stress hold, strain limited. (c) Tensile strain hold, stress relaxation. (d) Slow tensile straining rate. (e) Compressive stress hold, strain limited. (f) Compressive strain hold, stress relaxation. (g) Slow compressive straining rate. (h) Tensile and compressive stress hold, strain limited. (i) Tensile and compressive strain hold, stress relaxation. (j) Slow tensile and compressive straining rate. Source: Ref 3.27

gion. Whereas Miner's rule is computationally efficient, it is strictly related to gross-stress cycling as implicated by the *S-N* curve; it does not account for the load sequence effects that cause residual stresses. Such sequence effects may not be important if one is willing to pay a severe weight penalty for designing a structural part below the material's fatigue limit. As a result of these uncertainties and inaccuracies, much emphasis has been placed on describing the actual stress and strain history in the material immediately adjacent to the notch boundary. If the plastic deformation of this local volume can be assumed to be strain controlled by the mass of the surrounding elastically deformed material, then it is possible to define its stress-strain history by means of the classic Neuber relationship (Ref 3.40) as applied to cyclic stress-strain and hysteresis data. Damage can be related to results obtained for simple strain-controlled specimens and summed on a cycle-by-cycle basis. A distinct advantage of the strain-life method is its ability to deal with variable-amplitude loading through improved cumulative "damage" assessment. Cyclic plasticity responses are accounted for; load sequence effects are reflected in the analysis and results.

3.5.1 Constant-Amplitude ε-N Curves

A set of strain-life data is a plot of ε versus *N*, similar to *S* versus *N*. For convenience, an ε-*N* curve is shown as a log-log plot. An ε-*N* data point presents the result of one test; that is, the number of cycles to crack initiation (or final failure, whichever the case may be) corresponds to a finite strain level of that test. A typical ε-*N* curve is schematically shown in Fig. 3.44. A set of actual test data of low-carbon martensitic steel is shown in Fig. 3.45. Each data point on the "total" curve is the sum of two parts: the elastic strain amplitude and the plastic strain amplitude that come from a stabilized stress-strain hysteresis loop as defined in Fig. 3.3. Note that Fig. 3.44 is a plot of strain amplitude versus number of a pair of reversals ($2N_f$). It could have been plotted with strain range versus number of cycles (Fig. 3.45) or strain amplitude versus number of cycles. In the literature, $2N_f$ stands for two reversals, whereas N_f means one cycle. Therefore, interpreting the strain-life data, as well as applying the ε-*N* expressions that are presented below, can be very confusing.

Application of this method in its simplest form is to compare the total strain-amplitude

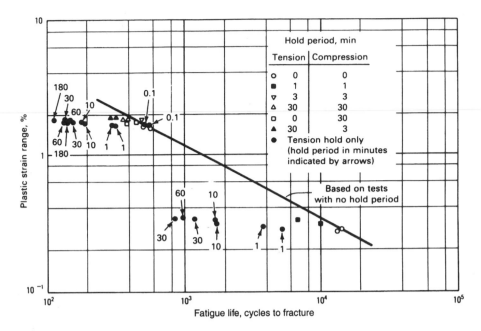

Fig. 3.43 Creep-fatigue interaction effects on the isothermal cyclic life of AISI type 304 stainless steel tested in air at 650 °C (1200 °F), normal straining rate of 4 × 10³/s. Source: Ref 3.38

($\Delta\varepsilon/2$) at a detail of the part to an ε-N curve having the necessary mean strain (stress) effects included. The assumption here is that the detail, perhaps in a highly constrained area, will respond identically to a specimen that is inherently a smooth bar in plane stress, although at the same strain level. The life, of course, corresponds to the intercept of the strain level at the ε-N curve. The result is typically an estimation of safe-life, or finite-life, so to speak.

3.5.2 Modeling the ε-N Curve

In a constant-amplitude fatigue test, the total strain range can be written as the sum of an elastic component and a plastic component:

$$\Delta\varepsilon = \Delta\varepsilon_e + \Delta\varepsilon_p \qquad \text{(Eq 3.6a)}$$

or, for zero mean stress (i.e., complete reverse loading):

$$\frac{\Delta\varepsilon}{2} = \frac{\Delta\varepsilon_e}{2} + \frac{\Delta\varepsilon_p}{2} \qquad \text{(Eq 3.6b)}$$

The local stress amplitude can be related to fatigue crack initiation life (or fatigue life) by:

$$\frac{\Delta\sigma}{2} = \sigma_f'(N_f)^b \qquad \text{(Eq 3.7)}$$

where σ_f' and b are material properties. As shown in Fig. 3.44, they are the intercept and the slop of the elastic component of the ε-N curve, and are called the fatigue strength coefficient and fatigue strength exponent, respectively. Here N_f is the number of cycles for crack initiation or complete failure; $2N_f$ will be used if the number of reversals is more desirable. The plastic strain amplitude can also be related to the

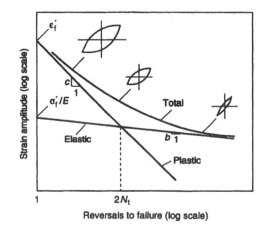

Fig. 3.44 Representation of total strain amplitude versus number of reversals to failure, including elastic and plastic portions as well as the combined curve. Note that two reversals constitute one cycle.

fatigue crack initiation life as:

$$\frac{\Delta\varepsilon}{2} = \varepsilon_f'(N_f)^c \qquad \text{(Eq 3.8)}$$

where ε_f' and c are the intercept and the slop of the plastic component of the ε-N curve, and are called the fatigue ductility coefficient and fatigue ductility exponent, respectively. Using Eq 3.7, the elastic component of strain can be written as:

$$\frac{\Delta\varepsilon}{2} = \frac{\Delta\sigma}{2E} = \sigma_f'(N_f)^b/E \qquad \text{(Eq 3.9)}$$

The strain-life equation is then obtained by adding Eq 3.8 and Eq 3.9 together, i.e.:

$$\frac{\Delta\varepsilon}{2} = \frac{\sigma_f'}{E}(N_f)^b + \varepsilon_f'(N_f)^c \qquad \text{(Eq 3.10)}$$

Note that Eq 3.10 applies to wrought metals only. When internal defects govern life (as is the case with castings, weldments, etc.), appropriate modification of the "internal micronotches" is required (Ref 3.7, 3.42).

The coefficients σ_f' and ε_f' in the above equations are empirical constants. They are the same coefficients previously used in Eq 3.1(a). When we treat these coefficients as measured values and those in Eq 3.1(b) and 3.1(d) as theoretical values, the correlations are fair (Ref 3.6). The slopes b and c are related to n' (the cyclic strain-hardening exponent). Through energy arguments, Morrow (Ref 3.2) has shown that:

$$b = -n'/(1 + 5n') \qquad \text{(Eq 3.11a)}$$

and

$$c = -1/(1 + 5n') \qquad \text{(Eq 3.11b)}$$

Thus:

$$n' = b/c \qquad \text{(Eq 3.11c)}$$

which allows a relationship between fatigue properties and cyclic stress-strain properties. Therefore, theoretically, we can construct a strain-life curve by using the properties of a cyclic stress-strain curve, n', ε_f', and σ_y', and the monotonic tensile fracture strength.

3.5.3 Cumulation of Damage

Miner's rule, used to compute the cumulation of fatigue damage for infinite life, is equally applicable here. However, just as in the S-N method of life assessment, several fundamental issues require close attention.

Effect of Mean Stress. Similar to the S-N curve, an ε-N curve is subjected to mean stress effects. Thus, to assess damage associated with a spectrum that contains multiple mean stress levels, one would need a family of strain-life curves corresponding to different R ratios.

An alternative approach that can greatly simplify this requirement has been proposed by Morrow (Ref 3.43): Mean stress can either increase or decrease fatigue strength. A positive (tensile) mean stress reduces fatigue strength, whereas a negative (compressive) mean stress

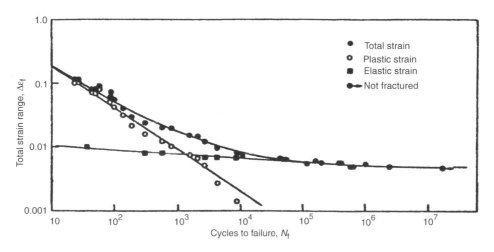

Fig. 3.45 Elastic, plastic, and total strain range as a function of life for a low-carbon martensitic steel. Source: Ref 3.41

increases fatigue strength. The curve for the "total strain amplitude" with a zero mean (i.e., Eq 3.10) can be modified by adjusting the elastic term in the strain-life equation by the mean stress. This concept is expressed in Eq 3.12 and is explained graphically in Fig. 3.46. That is, in terms of strain-life relationship, Eq 3.10 can be modified as:

$$\frac{\Delta \varepsilon}{2} = \frac{(\sigma_f' - \sigma_0)}{E}(N_f)^b + \varepsilon_f'(N_f)^c \qquad \text{(Eq 3.12)}$$

where σ_0 is the mean stress. As shown in Fig. 3.46, the curve for the zero mean stress has been transposed down by the amount of a tensile mean stress that carries a plus sign. If the mean stress is compressive, it will carry a minus sign; the elastic curve will be adjusted by $(\sigma_f' + \sigma_0)$, and the original "total" curve will be transposed upward. Note that at high strain amplitudes (0.5 to 1% or above), where plastic strains are significant, mean stress relaxation occurs, and the mean stress tends toward zero. This phenomenon is reflected in the modified ε-N curve in Fig. 3.46.

Recognizing that Eq 3.12 ignored the effect of mean stress on the plastic portion of the damage, Manson and Halford (Ref 3.44) and Smith et al. (Ref 3.45) modified both the elastic and plastic terms of the strain-life equation. As pointed out by Bannantine et al. (Ref 3.36), these proposed equations tend to overcorrect the contribution of mean stress to fatigue life. Though Eq 3.12 is not perfect, it generally does a better job of predicting mean stress effects.

Load Sequence Effects. To better understand the phenomenon of load sequence effect, consider a scenario that is given in Bannantine's book (Ref 3.36).* As shown in Fig. 3.47, Bannantine et al. considered the stress responses for two strain histories. Simply switching the order of a high tensile load with a compression load produces significantly different results. In history A, the tensile overload is followed by the compressive overload. In history B, these overloads occur in reverse order. Each of the corresponding stress-strain loops clearly show that these hysteresis responses result in different mean stress levels. More importantly, the resulting mean stress for history A has a positive value, while the resulting mean stress for history B is just the opposite. Because mean stresses significantly affect the fatigue life of a material, these two strain histories would have very different lives.

Stress and Strain at the Notch. An essential goal of a local stress-strain fatigue analysis model is to determine the detailed stress-strain history at the discontinuity (at the notch root, for instance) and track the changes in stress-strain states as the applied spectrum loads vary. The local stress $\Delta \sigma$ and local strain $\Delta \varepsilon$ at a discontinuity are related to the nominal stress ΔS and nominal strain Δe through Neuber's notch rule:

$$K_t = (K_\sigma K_\varepsilon)^{1/2} \qquad \text{(Eq 3.13a)}$$

where K_σ and K_ε are the stress concentration factors based on stress and strain, respectively. This relationship leads to:

$$K_t = \left(\frac{\Delta \sigma}{\Delta S}\frac{\Delta \varepsilon}{\Delta e}\right)^{1/2} \qquad \text{(Eq 3.13b)}$$

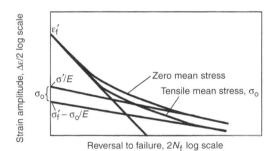

Fig. 3.46 Schematic mean stress modification to strain-life curve. Mean stress is tensile in this case. Source: Ref 3.43

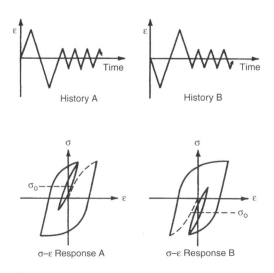

Fig. 3.47 Sample load sequence effects on material stress-strain response. Source: Ref 3.36

---

*Another similar example is given by Dowling (Ref 3.46).

or

$$K_t^2(\Delta S \cdot \Delta e) = \Delta\sigma \cdot \Delta\varepsilon \qquad \text{(Eq 3.14a)}$$

For a given nominal stress S in each load cycle and the associated geometric stress concentration factor, Eq 3.14(a) can be rewritten as:

$$\Delta\sigma \cdot \Delta\varepsilon = K_t^2(\Delta S^2/E) \qquad \text{(Eq 3.14b)}$$

where the right-hand side of the equation corresponds to the applied load and the left-hand side is the notch response. As mentioned before, the stress concentration factor K_t in these equations is interchangeable with the stress severity factor K_{SSF}.

Equations 3.14(a) and (b) relate the nominal stress and strain behavior of a notched member to the actual stress-strain behavior at the notch root. It is assumed that a smooth specimen undergoing a stress-strain history given by the right side of Eq 3.14(a) will develop a crack in the same lifetime as the notch root. Consequently, a smooth specimen undergoing the history given by the right side of Eq 3.14(a) may be used to simulate the life behavior of a notched member undergoing the history given by the left side of Eq 3.14(a). Thus, using a cyclic stress-strain curve and strain-life curve obtained from smooth specimen tests, we can construct a strain-life plot for the notched member. Since the smooth specimen test is conducted at $R = -1$ (i.e., complete reverse), mean stress adjustments (based on the mean notch stresses) as described previously are still needed.

Although Eq 3.14(a) and (b) are universally implemented in fatigue analysis, note that Eq 3.14(b) works in the strain range where the applied e is within the elastic limit. Equation 3.14(a) is applicable to strain ranges beyond the elastic limit where $S \neq Ee$. It has been observed that for small notches, the predicted fatigue life (which is based on a theoretical K_t) would be conservative. For this reason, Topper et al. considered that a K_t of another sort with a lower value should be implemented (Ref 3.47). They adopted Peterson's K_f factor to replace the K_t in Eq 3.14(a) and (b) to account for the size effect. Thus, these equations become:

$$K_f^2(\Delta S \cdot \Delta e) = \Delta\sigma \cdot \Delta\varepsilon \qquad \text{(Eq 3.15a)}$$

and

$$\Delta\sigma \cdot \Delta\varepsilon = K_f^2(\Delta S^2/E) \qquad \text{(Eq 3.15b)}$$

where K_f is called the fatigue strength reduction factor, or effective fatigue concentration factor. The expression for K_f is (Ref 3.48, 3.49):

$$K_f^2 = 1 + \frac{K_t - 1}{1 + \dfrac{\omega}{r}} \qquad \text{(Eq 3.16)}$$

where r is the root radius, and ω is an experimentally determined characteristic length parameter for sharply notched specimens and is material dependent. The methods that can be used to derive a ω value for a given configuration and material combination are given in Ref 3.46 to 3.49.

Application of Neuber's Rule. Fatigue damage occurs preferentially in regions of high stresses—areas adjacent to holes, notches, and defects, for example. We can assume that such a region is strain controlled by the surrounding mass of nonplastically deformed material. The material's actual hysteresis behavior is determined via the Neuber relationship, so that gross or reference area stresses are converted to local stresses and strains. Then the changes in the detailed stress-strain history in this local region are tracked, following every step of the applied load spectrum, as an essential part of any "sequence-accountable" fatigue analysis. A typical example is shown in Fig. 3.48. This hysteresis record shows that stress-strain states at points coinciding with the various load peaks and valleys are not necessarily elastic but can be plastic, and that significant residual stress-strain states can exist before the application of the next half-load cycle.

To explain how to obtain the stress-strain loops in Fig. 3.48, let us begin with Eq 3.14(b). The right side of Eq 3.14(b) represents the structural geometry and the global stress and strain away from the notch, which is the same as the stress and strain in a smooth specimen. The left side of Eq 3.14(b) represents the local stress and strain at the notch, as shown in Fig. 3.49. For a peak-to-peak change in load on a notched plate, values of K_t (or K_{SSF}, or K_f), ΔS, and E can be substituted in Eq 3.14(b) and the value of the left term (i.e., $\Delta\sigma \cdot \Delta\varepsilon$) calculated. This results in the equation of a rectangular hyperbola ($x \cdot y$ = constant) with vertical and horizontal asymptotes. That means there is a family of values of the product of local stress range, $\Delta\sigma$, and strain range, $\Delta\varepsilon$, that is equal to the constant, ($K_t \cdot \Delta S)^2/E$. Among them, there is a unique combination of $x \cdot y$ (i.e., $\Delta\sigma \cdot \Delta\varepsilon$) that satisfies the equation. This unique value occurs at the intersection of the cyclic stress-strain curve with the rectangular hyperbola.

Consider the first half-cycle where the far-field stress is raised from zero to S_1, and the value of $\Delta\sigma \cdot \Delta\varepsilon$ ($= K_{SSF} \cdot \Delta S_1$) is calculated and depicted as point 1 in Fig. 3.49. These would be the local stress and strain ranges if the material remains elastic. Iterations are made to locate a right combination of $\Delta\sigma \cdot \Delta\varepsilon$ (along the hyperbola) that intersects the cyclic stress-strain curve, and its magnitude is equal to $K_{SSF} \cdot \Delta S_1$. This is point 2 in Fig. 3.49. For a reversal such as ΔS_2 (shown in Fig. 3.49), the above procedure is repeated. However, the origin of the rectangular coordinate system used for the downloading step is now set at point 2. On unloading or for any subsequent events not starting at zero stress and strain, the cyclic stress-strain curve is magnified by a factor of two in order to trace the hysteresis loop. Thus, in this case, the intersect-

ing point is point 3. For clarification, numeric examples are shown in Fig. 3.48, which are computed for the first three points—A, B, and C—of that figure.

This point-by-point tracking procedure is obviously tedious. To streamline the computational routine, we can make use of Eq 3.2(b); that is,

$$\Delta\sigma\Delta\varepsilon = \frac{(\Delta\sigma)^2}{E} + \Delta\sigma\left(\frac{\Delta\sigma}{\zeta'}\right)^{1/n'} \qquad \text{(Eq 3.17a)}$$

By equating Eq 3.17(a) to Eq 3.14(b) we have:

$$\frac{(\Delta\sigma)^2}{E} + \Delta\sigma\left(\frac{\Delta\sigma}{\zeta'}\right)^{1/n'} = \frac{(K_t \cdot \Delta S)^2}{E} \qquad \text{(Eq 3.17b)}$$

where

$$\left(\frac{1}{\zeta'}\right)^{1/n'} = \varepsilon_f'\left(\frac{1}{\sigma_f'}\right)^{1/n'} \qquad \text{(Eq 3.17c)}$$

Equation 3.17(b) can be solved easily by iteration.

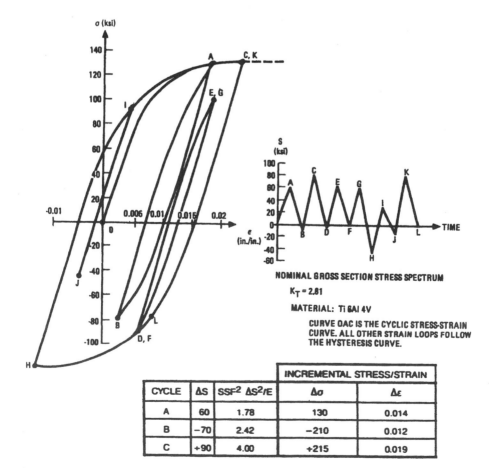

CYCLE	ΔS	SSF² ΔS²/E	INCREMENTAL STRESS/STRAIN	
			Δσ	Δε
A	60	1.78	130	0.014
B	−70	2.42	−210	0.012
C	+90	4.00	+215	0.019

NOMINAL GROSS SECTION STRESS SPECTRUM

$K_T = 2.81$

MATERIAL: Ti 6Al 4V

CURVE OAC IS THE CYCLIC STRESS-STRAIN CURVE. ALL OTHER STRAIN LOOPS FOLLOW THE HYSTERESIS CURVE.

Fig. 3.48 Sample idealized stress-strain history at a notch root during reversed load cycles. Courtesy of J.M. Waraniak

Local Stress-Strain Tracking. In order to assess the amount of fatigue damage in a complex random-amplitude load history, it is necessary to relate the variable-amplitude cycles to the corresponding constant-amplitude cycles. Then Miner's rule can be implemented as described earlier. In tracking the stress-strain history (with or without a stress raiser) in the form of stress-strain loops, it is necessary to define what type of loop causes fatigue damage. Generally, damage is assumed to occur only in fully closed loops of the material stress-strain hysteresis. The strain in each cycle corresponding to a closed hysteresis loop can be compared to the constant-amplitude strain-life data to determine the number of cycles to crack initiation (or failure), hence the fraction of damage. There are many variations in treating loop closures. One method of counting loop closures for a given spectrum is shown schematically in Fig. 3.50 and described below.

For each load cycle in the spectrum, the peaks and valleys are recorded in order of occurrence, so that sequence effects are preserved. Spectrum editing will not be needed as fatigue damage is calculated for each hysteresis loop closure, and the close-loop counting is an automatic part of the procedure. When loop closure occurs, the local stress and strain of the loop, calculated from Neuber's notch rule, are used to determine the number of cycles (or reversals) to crack initiation (or failure) from constant-amplitude strain-life data at the same local strain values.

Cycle counting begins at the first peak. Referring to Fig. 3.50, cycle counting starts with peak P_1. The next peak amplitude is compared to P_1. If it is greater than or equal to P_1 then the hysteresis loop in stress-strain space corresponding to the peak and valley P_1-V_1 is closed. Otherwise, it remains open and is examined for closure by subsequent peaks. Between peaks P_1 and P_2 are two complete cycles of the hysteresis loop corresponding to P_1-V_1-P_2, and they are counted as two cycles. Examining the next peak P_3, since $P_3 < P_2$, the loop corresponding to P_2-V_2 is still open. The next peak after P_3 closes the loop corresponding to P_3-V_3. Between peaks P_3 and P_4 are two cycles of P_3-V_3-P_3. However, the P_2-V_2 loop is still unclosed because the magnitude of P_4 is still lower than P_2. Examining P_5, since $P_5 > P_4$, another loop is closed, with three cycles of the type P_4-V_4-P_4. Finally, comparing P_5 to P_2, one finds that $P_2 = P_5$ and P_2-V_2-P_5 defines a closed loop. Notice that the peak P_5 actually has closed two loops. Also note that there are two closure points labeled as C_3 and C_4; these points indicate that closure occurs at the stress levels equal to the previous open peak levels while they are ascending to the next higher peak levels.

3.5.4 Summary

All the elements needed for strain-based fatigue life prediction have been described. The flowchart in Fig. 3.51 shows how all these elements are put together to perform a fatigue-life assessment analysis. Computer programs for tracking, and computation of local stresses and strains, and summation of fatigue damages are available in the public domain. The two computer codes given in Ref 3.50 to 3.52 are fundamentally similar except for some of their en-

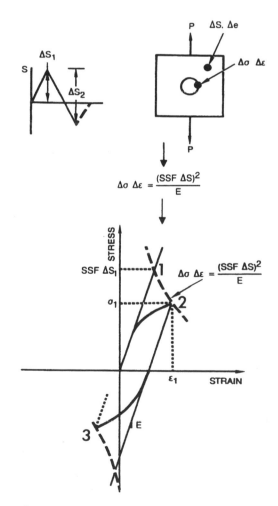

Fig. 3.49 Application of Neuber's rule. SSF, stress severity factor

gineering tools. For example, Potter uses an idealized elastic/perfectly-plastic cyclic stress-strain curve to compute the notch stresses and strains, whereas Porter's code uses a real cyclic stress-strain curve to monitor the local stresses and strains. The techniques used in these codes to account for the mean stress effects were different. For a detailed comparison of these two codes, see Ref 3.53.

3.6 Some Practical Fatigue Design Considerations

Figure 3.35 shows numerous material processing and design variables that can influence the fatigue performance of a final product. Many of these variables can be quantified and absorbed into the stress severity factor to help produce a more accurate fatigue life assessment. However, the bottom-line result for a finished product still depends on doing all that is necessary to improve fatigue quality. Many textbooks and reports have documented in detail the metallurgical aspects of fatigue, as well as the various conceivable sources of effects on fatigue performance. There is no need to cover all those subjects here.

One very clear fact concerning fatigue performance is the importance of the very local stress, its stress field or gradients, and its variation with time. For a given structural/machine element, the loads and environment that this piece of structure will experience are generally predetermined by the planned operational usage. From the stress engineer's or the designer's point of view, the fatigue life of an engineering structure to withstand a given loading spectrum will be controlled by the operational stress level and the quality of the detail design. Engineers are directly engaged in the detail design and the structure's subsequent manufacture. It is important to be able to visualize and identify the potential fatigue-critical areas in the structure and plan the design and manufacturing details. This section touches on several areas that deal with methods for reducing the chance of fatigue crack development.

3.6.1 Detail Design Considerations

The main thrust in improving fatigue performance of a part is reducing stress concentrations wherever possible. An obvious example is in Fig. 3.52, which simply shows that stress concentration can be reduced by increasing the curvature of the fillet. However, poor workmanship

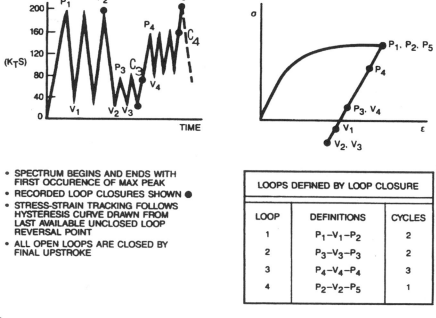

- SPECTRUM BEGINS AND ENDS WITH FIRST OCCURENCE OF MAX PEAK
- RECORDED LOOP CLOSURES SHOWN ●
- STRESS-STRAIN TRACKING FOLLOWS HYSTERESIS CURVE DRAWN FROM LAST AVAILABLE UNCLOSED LOOP REVERSAL POINT
- ALL OPEN LOOPS ARE CLOSED BY FINAL UPSTROKE

LOOPS DEFINED BY LOOP CLOSURE		
LOOP	DEFINITIONS	CYCLES
1	P_1–V_1–P_2	2
2	P_3–V_3–P_3	2
3	P_4–V_4–P_4	3
4	P_2–V_2–P_5	1

Fig. 3.50 Stress-strain loop closure tracking procedure. Courtesy of J.M. Waraniak

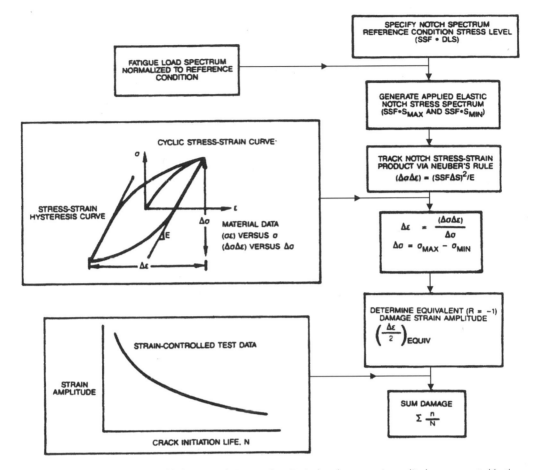

Fig. 3.51 Flow diagram of a strain-life fatigue analysis procedure. Equivalent damage strain amplitude was corrected for the mean stress effect.

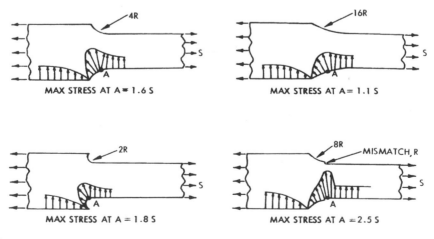

Fig. 3.52 Stress concentration at fillets

can produce adverse results. Reference books on the subject of fatigue contain many similar examples that more or less deal with structural members of monolithic geometry. With the aid of the modern computer technology, fatigue performance of complicated configurations in any type of structure can be analyzed. For design of complex structural assemblies, engineers still rely heavily on experience and structural testing. Several examples follow.

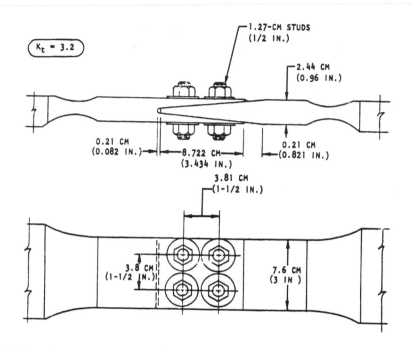

Fig. 3.53   Double-scarf joint specimen. Source: Ref 3.54

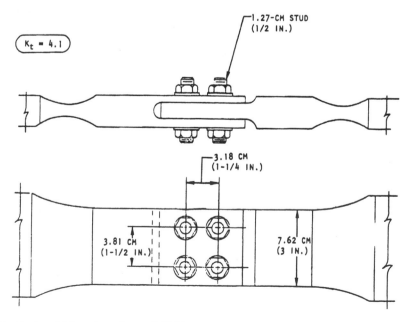

Fig. 3.54   Double-shear joint specimen. Source: Ref 3.54

The well-known tests of Hartman et al. (Ref 3.54, 3.55) are shown in Fig. 3.53 to 3.57, which present five joint configurations. The fatigue quality index (effective K_t) for these joints was determined from fatigue test results. The double-scarf joint specimen exhibited the best K_t value of 3.2, while the double-shear joint specimens ranged from 4.1 to 8.1, depending on the material. The worst case is the single-shear joint specimen, which exhibited a K_t value of 13.0. The better fatigue performance of the first three joint types is partly a result of configuration and

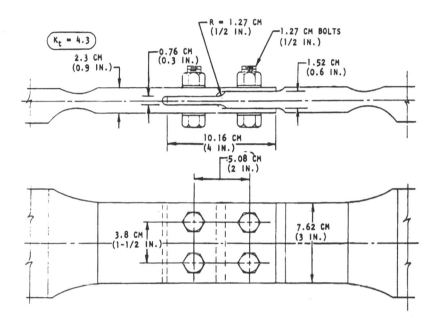

Fig. 3.55 Stepped double-shear joint specimen. Source: Ref 3.55

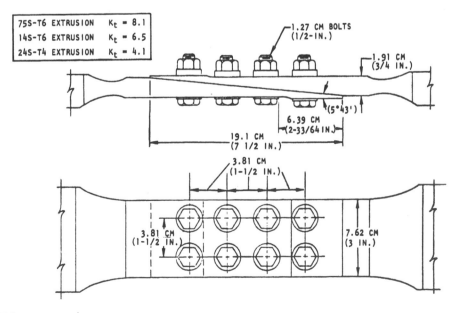

Fig. 3.56 Plain-scarf joint specimen. Source: Ref 3.54

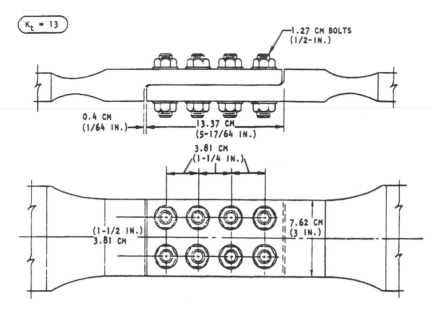

Fig. 3.57 Single-shear joint specimen. Source: Ref 3.54

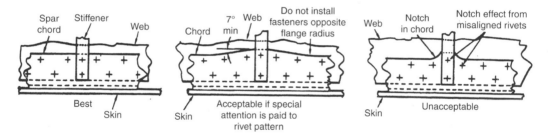

Fig. 3.58 Design of spar chord

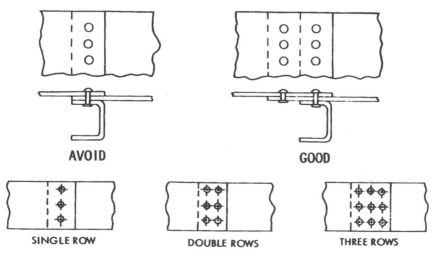

Fig. 3.59 Design of attachment joints

partly because these specimens contained fewer fasteners in a row.

Figure. 3.58 presents three designs for a spar chord that attaches the skin panel and the web. This example shows the importance of fastener pattern and notches in the sheet. Three degrees of success in these joints are cited: a best design, a conditionally acceptable design, and an unacceptable design. Figure 3.59 compares the fatigue performance of a sheet and various stringer attachments. The attachment design ranges from a single row of fasteners to three rows of fastener attachments. Based on fatigue test results, the three-row configuration is rated the best; the single-row configuration should be avoided.

3.6.2 Design against Fretting

Fretting is a type of wear that occurs at the contact area between two pieces of metal under load and subjected to oscillation at very small amplitude. In most assembled components and structures, fretting can occur at extremely small

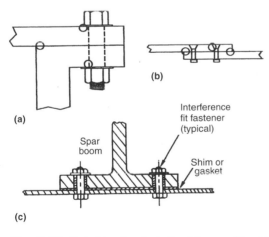

Fig. 3.60 Typical locations for fretting fatigue cracking. (a) Bolted flange. (b) Lap joint. (c) Interference-fit fastener, shims, or gaskets can reduce fretting.

relative amplitude ($<10^{-5}$ mm) of vibration (Ref 3.30). Fretting damage begins with local adhesion between mating surfaces. When adhered particles are removed from the surface, they may react with air or other corrosive environments. When corrosive agents are involved, the term "fretting corrosion fatigue" may be used. Affected surfaces show pits or grooves with surrounding products.

During fretting fatigue, cracks can initiate at very low stresses. In fatigue without fretting, the fatigue crack initiation time may represent 90% of the total component life. However, fretting fatigue can initiate a crack within the first several thousand cycles. Under fretting conditions, fatigue strength can be reduced by as much as 50 to 70% (Ref 3.56), or even 80% (Ref 3.30). Stress concentration is always involved. Some common fretting areas are shown in Fig. 3.60 to 3.62, where all the marked areas are sites of stress concentration. An interference-fit fastener is often used to eliminate fretting at fastener holes. A shim, gasket, or bushing can be used to reduce fretting at a tightly clamped area or where two pieces make a point contact. Thus, reduction of stress concentration is of primary concern. Table 3.1 lists recommended practices given in Ref 3.56.

3.6.3 Role of Residual Stresses

A brief overview of the mechanics of residual stress as it relates to static strength is given in Chapter 2. This section discusses fatigue aspects in the presence of residual stresses. Only a few essential points are addressed here, as comprehensive reviews of all aspects of this subject are readily available.

Stresses formed by quenching increase fatigue strength. Stresses derived from local deformations, particularly surface rolling, considerably reduce the injurious effects on the fatigue strength of notches, where corrosion or chafing

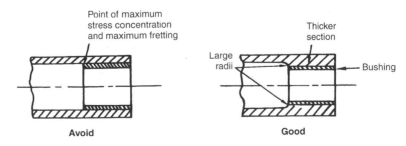

Fig. 3.61 Point of maximum stress concentration and maximum fretting in a pressurized tube fitting, with suggested possible fix

may be of concern. However, rolling should be done after heat treatment. Cold-worked holes benefit from negative residual stress, which increases the head room for yielding. Interference-fitted holes benefit from positive residual stress, which raises the level of minimum stress of a fatigue load cycle (constant or variable amplitude), which in turn closes up the range of ΔS.

Despite these benefits, residual stresses also can cause adverse effects. High compressive residual stresses generally are desirable on the surface of parts subjected to fatigue, stress corrosion, and fretting. However, these must be

balanced by tensile residual stresses within the part. In certain circumstances, these internal tensile residual stresses can be a problem. Figure 3.63 demonstrates how a layer of compressive residual stresses is formed on a part surface by shot peening. Needless to say, shot peening must be performed at the correct intensity in order to hold the tensile residual stress to an acceptably low level.

Another example is the residual stress in a surface-hardened part. The pattern of residual stress created by carburizing, nitriding, and so on is shown in Fig. 3.64. There is a layer of

Table 3.1 Reduction or elimination of fretting fatigue

Principle of abatement or mitigation	Practical method
Reduction in surface shear forces	Reduction in surface normal forces; reduction in coefficient of friction with coating or lubricants
Reduction/elimination of stress concentrations	Large radii; material removal (grooving); compliant spacers
Introduction of surface compressive stress	Shot or bead blasting; interference fit; nitriding/heat treatment
Elimination of relative motion	Increase in surface normal load; increase in coefficient of friction
Separation of surfaces	Rigid spacers; coatings; compliant spacers
Elimination of fretting condition	Drive oscillatory bearing; material removal from fretting contact (pin joints); separation of surfaces (compliant spacers)
Improved wear resistance	Surface hardening; ion implantation; soft coatings; slippery coatings
Reduction of corrosion	Anaerobic sealants; soft or anodic coatings

Source: Ref 3.56

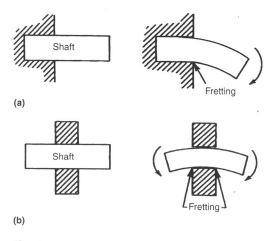

Fig. 3.62 Deformation of shaft/fixture indicating fretting locations. (a) Shrink or press-fit shaft. (b) Hub, wheel, or gear on a shaft

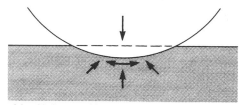

A hard ball pressed into a metal surface at point of greatest penetration. Note that the original surface (dashed line) is stretched (tension) into a spherical shape by the force on the ball. Radial reaction forces (below the ball) are compressive at this stage.

(a)

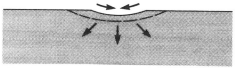

After the ball is removed, elastic recovery (or springback) causes a stress reversal: surface residual stresses in the cavity are now compressive, radial reaction forces are tensile.

(b)

Creation of numerous small indentations in surface, as by shot peening, forms a compressive residual stress barrier that resists cracking.

(c)

Fig. 3.63 Demonstration of the principle of mechanically induced residual stresses. (a) A hard ball pressed into a metal surface at the point of greatest penetration. Note that the original surface (dashed line) is stretched (tension) into a spherical shape by the force on the ball. Radial reaction forces (below the ball) are compressive at this stage. (b) After the ball is removed, elastic recovery (or springback) causes a stress reversal; surface residual stresses in the cavity are now compressive, and radial reaction forces are tensile. (c) Creation of numerous small indentations in the surface, as by shot peening, forms a compressive residual stress barrier that resists cracking. Source: Ref 3.15

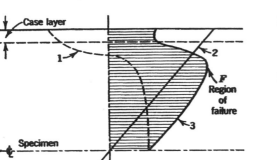

1—Residual stress due to case carburizing
2—Bending stress due to external loading
3—Resultant stress from 1 and 2

Fig. 3.64 Schematic stress distribution through the tension side of a case-carburized bending specimen

compressive stress on the part surface (over the case layer) and in the transition zone between the hardened and unhardened areas, along with some tensile residual stress deep inside the part to maintain equilibrium. When the part is subjected to bending fatigue, the bending stress superimposes with the residual stress; the resultant stress is shown in Fig. 3.64. Depending on the magnitudes of the applied stress and the residual stress, the point of maximum stress would have moved toward the mid-thickness of the plate, to point F, due to the presence of negative residual stress near the surface. Strictly from a mechanics point of view, the crack initiation site would be either on the surface or in the interior (i.e., point F), depending on which location has a higher stress. Usually, failure beneath the surface in surface-treated components is due to inadequate case depth. However, it should be noted that subsurface failure due to the stress distribution can occur without residual stresses. Point-loaded or line-loaded parts that involve contact pressure and friction, such as gears, ball bearings and their races, train wheels and rails, and so on, are good examples. Nonetheless, carburizing and nitriding produce a hard case on the part surface, thereby providing excellent wear resistance. However, the resulting fatigue performance must be watched.

REFERENCES

3.1. M.E. Fine and Y.-W. Chung, Fatigue Failure in Metals, *ASM Handbook*, Vol 19, *Fatigue and Fracture*, ASM International, 1996, p 63

3.2. J. Morrow, Cyclic Plastic Strain Energy and Fatigue of Metals, *Internal Friction, Damping, and Cyclic Plasticity*, STP 378, ASTM, 1965, p 45–87

3.3. R.W. Smith, M.H. Hirschberg, and S.S. Manson, "Fatigue Behavior of Materials Under Strain Cycling in Low and Intermediate Life Range," Report TN D-1574, NASA, April 1963

3.4. S.S. Manson and M.H. Hirschberg, Fatigue Behavior in Strain Cycling in the Low and Intermediate Cycle Range, *Fatigue—An Interdisciplinary Approach*, Syracuse University Press, 1964, p 133

3.5. F.R. Tuler and J. Morrow, "Cycle-Dependent Stress-Strain Behavior of Metals," TAM Report 239, University of Illinois, Urbana, March 1963

3.6. C.E. Feltner and R.W. Landgraf, Selecting Materials to Resist Low Cycle Fatigue, *J. Basic Eng. (Trans. ASME) D*, 1971, p 444–452

3.7. M.R. Mitchell, Fundamentals of Modern Fatigue Analysis for Design, *ASM Handbook*, Vol 19, *Fatigue and Fracture*, ASM International, 1996, p 227–249

3.8. F. Ellyin, *Fatigue Damage, Crack Growth and Life Prediction*, Chapman & Hall, 1997

3.9. R.W. Landgraf, J. Morrow, and T. Endo, Determination of the Cyclic Stress-Strain Curve, *J. Mater.*, Vol 4, 1969, p 176–188

3.10. W.A. Wood, *Bull. Inst. Met.*, Vol 3, 1955, p 5–6

3.11. A.H. Cottrell and D. Hull, *Proc. R. Soc. (London) A*, Vol 242, 1957, p 211–213

3.12. P. Neumann, *Acta Metall.*, Vol 17, 1969, p 1219

3.13. D.A. Ryder, "The Elements of Fractography," AGARDograph No. AGARD-AG-155-71, Advisory Group for Aerospace Research and Development of NATO, 1971

3.14. R.A. Lund and S. Sheybany, Fatigue Fracture Appearances, *ASM Handbook*, Vol 11, *Failure Analysis and Prevention*, ASM International, 2002, p 637

3.15. D.J. Wulpi, *Understanding How Components Fail*, ASM International, 1985

3.16. D.A. Meyn, *Trans. ASTM*, Vol 61 (No. 1), 1968, p 52

3.17. D.A. Meyn, *Metall. Trans.*, Vol 2, 1971, p 853

3.18. R.M.N. Pelloux, "An Analysis of Fracture Surfaces by Electron Microscopy," Report

D1-82-0169-R1, Boeing Scientific Research Laboratories, Dec 1963

3.19. R.M.N. Pelloux, *Met. Q.,* Nov 1965, p 34

3.20. J.M. Barsom and R.C. McNicol, Effect of Stress Concentration on Fatigue-Crack Initiation in HY-130 Steel, *Fracture Toughness and Slow-Stable Cracking,* STP 559, ASTM, 1974, p 183–204

3.21. Y.H. Kim, T. Mura, and M.E. Fine, Fatigue Crack Initiation and Microcrack Growth in 4140 Steel, *Metall. Trans. A,* Vol 9, 1978, p 1679–1683

3.22. R.G. Forman, "Study of Fatigue Crack Initiation from Flaws Using Fracture Mechanics Theory," Report AFFDL-TR-68-100, Air Force Flight Dynamics Laboratory, Sept 1968

3.23. P.G. Porter, private communication

3.24. N.E. Dowling, Notched Member Fatigue Life Predictions Combining Crack Initiation and Propagation, *Fatigue Eng. Mater. Struct.,* Vol 2, 1979, p 129–138

3.25. A.R. Jack and A.T. Price, The Initiation of Fatigue Cracks from Notches in Mild Steel Plates, *Int. J. Fract. Mech.,* Vol 6, 1970, p 401–409

3.26. W.G. Clark, Jr., Evaluation of the Fatigue Crack Initiation Properties of Type 403 Stainless Steel in Air and Steam Environments, *Fracture Toughness and Slow-Stable Cracking,* STP 559, ASTM, 1974, p 205–224

3.27. G.R. Halford, B.A. Lerch, and M.A. McGaw, Fatigue, Creep Fatigue, and Thermomechanical Fatigue Life Testing, *ASM Handbook,* Vol 8, *Mechanical Testing and Evaluation,* ASM International, 2000, p 686–716

3.28. D.F. Socie, J. Morrow, and W.C. Chen, A Procedure for Estimating the Total Fatigue Life of Notched and Cracked Members, *Eng. Fract. Mech.,* Vol 11, 1979, p 851–860

3.29. G.E. Dieter, Jr., *Mechanical Metallurgy,* McGraw-Hill, 1961

3.30. H.O. Fuchs and R.I. Stephens, *Metal Fatigue in Engineering,* John Wiley & Sons, 1980

3.31. D.K. Bullens, *Steel and Its Heat Treatment,* Vol 1, John Wiley & Sons, 1938, p 37

3.32. P.G. Forrest, *Fatigue of Metals,* Pergamon Press, 1962

3.33. D.J. McAdam, Corrosion Fatigue of Metals, *Trans. Am. Soc. Steel Treat.,* Vol 11, 1927, p 355

3.34. H.J. Grover, *Fatigue of Aircraft Structures,* NAVAIR 01-1A-13, U. S. Government Printing Office, 1966

3.35. "Metallic Materials and Elements for Aerospace Vehicle Structures," MIL-HDBK-5D, U.S. Department of Defense, May 1986

3.36. J.A. Bannantine, J.J. Comer, and J.L. Handrock, *Fundamentals of Metal Fatigue Analysis,* Prentice-Hall, 1990

3.37. A.F. Liu, unpublished data

3.38. V.M. Radhakrishnan, Life Prediction in Time Dependent Fatigue, *Advances in Life Prediction Methods,* ASME, 1983, p 143–150

3.39. S. Sehitoglu, Thermal and Thermalmechanical Fatigue of Structural Alloys, *ASM Handbook,* Vol 19, *Fatigue and Fracture,* ASM International, 1996, p 526–556

3.40. H. Neuber, Theory of Stress Concentration for Shear-Strained Prismatic Bodies with Arbitrary Nonlinear Stress-Strain Law, *J. Appl. Mech. (Trans. ASME) E,* Vol 28, 1961, p 544–550

3.41. P. Beardmore and C.E. Feltner, Cyclic Deformation and Fracture Characteristics of a Low Carbon Martensitic Steel, *Fracture 1969,* Chapman & Hall, 1969

3.42. M.R. Mitchell, *A Unified Predictive Technique for the Fatigue Resistance of Cast Ferrous-Based Metals and High Hardness Wrought Steels,* SP 442, SAE, 1979

3.43. J. Morrow, *Advances in Engineering,* Vol 4, *Fatigue Design Handbook,* SAE, 1968, p 21–29

3.44. S.S. Manson and G.R. Halford, Practical Implementation of the Double Linear Damage Rule and Damage Curve Approach for Treating Cumulative Fatigue Damage, *Int. J. Fract.,* Vol 17, 1981, p 169–172

3.45. K.N. Smith, P. Watson, and T.H. Topper, A Stress-Strain Function for the Fatigue of Metals, *J. Mater.,* Vol 5, 1970, p 767–778

3.46. N.E. Dowling, Estimating Fatigue Life, *ASM Handbook,* Vol 19, *Fatigue and Fracture,* ASM International, 1996, p 250–262

3.47. T.H. Topper, R.M. Wetzel, and J. Morrow, Neuber's Rule Applied to Fatigue of Notched Specimens, *J. Mater.,* Vol 4, 1969, p 200–209

3.48. R.E. Peterson, Notch-Sensitivity, *Metal*

Fatigue, G. Sines and J.L. Waisman, Ed., McGraw-Hill, 1959, p 293–306

3.49. R.E. Peterson, *Stress Concentration Factors,* John Wiley & Sons, 1974

3.50. J.M. Potter, The Effect of Load Interaction and Sequence on the Fatigue Behavior of Notched Coupons, *Cyclic Stress-Strain Behavior—Analysis, Experimentation, and Fatigue Prediction,* STP 519, ASTM, 1973, p 109–132

3.51. J.M. Potter and R.A. Noble, "A User's Manual for the Sequence Accountable Fatigue Analysis Computer Program," Report AFFDL-TR-74-23, Air Force Flight Dynamics Laboratory, 1974

3.52. P.G. Porter, "A Rapid Method to Predict Fatigue Crack Initiation, Vol. II, Computer Program User's Instructions," Report NADC-81010-60, Naval Air Development Center, 1983

3.53. P.G. Porter and A.F. Liu, "A Rapid Method to Predict Fatigue Crack Initiation, Vol. I, Technical Summary," Report NADC-81010-60, Naval Air Development Center, 1983

3.54. E.G. Hartman, M. Holt, and I.D. Eaton, "Static and Fatigue Strengths of High Strength Aluminum Alloy Bolted Joints," Report TN-2276, National Advisory Committee for Aeronautics, 1951

3.55. E.G. Hartman, M. Holt, and I.D. Eaton, "Additional Static and Fatigue Tests of High Strength Aluminum Alloy Bolted Joints," Report TN-3269, National Advisory Committee for Aeronautics, 1954

3.56. S.J. Shaffer and W.A. Glaeser, Fretting Fatigue, *ASM Handbook,* Vol 19, *Fatigue and Fracture,* ASM International, 1996, p 321–330

SELECTED REFERENCES

- W.E. Anderson, Fatigue of Aircraft Structures, *Int. Metall. Rev.,* Vol 17, 1972, p 240–263

- J. Bauschinger, On the Change of the Position of the Elastic Limit of Iron and Steel under Cyclic Variations of Stress, *Mitt. Mech. Tech. Lab.,* Vol 13, 1886, p 1

- N.E. Dowling, *Mechanical Behavior of Materials: Engineering Methods for Deformation, Fracture, and Fatigue,* 2nd ed., Prentice-Hall, 1999

- U.G. Goranson, J. Hall, J.R. Maclin, and R.T. Watanabe, Long-Life Damage Tolerant Jet Transport Structures, *Design of Fatigue and Fracture Resistant Structures,* STP 761, ASTM, 1982, p 47–90

- U.G. Goranson and J.T. Rogers, "Elements of Damage Tolerance Verification," presented at 12th Symposium of the International Committee on Aeronautical Fatigue (ICAF) (Toulouse, France), May 25–31, 1983

- L.E. Jarfall, "Optimum Design of Joints: The Stress Severity Factor Concept," presented at Fifth ICAF Symposium (Melbourne), May 1967

- M.A. Miner, Cumulative Damage in Fatigue, *Trans. ASME,* Vol 67, 1945, p A159

- A. Palmgren, Die Lebensdauer von Kugellagern, 2, *Verein Deutscher Ingenieure,* Vol 68, 1924, p 339–347

- J. Schijve, "The Accumulation of Fatigue Damage in Aircraft Materials and Structures," AGARDograph AGARD-AG-157, North Atlantic Treaty Organization, Jan 1972

- R.I. Stephens, A. Fatemi, R.R. Stephens, and H.O. Fuchs, *Metal Fatigue in Engineering,* 2nd ed., John Wiley & Sons, 2000

- S. Suresh, *Fatigue of Materials,* 2nd ed., Cambridge University Press, 1998

- R.M. Wetzel, Ed., *Advances in Engineering,* Vol 6, *Fatigue Under Complex Loading: Analysis and Experiments,* SAE, 1977

- A. Wohler, Tests to Determine the Forces Acting on Railway Carriage Axles and the Capacity of Resistance of the Axles, *Engineering,* Vol 11, 1871

CHAPTER 4

Static and Dynamic Fracture Toughness of Metals

FRACTURE TOUGHNESS was introduced in chapter 2 as the quality of a material that provides a means to measure the residual strength of a cracked body. Determination of residual strength (as opposed to the static strength of the same structural member without a crack) is very important in structural design and analysis.

Consider a plate loaded in tension (Fig. 4.1a). The axial directional stress acting at the cross section of an uncracked plate is equal to P/A, where P is the applied load and A is the cross-sectional area perpendicular to the applied load. Fracture of this plate occurs when P/A equals the material ultimate strength. If the plate contains a crack as shown in Fig. 4.1(b), it would normally be presumed that the applied load is supported by the remaining cross-sectional area. In other words, on the basis of conventional stress analysis, the plate will fail when the net section stress exceeds the material tensile ulti-

mate strength. For example, if the crack depth is one-half the original width of the plate, the fracture load will be $P/2$.

A graph depicting fracture stress as a function of crack size is presented in Fig. 4.2. In this figure, the stresses are computed using the applied load divided by the original (uncracked) cross-sectional area. The dotted line defines the load-carrying limit of the cracked plate, of which the net section stress of the plate is set to the tensile yield strength of the material. Note that net section yield stress (not ultimate) is customarily used in association with fracture strengths. Therefore, as shown in Fig. 4.2, the upper and lower limits of fracture strength are the tensile yield strength of the material (for an uncracked plate) and zero (when the crack size approaches the full width of the plate).

In reality, fracture testing of cracked plates has shown that, in general, fracture strengths are significantly below the net section yield limit (see Fig. 4.2). In other words, the remaining cross-sectional area of the cracked plate could not support the applied load as was assumed.

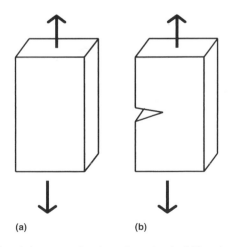

(a)   **(b)**

Fig. 4.1 Rectangular plate subjected to far-field tension. (a) Uncracked. (b) With a crack

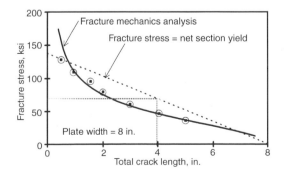

Fig. 4.2 Probable fracture strength of a cracked plate. The circles are presumed data points, not actual data.

Clearly, conventional stress analysis is not suitable for estimating the fracture strength of a cracked member. An approach that incorporates fracture toughness and stress analysis of cracks is needed. This technique is known as fracture mechanics. Currently, fracture mechanics is divided into three subcategories: linear elastic fracture mechanics, elastic-plastic fracture mechanics, and time-dependent fracture mechanics. The basic principles and applications of linear elastic fracture mechanics (LEFM) are dealt with in the remainder of this chapter and in Chapter 5. Nonlinear fracture mechanics, which includes the elastic-plastic and time-dependent methods, is discussed in Chapter 6.

Two terms frequently used throughout this book are "fracture toughness" and "residual strength." Fracture toughness is a material property. Residual strength is the fracture stress of a structural part. Like tensile strength, which can be used to calculate the fracture stress for an uncracked structural part, fracture toughness is used to calculate the part's residual strength. This chapter deals primarily with how to determine the various types of material fracture toughness. Chapter 5 then deals with how to determine structural residual strengths and the life of crack growth.

These concepts also lead to two other terms: "material allowable" and "structural allowable." Similar to tensile strength, fracture toughness is a material allowable. However, residual strength is a structural allowable. As previously demonstrated, the strength of a cracked part may be only a fraction (in magnitude) of a similar uncracked part. Therefore, "residual" can be regarded as "whatever is left" of a structure's load-carrying capability. Fracture stress (of an uncracked part) and residual strength (of a cracked part) are both design allowables.

4.1 Linear Elastic Fracture Mechanics

4.1.1 Griffith-Irwin Theory of Fracture

The Griffith crack theory was first published in 1920 (Ref 4.1) and subsequently modified in 1924 (Ref 4.2).* The basic idea of the modified Griffith theory is that at onset of unstable frac-

turing, one can equate the fracture work per unit crack extension to the rate of disappearance of strain energy from the surrounding elastically strained material (see Chapter 2, Eq 2.43). In other words, compare the work required to extend a crack (dW) with the release of stored elastic energy (dU_e) that accompanies crack extension, and each as associated with the increment of fracture area (dA). Thus, dU_e/dA equals dW/dA being the point of instability. That is, when dU_e/dA becomes slightly larger than dW/dA, rapid crack propagation will occur, driven by release of strain energy. Otherwise, no fracture will occur.

In ceramics, and especially in glasses, hardly any deformation accompanies fracture; this led to the surface-tension concept of brittle fracture. However, Griffith's surface-tension concept (which was validated by experimenting on glasses) is not suitable for metals, where plastic deformation always takes place. Both Orowan (Ref 4.3) and Irwin (Ref 4.4–4.7) independently concluded that the slight plastic flow that occurs in the brittle fracture zone absorbs a large amount of additional energy, and must be considered in determining the energy required to create a new fracture surface. But the Irwin version of the Griffith theory has the advantage over Orowan's version because it can be put to practical use. By applying the Griffith-Irwin theory of fracture, various methods can be established to predict the fracture stress of engineering structures in different shapes and/or under certain conditions by measuring the fracture toughness ($\mathcal{G}_C$). For this purpose, we must understand the stress field at the close neighborhood of the crack. For each case under consideration, it would be required to resolve the relationship between $\mathcal{G}_C$ and factors such as stress, specimen geometry, mechanical coefficients, and so on. These concepts are elaborated in this chapter and in Chapters 5 and 6.

The Irwin "strain energy release rate concept" was introduced in an attempt to explain the process of fracture. It encompasses the following definitions:

- *Strain energy release rate, $\mathcal{G}$,* is the quantity of stored elastic strain energy released from a cracking specimen as a result of extension of the advancing crack by a unit area (of crack surface just created). Actually, this energy rate can be regarded as composed of two terms: (1) The strain energy loss rate associated with extension of the fracture ac-

---

*A reprint of the 1920 paper appeared in the *Transitions Quarterly of the ASM* as part of the Metallurgical Classics Series in the late 1960s. It includes annotations showing the 1924 changes and commentaries of J.J. Gilman.

companied only by plastic strain local to the crack surface. This term is called $\mathcal{G}$. (2) The strain energy loss rate associated with non-recoverable displacements of the points of load application; however, this term is assumed zero.

- Fracture toughness, $\mathcal{G}_C$, is the component of work irreversibly absorbed in local plastic flow and cleavage surface tension to create a unit area of fracture.
- The condition for fracture is reached when the strain energy release rate $\mathcal{G}$ equals the fracture toughness $\mathcal{G}_C$.

It is assumed that any unit area of cleavage-type fracture surface of uniform appearance, which is produced at a given temperature in a specific material, requires the same amount of expended work $\mathcal{G}_C$. As will be shown later, it appears that $\mathcal{G}_C$ is a material property in the same sense as yield strength or tensile strength. Like the latter properties, $\mathcal{G}_C$ cannot be calculated on the basis of solid mechanics, but must be obtained experimentally. It is not yet clear just how complex a property $\mathcal{G}_C$ is. In order to obtain $\mathcal{G}_C$ experimentally, it is first necessary to obtain a mathematics expression for $\mathcal{G}$, in terms of crack dimensions, geometry of the test specimen, modulus of elasticity, Poisson's ratio, and the existing stress field. If the test specimen is then loaded, with the stress gradually increasing until a stress level is reached that results in fast propagation of the initial crack, $\mathcal{G}$ is obtained

from the applicable formula and equals the required experimental fracture toughness $\mathcal{G}_C$. In such manner $\mathcal{G}_C$ may be determined from a variety of test specimen configurations and test setups, provided that the test conditions result in brittle fracture of some sort.*

4.1.2 Fracture Indices and Failure Criteria

As discussed thus far, there is something called the strain energy release rate $\mathcal{G}$ operating in the system; $\mathcal{G}$ increases as the crack grows. The plate, or solid, will fail catastrophically when $\mathcal{G}$ reaches its critical value $\mathcal{G}_C$. It is possible to determine $\mathcal{G}_C$ by testing a specimen containing a crack (with an initial crack length of $2a_0$), and noting the stress level and crack length at failure, that is, S_c and $2a_c$ (see Fig. 4.3). Relating this to the Griffith solution, the strain energy release rate for each crack tip is:

$$\mathcal{G}_C = S_c^2 \pi a_c / E \qquad \text{(Eq 4.1a)}$$

for plane stress, and

$$\mathcal{G}_{IC} = S_c^2 \pi a_c (1 - v^2) / E \qquad \text{(Eq 4.1b)}$$

for plane strain. Also based on Irwin's calculation, $\mathcal{G}$ is directly related to the stress intensity factor K that defines the crack-tip stress and displacement fields. That is:

$$\mathcal{G} = K^2 / E \qquad \text{(Eq 4.1c)}$$

for plane stress, and

$$\mathcal{G}_I = K^2 \frac{(1 - v^2)}{E} \qquad \text{(Eq 4.1d)}$$

for plane strain.** Similarly, the critical value of K is regarded as the equivalent of the critical value of $\mathcal{G}$ (the material fracture toughness), that is:

$$\mathcal{G}_C = K_c^2 / E \qquad \text{(Eq 4.1e)}$$

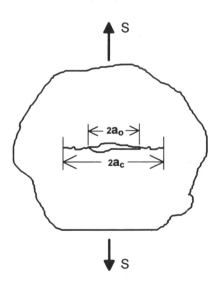

Fig. 4.3 Schematic view showing a through-thickness crack growing from an initial crack length to a critical crack length

---

*Here "some sort" means that a region of large plastic deformations may exist close to the crack but does not extend away from the crack by more than a small fraction of the crack length, as well as the physical dimensions of the part/test coupon. Discussion regarding this is presented later in this chapter in paragraphs that deal with plastic zone size and the ASTM requirements for validation of fracture toughness test results.

**The derivations of Eq 4.1(a) to (d) are omitted here because they can be found in many fracture mechanics books (e.g., Ref 4.8–4.10).

and

$$\mathcal{G}_{IC} = K_{IC}^2 \frac{(1 - v^2)}{E} \qquad \text{(Eq 4.1f)}$$

Comparing Eq 4.1(a) with 4.1(e) (and Eq 4.1b with 4.1f), it is seen that K is a function of stress and crack length. It is clear by now that the K and $\mathcal{G}$ relationship is the most important element in linear elastic fracture mechanics because, as will be evident later, K can be analyzed mathematically.

Chapter 1 discussed three different modes of crack-tip displacement. The notations for them include subscripts to K that are either a Roman numeral or an Arabic numeral. The Roman I and II designate the plane-strain condition. The Arabic 1, 2, and 3 are used for plane stress, or an unspecified state of stress (i.e., plane stress or plane strain, or somewhere in between the two). This notation system is directly applicable to the critical values of stress intensity factors in all three modes—that is, simply attach a second subscript "c" adjacent to the numeral subscript. The lowercase c and the capital C are used interchangeably. Again, a K-value (critical or otherwise) without a numeral subscript is mean for K_1. Incidentally, in this book, we will center our discussion on K_1, which is most common in structural failures. Limited discussions about mode 2 are given at the end of this chapter, and in Chapter 8.

In Chapter 1 it was also noted that there is no difference in the K-solution for the conditions of plane stress and plane strain. However, the critical stress intensity factor (i.e., the fracture index) for a material does vary with test specimen configuration and crack morphology. Among the two types of fracture indices, K_{IC} and K_C, the plane-strain fracture toughness K_{IC} is conceptually independent of structural geometry but normally associated with thick sections. However, as will be shown later in this chapter, the plane-stress fracture toughness K_C is a function of plate thickness and other geometric variables. Its values are obtained from testing of plate or sheet specimens of various thicknesses.

In structural analysis and fracture testing, K is expressed as a function of stress and crack size. More importantly, K can account for the influence of the details of structural geometry, crack morphology, and loading condition. Analytical and semiempirical tools are available for obtaining K-solutions. That is why we need to use K in lieu of $\mathcal{G}$. However, it is significant to note that K is really not a factor, although it is being used like a factor. It is a physical quantity with dimensional units of MPa$\sqrt{m}$ or ksi$\sqrt{in.}$. In this book, the term "stress intensity factor" is loosely defined; it might mean K, or a geometric factor such as those presented in Chapter 5. This is particularly true for K_C and K_{IC} (the critical stress intensity factors) because they are material properties. The remainder of this chapter discusses the various aspects of K_C and K_{IC}.

4.1.3 Role of Crack-Tip Plasticity

Consider the distribution of σ_y along $y = 0$ for a remotely loaded cracked sheet in a state of plane stress. The solid line in Fig. 4.4(a) was

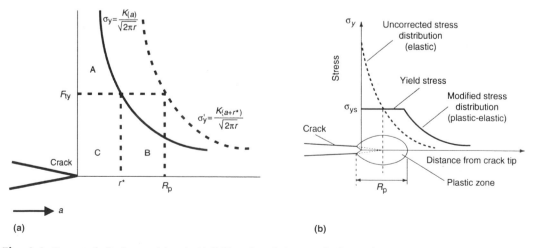

Fig. 4.4 Concept of effective crack length. (a) Shifting of crack-tip stress distribution due to material yielding at crack tip. (b) Interrelationship among elastic crack-tip stress distribution, shifted crack-tip stress distribution, and crack-tip plastic zone

obtained by using Eq 1.77 with $\theta = 0$. Examination of Eq 1.77 shows that $\sigma_y \to \infty$ when $r \to 0$. This is called the "stress singularity." Therefore, the material at the crack tip would have undergone plastic yielding at any applied stress levels. It seems logical to assume that the material in between the crack tip ($r = 0$) and some distance r^* is yielded, because σ_y over this area has exceeded the tensile yield strength of the material. Substituting F_{ty} (the material tensile yield strength) for σ_y, and r^* for r yields:*

$$F_{ty} = K_1/\sqrt{2\pi r^*} \qquad \text{(Eq 4.2a)}$$

or

$$r^* = \frac{1}{2\pi} (K_1/F_{ty})^2 \qquad \text{(Eq 4.2b)}$$

If we take r^* as the first estimate of the extent of the plastic zone, clearly the force produced by the stress acting over the length r^* will produce further yielding. For the purposes of maintaining equilibrium within the system, the whole curve for σ_y must be shifted (shown as the dotted line in Fig. 4.4a). In other words, area A must be balanced by area B. The area under the distribution curve of σ_y (i.e., area A plus area C) can be obtained by integrating σ_y from 0 to r^*. Since the size of area A is equal to that of area B, the area to be integrated is algebraically equal to area B plus area C:

$$\int_0^{r^*} \sigma_y \, dr = F_{ty}(r_p - r^*) + F_{ty} \cdot r^* \qquad \text{(Eq 4.3a)}$$

or

$$\int_0^{r^*} \frac{K_1}{\sqrt{2\pi r}} \, dr = F_{ty} \cdot r_p \qquad \text{(Eq 4.3b)}$$

After integration, we have:

$$\left(\frac{2}{\pi}\right)^{1/2} \cdot K_1 \sqrt{r^*} = F_{ty} \cdot r_p \qquad \text{(Eq 4.4a)}$$

$$\frac{2}{\pi} \left(\frac{K_1}{F_{ty}}\right)^2 \cdot r^* = r_p^2 \qquad \text{(Eq 4.4b)}$$

$$\frac{2}{\pi} \left(\frac{K_1}{F_{ty}}\right)^2 \cdot \frac{1}{2\pi} \left(\frac{K_1}{F_{ty}}\right)^2 = r_p^2 \qquad \text{(Eq 4.4c)}$$

*The material is assumed to be ideal elastoplastic with no strain hardening.

Therefore,

$$r_p = \frac{1}{\pi} \left(\frac{K_1}{F_{ty}}\right)^2 \qquad \text{(Eq 4.5)}$$

Comparing Eq 4.5 with Eq 4.2(b), we have:

$$r_p = 2r^* \qquad \text{(Eq 4.6)}$$

A schematic summary of the relationship between the crack-tip plastic zone and the modified crack-tip stress distribution is shown in Fig. 4.4(b).

From now on, r_p refers to the full plastic zone, or plastic zone width. If we imagine the crack-tip plastic zone as a circle, the radius of the circle will be $r_p/2$ ($= r^*$). The plastic zone radius is designated by r_y. The solution for r_y under plane strain has been derived by Irwin (Ref 4.11) as:

$$r_y^2 = \frac{1}{4\sqrt{2}\pi} \left(\frac{K_1}{F_{ty}}\right)^2 \cong \frac{1}{6\pi} \left(\frac{K_1}{F_{ty}}\right)^2 \qquad \text{(Eq 4.7)}$$

Therefore, the plane-strain plastic zone radius is equal to one-third of the plane-stress plastic zone radius. Similarly, the plane-strain plastic full zone width is equal to one-third of the plane-stress full zone.

Effective Crack Length. The shifting of σ_y to σ_y' (the dotted line in Fig. 4.4a) implies that the K-factor associated with this crack (of length $2a$) has become a function of ($a + r_y$). Thus, it can be said that the crack behaves as if it were of length ($a + r_y$). The quantity ($a + r_y$) is called the effective crack length. It is academically correct to use the effective crack length ($a + r_y$) in place of the physical crack length a to compute K. This new K is called the K_{eff}. However, in the above equations, the crack-tip plastic zone size (r_p or r_y) has been given as functions of K and F_{ty}. Now K is also a function of the effective crack length ($a + r_y$). Clearly, implementation of the effective crack length approach to fatigue crack growth prediction is difficult and cumbersome.

Other than making an adjustment to the effective crack length (from a physical crack length), there are two additional uses for the crack-tip plastic zone. The first is its contribution to the special meaning of plane stress and plane strain in fracture mechanics terminology. The second is its role in governing the behavior of fatigue crack propagation. These two uses are discussed in the next section and in Chapter 5, respectively.

Other Analytical and Physical Aspects of the Crack-Tip Plastic Zone. Research papers often depict the crack-tip plastic zone r^* shaped like a butterfly (Fig. 4.5). These are stress (or strain) contour maps computed on the basis of the von Mises yield criterion (i.e., Eq 1.27a to 1.28b). Combining Eq 1.77 with Eq 1.28(b), the resulting elastic-plastic boundary is given by:

$$r^* = \frac{1}{2\pi}\left(\frac{K_1}{F_{ty}}\right)^2 \cos^2\left(\frac{\theta}{2}\right)\left[1 + 3\sin^2\left(\frac{\theta}{2}\right)\right] \quad \text{(Eq 4.8)}$$

for plane stress. At $\theta = 0$, Eq 4.8 reduces to Eq 4.2(b). Therefore, the plastic zone contour maps are actually plotting r_y, not r_p.

It should be noted that the plastic zones just described are derived from the local stress distribution at the crack tip and are commonly referred to as the Irwin plastic zone. Other approaches have also been used to derive crack-tip plastic zone expressions; the most notable is the Dugdale plastic zone (Ref 4.12), and for plane stress its width equals:

$$\varpi = a\left[\sec\left(\frac{\pi S_y}{2F_{ty}}\right) - 1\right] \quad \text{(Eq 4.9)}$$

For small-scale yielding, this leads to

$$r_p = \frac{\pi}{8}\left(\frac{K_1}{F_{ty}}\right)^2 \quad \text{(Eq 4.10)}$$

and is known as the Rice plastic zone (Ref 4.13). The Dugdale zone represents a product of large-scale yielding that results from high stress levels. Physically, the Rice plastic zone is another form of the Irwin plane stress plastic zone, but approximately 23% larger. The Dugdale zone width is shaped like a long, thin strip in front of the crack tip, whereas the Irwin or Rice plastic zone is circular.

As mentioned in Chapters 1 and 2, yielding of a structural piece can be determined by elastic-plastic finite element analysis. Using the NASTRAN code and the finite element model of a cruciform specimen (Fig. 1.39), crack-tip plastic zone sizes have been determined for two aluminum alloys (7075-T7351 and 2024-T351) for crack lengths ($2a$) ranging from 13 to 50 mm (0.5 to 2 in.). Actual stress-strain curves of these alloys were used to handle crack-tip plasticity. The specimen models were subjected to a uniaxial tension stress ranging from 165 to 207 MPa (24 to 30 ksi). For data reduction, the dimension r^{fe} is defined as the largest distance between the crack tip and the tip of the supposedly butterfly-shaped plastic zone. Then an equivalent dimension r^{**} representing the diameter of a circle is computed. The area of such an imaginary circle is set equal to the total area of the butterfly. The results are presented in Table 4.1. Also in Table 4.1 are the plastic zone sizes computed by using Eq 4.5 (the Irwin full zone for plane stress) and Eq 4.10 (the Rice plastic zone), for comparison. It seems that the tip-to-tip dimension, r^{fe}, correlates better with the Irwin full zone size r_p, whereas the equivalent diameter, r^{**}, correlates better with the Rice plane stress zone size.

Crack-tip plastic zone sizes and shapes have been the subject of many experimental investigations. Etching of a metallographic sample is commonly used to reveal the plastic zone shape (under an optical microscope) and for quantitative measurement of its size. Other techniques include photostress and interferometry photography. Under normal circumstances, reliable results are difficult to obtain while the specimen is under load, due to testing machine vibration and so forth. Measuring the plastic zone from an unloaded specimen would not correlate to the loaded specimen. Regarding shape, the results of Hahn et al. (Ref 4.15, 4.16) show a basic difference between the shapes of the plane-stress and plane-strain plastic zones. The material's

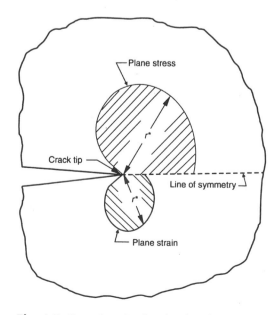

Fig. 4.5 Comparison of analytical mode 1 plastic zone sizes (plane stress versus plane strain)

Table 4.1 Comparison of classical and finite element solutions of crack-tip plastic zones

a		S		r^{fe}		r^{**}		$(K_{max}/F_{ty})^2$		Irwin (full zone)		Rice	
mm	in.	MPa	ksi	mm	in.	mm	in.	mm	in.	mm	in.	mm	in.
7075-T7351, cruciform specimen, $S_x = 0$													
6.4	0.25	207	30	1.3	0.05	1.73	0.068	4.97	0.196	1.58	0.0624	1.96	0.077
12.7	0.5	207	30	3.86	0.152	4.72	0.186	9.98	0.393	3.18	0.1251	3.92	0.1543
25.4	1	207	30	6.9	0.27	9.75	0.384	19.9	0.785	6.4	0.25	7.83	0.3083
25.4	1	173.8	25.2	4.6	0.181	5.94	0.234	14.07	0.554	4.48	0.1763	5.53	0.2176
2024-T351, cruciform specimen, $S_x = 0$													
6.4	0.25	207	30	2.8	0.11	4.37	0.172	7.19	0.283	2.3	0.09	2.82	0.1111
12.7	0.5	207	30	4.8	0.19	6.2	0.244	14.35	0.565	4.52	0.178	5.64	0.2219
12.7	0.5	165.5	24	4.13	0.1625	4.8	0.19	9.19	0.362	2.93	0.1152	3.61	0.1422
25.4	1	207	30	10.49	0.413	14.27	0.562	28.73	1.131	9.1	0.36	11.28	0.4441

Note: r^{fe} = maximum length of the plastic zone determined by finite element; r^{**} = diameter of a circle having an area equivalent to the total area of the finite element determined plastic zone. Source: Ref 4.14

strain-hardening exponent further influences the shape of the plastic zone of a given type (Fig. 4.6). For a nearly-elastic/perfectly-plastic material (i.e., the strain-hardening exponent $n \approx 0$), the plane-stress plastic zone becomes a very thin strip, resembling the shape of the Dugdale plastic zone (Fig. 4.7).

The foregoing discussion has undoubtedly complicated the general perspective of the crack-tip plastic zone. However, it helps to explain why it is often difficult to match field service life or experimental results with analytical prediction.

4.1.4 Ductile and Brittle Fracture

The state of stress throughout most bodies falls in between the limits defined by the states

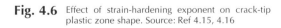

Fig. 4.6 Effect of strain-hardening exponent on crack-tip plastic zone shape. Source: Ref 4.15, 4.16

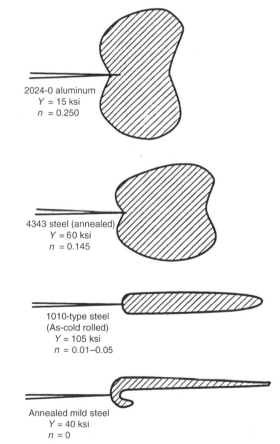

Fig. 4.7 Effect of strain-hardening exponent on crack-tip plastic zone shape for three low-alloy steels and one aluminum. Source: Ref 4.15, 4.16

of plane stress and plane strain. In the theory of elasticity, a state of plane stress exists when $\sigma_z = \tau_{xz} = \tau_{yz} = 0$. This two-dimensional state of stress is frequently assumed in practice when one of the dimensions of the body is small relative to the others. In a thin sheet loaded in the plane of the sheet (the X-Y plane), there will be virtually no stress acting perpendicular to the sheet. The remaining stress system will consist of two normal stresses, σ_x and σ_y, and a shear stress, τ_{xy}. The term "generalized plane stress" applies to cases of deformation where the through-thickness of a thin plate is subjected to external forces. Often when the term "plane stress" is used, generalized plane stress is actually implied. Similarly, plane strain is mathematically defined as $\varepsilon_z = \gamma_{xz} = \gamma_{yz} = 0$, which ensures that $\tau_{xz} = \tau_{yz} = 0$ and $\sigma_z = \nu(\sigma_x + \sigma_y)$. In such a system, all displacements can be considered limited to the X-Y plane (i.e., $w = 0$) so that strains in the Z-direction can be neglected in the analysis. Neither total plane stress nor total plane strain conditions are found in real structural configurations. However, in stress field analysis, employment of either of these constraints allows for two-dimensional solutions of three-dimensional problems, which provides the foundation for linear elastic fracture mechanics theory as it is applied today.

In fracture mechanics terminology, the terms plane stress and plane strain take on special, more restricted meanings. Instead of characterizing stress and strain states throughout a body, in fracture mechanics special concern is given to the crack tip and surrounding region. Because a bulk of material tends to deform in all directions, to develop a plane strain condition it is necessary to constrain the flow in one direction. In a region near the leading edge of a crack, the magnitude of stresses and strains is very high compared to those at some relatively greater distance away. Because of this high degree of tensile deformation, the material near the crack front tends to shrink (a Poisson's ratio effect) in a direction parallel to the leading edge. However, this shrinkage is constrained by the surrounding material, which is less deformed. If the leading-edge length is long compared to other dimensions of the plastically deformed zone, then it is highly constrained against shrinkage parallel to the leading edge. This is considered a state of localized plane strain. Conversely, a plane stress state exists if the plastic zone dimensions are large compared to the leading-edge length of the crack. By viewing the plastic zone

as a short cylinder with free ends it is bound to be relatively unconstrained or in a state of plane stress. Naturally, plane stress conditions would still exist at free surfaces of a thick plate containing a flaw, while localized plane strain conditions would prevail in the interior of the plate (Fig. 4.8). These considerations are equally relevant for three-dimensional cracks with curved crack fronts. Therefore, a part-through crack is usually considered to be in the state of plane strain, whereas a through-thickness crack is in the state of generalized plane stress unless the plate is very thick. Needless to say, a much thicker plate is required for a ductile material than for a not so ductile material. The proposed test procedure for obtaining a valid plane-strain fracture toughness value is given in ASTM Standard E 399 (Ref 4.17). All the details relating to specimen size and preparation requirements, test setup, data reduction method, and so on, are specified in this document.

A K_{IC} failure is considered brittle, whereas a K_C failure is somewhat ductile. To put it the other way, a K_{IC} value is used to determine the residual strength of a structural member containing a part-through crack or any geometry that is in a situation of localized plane strain. A K_C value would be used for determining the residual strength of a cracked sheet or plate, which is presumably in a state of generalized plane stress.

In order to understand the reasons for this apparently complex failure mechanism (i.e., plane stress versus plane strain), consider two small rectangular elements of material in a moderately thick plate, close to the crack tip. One of them (element A in Fig. 4.9) is on the free surface of

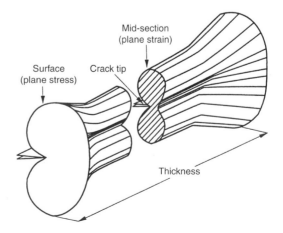

Fig. 4.8 Variation of crack-tip plastic zone through the thickness of a plate

the plate normal to the crack front. The other (element B) is at a similar position relative to the crack, but in the midthickness of the plate. If the test panel is thick enough to produce sufficient constraint, a plane strain condition exists in the interior. However, the condition of plane stress will still prevail at the surface, because it is free of stresses. As the remote loading increases, each of these elements will fail at some particular load level, either by shear or by tensile separation. A shear failure means one atomic plane is sliding over another and the mechanism of failure is governed by the von Mises or some other yield criterion. Tensile separation failure is direct separation of one plane of atoms from another that is caused by loading normal to the eventual fracture surface. Under pure hydrostatic loading, plastic deformation cannot occur in a material that can maintain a constant volume; therefore, a hydrostatic state of stress ($\sigma_x = \sigma_y = \sigma_z$) cannot produce a ductile fracture in metals. Element B (which is near-hydrostatic) may fail by tensile separation in the plane of the crack before it is able to achieve a critical shear stress level (a stress level that will cause ductile failure of element A). Thus we associate slant fracture with incomplete constraint and describe

its visual appearance as ductile, and flat fracture as macroscale brittle.

Fracture appearances are indicated by the percentage of shear lips visible on the rapid-growth portion of the fracture surfaces. A plane-strain state causes a flat fracture appearance. The fracture plane is perpendicular to the loading direction and is called the tensile mode failure. The plane-stress state causes a slant fracture appearance (having single or double shear lips 45° to the loading direction) and is called the shear mode failure. A combined state of plane strain and plane stress results in a mixed failure mode (Fig. 4.10). It is also evident that the tensile mode of failure is usually associated with a shorter crack or a lower K-level, whereas the shear mode occurs at higher K-levels or is associated with a longer crack. Refer back to the scenario for tensile separation and shear fracture discussed earlier. Should the applied load continue to increase, the plastic zone ahead of the crack tip will also become larger as the crack grows, allowing less through-thickness restraint on internal elements. Under these conditions, the proportion of flat fracture surface reduces as the crack extends and the proportion of shear lip increases. It might even approach plane stress right through the thickness. The resulting failure is illustrated in Fig. 4.10.

4.2 Plane-Strain Fracture Toughness: Static K_{IC}

In a given condition and at a given temperature and rate of loading, the plane-strain fracture

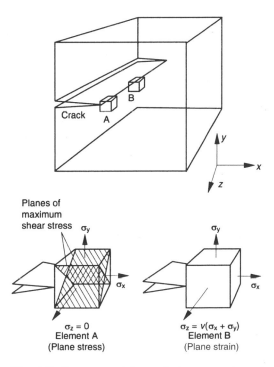

Fig. 4.9 Plane stress and plane strain elements near the crack tip

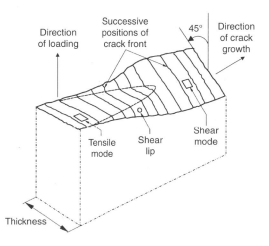

Fig. 4.10 Fatigue crack surface in a sheet (or plate), showing the transition from a tensile mode to a shear mode

toughness K_{IC} represents a practical lower limit to the fracture toughness of a material. From the point of view of having a single value representing the fracture toughness of a material, K_{IC} is conceptually independent of the specimen dimensions. However, this is true provided that the specimen is sufficiently large for a proper K_{IC} measurement, or that there is appreciable constraint in the crack front to create a plane-strain condition. Of course, materials exhibit nonuniformity and anisotropy with respect to K_{IC}, just as they do for other properties, and this must be taken into consideration in material evaluation. The most obvious way to measure K_{IC} would be to test a sufficiently thick plate specimen. This may not always be convenient, or even possible, and certainly noneconomical. In any event, the plane-strain fracture index K_{IC} is associated with a flat fracture appearance without appreciable amount of crack growth prior to fracture. K_{IC} is regarded as the minimum value of K_C and is an intrinsic material mechanical property.

Several test specimen configurations are available for obtaining plane-strain fracture toughness. The ASTM standards specify the specimen configuration and size, test setup, loading requirements, and data interpretation techniques (Ref 4.17). Stress intensity solutions for these specimens are also specified. If these K-solutions were correctly derived, and the fracture stress for a test is below 80% of the material tensile yield strength (the ASTM criterion for validation), the K_{IC} value for a given thickness and product form should be the same, regardless of which specimen type is used.

The problem in K_{IC} testing is to meet the valid K_{IC} requirement imposed by ASTM. To ensure a plane-strain condition, ASTM E 399 (Ref 4.17) specifies that both specimen thickness and crack length must meet a minimum size requirement. The following conditions:

$$B > 2.5 \cdot (K_{IC}/F_{ty})^2 \qquad \text{(Eq 4.11a)}$$

and

$$a > 2.5 \cdot (K_{IC}/F_{ty})^2 \qquad \text{(Eq 4.11b)}$$

must be met in order to obtain consistent K_{IC} values. Here B is the ASTM designation for specimen thickness. In this book, both t and B are used for thickness. The term $(K_{IC}/F_{ty})^2$ suggests that the required plate thickness B and crack length a are related to some measure of

crack-tip plastic zone size. Taking $r_p = (K_{IC}/F_{ty})^2/3\pi$, it is seen that the criteria for plane strain reflect a condition where B and a are both greater than, or equal to, $24r_p$.

To arrive at a valid K_{IC}, it is necessary to first compute a tentative value called K_Q based on a graphical construction of the load-displacement test record. If K_Q satisfies the conditions of Eq 4.11(a) and (b)—that is, if the quantity $(K_Q/F_{ty})^2$ is less than $0.4B$ (and also less than $0.4a$), then $K_Q = K_{IC}$. The graphical procedure for obtaining a K_Q value will not be discussed here because it is specified in ASTM E 399 (Ref 4.17). If the test result fails to satisfy Eq 4.11(a) and (b) and/or other requirements specified in ASTM E 399, it will be necessary to use a larger specimen to determine K_{IC}. For some relatively ductile materials, it will be very difficult to establish a valid K_{IC} value. One may have to compromise by using a K_Q value instead.

As an alternative to accepting K_Q as fracture toughness, ASTM provides another testing standard (E 992) for test data interpretation. The rationale is based on the "equivalent energy" concept, which has been discussed earlier in the section "Material Toughness" in Chapter 2. Test results that fail to meet the validity criteria of E 399 can be reinterpreted using the methods provided by E 992. Basically, a fictitious "maximum load," P_E, is located by extending the linear portion of the load-displacement curve until the area under the linear portion equals the area to the maximum load of the test. Then a fracture toughness value, $K\text{-}EE$, is recalculated by using P_E. In addition to reinterpreting E 399 test results, this method can be used to run tests using subsize specimens machined from materials of limited thickness and availability. The specimen geometries and test methods are the same as E 399, except that the dimensional restrictions are relaxed and the data reduction procedure is modified.

4.2.1 Pop-in K_{IC}

During monotonic loading, some crack extension may take place prior to rapid fracture. For some materials, the onset of slow crack growth occurs in a discontinuous manner termed "pop-in" failure. The pop-in load is identifiable in the load-displacement record of a fracture test. In addition, an audible sound can be heard during pop-in. A sudden crack extension associated with a slight load drop is usually observed during pop-in. After pop-in, the applied load is further increased until final fracture occurs at the expected fracture load. Occasionally, multiple pop-in can occur in a single test. Pop-in can occur in sheets or plates of any thickness. The crack growth process initiates from a local area in the midthickness region of the plate, and thus is plane strain in nature.

Experimental tests show that pop-in K_C values are relatively constant regardless of panel thickness and are equivalent to K_{IC} values obtained from thick panels. Or, more appropriately, their values are more or less equivalent to K_Q. Therefore, it seems logical to assume that it is economical to generate K_{IC} values by using thin sheet specimens because of lower material and fabrication costs, as well as a smaller testing machine. However, there is one problem: It is not known under what conditions pop-in will occur. Pop-in just happens during a test; its occurrence cannot be prearranged. In other words, pop-in does not reliably determine K_{IC}.

4.2.2 K_{IE} for Surface Flaws

For a partially exposed flaw such as a surface flaw or part-through crack (Fig. 4.11a and c) and a fully embedded flaw (Fig. 4.11b), the constraint at the leading border of the crack is very

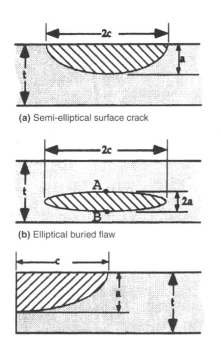

(a) Semi-elliptical surface crack

(b) Elliptical buried flaw

(c) Quarter-elliptical corner crack

Fig. 4.11 Three-dimensional cracks. (a and c) Partially exposed. (b) Fully embedded

high and the mechanism of crack propagation is controlled by the plane-strain condition. Therefore, the K_{IC} value customarily is used to predict the residual strength for a surface flaw. Incidentally, NASA is using an alternate fracture toughness value for routine engineering tasks in the Space Shuttle program. This fracture toughness value, K_{IE}, was specifically developed from fracture testing of a surface flaw specimen, because NASA workers found that the fracture behavior of the surface flaw is not entirely the same as K_{IC}. After testing a number of engineering alloys, including aluminum, titanium, steel, Inconel, magnesium, and beryllium-copper, Forman et al. (Ref 4.18)* found that K_{IE} is related to K_{IC} in the following manner:

$$K_{IE} = K_{IC} \cdot (1 + C_k \cdot K_{IC}/F_{ty}) \qquad \text{(Eq 4.12)}$$

where C_k is an empirical constant whose value equals $0.1984\sqrt{mm}$ ($1.0\sqrt{in.}$). However, the relationship of Eq 4.12 is not applicable to very ductile materials (which have a very high K_{IC}/F_{ty} ratio). The limitation specified by Forman et al. is $K_{IE} \leq 1.4 K_{IC}$.

When the crack depth for a surface flaw is large compared to the specimen thickness, the crack front constraint may be significantly diminished and the stress system could become plane stress. In Fig. 4.11(b), for example, point A in the center of the fully embedded elliptical flaw would have a higher constraint compared to point B, because point B is very close to the free surface of the plate. The points on or near the free surface of the part-through flaw would be nearly plane stress instead of plane strain. However, fracture of the part-through flaw is still initiated at the point of the highest constraint and thus controlled by K_{IC}. If the operating stress level is quite low (for the case of fatigue cycling), then the crack might grow through the thickness before rapid fracture. Once the crack has grown through the thickness, the crack propagation behavior is the same as that of a through-thickness crack. Whether it should be called plane-strain or plane-stress failure depends on the conditions.

4.2.3 Metallurgical Variables

Microstructural variables influence the mechanical properties of materials. In a review pa-

---

*The NASA computer code contains built-in material property values needed for space programs.

per, Ravichandran and Vasudevan (Ref 4.19) identified all the constituents in steels and in aluminum and titanium alloys, and determined how they affect the K_{IC} properties of these alloys. Their results are tabulated in Tables 4.2 to 4.4.

Heat treatment or other mechanical (or thermomechanical) means are known to cause strength variations in a given alloy. The fracture toughness K_C and K_{IC} of a given alloy are generally found to be inversely proportional to its tensile ultimate strength or tensile yield strength, at least for structural grades and high-strength grades of materials commonly used in engineering service because of their load-bearing capability. Among the various transformation products, tempered martensite exhibits the highest toughness, followed by bainite, followed by ferrite-pearlite structures. A typical example show-

ing K_{IC} as a function of tensile yield strength for steels is given in Fig. 4.12. Figure 4.13 shows a broad range of K_{IC} data as a function of tensile yield strength for the 2000 series and 7000 series aluminum alloys. It is also shown that the K_{IC} values are further affected by aging temperature. That is, an overaged alloy generally has a lower K_{IC} value when compared with an underaged alloy of the same yield strength. For a given alloy one might find a temper that can optimize the strength and toughness. Data of this sort for specific alloys can be found in Appendix 6.

Heat treatment alone may also affect fracture toughness values of an alloy, even if the heat treatments results in no change in tensile strength. To explain what this means, some data for D6AC steel are presented in Tables 4.5 and 4.6 and Fig. 4.14. These data indicate that the alloy underwent several different heat treatments but yielded similar tensile strengths. However, depending on heat treatment, its fracture toughness could have been different by a factor of two. These data were reported as K_{IC}, but their values still depend on thickness.

Another example cited here refers to the effect of constituent contents on fracture toughness. An experiment conducted on a high-strength steel, Fe-0.65Mn-0.35Si-0.8Cr-0.3Mo-0.1V (wt%), kept the carbon content constant at 0.35% in one case (Fig. 4.15a) and the nickel content at a level of 3% in the second case (Fig. 4.15b). By holding the amount of one element constant and varying the second element by a small amount, plane-strain fracture toughness changed appreciably despite the fact that all alloys had been heat treated to the same tensile yield strength.

The plane-strain fracture toughness for titanium can be improved by controlling the two

Table 4.2 Effects of microstructural variables on fracture toughness of steels

Microstructural parameter	Effect on toughness
Grain size	Increase in grain size increases K_{IC} in austenitic and ferritic steels
Unalloyed retained austenite	Marginal increase in K_{IC} by crack blunting
Alloyed retained austenite	Significant increase in K_{IC} by transformation-induced toughening
Interlath and intralath carbides	Decrease K_{IC} by increasing the tendency to cleave
Impurities (P, S, As, Sn)	Decrease K_{IC} by temper embrittlement
Sulfide inclusions and coarse carbides	Decrease K_{IC} by promoting crack or void nucleation
High carbon content (>0.25%)	Decrease K_{IC} by easily nucleating cleavage
Twinned martensite	Decrease K_{IC} due to brittleness
Martensite content in quenched steels	Increase K_{IC}
Ferrite and pearlite in quenched steels	Decrease K_{IC} of martensitic steels

Source: Ref 4.19

Table 4.3 Effects of processing/microstructural variables on fracture toughness of aluminum alloys

Variable	Effect on fracture toughness
Quench rate	Decrease in K_{IC} at low quench rates
Impurities (Fe, Si, Mn, Cr)	Decrease in K_{IC} with high levels of these elements
Grain size	Decrease in K_{IC} at large grain sizes
Grain-boundary precipitates	Increase in size and area fraction decrease K_{IC}
Underaging	Increases fracture toughness
Peak aging	Increases fracture toughness
Overaging	Decreases fracture toughness
Grain-boundary segregates (Na, K, S, H)	Lower fracture toughness in Al-Li alloys

Source: Ref 4.19

Table 4.4 Effects of microstructural variables on fracture toughness of titanium alloys

Variable	Effect on fracture toughness
Interstitials (O, H, C, N)	Decrease in K_{IC}
Grain size	Increase in grain size decreases K_{IC}
Lamellar colony size	Increase in colony size increases K_{IC}
β phase	Increases in β volume fraction, continuity increase K_{IC}
Grain boundary α phase	Increases in thickness and continuity increase K_{IC}
Shape of α phase	Increase in aspect ratio of α phase increases K_{IC}
Orientation	Crack oriented for easy cleavage along basal planes gives low K_{IC}

Source: Ref 4.19

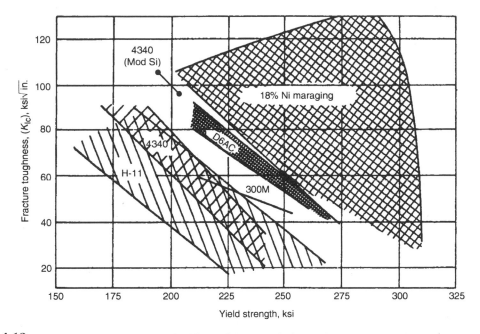

Fig. 4.12 Fracture toughness as a function of yield strength for various high-strength steels. Source: Ref 4.20

phases of crystal structure, that is, primary hexagonal close-packed (hcp) α and body-centered cubic (bcc) β. Of the two phases, β is more ductile and is preferable for increasing the fracture toughness of titanium alloys. The three broad classes of titanium alloys are near-α, α + β, and β alloys, grouped according to the level of α- or β-stabilizing elements. Typically, β content by volume is near-α, <10%; α + β, 10 to 25%; and β, >25%. Figure 4.16 shows a fracture toughness/strength relationship map for many types of titanium alloys. The metastable β alloys possess the highest combination of strength and toughness. This arises from a large volume fraction of β phase and fine aged-α precipitates. Like aluminum, of which the high-purity grade 2000, 6000, and 7000 series alloys have higher K_{IC} values,* titanium alloys can have higher K_{IC} values by keeping the interstitial at an extralow level. Better fracture toughness properties are expected in those alloys designated as extralow interstitial (ELI). For example, K_{IC} values of the Ti-6Al-4V alloy are appreciably higher in those having low oxygen or hydrogen contents (Ref 4.19).

Correlating toughness to strength is desirable. So far, some metallurgical elements that affect

alloy toughness and strength have been identified. Intuitively, these properties are controlled by microstructure, which is a result of alloying elements and heat treatment. However, simple rules that cover all materials are impossible due to the complexity involved. Based on a vast amount of room-temperature test data, it is strongly believed that fracture toughness of metals increases as material tensile strength decreases. One might ordinarily think that this trend could be extrapolated to the relation of fracture toughness with temperature. Strengths

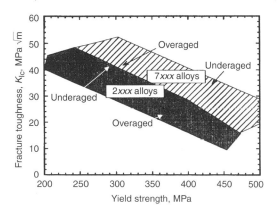

Fig. 4.13 Effects of alloy type and aged condition on the strength/fracture toughness relationship for aluminum alloys. Source: Ref 4.19

*An example comparing the 7475 alloy and conventional alloys is given in Fig. 4.17.

Table 4.5 Heat-treating methods versus heat-treating operations for D6AC steel

| Heat treating operation | Heat-treating methods | | | | | | | |
	A	B	C	D	E	F	G	H
Austenitizing	1700 ± 25 °F		1650 ± 25 °F			1650 ± 25 °F		
Ausbay (interrupted) quenching	Cooled from austenitizing temperature to 975 ± 25 °F in austenitizing furnace and held at 975 ± 25 °F until material is stabilized at this temperature. (Note: Cooling rate between 1350 and 1150 °F must not be less than 6 °F per minute)							
				Salt			Salt	
Quenching	140 °F oil		325 °F	325 °F	400 °F	400 °F	400 °F	375 °F
Tempering	Double tempered at 1025 °F; held at temperature for 2 h per cycle							

Source: Ref 4.21

Table 4.6 Summary of D6AC steel K_{IC} data

Thickness, in.	Product form	Heat treatment(a)	Avg K_{IC}, ksi√in.	Number of specimens
0.8	Forging	E, 400 °F salt	65.3	60
0.8	Plate	E, 400 °F salt	64.5	100
0.8	Plate	C, 325 °F salt + agitation	81.8	4
0.8	Plate	F, 400 °F salt	53.8	12
0.8	Plate	A, 140 °F oil	94.6	25
0.8	Forging	A, 140 °F oil	96.9	26
1.5–1.8	Forging	G, 400 °F salt	43.8	6
1.5–1.8	Forging	H, 375 °F salt + agitation	49.4	14
1.5–1.8	Plate	H, 375 °F salt + agitation	61.3	20
1.5–1.8	Plate	D, 325 °F salt + agitation	47.0	3
1.5–1.8	Plate	B, 140 °F oil	79.1	5
1.5–1.8	Forging	B, 140 °F oil	89.6	8

(a) See Table 4.5 for heat treatment designations. Source: Ref 4.21

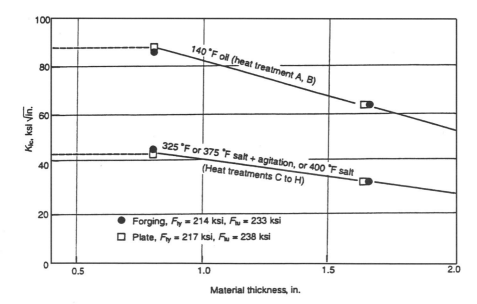

Fig. 4.14 MIL-HDBK-5 B-scale plane-strain fracture toughness values for D6AC steel as a function of heat treatment. Heat treatment designations refer to Table 4.5. Source: Ref 4.21

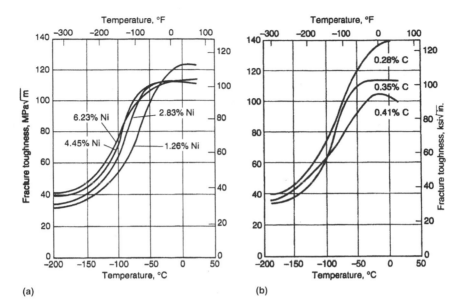

Fig. 4.15 Effects of alloying element content on plane-strain fracture toughness of high-strength steel (Fe-0.65Mn-0.35Si-0.8Cr-0.3Mo-0.1V). (a) Holding carbon at 0.35% and varying the amount of nickel. (b) Holding nickel at 3% and varying the amount of carbon. All steels were hardened and tempered to a room-temperature yield strength of approximately 1175 MPa (170 ksi). Source: Ref 4.20

of bcc and hcp metals normally increase with decreasing temperature, and vice versa; fracture toughness and ductility decrease accordingly. In other words, fracture toughness increases at temperatures higher than room temperature but decreases at temperatures below room temperature.*

However, this is not true for face-centered cubic (fcc) metals, because fcc metals do not normally lose ductility at subzero temperatures and remain ductile at all temperatures. In fact, fcc metals even gain greater elongation at subzero temperatures. The total area under the tensile stress-strain curve would then be larger at low temperature compared to room temperature. Consequently, fracture toughness of fcc metals increases with tensile strength and elongation while temperature decreases. Typical test data are presented in Table 4.7. Except for the short-transverse Al-Li, which shows negligible influence of temperature, all other materials in Table 4.7 gain higher fracture toughness values at low temperatures. More fracture toughness data for aluminum alloys (as functions of strength and temperature) are compiled in Appendix 6.

4.2.4 Effect of Environment

Environment, in a general sense, includes numerous conditions such as temperature, aqueous corrosion, oxidization, reducing atmosphere, radiation damage, and so on. However, only temperature and corrosion are discussed here. The effects of hydrogen embrittlement on subcritical crack growth will also be considered.

The effect of corrosive atmosphere probably is not applicable to monotonic load fracture test-

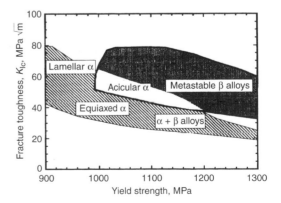

Fig. 4.16 Relationship between fracture toughness and strength for different classes of titanium alloys and microstructures. Source: Ref 4.19

---

*The temperature dependence in bcc and hcp metals is also associated with the ductile-to-brittle transition behavior discussed later in this chapter.

ing, but it does apply to fracture testing with a sustained load. The phenomenon of stress corrosion cracking (SCC) that is described below is the same as that for stress corrosion (described in Chapter 2). The major difference is that SCC concerns testing of a precracked specimen, whereas stress corrosion testing concerns testing of an uncracked specimen. Because fracture mechanics analysts are interested only in the analytical aspects of predicting the residual strength of structural components, it makes little difference to them whether the fracture path is inter-

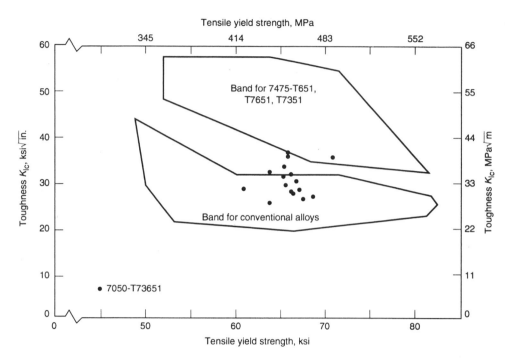

Fig. 4.17 K_{IC} versus F_{ty} plot showing the improvement of K_{IC} values for 7475 aluminum alloy plate (L-T orientation), accomplished by reducing iron and silicon contents, optimizing dispersoids, altering precipitates, and controlling quenching rate and grain size. Source: Ref 4.22

Table 4.7   Fracture toughness and tensile properties of two aluminum alloys and a nickel-base superalloy at room and cryogenic temperatures

Temperature, °C	Orientation	2090-T8E41 Al-Li plate				5083-O aluminum plate				Inco 718 Nickel-base superalloy			
		F_{ty}, MPa	F_{tu}, MPa	Elongation, %	K_{IC}, MPa$\sqrt{m}$	F_{ty}, MPa	F_{tu}, MPa	Elongation, %	K_{IC}, MPa$\sqrt{m}$	F_{ty}, MPa	F_{tu}, MPa	Elongation, %	K_{IC}, MPa$\sqrt{m}$
25	L	535	565	11									
−196	L	600	715	14									
−269	L	615	820	18									
25	LT				35								
−196	LT				51								
−269	LT				64								
25	SL				17								
−196	SL				15								
25	ST				16								
−196	ST				13								
24	TL					142			27				
−196	TL								43				
−269	TL								48				
24										1172	1404	15.4	96
−196										1342	1649	20.6	103
−269										1408	1816	20.6	112

Note: See Appendix 6 for designation of grain orientations. Heat treatment for Inco 718: 980 °C for 0.75 h, air cool; double age 720 °C, 8 h, furnace cool to 620 °C, hold 10 h, air cool. Source: Ref 4.23–4.25

granular or transgranular for a certain material in a certain environment. However, it is significant that cracks propagated by different modes progress at different rates. State-of-the-art fracture mechanics analysis methodology does not have a good handle on this problem. Therefore, this chapter only touches on a few key points that are relevant to fracture mechanics analysis.

In fracture mechanics terminology, stress corrosion cracking is not the same as stress corrosion. In the case of SCC, it is assumed that the structural part contains a preexisting crack. Certain environments can have a pronounced effect on subcritical crack growth. The effect may be thought of as the promotion of time-dependent crack extension at stress intensity levels below K_{IC}, or K_C. Processes commonly referred to as electrochemical and hydrogen embrittlement are examples of this. The scenario involves a cracked structure (or a test specimen containing a crack) that is submerged in a corrosive environment and subjected to a sustained load (constant load) for a prolonged period. At some point the crack will start to grow. When the crack grows to a critical length, rapid fracture will occur. The final absolute value of fracture toughness (computed using final crack length and the sustained load level) obtained in this manner will be the same as the fracture toughness (based on final crack length and fracture load) obtained from monotonically increasing load tests conducted in a nonaggressive environment.

Figure 4.18(a) illustrates sustained-load environmental crack growth behavior. This type of curve can be obtained by loading a series of specimens to various percentages of the monotonic fracture stress. In other words, each specimen is loaded to a certain percentage of the baseline critical stress intensity factor value for the same crack size. These loads are maintained until fracture (or retired after a very long time period, in the case of no failure). The initial stress intensity for a given specimen is calculated based on the initial precrack size and the sustained load applied to the specimen. The corresponding stress intensity factor is usually called K_i (or K_{Ii}). The time required to fracture depends on the applied stress level and the properties of the material. The times to failure for all specimens are recorded and plotted against their corresponding K_i values. The apparent threshold level, indicating no failure below this K-level, is called the K_{ISCC} (for plane strain) or K_{SCC} (for plane stress).

Some investigators believe that K_{SCC} does not exist. One possible reason is that thin sheet is not likely to promote flat fracture surfaces that are intrinsically associated with SCC. Figure 4.18(b) indicates that the susceptibility to SCC decreases with decreasing thickness and that there is a critical thickness below which SCC does not occur. This critical thickness depends on alloy composition, heat treatment, orientation, and loading rate. Apparently, the critical thickness relates to the transition from plane strain to plane stress. Only a small quantity of thin sheet test data is available, and cannot prove or disprove the existence of K_{SCC}. Nevertheless, K_{ISCC} is the only type of data being used today.

Results of tests conducted on various specimen geometries indicate that K_{ISCC} is a material property. Thus, K_{ISCC} can be used as a measure of a material's susceptibility to SCC in a given environment. Some actual examples (for Ti-6Al-4V and D6AC steel) are shown in Fig. 4.19. Certainly, K_{ISCC} depends on material processing. For a given material composition, thermomechanical treatment, and environment, K_{ISCC} may be different for different orientations of the test

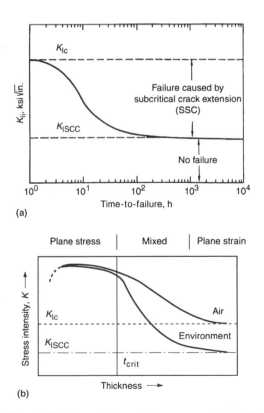

Fig. 4.18 Schematic time-to-failure curves. (a) General SCC behavior of precracked specimens. (b) Schematic plot showing the effect of specimen thickness on the SCC susceptibility of titanium alloys. Note: Stress corrosion cracking does not occur when $t < t_{crit}$. Source: Ref 4.26

specimen. For example, in aluminum alloys, susceptibility to SCC is much greater in the short-transverse direction than in any other grain direction. Therefore, extreme care should be used to ensure use of the correct value of K_{ISCC} for a specific application.

During an environmental sustained-load test, the increments of crack extension can be recorded to obtain a curve for crack length versus time. From this curve, a crack growth rate curve (da/dt versus K) can be plotted. It has been shown that this curve is a basic material property. The relationship between da/dt and K can be divided into three regions (Fig. 4.20a).

In stage 1, da/dt is highly sensitive to changes in K, exhibiting a threshold below which cracks do not propagate. This threshold value of K corresponds to K_{ISCC}. In this region, crack growth is controlled by the interactive effects of the mechanical and chemical driving forces, but is very sensitive to K.

Stage 2 represents crack growth above K_{ISCC}. Crack growth behavior in this region is bounded by the solid and the dotted lines shown in Fig. 4.20(a). The crack may continue to grow at a reducing rate (the solid line in Fig. 4.20a). Or, depending on the alloy/environment system, crack growth rate appears to be independent of K; that is, da/dt is nearly a constant regardless of K (the dotted line behavior). In such cases the primary driving force for crack growth is no

longer mechanical but is related to the chemical corrosion processes occurring at the crack tip (i.e., how fast the corrosive media can be transported to the crack tip). Crack growth rate is more sensitive to parameters such as pH level, pressure, temperature, and so on. In other words, this is a transport-controlled behavior. This behavior (independent of K in stage 2) was first reported by Wei et al. in the early 1970s (Ref 4.27). Other researchers since then have reported other types of behaviors, such as that indicated by the solid line in Fig. 4.20(a). For convenience, let's refer to the behaviors indicated by the dotted line and the solid line as type A and type B behaviors, respectively. The third type (type C behavior) shown in Fig.4.20(b), consists of a stretched linear portion in stage 1, and is more often used in the field of ceramics and glasses.*

In stage 3, crack growth rate is again strongly dependent on K, increasing rapidly as K approaches K_{IC}. In this stage, crack growth is mechanically controlled, unaffected by environment.

Actual test data for aluminum alloys and steels are shown in Fig. 4.21. It can be seen that most aluminum data match the type A behavior, but the da/dt behavior for steels may go either

---

*In the community of ceramics and glasses, the term K_{Ith} is sometimes used for the threshold value K_{ISCC}.

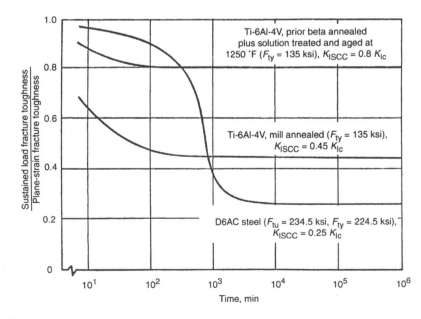

Ti-6Al-4V, prior beta annealed plus solution treated and aged at 1250 °F (F_{ty} = 135 ksi), K_{ISCC} = 0.8 K_{Ic}

Ti-6Al-4V, mill annealed (F_{ty} = 135 ksi), K_{ISCC} = 0.45 K_{Ic}

D6AC steel (F_{tu} = 234.5 ksi, F_{ty} = 224.5 ksi), K_{ISCC} = 0.25 K_{Ic}

Time, min

Fig. 4.19 Effect of environment (3.5% salt solution) and time at sustained load on fracture toughness for titanium alloys and steel. Source: Ref 4.21, 4.26

way. A second look shows another trend: Type B behavior is associated with the effect of hydrogen (i.e., the data labeled as "distilled water" in Fig. 4.21d and all data in Fig. 4.21b). Sodium chloride seems to contribute to da/dt behavior that is independent of K in stage 2, whether the material is aluminum or steel (except for the 7039-T64 alloy, where type C behavior is prominent). After comparing all the test data presented here for metals and the data in Chapter 7 for glasses with the models in Fig. 4.20, one might observe that type C behavior actually lies between types A and B. It also seems, depending on the combination of material and environment, that the extent of stage 2 may be extensive, narrow, or nil, as depicted by models A, C, and B, respectively. Therefore, perhaps types A and B alone are sufficient to bracket the entire range of SCC behavior. Actually, whether the slope in stage 1 is steep or not so steep, with or without a linear region, depends in part on the scale used to plot the data. The original Wei model used the linear scale on K. However, both linear and logarithmic scales are used by engineers today. There has been no systematic study to identify the source or reason for either type of these behaviors—for example, whether the crack growth mechanism is a function of material, environment, test method, and so on. Therefore, we cannot further elaborate.

Hydrogen embrittlement and stress corrosion cracking sometimes share a common cause, as briefly mentioned in Chapter 2. In fracture mechanics testing, test methods for determining the threshold allowable and the da/dt versus K curve are the same. In fact, Fig.4.21(b) and (d) demonstrate the effect of hydrogen on crack growth velocity. The same type of K_{Ii} versus time to failure curve is obtained for either SCC or hydrogen embrittlement. The terminologies for separating them are K_{ISCC} for stress corrosion cracking and K_{IHE} for hydrogen embrittlement. Test results of hydrogen embrittlement cracking of iron-nickel-cobalt steels are shown in Fig. 4.22. Note the similarities of the K versus time curves shown in Fig. 4.18, 4.19, and 4.22.

4.3 Dynamic K_{IC}

Many structural alloys (steels in particular) can fail in either a ductile or brittle manner, depending on service conditions such as temperature, loading rate, and constraint. Low tempera-

ture, fast loading rate, and triaxial state of stress contribute to brittle fracture, though all three of these factors need not be present at the same time. Toughness is defined as the ability of a material to absorb energy. It is usually characterized by the area under a stress-strain curve of a smooth (unnotched) tension specimen that is loaded slowly to fracture. Notch toughness represents the ability of a material to absorb energy and is measured by using a variety of specimens and test procedures. Engineers began doing this type of fracture testing in the World War II era. Tests in those days were performed by making a side impact to a specimen containing a V-notch (or some other type of notch). The tests were

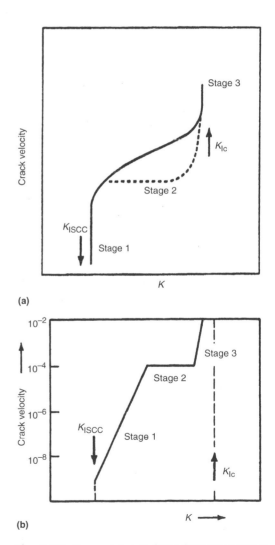

Fig. 4.20 Stress corrosion cracking velocity versus stress intensity factor. (a) Type A, dotted line; type B, solid line. (b) Type C behavior

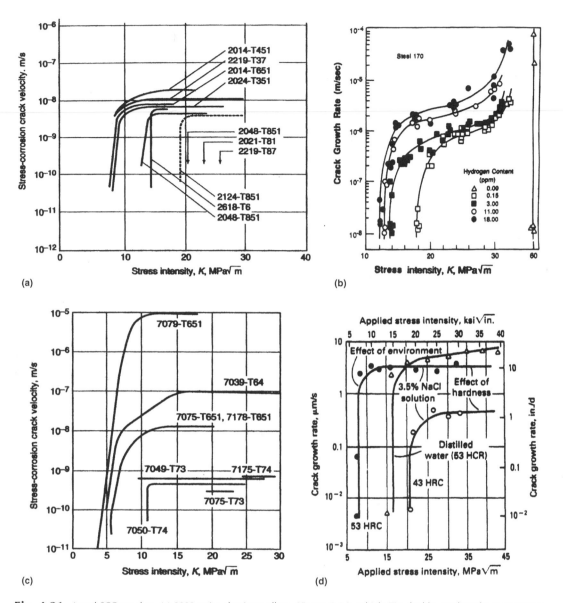

Fig. 4.21 Actual SCC test data. (a) 2000 series aluminum alloys, 25 mm (0.1 in.) thick (TL), double cantilever beam specimens, wet twice a day with an aqueous solution of 3.5% NaCl, 23 °C (73 °F). Source: Ref 4.22. (b) Effect of hydrogen content on crack growth velocity on an ultrahigh-strength steel. Source: Ref 4.28. (c) 7000 series aluminum alloys, double cantilever beam specimens, 25 mm (0.1 in.) thick, short-transverse orientation of die forging, long-transverse orientation of hand forging and plate. Specimens were subjected to alternate immersion tests in 3.5% NaCl solution, 23 °C (73 °F). Source: Ref 4.29. (d) Two AISI 4340 steels (43 and 53 HRC, respectively), contoured double cantilever beam specimens. The 53 HRC steel was subjected to distilled water or 3.5% NaCl solution. The 43 HRC steel was subjected to 3.5% NaCl solution only. Source: Ref 4.30

done at a range of temperatures to determine which alloys remain ductile in a cold environment, or at what temperature a given metal can remain ductile. This type of test, called the "pendulum impact test," is better known as the Charpy test or Izod test. These tests are not the same as K_{IC} testing. The basic difference is in the applied loading system (i.e., impact versus

monotonic) and high speed versus slow and gradual.

The significance of the conventional (static) K_{IC} properties also applies to the case of rapid loading. The plane-strain fracture toughness of certain materials (bcc, for example) is sensitive to loading rate; toughness may substantially decrease as the loading rate increases. Generally,

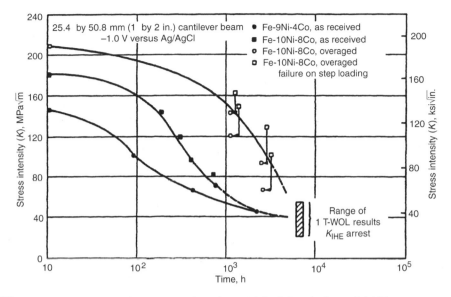

Fig. 4.22 Test results of hydrogen embrittlement cracking of iron-nickel-cobalt steels. Source: Ref 4.31

such materials also show a pronounced dependence of K_{IC} on test temperature. It is possible to measure dynamic toughness in a quasi-static fracture test, which basically involves applying a fast loading rate to conduct an ordinary K_{IC} test. Procedures for this "rapid-load $K_{IC}(t)$ test" are given in a special annex to ASTM Standard E 399 for plane-strain fracture testing. For conventional K_{IC} the maximum loading rate is limited to 2.75 MPa$\sqrt{m}$/s (2.5 ksi$\sqrt{in.}$/s). Any loading rate faster than that is considered rapid-load fracture toughness. Usually, these tests are done in the range of 10 to 10^4 MPa$\sqrt{m}$/s (9 to 9 × 10^3 ksi$\sqrt{in.}$/s). A typical data set of this type is shown in Fig. 4.23. Note that the term K_{Id} was used at that time; the subscript d stands for "dynamic." Alternatively, the value of K_{Id} was obtained by applying an impact load to a conventional K_{IC} test coupon. There is a general perception that all three fracture parameters—static K_{IC}, K_{Id}, and V-notch impact energy—are connected in some way so that correlation among any two parameters is feasible.

The apparatus and test procedure for the $K_{IC}(t)$ test are much the same as for the static K_{IC} test. Special instructions are given to ensure that the instrumentation can handle the rapidly changing signals. All of the criteria for static K_{IC} determination apply to the rapid-load test. The rapid-load fracture toughness property is denoted by $K_{IC}(t)$, where t is the time to reach the load corresponding to K_Q (in fractions of a second). To validate the test result, the 0.2% offset

tensile yield strength σ_{YS} is first used to determine the specimen size requirements. If the rapid-load value of K_Q is valid using a static tensile yield strength value determined at a temperature at or above that of the rapid-load test, no further yield strength considerations are necessary. If the test is invalid using such a yield strength, a recheck should be made using a dynamic yield stress, σ_{YD}. The value of σ_{YD} is determined by conducting a tension test on the test material at the temperature and loading rate of the $K_{IC}(t)$ test. That is, the time to reach the yield load in the tension test is calibrated approximately equal to the time t of the $K_{IC}(t)$ test. In the absence of a σ_{YD} value, using an estimated σ_{YD} is allowed. An equation is given in ASTM Standard E 399 for making such an estimation.

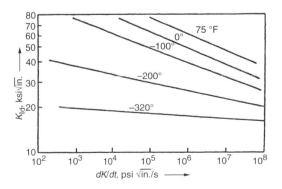

Fig. 4.23 Influence of loading rate and temperature on fracture toughness of A 302 B steel. Source: Ref 4.32

However, that equation is useful only for certain construction steels with room-temperature tensile yield strengths below 480 MPa (70 ksi). Therefore, the applicability of the ASTM equation is limited and will not be discussed here.

Actually, to this day, use of the term "dynamic fracture toughness" is quite confusing. ASTM does not clearly define this quantity. Yet the term K_{Id} appears in the literature, which might mean $K_{IC}(t)$ or values that were converted from data developed from notched-bar impact tests. A nearly true K_{Id} value probably comes from impacting the precracked Charpy specimen. An example of the dependence of K_{Id} on temperature, obtained from instrumented Charpy tests, is shown in Fig. 4.24.

4.3.1 Notched-Bar Impact Tests

Various types of notched-bar impact tests are used to determine the tendency of a material to behave in a brittle manner. The notched-bar impact test detects differences between materials that are not observable in a tension test. The data obtained from impact testing provide no values that can be used in component design, but do provide comparative information between different materials as well as between lots of materials and/or heat treatments. Many notched-bar test specimen designs have been used by investigators. During World War II, the catastrophic failure of many Liberty ships while sitting in the ocean resulted in development of the Charpy impact and Izod impact tests. These tests use small-size specimens, are low cost, and provide rapid detection of changes in material ductility. Improved versions of these tests and several similar types of tests were developed after the war. Three of them belong to the family of "drop-weight tests," which use larger specimens.

All these tests were developed at about the same time as fracture mechanics. However, the real engineering application of fracture mechanics did not begin until much later. Impact energy (measured from these earlier tests) was used to represent the fracture toughness of a material. The energy unit is not a useful parameter for structural design/analysis. However, it still is a good indicator of relative brittleness, allowing comparison of two or more materials. Most importantly, the primary purpose of this type of testing was to determine the ductile-to-brittle transition temperature in bcc metals, and it served this purpose nicely. Since the K-factor in fracture mechanics is closely related to the strain energy release rate, it is possible to convert impact test results to the more desirable parameter K_{IC}. Standardized test procedures for these tests are available (ASTM Standard E 23). Here are a few key points about these tests.

The Charpy specimen has a square cross section and contains a notch at the center of its length. Although either a V-notch, U-notch, or keyhole notch can be used, it has become common practice to use a V-notch or a precracked V-notch. The Charpy specimen (Fig. 4.25) is supported as a beam in a horizontal position. The load is applied by impact of a heavy swinging pendulum (approximately 5 m/s, or 16 ft/s, impact velocity) at the midspan of the beam on the opposite side of the notch. The specimen is thereby forced to bend and to split open the notch, and finally fracture at a strain rate on the order of 10^3 s^{-1}.

The Izod specimen (Fig. 4.25) is either circular or square in cross section and contains a

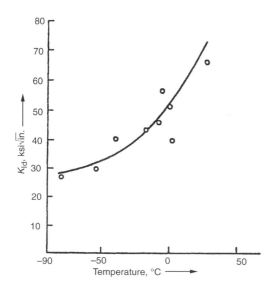

Fig. 4.24 Dynamic fracture toughness from instrumented Charpy tests. Source: Ref 4.33

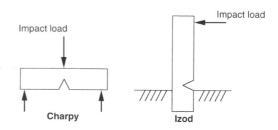

Fig. 4.25 Loading method in Charpy and Izod impact tests

V-notch at one end. The specimen is clamped vertically at the notched end like a cantilever beam and is struck with the pendulum at the opposite end. Note that in both types of impact tests, the notch is subjected to a tensile stress as the specimen is bent by the moving pendulum. Plastic constraint at the notch produces a triaxial state of stress. The relative values of the three principal stresses depend strongly on the dimensions of the bar and the details of the notch. Therefore, it is important to use standard specimens. Note that by inducing a fatigue crack in the Charpy specimen, the notch acuity and depth restrictions are eliminated. Test procedures for the V-notch Charpy and the precracked Charpy tests are available (with and without instrumentation, ASTM Standard E 23). There is no ASTM standard to cover the Izod test, which was developed in England.

The response of a specimen to the impact test is usually measured by the energy absorbed in fracturing the specimen. The energy absorbed is expressed in joules (foot-pounds), or normalized to joules (foot-pounds) per unit cross-sectional area. Often a measure of ductility, such as the percent contraction at the notch or lateral expansion of the back face opposite the notch, is used to supplement this information. It is also important to examine the fracture surface to determine whether it is fibrous (shear failure) or granular (cleavage fracture). Examples of each of these three parameters (as a function of temperature) for a low-carbon steel plate are shown in Fig. 4.26.

The transition-temperature curve developed from the fracture appearance method is based on the percentage of shear fracture on the fracture surface of a specimen (tested at a given temperature). This is done by carefully measuring the dimension of the brittle cleavage exhibited on the specimen fracture surface (Fig. 4.27) and then referring to Table 4.8. The percentage of shear can be plotted against test temperature, and the transition temperature can be ascertained using the shear percentage specified. The 50% point is called the fracture-appearance transition temperature (FATT).

The lateral expansion method is based on the observation that there is transverse strain (tension and contraction) directly behind the notch, and compression and expansion near the back side of the specimen. The expansion is easily measured. Shear lips are also created if constraint is incomplete; the thickness of the shear lips (B dimension) is an indication of increased toughness. The lateral expansion can be expressed as a measure of acceptable ductility at a given test temperature. The broken halves from each end of each specimen are measured (i.e., the dimensions of A_1, A_2, A_3, and A_4 in Fig. 4.28). The higher values from each side are added together, and this total is the lateral expansion value. These test results are then plotted against test temperature and a curve interpolated.

The notched-bar impact test is most meaningful when conducted over a range of temperature so that the temperature at which the ductile-to-brittle transition takes place can be determined. Figure 4.26 illustrates the types of curves obtained. The lower flat portion of the plot is generally referred to as the "lower shelf" region; the higher flat portion is the "upper shelf" region. Materials in these regions behave in a "brittle" and "ductile" manner, respectively. Some materials exhibit a very pronounced transition behavior, in contrast to those shown in Fig. 4.26. However, transition temperature sometimes is difficult to determine from the types of curves in Fig. 4.26 because the decrease in impact energy does not occur sharply at a certain temperature. Nevertheless, somewhere in that transition zone, between the lower and upper shelves (between -10 and $+20$ °C, or 15 and 70 °F, in this case), there is an energy value that can be defined as the transition temperature. It is desirable to define a single temperature within the transition range that reflects the behavior of the steel under consideration. Some suggest choosing a temperature that corresponds to 20 J (15 ft · lbf) fracture energy, or a 50% shear, as a target value: in this case, 0 °C (32 °F) and approximately 10 °C (50 °F), respectively. In either case, the selected transition temperature falls somewhere near the midpoint of the entire transition temperature range. That is fine for a material whose transition temperature range is very narrow and whose transition temperature is actually sharply defined. From a practical standpoint, engineers can just use the data (the S-shaped curve) to determine whether or not the material being tested has adequate impact strength for their design application. There is no need for a definition. Nevertheless, Barsom and Rolfe (Ref 4.36) suggest a minimum impact energy level of 20 J (15 ft · lbf) at the service temperature.

Regarding material selection from the standpoint of notch toughness or tendency for brittle failure, consider a hypothetical case by comparing the two transition curves shown in Fig. 4.29.

Despite the fact that material A has a higher impact toughness at room temperature, clearly material B is the preferred choice because it exhibits a lower transition temperature even though the transition temperatures in both materials are not sharply defined.

4.3.2 Drop-Weight Tests

Specimen size always presents a problem in any fracture testing. When conducting a series of notched-bar impact tests using specimens of different sizes, at some temperature the largest specimens will be completely brittle while the smallest specimens will be completely ductile. The fractures for in-between specimens will vary from almost fully brittle to almost fully ductile. Tests of ship hatch corners carried out at the University of California dramatically demonstrate the effect of specimen size. Full-scale, one-half scale, and one-quarter scale models were tested. Similar in all details, these mod-

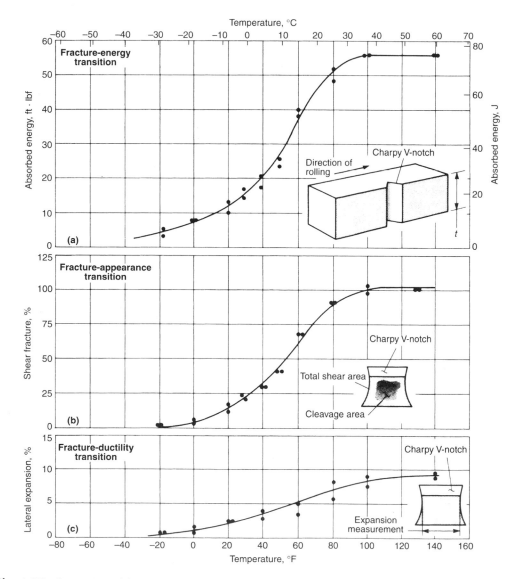

Fig. 4.26 Characteristics of the transition-temperature range for Charpy V-notch testing of semikilled low-carbon steel plate (0.18% C, 0.5% Mn, 0.07% Si), as determined by (a) fracture energy, (b) fracture appearance, and (c) fracture ductility. The drawing at lower right in each graph indicates (a) orientation of the specimen notch with plate thickness, t, and direction of rolling; (b) location of the total shear area on the fracture surface; and (c) location of the expansion measurement in this series of tests—all illustrated for a Charpy V-notch specimen. Percentages of shear fracture and lateral expansion were based on the original specimen dimensions. Source: Ref 4.34

els were made from the same material by the same welding procedures. When fracture strength was measured in terms of pounds per square inch of the net cross-sectional area, the full-size specimen exhibited only about one-half the strength of the quarter-scale model.

The problem of differing transition temperatures for full-size parts and test specimens was discovered when a series of full-size parts were tested using a giant pendulum-type impact machine. The results were compared with those determined using small standard test bars made from the same material. A partial solution to this problem was the development of the drop-weight test (DWT) and the drop-weight tear test (DWTT). A third type, which actually is a derivative of the DWTT, is called the dynamic tear test (DT). These tests produce transition temperatures similar to those found when testing full-size parts. However, such tests are adaptable only for plate specimens of limited sizes and have not become widely used. The essential features of these tests are discussed below. For details concerning test procedures, apparatus, and interpretation of test results, consult other authoritative documents on material testing (e.g., Ref 4.17).

Drop-Weight Test. The DWT specimen and setup are shown in Fig. 4.30 and described in ASTM Standard E 208 (Ref 4.37). The crack inducer is a bead of hardfacing metal. The weld bead, deposited on one side of the specimen at the center using a copper template, is purposely hard and brittle (the Murex-Hardex N electrode is recommended). A notch is made in the weld bead, but not in the specimen itself. After cooling to the desired temperature, the specimen is placed, weld down, on rounded-end supports and struck by a falling weight with sufficient energy to bend the specimen about 5° (the amount of weight and height depend on the strength of the material being tested; see Table 4.9). A cleavage crack forms in the bead as soon as incipient yield occurs (at about 3° deflection), thus forming the sharpest possible notch. When the crack reaches the specimen material, it will either propagate or stop. A series of tests are conducted over a range of temperatures to find the nil-ductility transition temperature (NDTT).

The specimen is then examined to see whether or not it has fractured. A specimen is considered broken if the crack extends to one or both sides of the surface with the weld bead. If the crack does not propagate to the edge, it is considered a "no break." If the weld notch is not visibly cracked, or if complete deflection does not occur (determined by mark transfer on the deflection stops), it is considered a "no test." Thus, this is a "go, no-go" type of test in that the specimen will either break or fail to break. The NDTT is the maximum temperature at which the specimen breaks. When minimum temperatures are set in material specifications, at least two specimens must be tested at the specified temperatures. All specimens tested must show a "no break" performance.

The drop-weight tear test can be used to determine the appearance of fractures in plain carbon or low-alloy pipe steels (yield strengths less than 825 MPa, or 120 ksi) over the temperature range where the fracture mode changes from brittle to ductile. It has been used to evaluate line-pipe material for natural-gas transmission lines. Specimens made from curved pipe may be flattened prior to testing.

This test method is similar in some ways to the DWT. The transition fracture appearance occurs at the same temperature as for full-size parts. It has the same sudden change from shear fracture to flat fracture as that observed in full-scale equipment. The test is relatively simple in terms of specimen preparation and lack of sensitivity to specimen preparation techniques. The results vary with specimen thickness in the same manner as actual parts.

A major shortcoming, as in the DWT, is that testing is confined to plate material thicknesses between 3 and 20 mm (0.125 and 0.75 in.). The test specimen is even larger than the DWT spec-

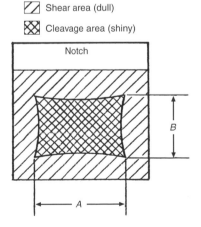

Fig. 4.27 Dimension of percent shear fracture. Average dimensions A and B are measured to the nearest 0.5 mm (0.02 in.), and the percent shear fracture is then determined using Table 4.8. Source: Ref 4.35

imen: 75 mm (3 in.) wide by 300 mm (12 in.) long. Tests are made with the same apparatus used for the DWT, but the test fixture for holding the specimen is altogether different. A large pendulum-type machine can also be used, but the vertical weight-dropping apparatus is more commonly used. Up to 2700 J (2000 ft · lbf) of energy may be required. Figure 4.31 shows the specimen configuration and test setup. Consult ASTM Standard E 436 (Ref 4.38) for test procedures and other details regarding specimen preparation, test apparatus, and so on.

The test result is evaluated by examining the broken pieces. The idea is to determine the percentages of the apparently flat and shear regions on the fracture surface, which are very different in appearance. The transition from one to the other is abrupt. There are two methods for making this evaluation. One is for percentages of shear from 45 to 100% and the other for percentages from 0 to 45%. The acceptance criterion is percentage of shear at a specific tempera-

ture. The temperature at which 50% shear occurs is sometimes considered the DBTT. The procedure for measuring the DWTT percent shear area is detailed in ASTM Standard E 436 (Ref 4.38).

The dynamic tear test belongs to the same family as the DWTT test, but uses a different specimen size. Like the DWTT, the use of either a drop-weight or a pendulum-type machine is allowed. Applicable thicknesses for the DT are 5 to 16 mm 0.2 to 0.6 in.), compared to 3 to 20 mm (0.125 to 0.75 in.) for the DWTT. The material to be tested (not necessary steel) should have a hardness not higher than 36 HRC. The hardness restriction is due to using a press-knife procedure for sharpening the specimen notch. Although test setups for both types of test are similar, the DT specimen is somewhat smaller than the DWTT specimen. Evaluation of the DT test result is also based on fracture appearance; however, the method used to estimate the fracture appearance is not the same as for the

Table 4.8 Conversion of measured fracture surface dimensions to percent shear of fracture appearance

Percent shear for measurements made in millimeters

Dimension B, mm	Dimension A, mm																		
	1.0	1.5	2.0	2.5	3.0	3.5	4.0	4.5	5.0	5.5	6.0	6.5	7.0	7.5	8.0	8.5	9.0	9.5	10
1.0	99	98	98	97	96	96	95	94	94	93	92	92	91	91	90	89	89	88	88
1.5	98	97	96	95	94	93	92	92	91	90	89	88	87	86	85	84	83	82	81
2.0	98	96	95	94	92	91	90	89	88	86	85	84	82	81	80	79	77	76	75
2.5	97	95	94	92	91	89	88	86	84	83	81	80	78	77	75	73	72	70	69
3.0	96	94	92	91	89	87	85	83	81	79	77	76	74	72	70	68	66	64	62
3.5	96	93	91	89	87	85	82	80	78	76	74	72	69	67	65	63	61	58	56
4.0	95	92	90	88	85	82	80	77	75	72	70	67	65	62	60	57	55	52	50
4.5	94	92	89	86	83	80	77	75	72	69	66	63	61	58	55	52	49	46	44
5.0	94	91	88	85	81	78	75	72	69	66	62	59	56	53	50	47	44	41	37
5.5	93	90	86	83	79	76	72	69	66	62	59	55	52	48	45	42	38	35	31
6.0	92	89	85	81	77	74	70	66	62	59	55	51	47	44	40	36	33	29	25
6.5	92	88	84	80	76	72	67	63	59	55	51	47	43	39	35	31	27	23	19
7.0	91	87	82	78	74	69	65	61	56	52	47	43	39	34	30	26	21	17	12
7.5	91	86	81	77	72	67	62	58	53	48	44	39	34	30	25	20	16	11	6
8.0	90	85	80	75	70	65	60	55	50	45	40	35	30	25	20	15	10	5	0

Percent shear for measurements made in inches

Dimension B, in.	Dimension A, in.																
	0.05	0.10	0.12	0.14	0.16	0.18	0.20	0.22	0.24	0.26	0.28	0.30	0.32	0.34	0.36	0.38	0.40
0.05	98	96	95	94	94	93	92	91	90	90	89	88	87	86	85	85	84
0.10	96	92	90	89	87	85	84	82	81	79	77	76	74	73	71	69	68
0.12	95	90	88	86	85	83	81	79	77	75	73	71	69	67	65	63	61
0.14	94	89	86	84	82	80	77	75	73	71	68	66	64	62	59	57	55
0.16	94	87	85	82	79	77	74	72	69	67	64	61	59	56	53	51	48
0.18	93	85	83	80	77	74	72	68	65	62	59	56	54	51	48	45	42
0.20	92	84	81	77	74	72	68	65	61	58	55	52	48	45	42	39	36
0.22	91	82	79	75	72	68	65	61	57	54	50	47	43	40	36	33	29
0.24	90	81	77	73	69	65	61	57	54	50	46	42	38	34	30	27	23
0.26	90	79	75	71	67	62	58	54	50	46	41	37	33	29	25	20	16
0.28	89	77	73	68	64	59	55	50	46	41	37	32	28	23	18	14	10
0.30	88	76	71	66	61	56	52	47	42	37	32	27	23	18	13	9	3
0.31	88	75	70	65	60	55	50	45	40	35	30	25	20	18	10	5	0

Note: 100% shear is to be reported when either A or B is zero. Source: Ref 4.35

DWTT. Details about the DT test are given in ASTM Standard E 604 (Ref 4.39).

4.3.3 Significance of the Transition Temperature

In concept, the transition temperature approach states that below this temperature the energy absorbed by a specimen in fracture is small. Fracture may occur at nominal stresses in a structure below the yield strength. Above the transition temperature, the energy absorbed by a specimen in fracture is large, and fracture will

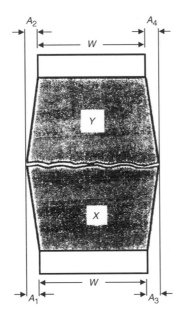

Fig. 4.28 Halves of broken Charpy V-notch impact specimen illustrating the measurement of lateral expansion. Dimension W is the original width. Source: Ref 4.35

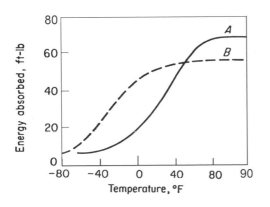

Fig. 4.29 Hypothetical transition-temperature curves for two steels, showing that room-temperature results can be fallible

not occur until the nominal stress in a structure exceeds the yield strength. In reality, we are dealing with two elements in the transition curve: temperature and loading rate, provided that all other parameters are equal: metallurgical factors, notch acuity, test setup, specimen size and thickness (which influences the triaxial state of stresses), and so forth. Test data showing the effects of loading rate on plane-strain fracture toughness are available. An example for ASTM type A36 steel is shown in Fig. 4.32, where it can be seen that in the lower shelf region, plane strain fracture toughness is independent of loading rate. However, at a given temperature, in the transition temperature region, fracture toughness values measured at higher loading rates are much lower than those measured at lower loading rates. A different interpretation of this graph is that the transition temperature increases as the loading rate increases. Note the β values in Fig. 4.32 pertain to the ASTM standard, which specifies that specimen thickness must be greater than 2.5 $(K_{IC}/F_{ty})^2$ so that the β value will not exceed 0.4. All the data points presented in the figure (below the cutoff lines) had a β value equal to or less than 0.4, thus meeting the ASTM requirement.

Figure 4.33 presents a generalization of this phenomenon. Recognizing that $K_{Id} \approx$ CVN (Charpy V-notch), and that CVN testing is traditionally used for determining transition temperature, one might wonder what is the real transition temperature in Fig. 4.33. The answer is: It all depends. Actually what this figure shows is that loading rate plays a role in shifting the transition temperature. Engineers must pay attention to how the part under consideration is loaded. The data in Fig. 4.33 further enforce the notion that high loading rates will cause a normally ductile material to become brittle. It shows that each transition temperature curve is

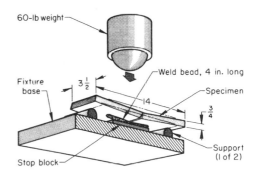

Fig. 4.30 Drop-weight test setup

divided into three regions: cleavage (brittle), transition (increasing shear), and shear (ductile). The subscripts s and d denote "static" and "dynamic," respectively. Impact fracture toughness is significantly lower than the static fracture toughness at a given temperature. The same material would start to become brittle at a much higher temperature when subjected to impact (i.e., higher transition temperature).

Another important feature of Fig. 4.33 is the vertical lines that are used to divide each transition temperature curve into three regions are significant. Two types of transitions occur in each transition temperature curve. The line that divides regions I and II indicates "ductility transition." The line that divides regions II and III indicates "fracture appearance transition." The ductility transition temperature relates to the fracture initiation tendencies of the material. Completely cleavage fracture occurs readily below this point. The fracture appearance transition temperature is related to the crack propagation characteristics of the material. Above this temperature, cracks do not propagate catastrophically because constraint is incomplete. Fracture occurs on planes of high shear stress, creating either slant fracture or mixed-mode fracture. In the region between these two transition temperatures (i.e., region II), fractures are difficult to initiate. Once the fracture does initiate, however, the crack propagates rather rapidly, with a little energy absorption.

Figure 4.33 can be generated by conducting CVN and slow-bend K_{IC} tests on many types of steels. For demonstration purposes, two actual data sets (for ABS-C and HY-80 steels) are shown in Fig. 4.34 and 4.35. Using these figures, Barsom and Rolfe (Ref 40) tried to determine the influence of material tensile yield strength and loading rate on transition temperature. According to Barsom and Rolfe, they used a somewhat arbitrary procedure to determine a reference point for the onset of dynamic transition. They defined this point as the intersection of two

Table 4.9   Standard DWT conditions

Type of specimen	Specimen size, mm (in.)	Span, mm (in.)	Deflection stop, mm (in.)	Yield strength level, MPa (ksi)	Drop-weight energy for given yield strength level(a)	
					J	ft · lbf
P-1	25.4 × 89 × 356 (1 × 3½ × 14)	305 (12.0)	7.6 (0.3)	210–340 (30–50)	800	600
				340–480 (50–70)	1100	800
				480–620 (70–90)	1350	1000
				620–760 (90–110)	1650	1200
P-2	19 × 51 × 127 (¾ × 2 × 5)	102 (4.0)	1.5 (0.06)	210–410 (30–60)	350	250
				410–620 (60–90)	400	300
				620–830 (90–120)	450	350
				830–1030 (120–150)	550	400
P-3	15.9 × 51 × 127 (⅝ × 2 × 5)	102 (4.0)	1.9 (0.075)	210–410 (30–60)	350	250
				410–620 (60–90)	400	300
				620–830 (90–120)	450	350
				830–1030 (120–150)	550	400

(a) Initial tests of a steel with a given strength level should be conducted with the drop-weight energy stated in this column. In the event that insufficient deflection is developed (no-test performance), an increased drop-weight energy should be employed for other specimens of the given steel. Source: Ref 4.37

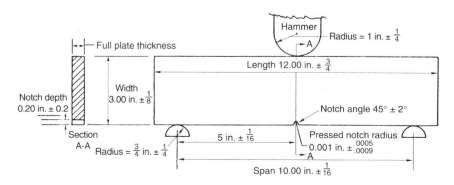

Fig. 4.31   Drop-weight tear test specimen and support dimensions

tangential lines: one drawn from the lower shelf level, another drawn from the transition region of the CVN test. This reference point is designated the "beginning of dynamic transition temperature." The reference points for the two steels in Fig. 4.34 and 4.35 are −18 and −123 °C (0 and −190 °F), respectively. Then they drew a

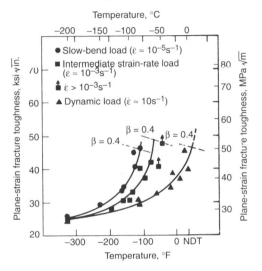

Fig. 4.32 Effect of temperature and strain rate on plane-strain fracture toughness behavior of ASTM type A36 steel. Source: Ref 4.36

horizontal line from the reference point across the data set, as depicted by an arrow pointing left (toward the slow-bend test line). The net difference of temperature (between the two types of tests) is defined as the indicator for transition temperature shifted due to changes in test methods. For these two cases, the amount of shift (ΔT) in transition temperature was −78 Δ°C (−140 Δ°F) for the ABS-C steel, and −36 Δ°C (−65 Δ°F) for the HY-80 steel. The onset of transition temperature for the ABS-C steel was shifted to −130 °C (−200 °F) from −50 °C (−60 °F). As for the HY-80 steel, the onset of transition temperature was shifted to −180 °C (−290 °F) from −143 °C (−225 °F). Table 4.10 presents data reduced from the test results of Barsom and Rolfe for nine steels.

On the basis of these data, it can be concluded that depending on the material, dynamic transition can occur at any temperature. Figure 4.36 shows the correlation between shifting of transition temperature as a function of material tensile yield strength. This graph has been constructed using data obtained from a number of steels with a yield strength between 250 and 1725 MPa (36 and 250 ksi), including those listed in Table 4.10. The key point about Fig. 4.36 is that while high-strength steels exhibit a

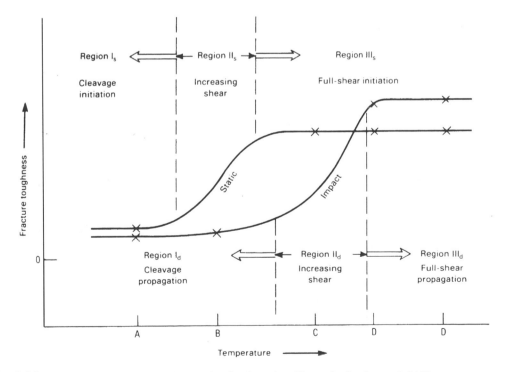

Fig. 4.33 Fracture toughness transition behavior of steel under static and impact loading. Source: Ref 4.36

small shift in transition temperature, steels that have lower strengths show a large shift in transition temperature. It should be noted that the data presented in Table 4.10 basically represent the lower shelf region. The term "transition temperature" is not exactly the same as in the normal sense, which implies the midpoint (or somewhere near the midpoint) of the entire transition range. That might actually be all right when taking a conservative approach.

4.3.4 Correlation between Static K_{IC} and Dynamic K_{IC}

Researchers have sought to quantitatively assess critical flaw size and permissible stress levels by correlating Charpy impact energy and K_{IC}. Most of these correlations were dimensionally incompatible and ignored the differences between the two measures of toughness (loading rate and notch acuity, in particular). The corre-

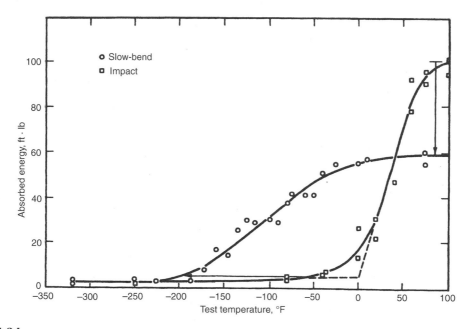

Fig. 4.34 Slow-bend and impact CVN test results for ABS-C steel. Source: Ref 4.40

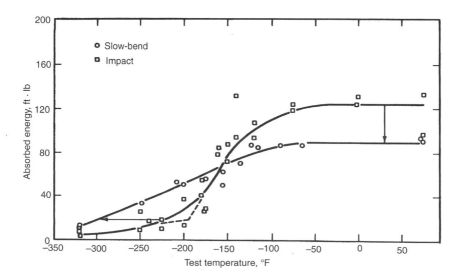

Fig. 4.35 Slow-bend and impact CVN test results for HY-80 steel. Source: Ref 4.40

lations were made only for limited types of materials and ranges of data, and data scattering was sometimes extensive. Also adding to the confusion was that the two types of transition temperatures (i.e., the fracture appearance transition and the ductility transition) were not always distinguished in the data used. Comparisons should not be made between test results where the two criteria have been mixed.

A review of the literature shows that the term K_{Id} was often used for dynamic fracture toughness. However, how it was determined, the type(s) of specimen used, and the significance of its value are not always clear. We have shown sample data in Fig. 4.23 and 4.24, taken from two different sources, to which two different test methods were applied. Despite the difficulties, it is desirable to correlate CVN with K_{Id}, K_{Id} with

K_{IC}, and, better yet, CVN with K_{IC}, because this information is especially important in the upper shelf region, where the LEFM method that is normally used for determining fracture toughness becomes impractical. However, note that such a correlation is not precise due to differences in specimen size and notch geometry between Charpy and fracture mechanics specimens. For whatever it is worth, some examples are given below.

Two earlier investigations focused on rotor steel alloys (NiCrMoV, NiMoV, CrMoV) (Ref 4.41, 4.42). The following equations apply to the entire transition temperature regime:

$$K_{IC,0} = 0.5\sigma_{YS} \qquad \text{(Eq 4.13a)}[9]$$

$$K_{IC,100} = \sigma_{YS}\left[5\left(\frac{CVN}{\sigma_{YS}} - 0.05\right)\right]^{1/2} \qquad \text{(Eq 4.13b)}$$

where σ_{YS} is the same as F_{ty}. The second subscripts in these equations (i.e., 0 and 100) imply that the K_{IC} values correspond to 0% shear fracture and 100% shear fracture, respectively. A K_{IC} value at 50% shear fracture can be estimated by taking an average of Eq 4.13(a) and (b). Good agreement between the predicted K_{IC} versus temperature curves and those actually determined from tests for the rotor steels is shown in Fig. 4.37. Later, more data for steels in the 760 MPa (110 ksi) tensile yield strength range were accumulated. These data show good correlation with the upper shelf (100% shear) equation (Fig. 4.38). Many more empirical equations for steels are available (Ref 4.43–4.46).

The room-temperature dynamic fracture toughness for the steels in Fig. 4.38 can be computed from Eq 4.13(b) by substituting dynamic yield strength σ_{YD} for static tensile yield strength. It has been postulated that the effect of loading rate is to elevate the yield stress by 172 MPa (25 ksi). In other words,

$$K_{Id} = (25 + \sigma_{YS})\left[5\left(\frac{CVN}{25 + \sigma_{YS}} - 0.05\right)\right]^{1/2} \qquad \text{(Eq 4.13c)}$$

Note that the empirical equations presented here are not intended for all materials or all design

Table 4.10 Shifting of transition temperatures(a)

Steel	Shift in transition temperature		Beginning of dynamic transition temperature	
	°C	°F	°C	°F
A36	−107	−160	−4	25
ABS-C	−96	−140	−18	0
A302-B	−90	−130	10	50
HY-80	−62	−80	−123	−190
A517-F	−51	−60	−87	−125
HY-130	−18	0	−134	−210
10Ni-Cr-Mo-V	−18	0	−146	−230
18Ni(180)	−18	0	N/A	
18Ni(250)	−18	0	N/A	

(a) Shift is measured from the CVN impact curve to the slow-bend curve; see Fig. 4.34 and 4.35. Source: Ref 4.40

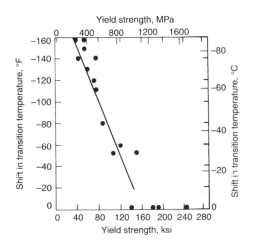

Fig. 4.36 Effect of yield strength on shift in transition temperature between impact and static plane-strain fracture toughness curves. Source: Ref 4.36

[9]A factor of 0.45 (instead of 0.5) is labeled on the figures in Volumes 8 and 19 of the *ASM Handbook*, conflicting with what is in the original paper. The 0.45 factor may have been an updated version of Eq 4.13(a) or merely a typographical error.

applications. Any CVN and K_{IC} correlation is not universal, and may vary with product form, temper, and many other material parameters. Even when the material and parameters are equal, the correlated data points can vary greatly and should be used with caution. Also remember that these equations were developed purely for showing a correlation. The physical units on both sides of the equations are not balanced. That is, the physical unit on the right side is not the same as those on the left side.

4.4 Plane-Stress Fracture Toughness, K_C

Consider the case of a sheet or plate containing a through-thickness crack. If loads are applied perpendicular to the crack so that tensile stresses act to open it, the level of K increases linearly with the level of the tension stress component normal to the crack. As the level of K increases, some point will be reached at which the crack length will start to increase. Then the crack will grow to a critical size, resulting in onset of rapid crack propagation (fracture). Fig-

ure 4.39 shows this schematically. In practice, the stress at which slow crack growth starts (point 0 in Fig. 4.39) is usually not very well defined. For practical purposes, the stress at the onset of rapid crack propagation (point 1 in Fig. 4.39) may be taken as the maximum stress reached in a test. The critical crack length at rapid fracture is not sharply defined, since crack length increases rapidly up to the length at failure. However, it can be measured to a useful degree of accuracy by observing the fracture appearance of the specimen, or by taking a high-speed motion picture during the test.

Strictly speaking, fracture toughness K_C is computed using critical crack length and fracture load. For engineering purposes, however, a critical K-value, i.e., K_C, can be computed based on initial crack length and maximum load (point N in Fig. 4.39). This engineering value is often called the K_{app} (K-apparent). Similarly, the terminology for critical crack length of a structural part is also not clearly defined. Most likely it refers to the criticality of the structural part under consideration. Any critical crack length that is computed using a single value of K_C (which is probably a K_{app} value) is actually the crack length at onset of slow crack growth, which has

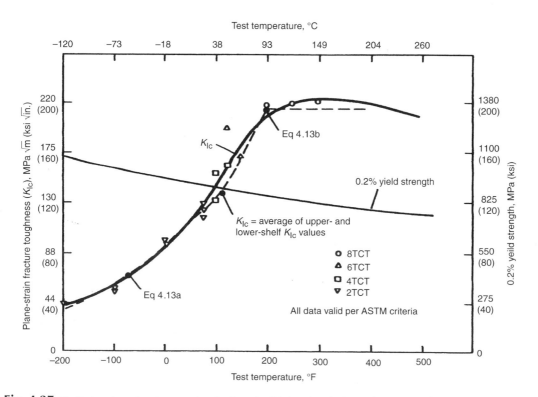

Fig. 4.37 Begley-Logsdon estimation procedure for K_{IC} and validation of results for a turbine rotor steel. Source: Ref 4.42

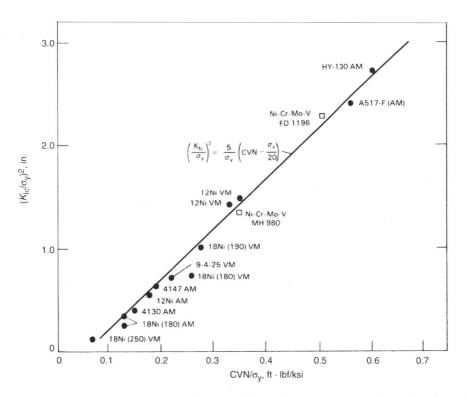

Fig. 4.38 Relation (Eq 4.13b) between plane-strain fracture toughness and CVN impact energy. Tests conducted at 27 °C (81 °F). VM, vacuum melted; AM, air melted. Source: Ref 4.36, 4.40

been literally taken as onset of rapid fracture. However, if desired, the real critical crack length can be determined analytically. The technique for doing that is discussed later in the section "Crack Growth Resistance Curve."

If the material is ductile and/or the test specimen is in a state of generalized plane stress, slow crack growth is expected to occur. The amount of slow stable crack growth depends on the ductility of the material and the extent of plastic constraint at the crack tip. On the other hand, when the environment is inert, and the temperature, thickness, and other characteristics of the material are such that it is quite brittle, the start of slow crack growth will be followed immediately by the onset of rapid fracture; that is, the amount of slow stable crack growth will be negligible. The obtained fracture toughness will be equivalently plane strain in magnitude. This may happen even in a very thin sheet (i.e., in the state of plane stress). For example, 7178-T6 aluminum sheet made in the 1960s (in its original composition) actually exhibit no ductility or slow growth (Ref 4.10).

Look again at Fig. 4.2, which is actually a fracture map relating fracture stress to crack size. It merely shows that fracture mechanics provides a realistic solution to the crack problem. It also shows that the new technology also has limitations. Notice that when the crack is either too short or too long, the fracture me-

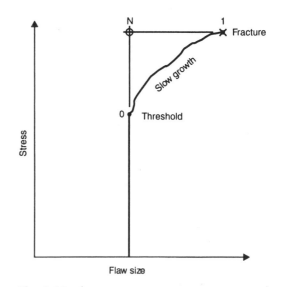

Fig. 4.39 Crack extension under monotonic increasing load

chanics predicted fracture stresses become unrealistic. The predicted fracture stress exceeds the material tensile yield (or even ultimate) strength for the short crack. The predicted fracture stress does not drop to zero as the crack approaches the edge of the panel. It has been suggested that the problem can be solved by graphical means (Ref 4.47). This method involves drawing two lines from each end of the scale: a line from F_{ty} at zero crack length, and another line from the edge of the panel at zero stress. Each line is drawing toward, and tangent to, the fracture mechanics line. The fracture map thus becomes a graph that contains three segments. An example of the suggested fracture map is shown in Fig. 4.40 along with a set of actual test data (taken from Ref 4.48) for comparison. The dotted lines in the figure are the two tangential segments meant to replace the analytical fracture mechanics solution in the two extreme areas.

In reality, if any fracture stress cannot be higher than net section yield, it is conceivable that net section yield would be the limit for all crack lengths. A sensible solution would be to limit the fracture stress at net section yield by performing such a check on a crack size that is either too small or too long. The net section yield limits are represented by two solid lines in Fig. 4.40. This way, we are not determining residual strengths graphically because net section stress is readily determined by geometry.

In summary, the fracture indices are, respectively, K_C for the plane-stress (or mixed-mode) fracture and K_{IC} for the plane-strain fracture. Plane-stress (and/or ductile) fracture is associated with a slant fracture appearance with shear tips and slow stable growth prior to fracture. In reality, it is too complicated to calculate the residual strength by using a real critical crack length for crack growth life prediction, even with the aid of a sophisticated computer program. It actually makes more sense to use a K_{app} value, because the crack length at onset of structural failure (i.e., initial crack length) is what is important in structural life/structural damage tolerance prediction. Note that K_{app} is not a universally adopted term, but is sometimes used as an engineering version of K_C. Unless stated otherwise, in the rest of this book K_C generally means K_{app}. The terminology for plane-strain (and/or brittle) fracture is quite straightforward because the amount of slow growth should be negligible, or nil.

4.4.1 Thickness Effect

An alloy's static strengths often vary with its product form. Therefore, the product form—the original thickness of the plate, for example—is identified in the report of tensile test results for a given alloy. In fracture mechanics, the thickness of a plate refers to the actual finished thickness of a part or test coupon. The dependence of K_C on plate thickness is one of the better known phenomena in fracture toughness testing. Figure 4.41 shows that K_{IC} and pop-in K_{IC} are independent of panel thickness. Beyond a certain thickness B_1, a state of plane strain prevails and fracture toughness reaches the plane-strain value K_{IC}, independent of thickness as long as $B > B_1$.

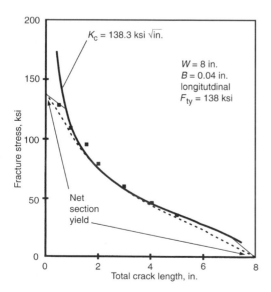

Fig. 4.40 Display of residual strength data for Ti-6Al-4V, mill annealed. Source: Ref 4.10

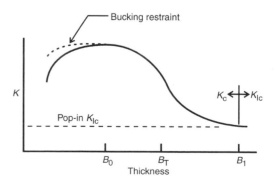

Fig. 4.41 Schematic representation of fracture toughness as a function of plate thickness. Source: Ref 4.10

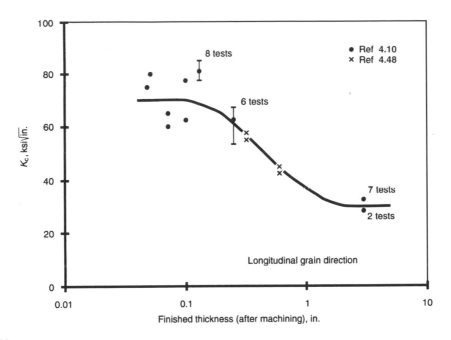

Fig. 4.42 Fracture toughness properties of 7075-T6 aluminum sheet and plate. Source: Ref 4.10, 4.48

According to Irwin (Ref 4.49), K_{IC} will be reached at a thickness of $B \geq 2.5(K_{IC}/F_{ty})^2$. Incorporating this thickness requirement into Eq 4.7, one concludes that the size of a plane-strain crack-tip plastic zone width has to be smaller than 4% of the plate thickness. There is an optimum thickness B_0 where the toughness reaches its highest level. This level is usually considered to be the real plane-stress fracture toughness. According to Broek (Ref 4.8), r_p has to reach the size of the sheet thickness in order for the full plane stress to develop. In the transition region, $B_0 < B < B_1$, the toughness has intermediate values. Actual examples showing the effect of (geometric) thickness on K_C for an aluminum alloy (in two tempers, 7075-T6 and 7075-T76) and a titanium alloy (Ti-6AL-4V) are given in Fig. 4.42 and 4.43. Again, according to Irwin (Ref 4.49), the inflection point B_T (in the transition region) occurs at approximately $B = (K_{IC}/F_{ty})^2$. Therefore, from the standpoints of design and material selection, it is desirable to design a part with $B < (K_{IC}/F_{ty})^2$, or to select a material with a K_{IC} value higher than $F_{ty}\sqrt{B}$.

For thicknesses below B_0, there is uncertainty about toughness. In some cases a horizontal level is found; in other cases a decreasing K_C value is observed. Recall the K_C versus B curves in Fig. 4.42 and 4.43; thin sheets of the 7075 aluminum (in either overaged or peak-aged con-

ditions) are insensitive to thickness variation. By contrast, the Ti-6Al-4V alloy exhibits a significant drop in K_C values for thicknesses below B_0. Such a phenomenon might be attributed to progress of the extremely large plastic zone during slow stable crack growth, leading to very high local strains at the crack tip, and possibly yielding across the net section of the specimen. Sheet buckling is also responsible for the drop in K_C

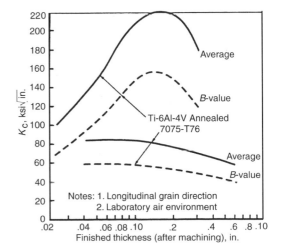

Fig. 4.43 Fracture toughness properties of Ti-6Al-4V and 7075-T76 aluminum sheet and plate. Source: Ref 4.50

value. Sheet buckling occurs above and below the crack line and induces lateral compressive stress ($-\sigma_x$) at the crack tip. During sheet buckling some additional normal stress, σ_y, is induced at the crack tip in order to maintain a balance of the stress system. The K level might actually have been higher than expected. Using a buckling constraint device in a test setup can help to reduce out-of-plane bulging, thereby minimizing the amount of the induced σ_y at the crack tip, and finally allowing the test panel to sustain a higher applied load level before failure. The end result is a higher computed K_C value. An antibuckling device can be as simple as four horizontal bars loosely hanging on both surfaces of the sheet, resting above and below the crack line. It also can be as sophisticated as the one described in Ref 4.10.

The foregoing discussion leads to the conclusion that large plastic zones tend to allow plane-stress conditions to develop. An enlarging plastic zone under increasing load would tend to reduce the through-thickness constraint, allowing more of the plastic zone to deviate from the plane-strain condition. Thus, the ratio of plastic zone size to specimen thickness must be much less than unity for plane strain to predominate. According to ASTM-established criteria, this ratio should be 0.04 or smaller. Since it is the plane-strain condition that produces the lower limiting value of fracture toughness, very thick plates may be required for plane-strain fracture toughness testing. This is particularly true for a material with a high K_{IC} value and a low F_{ty} value, because it would allow the specimen to reach a high K/F_{ty} ratio (and a high ratio of plastic zone to specimen thickness) before failure. In fracture mechanics, the effect of plate thickness on K_C refers to the thickness of the test panel, not the thickness of a product form. A test panel might have been fabricated from an orig-inally thicker plate. Due to microstructure variation through the thickness, some variation in K_C is expected depending on how much material was machined off the original plate and whether the machining was done on one side or both sides. Therefore, it is a good practice to identify the product forms of all the materials used in the testing problem. A map showing the exact original location of the test specimen in the plate or billet is also desirable.

4.4.2 Crack Growth Resistance Curve

As discussed earlier, the fracture process in a metal plate can be characterized by the type of curve shown in Fig. 4.39, where load (or stress) is plotted against crack length. For some materials, considerable slow stable crack growth takes place (along with increasing load) prior to catastrophic failure. Because the stress intensity factor, K, which is determined from linear elastic stress analysis of a cracked plate, uniquely defines the stresses in the crack-tip vicinity, tests on different panels can be correlated by the use of this parameter. Evidently, the S versus a curve can be converted to a K versus a curve. By subtracting the initial crack length from a, the K versus a curve becomes a K versus Δa curve (Fig. 4.44). The last curve in Fig. 4.44 is called the crack growth resistance curve, or simply an R-curve, or K_R-curve, because the K-parameter here is specifically designated as R or K_R.

We know that K_{IC} is a constant, and so a single value of K_{IC} can apply for a material and temper. However, K_C is not a constant, as will be shown later. The goal now is to find a single empirical parameter, or something, that can replace K_C. It would be ideal to show that a material and temper and thickness, in a generalized state of plane stress, can be characterized by a single R-curve. Everything else being equal (i.e., specimen ge-

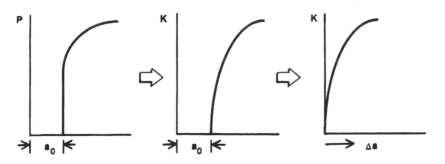

Fig. 4.44 Step-by-step procedure for reducing a load versus crack length record to an R-curve

ometry, initial crack length, loading condition, temperature, etc.), a more ductile material will have a longer period for slow stable growth (a longer *R*-curve).

In reality, the amount of slow stable tear also depends on structural configuration (or the geometry of the test specimen). For example, the *R*-curve obtained from a compact specimen has a longer length compared to one obtained from a center-cracked specimen, because the gradient of *K* in a compact specimen is decreasing whereas the gradient of *K* in a center-cracked specimen is increasing. If length is the only variable in an *R*-curve, it would mean that the shape of an *R*-curve is independent of geometric variables (e.g., panel width, the ratio of crack length to panel width, etc.).

Once an *R*-curve representing the fracture behavior of a given material is determined, it is regarded as a material property. An actual example of this type of curve for an aluminum alloy is shown in Fig. 4.45. However, the data in the figure show some indication that the *R*-curve for the narrower panels (20 and 30 cm, or 8 and 12 in., wide) had a steeper initial slope compared to the 60 cm (24 in.) wide panel; it may have been affected by the *K*-gradients in different panels. Currently, insufficient data are available in the literature to determine the effect of geometry on *R*-curves. For now, simply note that the original concept may not always hold.

An *R*-curve can be obtained by conducting a fracture test of a center-precracked panel, or any other type of specimen such as those recommended by ASTM (Ref 4.17), provided the load and crack length increments can be correlated on a time basis. Figure 4.46 provides an example of such data. The data were obtained by testing a precracked 2024-T3 Alclad aluminum sheet, 1.9 mm (0.0756 in.) thick, 122 cm (48 in.) wide, and 254 cm (100 in.) long. First, a 25.4 cm (10

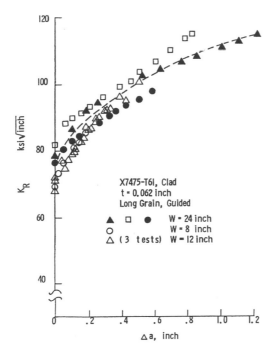

Fig. 4.45 *R*-curve for X7475-T61 aluminum alloy. "Guided" refers to use of a buckling guide. Source: Ref 4.51

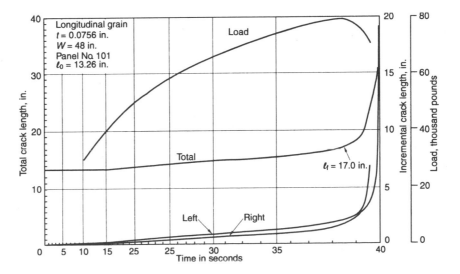

Fig. 4.46 Crack growth behavior of 2024-T3 aluminum sheet. Source: Ref 4.52

in.) sawcut was made at the center of the panel. Very low tension-tension cyclic loads were applied to the panel to produce a simulated fatigue crack approximately 33 cm (13 in.) long, symmetric to the vertical centerline of the panel. During monotonic increasing loading, close-up motion pictures (film speed: 200 frames per second) and electronically recorded load data were taken. Superposition of the load versus time and the crack length versus time curves yielded a load versus crack length plot similar to that shown in Fig. 4.39. From these data, K-values were computed for each consecutive crack length and the corresponding stress levels using an appropriate stress intensity factor equation. The stress intensity factor equation for the cen-

ter-cracked panel is given in Chapter 5. By setting K equal to K_R, an R-curve was constructed (with K_R versus a or K_R versus Δa). The result is shown in Fig. 4.47.

Perhaps the best way to explain an R-curve is to go through a hypothetical exercise that demonstrates how to use an R-curve to determine the residual strength of a piece of structure. For simplicity, consider a center-cracked panel having an initial crack length $2a_0$. To predict the failure load, we need to construct a diagram such that the R-curve of the *same material* is placed at $a = a_0$, as shown in Fig. 4.48. The material R-curve need not be developed from a specimen configuration identical to the one that is being analyzed. That is, the R-curve could have come from a compact specimen, or another type of specimen, as long as the thickness of the specimen and the part in question are the same.

The three dashed lines in Fig. 4.48, labeled P_1, P_2, and P_3, are K-curves hypothetically computed for three stress levels (as a function of a). Each K-curve corresponds to a stress level applied to the structure (with $P_1 < P_2 < P_3$). Thus, these are the applied K-curves. Let's focus on the points on P_1 and P_2; the applied K-levels on the right of the R-curve are lower than the K_R of the R-curve. This means that at stress levels P_1 and P_2 the material has more energy ($K_R \cong \mathscr{G}_R$) available for release against the input (the applied K). When the crack starts to extend at point 1 (stress level P_1), we can imagine that the K-value in the structure follows the applied K-curve and moves to the right of the R-curve. That makes the applied K-level fall beneath the R-curve, i.e., $K < K_R$. That is, the material is capable of taking a higher load. The same is true for point 2. But as the stress level reaches P_3, the applied K-levels corresponding to subse-

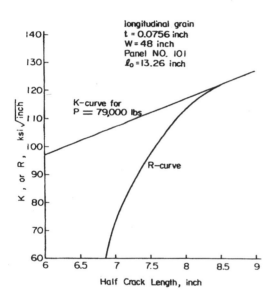

Fig. 4.47 *R*-curve for the 2024-T3 aluminum sheet. "Guided" refers to use of a buckling guide. Source: Ref 4.52

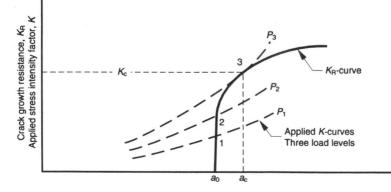

Fig. 4.48 Schematic representation of superposition of *R*-curve and applied *K*-curves to predict instability. $P_1 < P_2 < P_3$

quent crack extension (i.e., beyond point 3) are higher than K_R. This means that the material can no longer sustain more loads. Therefore, the failure criterion based on the crack growth resistance curve concept is that crack growth will be stable as long as the increase in resistance K_R, as the crack grows, is greater than the increase in applied K. Otherwise, unstable fast fracture will occur. That is, failure will occur when

$$K = K_R \qquad \text{(Eq 4.14a)}$$

and

$$\partial K/\partial a \geq \partial K_R/\partial a \qquad \text{(Eq 4.14b)}$$

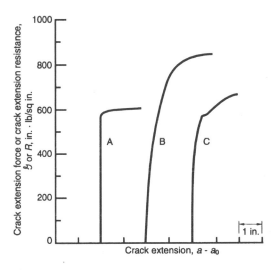

Fig. 4.49 Some conceivable types of R-curves. Source: Ref 4.53

The failure point can be graphically determined when the applied K-curve for a given load is tangent to the R-curve. This tangential point defines the K_C value, or the fracture load (P_3 in this case), of the structure (or the test panel). Recall that for panel no. 101 (Fig. 4.47), the failure load was 35,835 kg (79,000 lb). The tangential point where the applied K-curve merges with the R-curve gives a K_C value of 134 MPa$\sqrt{\text{m}}$ (122 ksi$\sqrt{\text{in.}}$). However, on the basis of initial crack length and final load, a K_{app} value of 114 MPa$\sqrt{\text{m}}$ (104 ksi$\sqrt{\text{in.}}$) is obtained.

An R-curve can now be used to explain the phenomenon of K_{IC}, and the effect of geometry on K_C. We can begin by showing several possible types of R-curves (Fig. 4.49). Obviously, the one with a sharp point (i.e., type A) is representative of the plane-strain K_{IC} behavior. The smooth curve (type B) represents most of the thin sheet (or mixed-mode K_C) behavior. As shown in Fig. 4.50, the value of K_{IC} for a given material is independent of initial crack lengths. The residual strength σ_C is higher when the initial crack length is shorter. The tangential point is right at the start of crack extension. In contrast to the K_{IC} behavior, the type B material exhibits a geometry dependency, as schematically shown in Fig. 4.51 and 4.52. Figure 4.51 illustrates K_C as a function of panel width, whereas Fig. 4.52 shows K_C as a function of $2a/W$ ratio for a fixed specimen width. It is interesting to note that for a given panel width there is an optimum K_C value that can be obtained at $2a \cong W/3$. Although the general trend for the panel width ef-

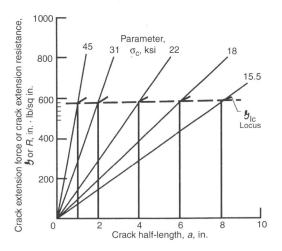

Fig. 4.50 Crack extension instability condition for various crack lengths in a brittle material. Source: Ref 4.53

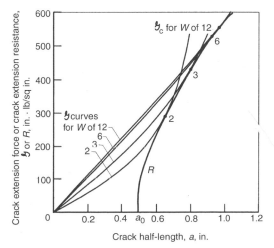

Fig. 4.51 Dependence of fracture toughness on specimen width, W, for center-cracked plate specimens having the same initial half-crack length. Source: Ref 4.53

fect is that K_C increases with panel width, the degree of susceptibility depends on the ductility of the material. As shown in Fig. 4.53 and 4.54, which present K_C values for two aluminum alloys (7075-T6 and 2024-T3), the more ductile 2024-T3 alloy exhibits a higher degree of panel width susceptibility.

At this point, one might wonder about showing, matter of factly, the geometric effects in Fig. 4.51 and 4.54, because it was stated earlier that R-curves are somewhat independent of specimen configuration. Bear in mind that we are dealing with two types of activities: constructing

an R-curve as a material fracture property, and using the R-curve for structural analysis. Perhaps this confusion can be cleared up by calling the tangential point K_C when dealing with material fracture property development (see Fig. 4.47) and referring to the tangential point as residual strength whenever a structural allowable prediction is involved. Although the test data points in Fig. 4.53 and 4.54 are K_{app} values, these figures show that the trend lines of these data sets agree with those shown in Fig. 4.51, the analytical trend developed by using an R-curve.

In light of the foregoing discussion, we can conclude that using an R-curve is perhaps the most efficient and economical method for residual strength assessment. Theoretically, a material R-curve of any origin can be used to predict the residual strength of a structural member. The structural member in question can be of any configuration and subjected to any type of loading, as long as a K-solution for such a combination is available.

4.5 Fracture under Mixed Modes 1 and 2

Thus far, discussions have centered around the tensile mode configuration, where the applied loading was perpendicular to the crack. Under a general in-plane loading condition, the crack faces will open or slide, or both, depend-

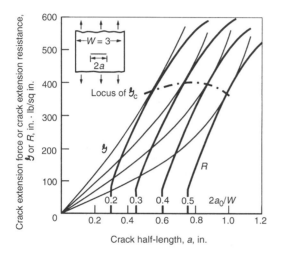

Fig. 4.52 Dependence of fracture toughness on relative initial crack length for a finite-width specimen having an R-curve identical to that of Fig. 4.51. Source: Ref 4.53

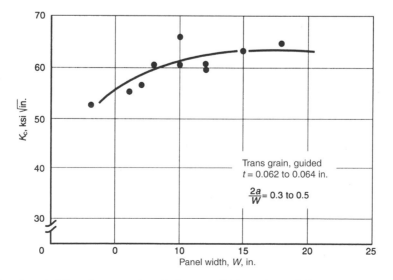

Fig. 4.53 Effect of panel width on fracture toughness for bare 7075-T6 sheets. Source: Ref 4.10

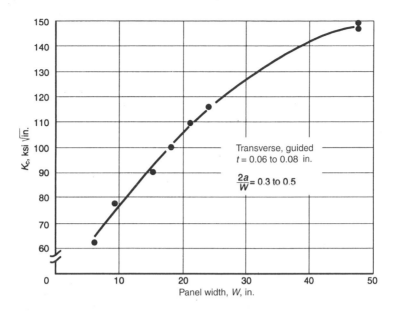

Fig. 4.54 Effect of panel width on fracture toughness for bare 2024-T3 sheets. Source: Ref 4.10

ing on the loading and the crack orientation. That is, the loading may be unidirectional tension, biaxial tension, or pure shear (Fig. 4.55–4.57). The crack may lie on an oblique angle with respect to the loading direction. Depending on the situation, the crack-tip displacement field will be pure mode 1, pure mode 2, or combined modes 1 and 2. As shown in Fig. 4.55 to 4.57, a mixed-mode condition exists whenever both normal and shear stress components act on the crack. In Fig. 4.55, for example, pure tension prevails when the crack lies perpendicular to the applied load ($\beta = 90°$). The percentage of shear increases, while β decreases. A pure shear condition is supposedly reached at $\beta = 0°$; obviously, this is impossible. A pure mode 2 crack exists in the other two loading conditions.

The failure criterion for a generalized loading condition is simply the total strain energy release rate in the body (Ref 4.9). That is:

$$\mathcal{G} = \mathcal{G}_1 + \mathcal{G}_2 + \mathcal{G}_3 \qquad \text{(Eq 4.15)}$$

Without $\mathcal{G}_3$, Eq 4.15 can be translated to:

$$\frac{\mathcal{G}_1}{\mathcal{G}_{1C}} + \frac{\mathcal{G}_2}{\mathcal{G}_{2C}} = 1 \qquad \text{(Eq 4.16)}$$

or

$$\left(\frac{K_1}{K_{1C}}\right)^2 + \left(\frac{K_2}{K_{2C}}\right)^2 = 1 \qquad \text{(Eq 4.17)}$$

These equations present a few problems. One would need to determine the K_{2C} value for the material and the K_2 value for the structure under

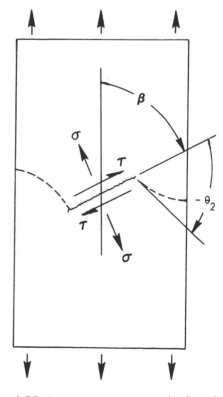

Fig. 4.55 Stress components on a crack subjected to in-plane uniaxial tension

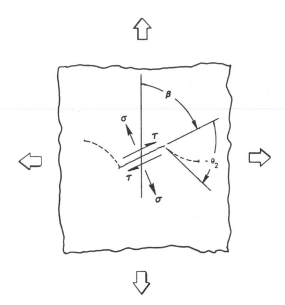

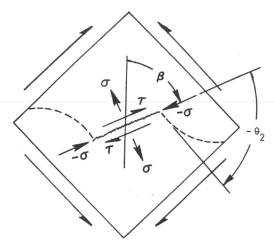

Fig. 4.57 Stress components on a crack subjected to in-plane shear

Fig. 4.56 Stress components on a crack subjected to in-plane biaxial tension

consideration. Handbook solutions for K_1 are available for common geometries; however, almost none exist for K_2. Therefore, determining K_2 (by finite element or other means) often is a necessary step in the process, whether for a fracture test specimen or a structure of specific geometry. One might wonder if there is a unique relationship between K_{1C} and K_{2C}, so that the step that determines K_{2C} can be skipped. One way is to assume that $K_{1C} = K_{2C}$ when no known data are available. In fact, the maximum stress criterion used by Erdogan and Sih (Ref 4.54) has predicted that $K_{2C} = 0.866K_{1C}$. On the other hand, on the basis of the strain energy density criterion, Sih and MacDonald (Ref 4.55) have predicted that:

$$\frac{K_{2C}}{K_{1C}} = \sqrt{\frac{3(1 - 2v)}{2(1 - v) - v^2}} \qquad \text{(Eq 4.18)}$$

which gives $K_{2C} = 1.021K_{1C}$ for $v = 0.25$, and $K_{2C} = 0.905K_{1C}$ for $v = 1/3$. Another prediction (Ref 4.56), based on the maximum energy release rate criterion, finds that the ratio of K_{2C} to K_{1C} equals 0.816.

Table 4.11 presents a compilation of available fracture test data. The K_{1C} and K_{2C} values shown

Table 4.11   Mode 1 and mode 2 fracture toughness values for various materials

Material	Thickness, mm	K_{1C}, MPa√m	K_{2C}, MPa√m	K_{2C}/K_{1C}	Specimen type (for mode 2)	Loading (for mode 2)	Source
Aluminum							
7075-T7651	7.62	89.2	89.2	1.0	Picture frame	Shear	Ref 4.57
7075-T651	7.62	70.0	70.0	1.0	Edge angle crack	Tension	Ref 4.58
2024-T351	7.95	99.4	74.2	0.75	Picture frame	Shear	Ref 4.57
2024-T3	...	...	...	0.75	Center angle crack	Tension	Ref 4.59
2519-T87 (LS)	...	38.2	41.6	1.09	Charpy V-notch	Shear	Ref 4.60
2519-T87 (LT)	...	30.0	38.8	1.29	Charpy V-notch	Shear	Ref 4.60
5083-H131 (LS)	...	32.2	28.3	0.88	Charpy V-notch	Shear	Ref 4.60
5083-H131 (LT)	...	33.3	27.3	0.82	Charpy V-notch	Shear	Ref 4.60
5083-H131 (TS)	...	31.1	31.2	1.0	Charpy V-notch	Shear	Ref 4.60
DTD 5050	10.2, 12.7	31.5	25.0(a)	0.79	Center angle crack	Tension	Ref 4.61
Steels							
4340 (at −200 °F)	6.35	45.7	49.7	1.09	Thin-wall tube	Torsion	Ref 4.58
AISI 01	7.11	26.4	27.4	1.04	Compact shear	Shear	Ref 4.62
Plastics							
PMMA	10.0	1.06(b)	0.94	0.89	Compact shear	Shear	Ref 4.63
	3.0	0.52	0.46	0.89	Center crack	Shear	Ref 4.54

(a) Extrapolated value. Note: K_{1C} values came from conventional center-crack specimens, except (b) in which compact specimens were used.

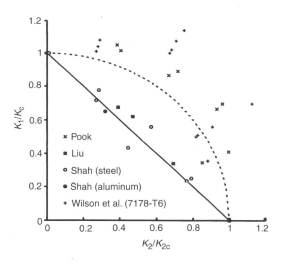

Fig. 4.58 Behavior of K_1 and K_2 interactions. Solid line: $\xi = \zeta = 1$. Dotted line: $\xi = \zeta = 2$

in the table are reportedly based on initial crack length and fracture load. On the basis of these data, the K_{2C}/K_{1C} ratio varies from 0.75 to 1.29. In comparison, the theoretical value for K_{2C}/K_{1C} falls between 0.816 and 1.021. The test data listed in Table 4.11 include high- and moderate-strength aluminum alloys, steels, and plastics. For a given material, the experimentally developed K_{2C} to K_{1C} ratios are consistent, even though different specimen types were used by different investigators. Therefore, it seems that the K_{2C}/K_{1C} ratios compiled in Table 4.11 are material dependent but specimen geometry and test procedure independent. In reality, experimental determination of a pure K_{2C} value is very difficult. Researchers realize that under any circumstance, some amount of K_1 always exists at the crack tip. Therefore, the K_{2C} values listed in Table 4.11 should be regarded only as close approximations.

Many investigators have developed various types of specimens and devices for the purpose of determining a pure K_{2C} value. The disk-type specimen developed by Banks-Sills et al. (Ref 4.62, 4.63) is considered good, because in this configuration the magnitude of K_2 is 25 times that of K_1 at the crack tip. In other words, they were able to keep the K_1 level down to a very minimum. Their test results are also included in Table 4.11. Many other specimen designs are available in the literature. Unfortunately, our database cannot be expanded because these papers did not report any K_{1C} value for the material tested, and thus the K_{2C}/K_{1C} ratio is not known.

Currently, Eq 4.16 and 4.17 are recognized as the criteria for mixed-mode failure. Based on experience, these equations do not always correlate well with test data. To make these equations work, it has been suggested that the exponents in Eq 4.17 be treated as empirical constants (Ref 4.64). That is:

$$\left(\frac{K_1}{K_{1C}}\right)^{\xi} + \left(\frac{K_2}{K_{2C}}\right)^{\zeta} = 1 \qquad \text{(Eq 4.19)}$$

Here ξ and ζ are empirical constants. They need not be the same value for a given material; theoretically, $\xi = \zeta = 2$.

Although while Eq 4.19 is widely accepted in the fracture mechanics community, mixed results persist—as shown by the test data in Fig. 4.58. The test data of Liu (Ref 4.57) and Shah (Ref 4.58) fit perfectly using Eq 4.19 with $\xi = \zeta = 1$. On the other hand, there is no clear correlation between K_{2C}/K_{1C} and K_{2C}/K_{1C} among the test data reported by Pook (Ref 4.61) and Wilson et al. (Ref 4.65). We will see more data of this nature later in the chapters on ceramics and composites.

REFERENCES

4.1. A.A. Griffith, The Phenomena of Rupture and Flow in Solids, *Philos. Trans. R. Soc.(London) A,* Vol 221, 1920, p 163–198

4.2. A.A. Griffith, The Theory of Rupture, *Proceedings, First International Congress of Applied Mechanics* (Delft, The Netherlands), 1924, p 55–63

4.3. E. Orowan, Fundamentals of Brittle Behavior in Metals, in *Fatigue & Fracture of Metals: A Symposium,* W.M. Murray, Ed., John Wiley & Sons, 1952, p 139–167

4.4. G.R. Irwin, Fracture Dynamics, *Fracturing of Metals,* American Society for Metals, 1948, p 147–166

4.5. G.R. Irwin, Relation of Stresses Near a Crack to the Crack Extension Force, *Ninth International Congress of Applied Mechanics* (Brussels, Belgium), Vol 8, 1957, p 245–251

4.6. G.R. Irwin, Fracture, *Hanbuch der Physik,* Vol VI, Springer-Verlag, 1958, p 551–590

4.7. G.R. Irwin, Analysis of Stresses and Strains Near the End of a Crack Transversing a Plate, *J. Appl. Mech. (Trans. ASME), Ser. E,* Vol 24, 1957, p 361

4.8. D. Broek, *Elementary Engineering Fracture Mechanics,* Noordhoff International Publishing, 1974

4.9. H. Tada, P.C. Paris, and G.R. Irwin, *Stress Analysis of Cracks Handbook,* 3rd ed., ASME, 2000

4.10. A.F. Liu, *Structural Life Assessment Methods,* ASM International, 1998

4.11. G.R. Irwin, Plastic Zone Near a Crack and Fracture Toughness, *Mechanical and Metallurgical Behavior of Sheet Materials,* Proceedings of Seventh Sagamore Ordnance Materials Conference, Syracuse University Research Institute, 1960, p IV-63 to IV-78

4.12. D.S. Dugdale, Yielding of Steel Sheets Containing Slits, *J. Mech. Phys. Solids,* Vol 8, 1960, p 100–104

4.13. J.R. Rice, Mechanics of Crack Tip Deformation and Extension by Fatigue, *Fatigue Crack Propagation,* STP 415, ASTM, 1967, p 247–306

4.14. A.F. Liu, Crack Growth Under Variable Amplitude Biaxial Stresses, *Proceedings of the 1982 Joint Conference on Experimental Mechanics,* Part 2, Society for Experimental Stress Analysis, 1982, p 907–912

4.15. A.R. Rosenfield, P.K. Dai, and G.T. Hahn, Crack Extension and Propagation Under Plane Stress, *Proceedings, First International Conference on Fracture,* 1966, p 223–258

4.16. G.T. Hahn and A.R. Rosenfield, Sources of Fracture Toughness: The Relation Between K_{IC} and the Ordinary Tensile Properties of Metals, *Symposim on Applications Related Phenomena in Titanium and Its Alloys,* STP 432, ASTM, 1968, p 5

4.17. Metals Test Methods and Analytical Procedures, *Annual Book of ASTM Standards,* Vol 03.01, *Metals—Mechanical Testing; Elevated and Low Temperature Tests; Metallography,* ASTM

4.18. R.G. Forman, V. Shivakumar, and J.C. Newman, Jr., "Fatigue Crack Growth Computer Program NASA/FLAGO Version 2.0," Report JSC-22267A, NASA, May 1994

4.19. K.S. Ravichandran and A.K. Vasudevan, Fracture Resistance of Structural Alloys, *ASM Handbook,* Vol 19, *Fatigue and Fracture,* ASM International, 1996, p 381–392

4.20. Fracture Mechanics Properties of Carbon and Alloy Steels, *ASM Handbook,* Vol 19, *Fatigue and Fracture,* ASM International, 1996, p 614–654

4.21. C.E. Feddersen and D.P. Moon, "A Compilation and Evaluation of Crack Behavior Information on D6AC Steel Plate and Forging Materials for the F-111 Aircraft," Defense Metals Information Center, Battelle Columbus Laboratories, 25 June 1971

4.22. R.J. Bucci, G. Nordmark, and E.A. Starke, Jr., Selecting Aluminum Alloys to Resist Failure by Fracture Mechanisms, *ASM Handbook,* Vol 19, *Fatigue and Fracture,* ASM International, 1996, p 771–812

4.23. K.T. Venkateswara Rao, H.F. Hayashigatani, W. Yu, and R.O. Ritchie, On the Fracture Toughness of Aluminum-Lithium Alloy 2090-T8E41 at Ambient and Cryogenic Temperatures, *Scr. Metall.,* Vol 22, 1988, p 93–98

4.24. J.G. Kaufman, F.G. Nelson, and R.H. Wygonik, Large Scale Fracture Toughness Tests of Thick 5083-O Plate and 5183 Welded Panels at Room Temperature, −260 and −320 °F, *Fatigue and Fracture Toughness—Cryogenic Behavior,* STP 556, ASTM, 1974

4.25. R.L. Tobler, Low Temperature Effects on the Fracture Behavior of a Nickel Base Superalloy, *Cryogenics,* Vol 16, 1976, p 669–674

4.26. D.E. Piper, S.H. Smith, and R.V. Carter, *Metall. Eng. Q.,* Vol 8, 1968, p 50

4.27. R.P. Wei, S.R. Novak, and D.P. Williams, *Material Research and Standards,* TMRSA, Vol 12, ASTM, 1972, p 25

4.28. S. Yoshizawa, K. Yamakawa, and S. Yoshida, *Seventh International Congress on Metallic Corrosion,* ICMC 7 (Rio De Janeiro), 1978, p 863

4.29. M.O. Speidel, *Metall. Trans.,* Vol 6A, 1975, p 631

4.30. D.L. Dull and L. Raymond, Stress History Effect on Incubation Time for Stress Corrosion Crack Growth in AISI 4340 Steel, *Metall. Trans.,* Vol 3, 1972, p 2943–2947

4.31. C.A. Zanis, P.W. Holsberg, and E.C. Dunn, Jr., Seawater Subcritical Cracking of HY-Steel Weldments, *Weld. J. Res. Suppl.,* Vol 59, 1980

4.32. J.M. Krafft and A.M. Sullivan, Effects of Speed and Temperature on Crack Toughness and Yield Strength, *ASM Trans.,* Vol 56, 1963, p 160–175

4.33. W. Server and A.S. Telelman, The Use of Pre-Cracked Charpy Specimens to Determine Dynamic Fracture Toughness, *Engineering Fracture Mechanics,* Vol 4, Pergamon Press, 1972, p 367–375

4.34. Fracture Toughness and Fracture Mechanics, *ASM Handbook,* Vol 8, *Mechanical Testing and Evaluation,* ASM International, 2000, p 563–575

4.35. "Standard Test Methods for Notched Bar Impact Testing of Metallic Materials," E 23, *Annual Book of ASTM Standards,* ASTM

4.36. S.T. Rolfe and J.M. Barsom, *Fracture and Fatigue Control in Structures—Application of Fracture Mechanics,* Prentice-Hall, 1977

4.37. "Standard Test Method for Conducting Drop-Weight Tests to Determine Nil-Ductility Transition Temperature of Ferritic Steels," E 208, *Annual Book of ASTM Standards,* ASTM

4.38. "Standard Test Method for Drop-Weight Tear Tests of Ferritic Steels," E 436, *Annual Book of ASTM Standards,* ASTM

4.39. "Standard Test Method for Dynamic Tear Testing of Metallic Materials," E 604, *Annual Book of ASM Standards,* ASTM

4.40 J.M. Barsom and S.T. Rolfe, Correlations Between K_{IC} and Charpy V-Notch Test Results in the Transition-Temperature Range, *Impact Testing of Metals,* STP 466, ASTM, 1970, p 281–302

4.41 S.T. Rolfe and S.R. Novak, Slow-Bend K_{IC} Testing of Medium-Strength High-Toughness Steels, *Review of Developments in Plane-Strain Fracture Toughness Testing,* STP 463, ASTM, 1970, p 124–159

4.42. J.A. Begley and W.A. Logsdon, "Correlation of Fracture Toughness and Charpy Properties for Rotor Steels," WRL Scientific Paper 71-1E7-MSLRF-P1, Westinghouse Research Laboratory, July 1971

4.43. R.H. Sailors and H.T. Corten, Relationship Between Material Fracture Toughness Using Fracture Mechanics and Transition Temperature Tests, *Fracture Toughness, Proceedings of the 1971 National Symposium on Fracture Mechanics, Part II,* STP 514, ASTM, 1972, p 164–191

4.44. J.D. Landes, Fracture Toughness Testing, *ASM Handbook,* Vol 19, *Fatigue and Fracture,* ASM International, 1996, p 393–409

4.45. Impact Toughness Testing, *ASM Handbook,* Vol 8, *Mechanical Testing and Evaluation,* ASM International, 2000 p 602

4.46. R. Viswanathan, *Damage Mechanisms and Life Assessment of High-Temperature Components,* ASM International, 1989

4.47. C.E. Feddersen, Evaluation and Prediction of the Residual Strength of Center Cracked Tension Panels, *Damage Tolerance in Aircraft Structures,* STP 486, ASTM, 1971, p 50–78

4.48. F.C. Allen, Effect of Thickness on the Fracture Toughness of 7075 Aluminum in the T6 and T73 Conditions, *Damage Tolerance in Aircraft Structures,* STP 486, ASTM, 1971, p 16–38

4.49. G.R. Irwin, Fracture Mode Transition of a Crack Traversing a Plate, *J. Basic Eng. (Trans. ASME), Ser. D,* Vol 82, 1960, p 417–425

4.50. J.C. Ekvall, T.R. Brussat, A.F. Liu, and M. Creager, Preliminary Design of Aircraft Structures to Meet Structural Integrity Requirements, *J. Aircraft,* Vol 11, 1974, p 136–143

4.51. A.F. Liu, unpublished data

4.52. A.F. Liu and M. Creager, On the Slow Stable Crack Growth Behavior of Thin Aluminum Sheet, *Mechanical Behavior of Materials,* Vol I, *Deformation and Fracture of Metals,* Society of Material Science, Japan, 1972, p 558–568

4.53. J.E. Srawley and W.F. Brown, Fracture Toughness Testing, *Fracture Toughness Testing and Its Applications,* STP 381, ASTM, 1965, p 133–198

4.54. F. Erdogan and G.C. Sih, On the Crack Extension in Plates Under Plane Loading and Transverse Shear, *J. Basic Eng. (Trans. ASME), Ser. D,* Vol 85, 1963, p 519–527

4.55. G.C. Sih and B. MacDonald, Fracture Mechanics Applied to Engineering Problems—Strain Energy Density Fracture Criterion, *Eng. Fract. Mech.,* Vol 6, 1974, p 361–386

4.56. K. Palaniswamy and W.G. Knauss, Propagation of a Crack Under General, In-Plane Tension, *Int. J. Fract. Mech.,* Vol 8, 1972, p 114–117

4.57. A.F. Liu, Crack Growth and Failure of Aluminum Plate under In-Plane Shear, *AIAA J.,* Vol 12, 1974, p 180–185

4.58. R.C. Shah, Fracture Under Combined Modes in 4340 Steel, *Fracture Analysis,* STP 560, ASTM, 1974, p 29–52

4.59. B.C. Hoskin, D.G. Graff, and P.J. Foden; see D. Broek, *Elementary Engineering Fracture Mechanics,* Noordhoff International Publishing, 1974, p 332, or A.P. Parker, *The Mechanics of Fracture and Fatigue,* E. & F. N. Spon Ltd., 1981, p 90

4.60. N. Tsangarakis, All Modes Fracture Toughness of Two Aluminum Alloys, *Eng. Fract. Mech.,* Vol 26, 1987, p 313–321

4.61. L.P. Pook, The Effect of Crack Angle on Fracture Toughness, *Eng. Fract. Mech.,* Vol 3, 1971, p 483–486

4.62. L. Banks-Sills, M. Arcan, and H. Gabay, A Mode II Fracture Specimen—Finite Element Analysis, *Eng. Fract. Mech.,* Vol 19, 1984, p 739–750

4.63. L. Banks-Sills and M. Arcan, A Compact Mode II Fracture Specimen, *Fracture Mechanics: 17th Volume,* STP 905, ASTM, 1986, p 347–363

4.64. E.M. Wu, Application of Fracture Mechanics to Anisotropic Plates, *J. Appl. Mech. (Trans. ASME), Ser. E,* Vol 34, 1967

4.65. W.K. Wilson, W.G. Clark, Jr., and E.T. Wessel, "Fracture Mechanics Technology for Combined Loadings and Low-to-Intermediate Strength Metals," Technical Report 10276 (Final), TACOM (Vehicular Components and Materials Laboratory, U.S. Army, Tank Automotive Command), Nov 1968

SELECTED REFERENCES

- J.M. Barsom and S.T. Rolfe, *Fracture and Fatigue Control in Structures—Application of Fracture Mechanics,* 3rd ed., ASTM, 1999
- D. Broek, *Elementary Engineering Fracture Mechanics,* 4th ed., Martinus Nijhoff, 1986
- D. Broek, *The Practical Use of Fracture Mechanics,* Kluwer Academic Publishers, 1988
- G.E. Dieter, Jr., *Mechanical Metallurgy,* 3rd ed., McGraw-Hill, 1986
- R.W. Hertzberg, *Deformation and Fracture Mechanics of Engineering Materials,* 4th ed., John Wiley & Sons, 1996
- J.F. Knott, *Fundamentals of Fracture Mechanics,* John Wiley & Sons, 1973
- H. Liebowitz, Ed., *Fracture,* Academic Press, 1971
- A.P. Parker, *The Mechanics of Fracture and Fatigue,* E. & F. N. Spon, Ltd., 1981
- E.R. Parker, *Brittle Behavior of Engineering Structures,* John Wiley & Sons, 1957
- N. Perrone, H. Liebowitz, D. Mulville, and W. Pilkey, Ed., *Fracture Mechanics,* University Press of Virginia, 1978
- A.S. Tetelman and A.J. McEvily, *Fracture of Structural Materials,* John Wiley & Sons, 1967

CHAPTER 5

Damage Tolerance of Metals

STRUCTURAL FAILURE can result from:

- Stressing an undamaged part to a stress level that exceeds its material allowable
- Premature failure of a part that contains a preexisting crack (or cracks)
- Final failure (fracture) following the formation of a fatigue crack that developed under repeated loads at subcritical stress levels (with or without preexisting damage)

Typically, damage tolerance design/analysis involves assessing the longevity of a structure that contains a preexisting flaw (or crack) in a critical location. Assessing the damage tolerance of a given design against its anticipated use has become an integral step during design/development and structural sizing. For parts already in service, periodic reassessment during the service life is essential. In the case of an aircraft, for example, fatigue loads spectra are updated periodically using loads data collected from the in-flight load recorder. If no crack is found during a scheduled inspection of a given part, the structural life of that part will be reassessed using the updated fatigue loads spectrum (for that part) and the original assigned initial crack size. If a crack of any size is found, the newly detected crack size will be used as the analytical initial crack size. The useful life of that part will be the recalculated life plus the life already in-service. This revised life is used to check against the intended service life. Corrective actions should be taken as required.

A fracture-mechanics-oriented damage tolerance design/analysis procedure consists of the following steps:

- Establish design criteria suitable to the expected use of the vehicle, machinery, etc.
- Identify structural elements and the expected loading conditions.

- Develop a systematic means of identifying the criticality of these structural elements.
- Check the actual criticality of the questionable structural elements by using the best available fatigue and fracture mechanics methodology, and perform trade studies regarding safety, weight, and cost. The fracture mechanics analysis also helps to establish reliable in-service inspection intervals and to periodically update the fracture critical parts list.

An actual application requires specifying the extent and type of damage and the load level to be achieved with the damage present. These basic criteria are assigned on the basis of engineering judgment. The goal is to ensure that damage is readily detectable before the strength is impaired beyond the point of safety. Once the basic criteria are defined, the structures must be designed to meet those criteria. Fracture-mechanics-based analytical tools include (a) methods to determine residual strength and (b) methods to determine fatigue crack growth life. The former have been presented in Chapter 4. This chapter presents tools that complement the former and methodologies needed to accomplish the tasks of the latter. Methods for handling environment-induced crack growth such as corrosion-fatigue and creep-fatigue crack growth are also discussed. Linear-elastic fracture mechanics is the base for all the methods presented in these two chapters. Chapter 6 discusses methods based on nonlinear fracture mechanics, which are suitable for fatigue crack growth and/or fracture due to excessive yielding at the crack tip.

5.1 Stress-Intensity Factor and Damage Tolerance Analysis

As discussed in Chapter 4, $\mathcal{G}_C$ (or $\mathcal{G}_{IC}$) is a real material property that provides a means for determining the structural allowable for a given design configuration. In order to obtain $\mathcal{G}_C$ experimentally, it is first necessary to obtain a mathematics expression for $\mathcal{G}$ in terms of crack dimension geometry. An expression for the structural geometry in question is also required. The stress-intensity factor K, which is directly related to $\mathcal{G}$, can be used in place of $\mathcal{G}$ because K can be mathematically analyzed.

Failed structural members may contain either through-thickness cracks (Fig. 5.1) or part-through cracks (Fig. 5.2). A through-thickness crack is also referred to as a two-dimensional crack or a line crack. A part-through crack is sometimes called a three-dimensional-crack, surface crack, or thumbnail crack (owing to its physical appearance). Note that in some literature, the through-thickness crack and the part-through crack are referred to as the one-dimensional crack and the two-dimensional crack, respectively.

Performing fracture mechanics analysis on a cracked structure (or machine part) requires a set of stress-intensity factors that appropriately represent the structure's global and local geometries, including crack morphology and loading condition. Stress-intensity factors for generic geometries in combination with loading types have been compiled in several handbooks (Ref 5.1–5.5). Stress-intensity factors for configurations commonly used for materials characterization,

structural design, life assessment, and failure analysis are available in Ref 5.6 and 5.7. This section presents and discusses some frequently used mode 1 (crack opening mode) stress-intensity factors. Several crack geometries simulate those found in structural/machine parts, or used as laboratory test specimens for fatigue crack growth and fracture testing.

5.1.1 Two-Dimensional Crack

For a given structural configuration and applied stress, the mathematical expression for K can be determined by a number of analytical methods. For purposes of illustration and a simple introduction, however, first consider a plate of width W containing a through-thickness crack of length $2a$ (see the inset in Fig. 5.3). As long as the plate response to the stress is elastic, the local crack-tip stress at any point is proportional to the applied stress σ (i.e., $\sigma_y \propto \sigma$). It must be expected that the crack-tip stress will also depend on the crack size. Because σ_y depends on $1/\sqrt{r}$ according to Eq 1.77, it is inevitable that it depends on $\sqrt{a}$; otherwise, the dimensions would be wrong. Hence, a simple argument shows that for $\theta = 0$:

$$\sigma_y \propto \sqrt{a}/\sqrt{2\pi r} \qquad \text{(Eq 5.1)}$$

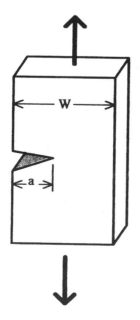

Fig. 5.1 Through-thickness crack at the edge of a plate

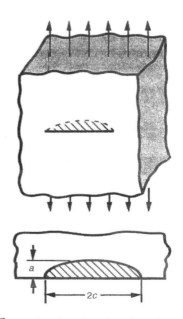

Fig. 5.2 Part-through crack on the surface of a plate

This proportional relation can be used to define an explicit relation between the applied stress σ and the crack-tip stress σ_y such that:

$$\sigma_y = \beta\sigma\sqrt{\pi a}/\sqrt{2\pi r} \qquad \text{(Eq 5.2)}$$

where β is a dimensionless constant. The crack-tip stresses will be higher when W is narrower. Thus, β must depend on W. It is known that β must be dimensionless, yet β cannot be dimensionless and depend on W at the same time unless β depends on W/a or a/W; that is, $\beta = f(a/W)$. Comparison of Eq 5.2, then, shows that:

$$K = \sigma \cdot \sqrt{\pi a} \cdot \beta\left(\frac{a}{W}\right) \qquad \text{(Eq 5.3)}$$

where the third term β on the right is expressed as a function of a/W. If the crack-tip stress is affected by other geometric parameters—for example, if a crack emanates from a hole, the crack-tip stress will depend on the size of the hole—the only effect on stress-intensity factor (and the crack-tip stress) will be in β. Consequently, β will be a function of all geometric factors affecting the crack-tip stresses: $\beta = f(a/W, a/L, a/D,$ etc.$)$, where W, L, and D are plate width, plate length, and diameter of a circular

hole, respectively. Crack-tip stresses are always given by Eq 1.77; the value of K in Eq 1.77 is always given by Eq 5.3 (or its alternate form, Eq 5.4). All effects of geometry are reflected in the geometric parameter β.

In this chapter, we will write a general expression for stress intensity as:

$$K = \sigma \cdot Y \cdot \sqrt{\pi a} \qquad \text{(Eq 5.4)}$$

For a given structural configuration, the definitions of σ and a depend on the problem under consideration. Y represents the complete geometric function such as $\beta = f(a/W)$. Therefore, Y is a product of all the geometric influence factors:

$$Y = \Pi\beta \qquad \text{(Eq 5.5)}$$

or

$$Y = \beta_1 \cdot \beta_2 \cdot \beta_3 \cdot \beta_4 \cdot \ldots \qquad \text{(Eq 5.5a)}$$

Each β factor is a correction factor for one boundary condition.

The value of K can be determined by a number of analytical methods (Ref 5.1, 5.2), such as finite element, boundary integral equation,

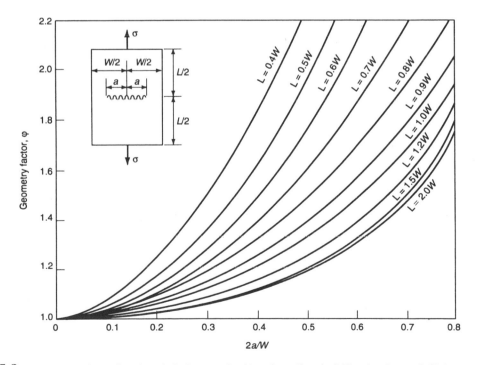

Fig. 5.3 Stress-intensity factors for a through-thickness crack subjected to uniform far-field tension. Source: Ref 5.8

weight function, boundary collocation, and so on. This chapter does not delve into the technicality of these methods, but instead presents only the results (mainly extracted from the open literature). However, keep in mind that each K value is determined for a specific crack size in an explicit structural (or test specimen) configuration. Each K value is obtained by solving one problem at a time. Then, a dimensionless factor Y is equal to K normalized by σ and $\sqrt{\pi a}$. That is, Y is a lumped factor that might represent the total contribution of several β factors. After a series of K values for various crack sizes across the entire crack plane are determined, a close-form equation can be obtained by fitting a curve through all the Y values. When the geometry under consideration contains only one boundary condition, the Y factor is the same as a single β factor.*

The following paragraphs provide short summaries of some mode 1 stress-intensity factors for crack geometries commonly found in structural parts.

Uniform Far-Field Loading. The term center crack literally means that the crack is located in the center of a plate (see the inset in Fig. 5.3). There are many stress-intensity factor solutions for a center crack subjected to far-field uniform tension stress. Among them, the Isida solution (Ref 5.8) is regarded as the most accurate. It accounts for the boundary effect (the free edges that define the width and length of the panel). The stress-intensity expression can be written in the form of Eq 5.4 as:

$$K = \sigma\sqrt{\pi a} \cdot \varphi \qquad \text{(Eq 5.6)}$$

where σ is gross cross-sectional area stress (i.e., ignoring the crack). The geometric factor φ is a dimensionless function of the ratio of the crack length to panel width and the ratio of the panel length to panel width, as given in Fig. 5.3. For a panel with a length greater than two times its width, the effect of panel length on K vanishes. According to Ref 5.9, the finite width correction factors (i.e., the curve labeled as $L \geq 2W$ in Fig. 5.3) can be represented by a secant function:

$$\varphi_W = \sqrt{\sec(\pi a/W)} \qquad \text{(Eq 5.6a)}$$

---

*It is necessary to use many symbols other than Y and β throughout this book due to the involvement of many different cracked body configurations. It varies unsystematically from one case to another. There is no significance as to the connection between symbols and geometries, or loading conditions.

Here the secant function, in radians, is commonly known as the width correction factor for the center crack panel configuration. It is a curve-fitted expression of the Isida solution for a long strip.

If the crack is right at the edge of the plate (see Fig. 5.1), the φ factor is given by Ref 5.2 as:

$$\varphi_e = \sec\beta\,[(\tan\beta)/\beta]^{1/2} \cdot [0.752 + 2.02\,(a/W) + 0.37\,(1 - \sin\beta)^3] \qquad \text{(Eq 5.7)}$$

where a is the total crack length measured from the edge of the plate across the width W, and β is a specific geometric factor for this case such that $\beta = \pi a/(2W)$. The subscript "e" stands for the edge crack; $\varphi_e = 1.122$ at $a/W = 0$.

For a crack that originates at the edge of a circular hole inside an infinitely wide sheet:

$$K = \sigma\sqrt{\pi c} \cdot \varphi_n \qquad \text{(Eq 5.8)}$$

where σ is the gross cross-sectional area stress (i.e., again ignoring the hole and the crack), c is the crack length measured from the edge of the hole (not from the center of the hole), and φ_n is the Bowie solution for cracks emanating from a circular hole (Ref 5.6, 5.10). According to Ref 5.11, the Bowie solution for uniaxial tension (Fig. 5.4) can be fitted by the following equations:

$$\varphi_1 = 0.707 - 0.18\chi + 6.55\chi^2 - 10.54\chi^3 + 6.85\chi^4 \qquad \text{(Eq 5.9a)}$$

and

$$\varphi_2 = 1 - 0.15\chi + 3.46\chi^2 - 4.47\chi^3 + 3.52\chi^4 \qquad \text{(Eq 5.9b)}$$

where $\chi = (1 + c/r)^{-1}$ (r is the hole radius); the subscripts 1 and 2 stand for a single crack and two symmetric cracks, respectively. If the crack(s) are subjected to equi-biaxial tension (i.e., equal tension force in both the X and Y directions), the Bowie factors can be expressed as:

$$\varphi_1 = 0.707 + 0.2744\chi + 2.765\chi^2 - 3.3824\chi^3 + 1.893\chi^4 \qquad \text{(Eq 5.9c)}$$

and

$$\varphi_2 = 1 - 0.0436\chi + 3.1256\chi^2 - 4.46077\chi^3 + 2.6281\chi^4 \qquad \text{(Eq 5.9d)}$$

The Paris and Sih solutions along with the fitted lines for all four configurations are plotted in Fig. 5.4.

To include the finite width effect:

$$K = \sigma\sqrt{\pi c} \cdot \varphi_n \cdot f_n \qquad \text{(Eq 5.10)}$$

where $n = 1$ for a single crack, $n = 2$ for two symmetric cracks, and the hole is located in the center of the plate. The width correction factors, according to Newman (Ref 5.12), are:

$$f_1 = \sqrt{\sec\left[\pi(2r \cdot c)/2(W - c)\right] \cdot \sec(\pi r/W)} \qquad \text{(Eq 5.11a)}$$

and

$$f_2 = \sqrt{\sec\left[\pi(r + c)/W\right] \cdot \sec(\pi r/W)} \qquad \text{(Eq 5.11b)}$$

Crack Line Loading. A crack of length $2a$, subjected to forces per unit thickness P, acting at the center of the crack surfaces, is located centrally in a rectangular plate of height $2H$ and width $2b$ (Fig. 5.5). The expression for the stress-intensity factor in this case is:

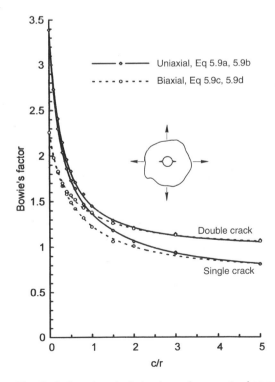

Fig. 5.4 Stress-intensity factors for cracks emanating from a circular hole. Data points: Paris and Sih. Solid lines: uniaxial tension (Newman's fit). Dotted lines: equi-biaxial tension (Liu's fit). Source: Ref 5.6, 5.7, 5.11

$$K = P \cdot Y/\sqrt{\pi a} \qquad \text{(Eq 5.12)}$$

where Y is the geometric factor, as shown in Fig. 5.5.

When forces are applied at a point on the surface of an edge crack* (Fig. 5.6), the stress-intensity factor is defined as:

$$K = 2P \cdot Y/\sqrt{\pi a} \qquad \text{(Eq 5.13)}$$

where the geometric factor Y is defined by the Tada solution (Ref 5.2) as:

$$Y = \alpha_1 - \alpha_2 + \alpha_3 \cdot \alpha_4 \qquad \text{(Eq 5.14a)}$$

where

$$\alpha_1 = 3.52(1 - \lambda)/(1 - \beta)^{3/2} \qquad \text{(Eq 5.14b)}$$

$$\alpha_2 = (4.35 - 5.28\lambda)/(1 - \beta)^{1/2} \qquad \text{(Eq 5.14c)}$$

$$\alpha_3 = [(1.3 - 0.3\lambda^{3/2})/(1 - \lambda^2)^{1/2}] + 0.83 - 1.76\lambda \qquad \text{(Eq 5.14d)}$$

$$\alpha_4 = 1 - (1 - \lambda)\beta \qquad \text{(Eq 5.14e)}$$

where $\lambda = c/a$ and $\beta = a/W$. When the load is applied right at the edge of the plate (i.e., $\lambda = 0$), Eq 5.14(a) reduces to:

$$Y = 3.52/(1 - \beta)^{3/2} - 4.35/(1 - \beta)^{1/2} + 2.13(1 - \beta) \qquad \text{(Eq 5.15)}$$

Compact Specimen. Although not commonly used for structural analysis, the compact specimen (Fig. 5.7) is specified in ASTM Standards E 399 and E 647 as the standard specimen geometry for generating material K_{IC} and fatigue crack growth rate data. Therefore, for the convenience of readers, the ASTM recommended stress-intensity equation for the compact specimen is included here. The equation given in the 1995 *Annual Book of ASTM Standards* (Ref 5.13, 5.14), is:

$$K = \frac{P}{B\sqrt{W}} \cdot F(\alpha_1) \cdot F(\alpha_2) \qquad \text{(Eq 5.16)}$$

---

*The equations presented herein are adopted from Ref 5.2. The accuracy for these equations is 2%. The updated equations that are now available in Ref 5.1 have an accuracy of 1%.

where

$$F(\alpha_1) = 0.886 + 4.64\alpha - 13.32\alpha^2$$
$$+ 14.72\alpha^3 - 5.6\alpha^4 \quad \text{(Eq 5.17a)}$$

and

$$F(\alpha_2) = (2 + \alpha)/(1 - \alpha)^{3/2} \quad \text{(Eq 5.17b)}$$

where $\alpha = a/W$. Equation 5.16 is accurate within 0.5% over the range of a/W from 0.2 to 1. As shown in Fig. 5.7, the height of the specimen (H), the length of the chevron notch, and the loading pin location (h) are functions of the specimen width (W). The ASTM recommended dimension for the thickness B is such that:

- For K_{IC} tests, the preferred thickness is $W/2$. Alternatively, $2 \leq W/B \leq 4$ (with no change in other proportions) is allowed.
- For fatigue crack growth rate tests, B can be in the range of $W/20$ to $W/4$.

For test data reported in the earlier literature, an older version of the K-equation (e.g., one from the 1972 edition of the ASTM standard) may have been used. That is:

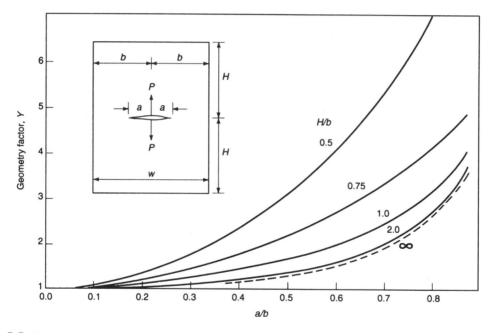

Fig. 5.5 Stress-intensity factors for central crack loaded in a retangular plate with opposite forces at the center of the crack, where P is the force (load) per unit thickness. Source: Ref 5.12

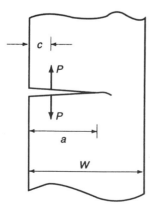

Fig. 5.6 Edge crack loaded by a pair of point forces

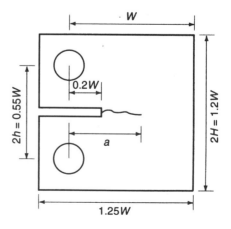

Fig. 5.7 Compact specimen configuration (pinhole diameter, 0.25 W)

$$F(\alpha_1) = 29.6\alpha^{1/2} - 185.5\alpha^{3/2} + 655.7\alpha^{5/2}$$
$$- 1017.0\alpha^{7/2} + 754.6\alpha^{9/2} \quad \text{(Eq 5.17c)}$$

for specimens where $H/W = 0.6$, or

$$F(\alpha_1) = 30.96\alpha^{1/2} - 195.8\alpha^{3/2} + 730.6\alpha^{5/2}$$
$$- 1186.3\alpha^{7/2} + 638.9\alpha^{9/2} \quad \text{(Eq 5.17d)}$$

for specimens where $H/W = 0.486$. The second part of the geometric correction factor, $F(\alpha_2)$, did not exist in the earlier version of the equation.

Bend Specimens. Before the compact specimen became available, test engineers used the three-point bend specimen to determine K_{IC} values. This specimen configuration (Fig. 5.8), cur-

rently included in the ASTM standard, is still being used in the fracture mechanics community. The K-expression listed here is not exactly the same as the previous version given in Ref 5.9. The current version of the equation (taken from ASTM Standard E 399) is:

$$K = \frac{P}{B}\frac{S}{W^{3/2}} \cdot F(\alpha_1) \cdot F(\alpha_2) \quad \text{(Eq 5.18)}$$

where

$$F(\alpha_1) = \frac{3}{2}\sqrt{a} \cdot [(1 + 2\alpha) \cdot (1 - \alpha)^{3/2}]^{-1}$$
$$\text{(Eq 5.19a)}$$

and

$$F(\alpha_2) = 1.99 - \alpha \cdot (1 - \alpha)$$
$$\cdot (2.15 - 3.93\alpha + 2.7\alpha^2) \quad \text{(Eq 5.19b)}$$

where $\alpha = a/W$ and B is the thickness. ASTM Standard E 399 specifies that the total span ($2L$) should have a length equal to $4W$, and the total specimen length should be at least 20% longer than $2L$.

The four-point bend specimen (Fig. 5.9), although not included in any ASTM standard, is

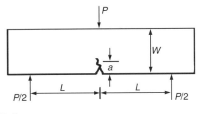

Fig. 5.8 Three-point bend specimen

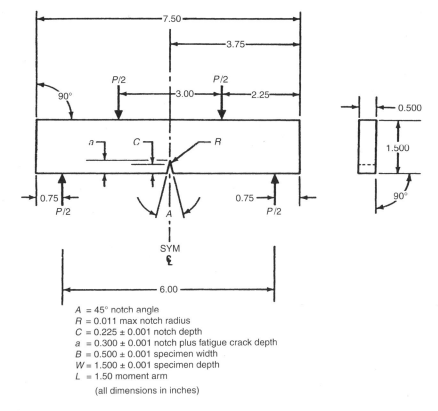

A = 45° notch angle
R = 0.011 max notch radius
C = 0.225 ± 0.001 notch depth
a = 0.300 ± 0.001 notch plus fatigue crack depth
B = 0.500 ± 0.001 specimen width
W = 1.500 ± 0.001 specimen depth
L = 1.50 moment arm

(all dimensions in inches)

Fig. 5.9 Four-point bend specimen

included here because the literature contains a large quantity of K_{IC} data that were developed from this type of specimen. The stress-intensity factor as given in Ref 5.9 is:

$$K = \frac{PL}{BW^{3/2}} \left[\left(\frac{1}{1 - \nu^2} \right) \left(34.7 \, \frac{a}{W} - 55.2 \left(\frac{a}{W} \right)^2 + 196 \left(\frac{a}{W} \right)^3 \right)^{1/2} \right]$$

(Eq 5.20)

Crack at a Pin-Loaded Lug. The stress-intensity factors for cracks at the bore of a pin hole that are presented in this section came from a Lockheed report (Ref 5.15). There are two good reasons for choosing the Lockheed data. First, among the available literature, the Lockheed data are most complete. They cover both straight and tapered lugs (Fig. 5.10), with cracks oriented in many loading directions. Second, for a given lug geometry and loading angle combination, the pin-lug contact pressure distribution was determined for each crack length. However, a few words of caution are in order. Many investigators have addressed the presence of K_2 at the crack tip even though the loading condition

is primarily K_1 (Ref 5.16–5.20). It is not clear whether the Lockheed investigators included K_2 in the course of their calculations. It is conceivable that K_2 was part of the formula that constitutes the special crack-tip element they used.

In any event, the ratio of K_2 to K_1 for short cracks is quite low. That means the contribution of K_2 to the total K may be insignificant. However, the inclusion of K_2 may become important when the crack is long. Experimental test programs conducted at Lockheed have shown satisfactory correlations between test data and predicted lives (Ref 5.21). Therefore, for all practical purposes, the Lockheed lug solutions are considered reasonably accurate, with or without K_2 included.

Figure 5.11 presents the finite element solutions of stress-intensity factors for an axially loaded straight attachment lug. The pin load is applied at the direction normal to the base of the lug ($\theta = 0°$). Either the bearing stress, σ_{br}, or the gross area stress, σ_o, can be used to compute stress intensity. That is, for a single through-thickness crack on one side of the pinhole (perpendicular to the direction of loading, P) according to Ref 5.15:

$$K = \sigma_{br} \cdot F_{RB} \cdot \sqrt{\pi c}$$

(Eq 5.21a)

Because $\sigma_{br} = P/2R_i t$, and $\sigma_o = P/2R_O t$, Eq 5.21(a) can be rewritten as:

$$K = \sigma_o \cdot F_{RB} \cdot R_O/R_i \cdot \sqrt{\pi c}$$

(Eq 5.21b)

The values of F_{RB} for five R_O/R_i ratios are given in Fig. 5.11.

All the equations for a straight lug are also applicable to a tapered lug (with a new set of F_{RB} factors). The F_{RB} factors for the 0° loading are presented in Fig. 5.12. Limited finite element data are available for tapered lugs loaded in other directions. The through-crack solutions for $\theta = 180°$, $-45°$, and $-90°$ are presented in Fig. 5.13 to 5.15, respectively.

5.1.2 Three-Dimensional Crack

Semielliptical Crack on the Surface of a Semi-Infinite Solid. A part-through crack originating on the surface of a semi-infinite solid, or a plate, is usually modeled as one-half of an ellipse (Fig. 5.16). Either the major or the minor axis of the ellipse may be placed on the front surface of the plate. As shown in Fig. 5.16, the length and depth of the crack are designated as

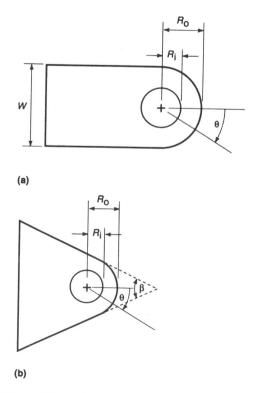

(a)

(b)

Fig. 5.10   Attachment lug configurations. (a) Straight. (b) Tapered

$2c$ and a, respectively. The crack shape is described by an aspect ratio of $a/2c$ (or a/c).

When tensile loads are applied normal to a semielliptical crack in a semi-infinite solid, the expression for K, for some point on the periphery of the crack, is given by Ref 5.22 as:

$$K = \sigma\sqrt{\pi a/Q} \cdot f_\varphi \cdot \alpha_f \qquad \text{(Eq 5.22)}$$

where the applied stress σ is equated to the applied load (force) over the full cross section of the plate (ignoring the crack). Both Q and f_φ are parameters accounting for the shape of the ellipse. Initially, Q was presented in the literature as a function of Φ and the σ/F_{ty} ratio, where F_{ty} is the material tensile yield strength, and Φ is a crack shape parameter, the complete elliptical integral of the second kind (dimensionless). That is:

$$\Phi = \int_0^{\pi/2} \sqrt{(1 - \kappa^2 \sin^2 \theta)}\, d\theta \qquad \text{(Eq 5.23a)}$$

and

$$\kappa = (1 - a^2/c^2)^{1/2} \qquad \text{(Eq 5.23b)}$$

To compute the elliptical integral of the second kind, i.e., Eq 5.23(a), the following approximate formula can be used:

$$\Phi = 1 + 4.593(a/2c)^{1.65} \qquad \text{(Eq 5.23c)}$$

A graphical solution of Φ is given in Fig. 5.17. As explained in Chapter 4, and in the next section regarding a surface flaw in a plate, the term

σ/F_{ty} is used for converting the physical crack length to the effective crack length (to include the effect of crack-tip plasticity). If this term is deleted from Q, the elastic K is solely related to Φ. In the remainder of this chapter, Φ and $\sqrt{Q}$ are regarded as the same (without plasticity correction).

The angular function f_φ has the following form (dimensionless):

$$f_\varphi = [(a/c)^2 \cdot \cos^2 \varphi + \sin^2 \varphi]^{1/4} \qquad \text{(Eq 5.24a)}$$

for $a/c \leq 1$, and:

$$f_\varphi = [(c/a)^2 \cdot \sin^2 \varphi + \cos^2 \varphi]^{1/4} \qquad \text{(Eq 5.24b)}$$

for $a/c > 1$.

In Eq 5.24(a) and (b), φ is a parametric angle measured from the plate surface toward the center of the crack (i.e., $\varphi = 0°$ is on the plate surface, $\varphi = 90°$ is at the maximum depth of the crack). This terminology is used in all open literature for defining the position of a point on the ellipse. However, φ is not an angle that actually connects the center of the ellipse to a specific point on the physical crack periphery. To translate φ to β (the angle between the plate surface and a specific point on the periphery of the ellipse, Fig. 5.18), the following relationship between φ and the geometric angle β can be used:

$$\beta = \tan^{-1}[(a/c) \cdot \tan \varphi] \qquad \text{(Eq 5.24c)}$$

The parameter α_f in Eq 5.22 is called the front face influence factor (dimensionless), accounting for the influence of the free surface coinci-

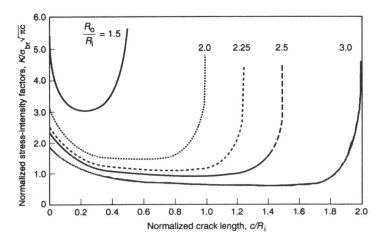

Fig. 5.11 Normalized stress-intensity factors for single through-thickness cracks emanating from a straight lug subjected to a pin loading applied in the 0° loading direction. Source: Ref 5.15

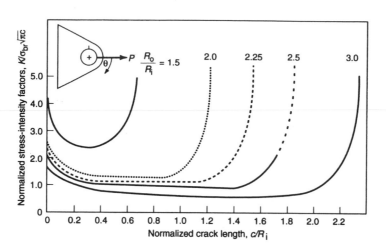

Fig. 5.12 Normalized stress-intensity factors for single through-thickness cracks emanating from a tapered lug subjected to a pin loading applied in the 0° loading direction. Source: Ref 5.15

dent with the visible length of the crack. It is a function of a/c and φ. For $a/c = 0$ (i.e., a scratch on the surface), one can idealize it as equivalent to an edge crack in a semi-infinite sheet. Therefore, $\alpha_f = 1.122$ for all points along the entire crack front. For a semicircular crack ($a/c = 1$), the solution given by Smith (Ref 5.23) is shown in Fig. 5.19. The α_f values for all intermediate a/c ratios can be obtained by interpolation.

For a given a/c ratio, f_φ is a function of φ (radians). Therefore, the combination of α_f, f_φ, and $1/\sqrt{Q}$ is the source of the variance in K values along the crack periphery. During crack propagation, each point along the crack front grows a different amount in a different direction. As a result, the crack shape continuously changes as the crack extends. In making struc-

tural life prediction, a minimum of two K values (i.e., at the maximum depth and on the surface) is required for each crack size and its corresponding aspect ratio. Figure 5.20 illustrates the fundamentals of K variation as a function of a/c and φ. The stress-intensity factor F, which is equal to $f_\varphi \cdot \alpha_f \cdot \sqrt{1/Q}$ (according to Eq 5.22), has been computed for several points on the crack periphery (for several a/c ratios) and is plotted in Fig. 5.20. Conceptually, the location

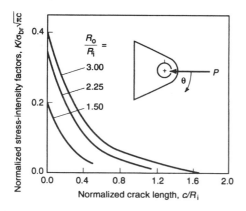

Fig. 5.13 Normalized stress-intensity factors for single through-thickness cracks emanating from a tapered lug subjected to a pin loading applied in the 180° loading direction. Source: Ref 5.15

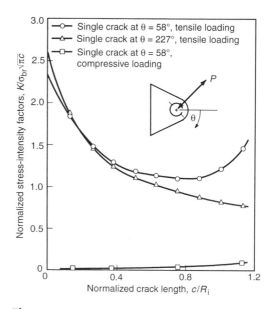

Fig. 5.14 Normalized stress-intensity factors for single through-thickness cracks emanating from a tapered lug subjected to a pin loading applied in the −45° loading direction and its reversed direction, $R_o/R_i = 2.25$. Source: Ref 5.15

of the highest K value is at the maximum depth (for $a < c$). However, the highest K value is on the surface of the plate when $a \geq c$. As shown in Fig. 5.20, the switching actually takes place at $a/c \cong 0.8$. For this flaw shape (i.e., $a/c \cong 0.8$), all the F values along the crack boundary are approximately the same. Thus, crack growth rates at each point along the crack boundary are approximately the same. Therefore, whether the crack starts as a scratch (having $a \ll c$) or a deep cavity (having $c \ll a$), given time its flaw shape will eventually stabilize at $a/c \simeq 0.8$.

We have worked out two hypothetical cases to demonstrate the flaw shape change phenomenon. The initial flaw shape in one case is shallow ($a/2c = 0.3$). The initial flaw in another case is a semicircle ($a/2c = 0.5$). We used a hypothetical material fatigue crack growth rate curve, then computed and recorded the history of crack growth rate increments and the change of flaw shape corresponding to each step of crack growth increment. Two crack growth history maps (one for each initial flaw shape) are plotted in Fig. 5.21 and 5.22. The method for computing fatigue crack growth is described in a later section. Since the exact amount of crack growth in each step of the entire crack growth history is not the issue here, we only show the

flaw shape changes in these maps. It is seen that the shallow flaw grew from an initial $a/2c = 0.3$ to a stabilized flaw shape, $a/2c = 0.429$. On the other hand, the initially semicircular flaw also grew to a stabilized flaw shape of $a/2c = 0.429$.

Historically, prior to the availability of Smith's front face solution (Ref 5.23), there was a perception that the highest K is always at the maximum depth of the shallow flaw (including $a/2c = 0.5$). Fatigue crack propagation and fracture were computed based entirely on the K values at the A-tip of the crack. This assumption is apparently all right for fracture because K_{IC} failure happens fast without slow stable crack growth, and that flaw shape change during and after fracture is not of concern here. However, in the case of fatigue crack propagation, that would mean flaw shape change is ignored along

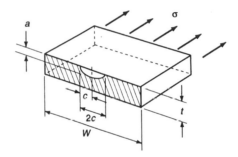

Fig. 5.16 Configuration of a semielliptical surface crack. The C-tip is at either end of the major axis (along the surface). The A-tip is at the maximum depth of the minor axis. In this book, C-tips are always on the surface; the A-tip is always at the maximum depth, whether c is greater than a, or vice versa.

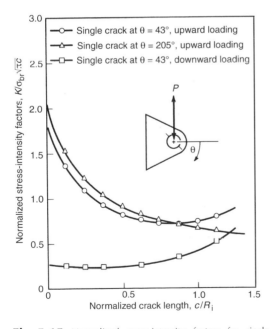

Fig. 5.15 Normalized stress-intensity factors for single through-thickness cracks emanating from a tapered lug subjected to a pin loading applied in the $-90°$ loading direction and its reversed direction, $R_o/R_i = 2.25$. Source: Ref 5.15

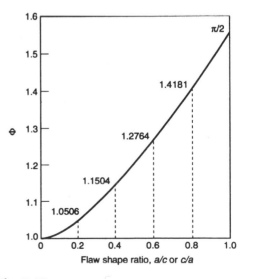

Fig. 5.17 Elliptical integral Φ

the way in each step of calculation. According to Eq 5.22, all the computed K values would be erroneous. That would cause a serious error in the computed life. After a period of fatigue crack propagation, the fracture stress would also be in serious error because the initial and the final flaw shapes are not the same. For comparison purposes, a dotted line is drawn in Fig. 5.21 and 5.22. Each dotted line in these figures represents the initial flaw shape at the crack's final position. That is, it shows what the final size and shape of the crack would have been like if the changes in flaw shape were not taken into account.

Figure 5.23 compares the analytical stable flaw shape with test data. The data sample consists of nine specimens made of 7075-T6 aluminum. These specimens were either 1.27 or 0.635 cm (0.5 or 0.25 in.) thick and 7.62 cm (3 in.) wide, with initial flaw depths approximately equal to 5 to 10% of the specimen thickness. The initial flaw shape $a/2c$ ranged from 0.262 to 0.533. The specimens were subjected to constant-amplitude cyclic loading, with the minimum to maximum applied stress ratio equal to 0.1. The maximum applied stress levels were not the same on every specimen. It covers a wide range of applied stress levels, $13 \leq \sigma_{max} \leq 50$ ksi. Examination of the broken halves of the specimens after testing revealed that they had a stabilized final flaw shape of $a/2c = {\sim}0.43$. The vertical dotted line in Fig. 5.23 represents the final flaw shape value based on Fig. 5.20. The flaw shape data points would have followed the 45° line if they had remained unchanged.

Semielliptical Crack on the Surface of a Plate. For a crack in a rectangular plate of finite thickness and width, the solution for K with a given crack size and shape should account for the influence of the width and the front and back faces of the plate. Irwin's original equation for K at the maximum depth of a shallow crack is (Ref 5.22):

$$K = 1.95\sigma\sqrt{a}/\Phi \qquad \text{(Eq 5.25)}$$

Here the value of 1.95 includes $\sqrt{\pi}$ and a 1.1 factor for the front face correction. Although width correction is not included, Irwin did try to account for the back face and the crack-tip plasticity. Equation 5.25 becomes:

$$K = 1.95\sigma\sqrt{a/Q} \qquad \text{(Eq 5.26)}$$

Using a plane-strain plastic zone radius correction factor, Eq 4.7, an approximate solution for the Q parameter is given as:

$$Q = \Phi^2 - 0.212(\sigma/F_{ty})^2 \qquad \text{(Eq 5.27)}$$

When σ/F_{ty} is very small, Eq 5.26 reduces back to Eq 5.25. Note that Eq 5.27 is not applicable to locations other than the maximum depth of the shallow crack. This is particularly true on the plate surface, where the state of stress is plane stress. It has also been assumed that Eq 5.27 is valid for less than one-half the plate thickness.

Finite element solutions for cracks subjected to tension or bending have been developed by Newman and Raju (Ref 5.24, 5.25). These solutions were subsequently updated by Raju et al. (Ref 5.26). The new data, the values of $K/(\sigma\sqrt{\pi a/Q})$, are presented in Tables 5.1 and 5.2. Note that all the geometric factors have been

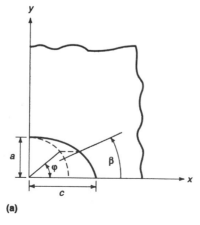

(a)

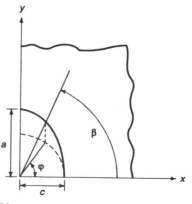

(b)

Fig. 5.18 Definition of φ and β for an elliptical crack

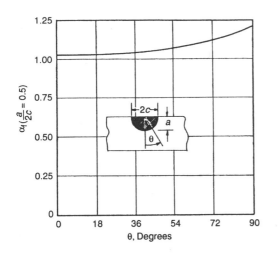

Fig. 5.19 Front-face geometric coefficient for a semicircular part-through flaw. Source: Ref 5.23

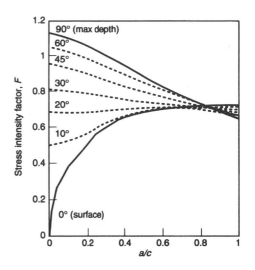

Fig. 5.20 Variation of stress-intensity factors for a shallow crack in a semi-infinite solid, according to Eq 5.22 with $F = f_\varphi \cdot \alpha_f \cdot \sqrt{1/Q}$

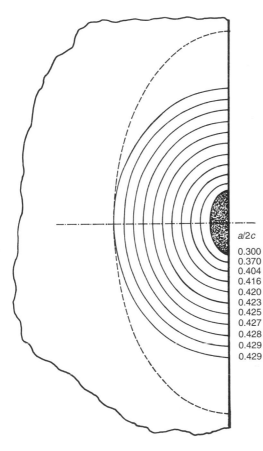

a/2c

0.300
0.370
0.404
0.416
0.420
0.423
0.425
0.427
0.428
0.429
0.429

Fig. 5.21 Propagation of surface flaw under uniform tension for initial flaw shape $a/2c = 0.3$. Courtesy of T.M. Hsu

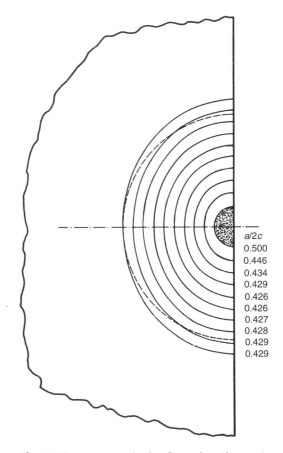

a/2c

0.500
0.446
0.434
0.429
0.426
0.426
0.427
0.428
0.429
0.429

Fig. 5.22 Propagation of surface flaw under uniform tension for initial flaw shape $a/2c = 0.5$. Courtesy of T.M. Hsu

lumped into one. These data have been built into the crack library in the NASA/FLAGRO computer program (Ref 5.27), with which interpolations are accomplished by using a nonlinear table lookup routine to obtain stress-intensity factors that are not available in the tables.

A final note on part-through crack growth behavior is directed at the estimate of stress-intensity factors while the crack front is approaching the back face of the plate. Direct use of the above equations would lead to a discontinuity where the calculation of K is suddenly switched from the part-through crack solution to the through-crack solution. This is commonly known as the "transition phenomenon." Techniques for making a smooth transition are available (Ref 5.15, 5.28–5.30). Plastic yielding at the crack tip may become excessive when the crack approaches the back face of the plate. The crack tip stress field may no longer be linear elastic. Back face yielding might have completely altered the crack growth and fracture behavior of the plate. Reference 5.28 offers some insight to this problem.

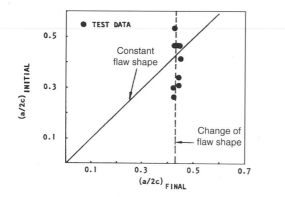

Fig. 5.23 Comparison of final flaw shape: test data versus theories. Courtesy of T.M. Hsu

Table 5.1 Correction factors ($K/\sigma\sqrt{\pi a/Q}$) for stress intensity at shallow surface cracks under tension

2c/W	a/c	a/t				
		0.0	0.20	0.50	0.80	1.0
At the C-tip: tensile loading						
0.0	0.20	0.5622	0.6110	0.7802	1.1155	1.4436
0.0	0.40	0.6856	0.7817	0.9402	1.1583	1.3383
0.0	1.00	1.1365	1.1595	1.2328	1.3772	1.5145
0.1	0.20	0.5685	0.6133	0.7900	1.1477	1.5014
0.1	0.40	0.6974	0.7824	0.9456	1.2008	0.4256
0.1	1.00	1.1291	1.1544	1.2389	1.3892	1.5273
0.4	0.20	0.5849	0.6265	0.8438	1.3154	1.7999
0.4	0.40	0.7278	0.8029	1.0127	1.4012	1.7739
0.4	1.00	1.1366	1.1969	1.3475	1.5539	1.7238
0.6	0.20	0.5939	0.6415	0.9045	1.5056	2.1422
0.6	0.40	0.7385	0.8351	1.1106	1.6159	2.1036
0.6	1.00	1.1720	1.2855	1.5215	1.8229	2.0621
0.8	0.20	0.6155	0.6739	1.0240	1.8964	2.8650
0.8	0.40	0.7778	0.9036	1.3151	2.1102	2.9068
0.8	1.00	1.2630	1.4957	1.9284	2.4905	2.9440
1.0	0.20	0.6565	0.7237	1.2056	2.6060	4.2705
1.0	0.40	0.8375	1.0093	1.6395	2.9652	4.3596
1.0	1.00	1.3956	1.8446	2.6292	3.6964	4.5865
At the A-tip: tensile loading						
0.0	0.20	1.1120	1.1445	1.4504	1.7620	1.9729
0.0	0.40	1.0900	1.0945	1.2409	1.3672	1.4404
0.0	1.00	1.0400	1.0400	1.0672	1.0883	1.0800
0.1	0.20	1.1120	1.1452	1.4595	1.7744	1.9847
0.1	0.40	1.0900	1.0950	1.2442	1.3699	1.4409
0.1	1.00	1.0400	1.0260	1.0579	1.0846	1.0820
0.4	0.20	1.1120	1.1577	1.5126	1.8662	2.1012
0.4	0.40	1.0900	1.1140	1.2915	1.4254	1.4912
0.4	1.00	1.0400	1.0525	1.1046	1.1093	1.0863
0.6	0.20	1.1120	1.1764	1.5742	1.9849	2.2659
0.6	0.40	1.0900	1.1442	1.3617	1.5117	1.5761
0.6	1.00	1.0400	1.1023	1.1816	1.1623	1.0955
0.8	0.20	1.1120	1.2047	1.6720	2.2010	2.5895
0.8	0.40	1.0900	1.1885	1.4825	1.6849	1.7727
0.8	1.00	1.0400	1.1685	1.3089	1.2767	1.1638
1.0	0.20	1.1120	1.2426	1.8071	2.5259	3.0993
1.0	0.40	1.0900	1.2500	1.6564	1.9534	2.0947
1.0	1.00	1.0400	1.2613	1.4890	1.4558	1.3010

Note: These values are built into the NASA/FLAGRO program (Ref 5.27).

However, so far, a good method of handling is non-existent.

5.2 Determination of Stress-Intensity Factors for Nonstandard Configurations

The value of K can be determined by a number of analytical methods. This chapter touches on two popular methods: an engineering method that is based on superposition and/or compounding of several known solutions, and the finite element method.

In general, regardless of what method is used, the stress-intensity factor is determined for a stationary crack having a specific crack size in an explicit structural (or test specimen) configura-tion. A center-crack panel, for example, would need a number of analyses (a dozen crack lengths, or more) to provide enough data points to cover the entire width of the panel. Acquiring a large quantity of K values for a single config-uration is necessary because in damage tolerance analysis the initial damage may start as a very small crack (near zero). When the situation al-lows (e.g., at very low stresses), the crack can propagate across the entire panel before fracture. For convenience, engineers may obtain a close-form equation by fitting a curve through all the K values, or otherwise present them in tabulated or graphical form. All the stress-intensity ex-pressions presented in the previous section were practically developed in this manner. The K val-ues listed in Tables 5.1 and 5.2 are perfect ex-amples. Many more stress-intensity solutions for a variety of geometric and loading combinations

Table 5.2 Correction factors ($K/\sigma\sqrt{\pi a/Q}$) for stress intensity at shallow surface cracks in bending

2c/W	a/c	a/t 0.0	0.20	0.50	0.80	1.0
At the C-tip: bending loading						
0.0	0.20	0.5622	0.5772	0.6464	0.7431	0.8230
0.0	0.40	0.6856	0.7301	0.7694	0.7358	0.6729
0.0	1.00	1.1365	1.0778	1.0184	0.9716	0.9474
0.1	0.20	0.5685	0.5809	0.6524	0.7646	0.8624
0.1	0.40	0.6974	0.7315	0.7856	0.8008	0.7895
0.1	1.00	1.1291	1.0740	1.0114	0.9652	0.9435
0.4	0.20	0.5849	0.5981	0.6934	0.8654	1.0249
0.4	0.40	0.7278	0.7519	0.8327	0.9312	1.0068
0.4	1.00	1.1366	1.1079	1.0634	1.0358	1.0268
0.6	0.20	0.5939	0.6158	0.7438	0.9704	1.1802
0.6	0.40	0.7385	0.7816	0.8906	1.0215	1.1211
0.6	1.00	1.1720	1.1769	1.1759	1.1820	1.1900
0.8	0.20	0.6155	0.6446	0.8320	1.1794	1.5113
0.8	0.40	0.7778	0.8386	1.0150	1.2791	1.5073
0.8	1.00	1.2630	1.3633	1.4785	1.5360	1.5431
1.0	0.20	0.6565	0.6848	0.9593	1.5053	2.0518
1.0	0.40	0.8375	0.9232	1.2285	1.7607	2.2637
1.0	1.00	1.3956	1.6821	2.0140	2.1482	2.1446
At the A-tip: bending loading						
0.0	0.20	1.1120	0.8825	0.6793	0.3063	−0.0497
0.0	0.40	1.0900	0.8292	0.5291	0.1070	−0.2489
0.0	1.00	1.0400	0.7411	0.3348	−0.1149	−0.4396
0.1	0.20	1.1120	0.8727	0.6697	0.3071	−0.0348
0.1	0.40	1.0900	0.8243	0.5170	0.1047	−0.2336
0.1	1.00	1.0400	0.7398	0.3322	−0.1172	−0.4408
0.4	0.20	1.1120	0.8683	0.6794	0.3439	0.0291
0.4	0.40	1.0900	0.8330	0.5270	0.1257	−0.1989
0.4	1.00	1.0400	0.7602	0.3572	−0.1080	−0.4543
0.6	0.20	1.1120	0.8904	0.7248	0.4033	0.0915
0.6	0.40	1.0900	0.8625	0.5803	0.1678	0.1874
0.6	1.00	1.0400	0.7982	0.4072	−0.0856	−0.4750
0.8	0.20	1.1120	0.9191	0.7925	0.5102	0.2254
0.8	0.40	1.0900	0.8987	0.6619	0.2524	−0.1300
0.8	1.00	1.0400	0.8556	0.4981	−0.0329	−0.4960
1.0	0.20	1.1120	0.9545	0.8827	0.6666	0.4351
1.0	0.40	1.0900	0.9417	0.7723	0.3810	−0.0250
1.0	1.00	1.0400	0.9323	0.6312	0.0505	−0.5249

Note: These values are built into the NASA/FLAGRO program (Ref 5.27).

are available in various handbooks (Ref 5.1–5.7).

Damage tolerance analysis (on a real structural member under a specific loading condition) usually treats the local structural area under consideration as if it were the same as a simplified model for which known stress-intensity solutions exist. For complex structural configurations where handbook solutions are not available, determining K becomes an integral and critical step in performing a structural life predictive analysis. Now that high-speed, workstation-based commercial structural analysis computer codes are available, the finite element technique has become a popular tool for both structural analysis and stress-intensity factor determination.

5.2.1 Principles of Superposition and Compounding

The original (raw) data of a stress-intensity factor represents a total sum of stress intensities attributed to a configuration that contains several boundary conditions. Engineers often try to break the lumped factor into pieces, each representing the effect attributed to a single boundary condition. The part-through flaw equations of Ref 5.24 and 5.25 are perfect examples. The advantage of separating a lumped geometric factor into several individual segments is obvious. The fracture mechanics analysts can build around a compounded equation to suit the crack model under consideration by picking and choosing several known solutions from different sources. Superposition and compounding are the most common and simplest techniques for obtaining stress-intensity factors. The methods are quite simple, and are discussed below.

A complex structural configuration may encompass complicated geometry and/or loading by more than one source of input loads. It can be treated as if it were a group of several simple configurations having separate boundary conditions. Known stress-intensity solutions are available to each of these idealized segments. The K-factors for the simple configurations are then added or multiplied together to obtain the required solution. This is less sophisticated but perhaps more expedient than other analytical methods. The advantage is that the compounded K-factors can be conveniently expressed in a parametric closed-form as functions of various dimensionless geometric variables (e.g., the ratio of crack length to panel width, the ratio of crack length to hole radius, etc.), useful to designers and analysts. This engineering equation can be easily incorporated into an operative crack growth analysis computer code. Stress-intensity factors for any combination of structural geometry, loading condition, crack length, and crack shape (even changes in crack length and crack shape while the crack growth is in progress) can be rapidly determined. Furthermore, once a set of workable engineering equations is determined, the cost of calculating a K value is quite economical.

Referring to Eq 5.3 and 5.4, when Y (or β) is a function of more than one variable, it is desirable to separate the β function into a series of dimensionless parametric functions, where each segment represents an explicit boundary condition. Therefore, Eq 5.4 can be rewritten as:

$$K = \sigma\sqrt{\pi a} \cdot \Pi\alpha \qquad \text{(Eq 5.28)}$$

where $\Pi\alpha$ is the product of a series of dimensionless parametric functions, or factors, accounting for the influence of the part and crack geometries and loading condition. In the absence of any geometric influence—for example, a through-thickness crack in an infinitely wide sheet under uniform far-field tension—$\Pi\alpha$ approaches unity ($\beta = 1$) and Eq 5.6 reduces to the basic stress-intensity expression:

$$K = \sigma\sqrt{\pi a} \qquad \text{(Eq 5.29)}$$

where σ is the applied stress and a is the half-length of the through-thickness crack.

Consider the case of two cracks emanating from the opposite sides of a hole in a finite-width sheet. The stress-intensity expression would be that of Eq 5.10, which is a compounded solution that consists of Bowie's hole factor for two cracks and a Newman's width correction factor. What if the Newman correction factor is not available? The natural thing to do is to use the Isida width correction factor. The compounded stress-intensity expression that we just created would be:

$$K = \sigma\sqrt{\pi c} \cdot \varphi_2 \cdot \varphi_w \qquad \text{(Eq 5.30a)}$$

where φ_2 comes from Eq 5.9(b). Modifying Eq 5.6(a) to include the hole radius, we have:

$$\varphi_w = \sqrt{\sec\left(\frac{\pi(c + r)}{W}\right)} \qquad \text{(Eq 5.30b)}$$

Comparison of Eq 5.30(b) with Eq 5.11(b) shows that the former is short by a factor of $[\sec(\pi r/W)]^{1/2}$. This would mean a 4% error for a moderate size hole ($r = W/8$). The error would have been less if the hole were smaller. However, the error increases to 19% for a relatively large hole, i.e., $W = 2D$ ($r = W/4$), which is not a common configuration.

Techniques are available for making up fitting functions for stress-intensity factors. For life assessment, the fracture mechanics analyst needs to use stress-intensity factors across the entire cross section of the part under consideration. Accurate stress-intensity solutions are difficult to obtain for cracks that are very small or close to the free boundary of the part, but this hurdle can be overcome via sound engineering judgment. Except for the center crack, most cracks are exposed at a free surface—that is, starting at an edge of some sort. A crack emanating from a hole can be considered an edge crack at a site of stress concentration.

This notion leads to a compounding solution for a very small crack at the edge of a circular hole. The compounding solution consists of two components: the stress-concentration factor of the hole and the stress-intensity factor for an edge crack with zero crack length. The stress-concentration factor for a circular hole in an infinite sheet is 3.0 (the K_t based on gross area stress); the edge crack geometric factor is 1.122. Therefore, the Bowie factor for $c/r = 0$ should be in the neighborhood of 3.37 (3 × 1.122). A value of 3.39 was chosen by Paris and Sih (Ref 5.6). When using Eq 5.9(a) and (b), a pair of values—3.39 and 3.36—will be obtained. For the equi-biaxial tension loading, the value for K_t is 2. The Bowie factor for $c/r = 0$ would then be in the neighborhood of 2.244 (2 × 1.122), as compared to the handbook solution of 2.26.

Next, consider the case of a long crack. The local stress distribution decreases from K_t at the hole edge to unity at locations far away from the hole, and the influence of being an edge crack vanishes; therefore, the total crack (including the hole) can be considered as a center crack. For the double crack configuration, the total crack length can be approximated by $2c$ because the hole size becomes negligible at $c/r \to \infty$. Therefore, $\varphi_2 = 1$ for $c/r \to \infty$, as given by Eq 5.9(b). For the single crack configuration, the crack length c is treated as the total crack length. It means that Eq 5.9 would become $K = \sigma\sqrt{\pi c/2}$, or $K = \sigma\sqrt{\pi c} \cdot \sqrt{1/2}$. Consequently, this leads to $\varphi_1 = 0.707$, as given by Eq 5.9(a).

As a rule of thumb, superposition is usually involved with combinations of loading conditions, and compounding is used for obtaining a solution for the combination of geometric variables. It should be noted that many engineers use "compounding" as a general term; it might mean superposition or compounding, or both. A compounded K-expression is usually formulated by multiplication of a number of individual K-expressions for each geometric boundary involved in a general configuration under consideration. For example, the K-expression for the center-cracked panel (the combination of Eq 5.6 and Eq 5.6a) can be regarded as a compounded solution that combines the basic K-solution for an infinite sheet and a so-called width correction factor. The K-expression for the surface flaw (Eq 5.22) is another well-represented example. In fact, most of the stress-intensity factors presented in the last section are compounded factors.

Certain known solutions for a partially loaded crack may be superimposed, by adding or subtracting one to another, to obtain approximate solutions for cracks in arbitrary stress fields. The summation may consist of many idealized individual conditions. The only restriction is that the stress-intensity factors must be associated with the same structural geometry (including crack geometry). Illustration of this principle is given in the following examples. For detailed descriptions of the principle and implementation of superposition and compounding, see Ref 5.15 and 5.31.

Example 5.1. Consider a sheet that contains a pin-loaded hole. The pin load is balanced by the reaction stress uniformly distributed at the other end of the sheet (Fig. 5.24). To keep the

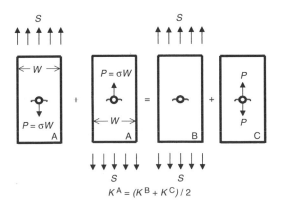

Fig. 5.24 Stress-intensity factor for a pin-loaded hole obtained by superposition

problem simple, assume that the size of the pin-loaded hole is very small so that the actual contact pressure distribution between the pin and the hole is not an issue. In addition, ignore the small amount of mode 2 deformation at the crack tip that would have been caused by this particular loading condition.

We know that there are solutions available for the center crack subject to either uniform far-field loading or point loading applied on the crack line. We also know that a stress-intensity factor for an open hole subjected to uniform far-field loading (the Bowie factor, Fig. 5.4) is available. With this knowledge in hand, we can set up a superposition scheme (as depicted in Fig. 5.24) to derive an approximate stress-intensity solution. The trick is to set up a balanced configuration so that the sum of stress-intensity factors on one side is equal to the sum of stress-intensity factors on the other side. Here, K^C (K for part C) is the K for a pair of concentrated loads; K^B (K for part B) is the K for Bowie's open hole solution. These stress-intensity factors are already given above. The unknown, which is on the left side of Fig. 5.24, has two sketches of part A. The loading conditions for these two sketches are identical. They are positioned upside down to balance with the loads of the right side. Thus, the final formulation for one of the part A configurations would be:

$$K^A = (K^B + K^C)/2 \qquad \text{(Eq 5.31)}$$

The P term in Fig. 5.24 is the applied force per unit thickness.

Example 5.2. The problem in Example 5.1 can be extended to include load transfer. As shown in Fig. 5.25, the stress field can be treated as if it were in two portions: a uniformly stretched strip (open hole) and a concentrated load at the fastener hole reacted by uniform stresses at the other end (loaded hole). In Fig. 5.25, α is the fraction of load transfer. Here, the open hole part is equivalent to K^B (with load = $(1 - \alpha)P$), and the loaded hole part is equivalent to K^A (with load = αP).

5.2.2 Finite Element Methods

Among the many analytical techniques used for determining K, the methods in the finite element category are most suitable to structural analysis application. Using finite element analysis, the crack is treated as an integral part of the structure, which can then be modeled in as much detail as necessary to accurately reflect the structural load paths, both near to and far from the crack tip. This technique is particularly useful in modeling three-dimensional (solid-type) structures with flaws. Therefore, in recent years, a great deal of attention has been given to its development.

Conventional Element versus Special Element. The approach to determining K using finite elements is to match the nodal point displacements (or element stresses) with classical crack-face displacement or crack-tip stress solutions (i.e., Eq 1.77 to 1.82). When performing a finite element analysis, one method of obtaining accuracy in determining the singular stress field ahead of the crack tip is to provide a fine mesh of elements around the crack. Using a conventional (h-version) finite element code, a model (Fig. 5.26) containing an extremely large number of elements in the vicinity of the crack tip is required; hence, setup and computer run time increase.

Special elements, such as the cracked element and the quarter point/collapse element, have been developed to account for crack-tip singularities. In the past 30 years, many types of special elements have been developed and published in the literature. Hundreds of documents that deal with theories, optimization, and application are available. These were thought to obtain a direct measure of the crack-tip stress singularity, or to provide a more accurate account of the stress and displacement fields at the crack tip, or both. The majority of this work has dealt with plane elements, that is, two-dimensional geometry. Published work on modeling with solid elements is limited.

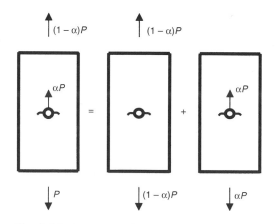

Fig. 5.25 Superposition of stress fields for a fastener hole with load transfer

The Williams, or Westergaad, stress function (Ref 5.32, 5.33) is often built into the element for determining the stress-intensity factor. At times, other stress functions have also been used. In general, whether it is a cracked element or a special element of any type, the computational accuracy is maximized by experimenting with the number of nodes, the positions of the nodes, and the position of the crack tip. Manipulating the finite element notes with crack line stress distributions can accomplish the desired results. Such elements not only enhance the reliability of accurate stress-intensity factor calculation but also streamline the analysis procedure. With a cracked element, the fine grids around the crack tip are no longer needed. Using a cracked element (or some other special element), data reduction for establishing the crack-front stress and displacement fields is not required. Interpretation of the reduced data to come up with a *K* value is also not required. A computation routine for determining *K* is built in with the finite element code.

In earlier versions of the finite element codes, the interelement displacement compatibility between the cracked element and the conventional elements is not satisfied. However, this problem has been resolved with the newer hybrid elements. Sometimes, the size of the cracked element is also important (smaller is not always better). It is the user's responsibility to optimize its usage by making preliminary checkout runs. When the analysis is for a mode 1 crack, a symmetric half-element can be used.

The key feature of a cracked element is that it can have any shape, with many nodes on any boundary. There is always a node directly in front of the crack tip, but no node at the crack tip. The crack is simulated by two opposite nodes splitting about the crack line. A typical example is shown in Fig. 5.27. Like the other special elements, the cracked element can be incorporated into an existing structural analysis finite element program. An example showing the connections between the cracked element and the surrounding elements is given in Fig. 5.28. Figure 5.29 is another example showing a general finite element mesh of a wing plank having a cracked element placed at the wing root. The resulting stress-intensity distribution is also shown.

Implementation of a p-Version Code. In this section we will demonstrate how to determine stress-intensity factors for a center-cracked panel by using a p-version code. The PFEC used this time is the same one used for a demonstration problem in Chapter 1. The PFEC was an older version (version 2.1) that runs on an IBM workstation. It does not contain any fracture mechanics extraction routine. Users must rely on matching the PFEM solutions with the classical solutions for crack-tip stress or displacement components—that is, σ_y of Eq 1.77 for $\theta = 0°$, or v (using Eq 1.80 or 1.82) for $\theta = 180°$. The

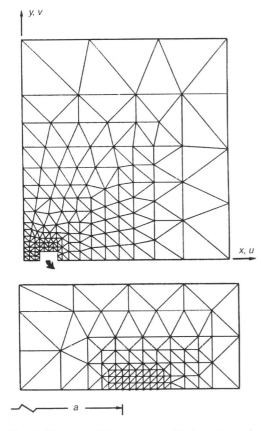

Fig. 5.26 Sample finite element model of a center-cracked panel, using conventional h-version elements

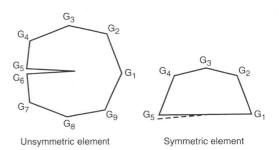

Fig. 5.27 The M.I.T. element for representation of crack-tip singularity. Source: Ref 5.34

procedures that are implemented below are equally applicable when using an h-version finite element mesh model, or the crack-tip stress and displacement distributions obtained by any other means. Therefore, this example is most appropriate for showing the practical meaning of classical fracture mechanics.

PFEM for a Center-Cracked Panel. Figure 5.30 shows a typical two-dimensional finite element mesh (which is made of plane elements) for a rectangular panel containing a through-thickness crack. The panel is geometrically symmetrical, so only one-quarter is shown. Comparing Fig. 5.30 with Fig. 5.26, clearly the new model has only a few elements. Due to the singular nature of the stresses near the crack tip, it might require a very high level of polynomial to converge. As shown in Fig. 5.30, a box of very small elements (0.254 mm, or 0.01 in., in radius) is placed around the crack tip. This group is designated as sacrificial elements. They are not special elements of any kind, except that verification of convergence inside these elements will not be required. Thus, computer space will not be exhausted as a result of running the program at p-levels associated with extraordinarily high degrees of freedom. Parametric evaluation was not performed to obtain an optimized size for the sacrificial elements, but experience has revealed that stresses or displacements could reach reasonably good convergence at p-levels as low

as 6 or 7. Beyond that, only insignificant improvements on minimizing the computational error could be achieved. Thus, quite often, sacrificial elements are unnecessary.

Six models of the center-cracked panel (Fig. 5.30) were constructed. The total width, W, was 50.8 mm (2 in.). The crack length to panel width ratios, $2a/W$, were 0.05, 0.25, 0.4, 0.55, 0.7, and 0.9. The material properties used were $E = 71,000$ MPa (10,300 ksi) and $v = 0.33$. Each panel was subjected to 68.9 MPa (10 ksi) uniform tension stress (i.e., the applied load divided by the product of the full width times the thickness of the panel).

Data Reduction and Interpretation. For each panel, convergence was separately checked for convergence on stress and convergence on displacement. Based on the reduced data, the impact of convergence criteria on the solutions for stresses, or displacements, was insignificant. Therefore, convergence criteria were not an issue here. For data reduction, the PFEC allows the user to extract 10 pairs of data points (of stress and displacement) along any preselected line connecting any two nodal points of a given element (e.g., along an edge of the element). The crack-tip stress (σ_y) and crack-face displacement (v) data for all six panels are reduced from the finite element program output and presented in Fig. 5.31 and 5.32. In these figures, r is the absolute distance from the crack tip, because the

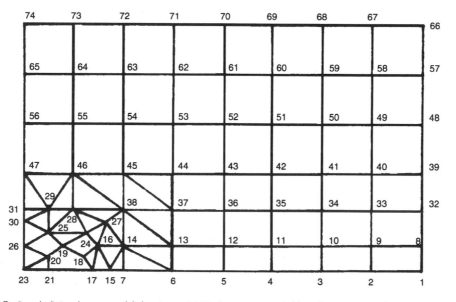

Fig. 5.28 Sample finite element model showing an M.I.T. element surrounded by other conventional rectangular and triangular elements. The five nodes for the cracked element are 17, 18, 19, 20, and 21. The crack line starts from node 23 and ends in the middle between nodes 21 and 17

K values for a given panel would be determined by using σ_y of Eq 1.77 (with $\theta = 0°$) and v of Eq 1.82 (with $\theta = 180°$). The displacement and stress distributions were computed by assigning a K value into these equations. Iterations were made, assigning a different K value each time, until the computed v, or σ_y, distribution fit the PFEM data. As shown in Fig. 5.31 and 5.32, the analytical solution (for v or σ_y) only fits the PFEM data over a narrow range of r that is not

too close to, and not too far away from, the crack tip. This is due, in part, to omission of the higher-order terms in the analytic equations and the limitation of the finite element program in producing a proper stress singularity at the crack tip.

Theoretically, the K value for a given crack can be directly calculated from the PFEM data (v or σ_y), with Eq 1.77 to 1.82. Within a range of r where the analytical solution matches the

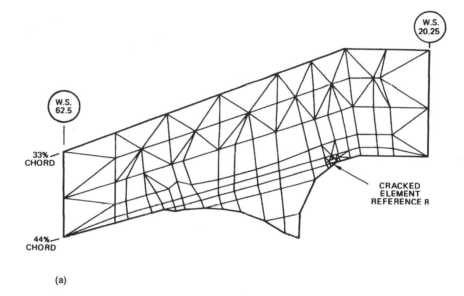

(a)

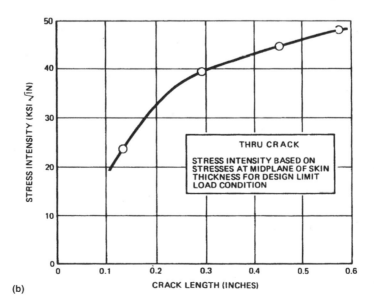

(b)

Fig. 5.29 NASTRAN finite element model of a wing of the F-5 fighter. (a) Cracked element is placed at the wing root. (b) Stress-intensity factor solutions. Source: Ref 5.35

finite element solution, K should be a single value. When K is plotted against r, K should be a constant in that region. The plots for this group of center-cracked panels are shown in Fig. 5.33 and 5.34. The K_V and K_S terms in these figures imply that the K values are determined based on finite element displacement and finite element stress, respectively. Also shown are the characteristics of the K_V (or K_S) versus r distribution, in which the K values are linear across the region where the analytical and the PFEM solutions correlate. This feature is typical of any finite element analysis results reported in the literature (e.g., Ref 5.36–5.38). In areas outside the linear region, the computed K values deviate from the regression line and are considered invalid. The data points inside the linear region can be used as a basis for determining K.

According to Eq 1.77 to 1.82, all K values on the regression line should have the same value (i.e., K is a constant). Although not shown here, the K_V or K_S values for many other crack configurations studied using this PFEC were indeed a single value. When K_V or K_S is not a constant, the real K values can be obtained by rewriting Eq 1.77 as:

$$K_s = \sqrt{2\pi r} \lim_{r \to 0} \sigma_y \qquad \text{(Eq 5.32)}$$

that is, by extrapolating the regression line to $r = 0$. The intercept point can be taken as the real K (Ref 5.36). The same approach can be applied to determine K_V.

Conceptually, the four methods illustrated in Fig. 5.31 to 5.34 should give an identical K value for a given geometry. In reality, due partly to the computational characteristics of a finite element program and partly to the configuration under consideration for which the missing higher-order terms in the analytical solution might become significantly important, one or more of these methods may yield more dependable results than the others.

The K values for all six panels, derived from each of the four data reduction methods, are labeled in Fig. 5.31 to 5.34 along with the fitted curves. From these figures, it appears that the displacement extrapolation method (i.e., those shown in Fig. 5.33) is consistently better than the others.

There are many stress-intensity solutions for the center-cracked configuration. Among them, the Isida solution (Fig. 5.3) is regarded by the fracture mechanics community as being the most accurate. For comparison purposes, the Is-

ida solution is plotted in Fig. 5.35 together with the finite element solution, which is based on the intercept point of the K_V versus r curve. The agreement between the two solutions is excellent.

5.3 Fatigue Crack Growth in Room-Temperature Air

Many structural failures are the result of the growth of preexisting subcritical flaws or cracks to a critical size under fatigue loading. The growth of these flaws will occur at load levels well below the ultimate load that can be sustained by the structure. A quantitative understanding of this behavior is required before the performance of the structure can be evaluated. Information on the crack propagation behavior of metals is needed for selecting the best performing material, evaluating the safe-life capability of a design, and establishing inspection intervals.

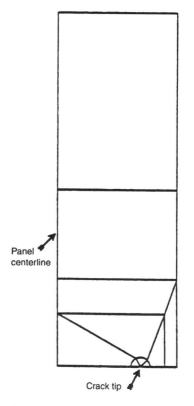

Fig. 5.30 Finite element mesh for one-quarter of a center-cracked panel, using p-version elements. Source: Ref 5.7

Fatigue crack propagation is a phenomenon where the crack extends at every applied stress cycle, clearly illustrated by a fractograph (Fig. 5.36) taken from the fracture surface of a specimen after termination of a cyclic crack growth test (Ref 5.39). The amount of crack extension due to a stress cycle can be easily calculated by counting the number of striations and correlating the width of each striation with the cyclic stress profile shown Fig. 5.36.

In this example, a loading block contains a total of 24 constant-amplitude stress cycles. All 24 cycles are of the same amplitude; however, three of these cycles have lower stress magnitudes. A crack growth test was conducted by repeatedly applying this loading block until the

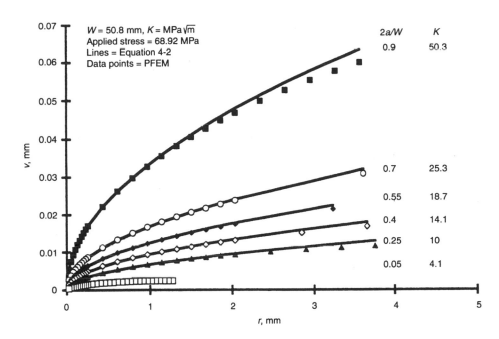

Fig. 5.31 Crack-face displacements for cracks in a center-cracked panel. Source: Ref 5.7

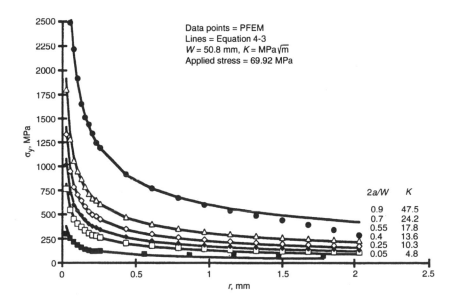

Fig. 5.32 Crack-tip stress distributions for cracks in a center-cracked panel. Source: Ref 5.7

crack grew to a preselected length (or the specimen failed). In each loading block, the biggest jump in crack length (i.e., the largest striation width) corresponded to the stress cycle for which the load was increased from point 4 (the lowest valley) to point 1 (the highest peak). This widest striation was followed by 20 medium-size striations and three tiny striations, corre-

sponding to 20 (remaining) higher stress cycles and three lower stress cycles, respectively. Therefore, the amount of crack growth (Δa) per stress cycle is simply denoted as da/dN (or $\Delta a/\Delta N$).

In the early 1960s, Paris was the first to demonstrate that each small increment of crack extension is actually governed by the stress inten-

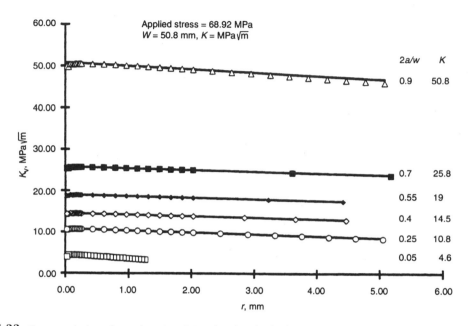

Fig. 5.33 Center-cracked panel stress-intensity solutions based on the displacement extrapolation method. Source: Ref 5.7

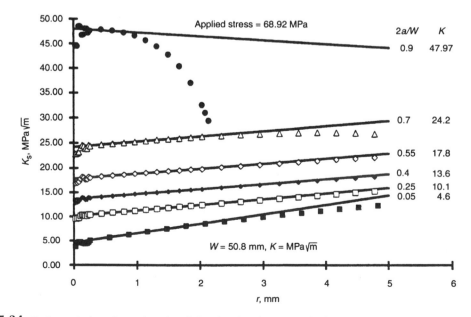

Fig. 5.34 Center-cracked panel stress-intensity solutions based on the stress extrapolation method. Source: Ref 5.7

sity at the crack tip (Ref 5.40, 5.41). In other words, he showed that the Irwin theory of fracture is also applicable to incremental subcritical crack growth. The following section shows that da/dN (of a given loading cycle) is a function of crack-tip stress intensity, K.

5.3.1 Driving Force

The mechanics of fatigue crack propagation are similar to those for rapid fracture. From the definition of $K = S\sqrt{\pi a} \cdot \beta$ (Eq 1.83), it follows that a cyclic variation of S will cause a similar cyclic variation of K, that is, K_{max} and K_{min} (Fig. 5.37). Following the similarity rule—comparing two specimens having different crack lengths and applied stresses—one may have a short crack subjected to higher stress, whereas another may have a longer crack subjected to a lower stress. If these combinations lead to the same stress-intensity level, an equal amount of crack growth should result. In other words, the stress intensity at the crack tip controls the rate of fatigue crack growth.

To further clarify this concept, consider two center-cracked panels of the same size, subjected to different maximum sinusoidal stress levels S_1 and S_2, with $S_1 > S_2$ and $R_1 = R_2$, where R is the minimum to maximum stress ratio. The fatigue crack length versus cycles histories are schematically shown in Fig. 5.38(a). Reduction of these data to da/dN versus ΔK will result in a graph such as that shown in Fig. 5.38(b). The reason for the shift in the curves in Fig. 5.38(b) is self-explanatory in Fig. 5.38(a). Because both tests have the same initial and final crack lengths, a lower K level is expected at the beginning and the end of the test that has a lower applied stress. In the center portion of these da/dN curves, crack growth rates generated at S_1 and S_2 will coincide, because the conditions at both crack tips are identical. That is, the same maximum stress intensity and fluctuation will produce identical crack growth rates.

A generalization of this concept is further illustrated in Fig. 5.39. The applicability of K is demonstrated by comparing three crack geometry and loading combinations: a center crack or an edge crack subjected to far-field applied stresses, and a center crack loaded by a pair of concentrated forces. Despite the fact that K increases with crack length in far-field loading and decreases with crack length in crack-face loading, there is full agreement among the crack growth rates for all three cases when $\Delta a/\Delta N$ is

plotted as a function of ΔK (or K_{max}). A material will exhibit the same rate of crack growth under the same K level, regardless of the loading condition and/or geometry that produced the K value. In other words, the result of one configuration can be predicted if the result of another configuration is known, provided that correct K-expressions for both configurations are available. Again, Paris (Ref 5.41) was the first to provide actual experimental data in validation of this concept (Fig. 5.40). Therefore, using a stress-intensity factor that correctly represents the geometry and loading under consideration is extremely important in performing crack growth analysis.

5.3.2 Constant-Amplitude Loading

The methods for conducting fatigue crack growth testing as well as data reduction procedures are specified in ASTM Standard E 647 (Ref 5.14). An excellent article that provides a concise summary of the techniques appears in Volumes 8 and 19 of the *ASM Handbook* (Ref 5.43, 5.44). This section discusses a general procedure for obtaining material crack growth rate data, serving as a stepping stone to discussion of the fatigue crack growth mechanism.

Fatigue crack growth testing can be conducted on any type of test specimen or structural component. The ASTM standards (Ref 5.14) provide several specimen configurations along with specific test procedures. Among them, the compact specimen is the most popular for reasons of economy. The test is run under constant-amplitude loading; that is, the maximum and minimum load levels are kept constant for the duration of the test. Either a sinusoidal or saw-tooth waveform is used for the applied loading.

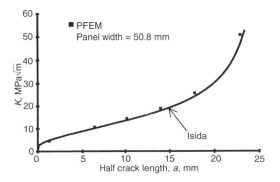

Fig. 5.35 Comparison of PFEM and Isida solutions for a center-cracked panel. Data points: PFEM. Solid line: Feddersen's fit. Source: Ref 5.7

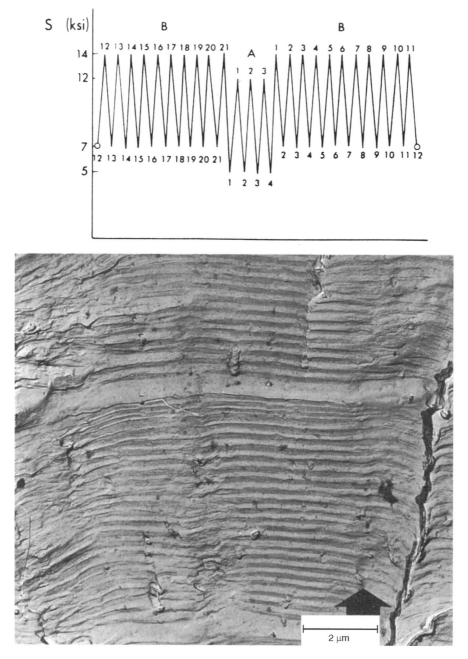

Fig. 5.36 Typical fractograph showing striation markings corresponding to fatigue load cycles. Note the large striation spacing due to the load amplitude A4-B1, preceded by three smaller striations corresponding to the load cycles of block A. The arrow indicates the crack propagation direction. Source: Ref 5.39

The data recorded from the test include the testing parameters as well as the crack length as a function of the number of cycles. The crack growth data are usually obtained by visually measuring the crack length and noting the number of load cycles on a counter. Crack gages or compliance techniques can be used in lieu of visual measurement to monitor the progress of crack propagation (Ref 5.14, 5.43, 5.44). This gives a series of points that describe the crack length as a function of load cycles for a given test.

Crack growth rates are computed on every two consecutive data points. The last point of

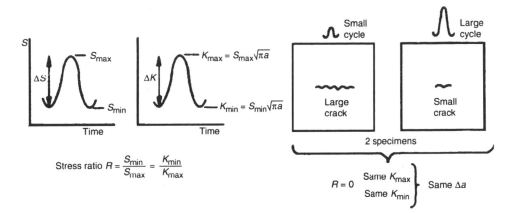

Fig. 5.37 Schematic illustration of K being the driving force of fatigue crack propagation. Source: Ref 5.42

each pair of the a versus N data points is used as the first point for the following pair of data points. The amount of each crack growth increment is Δa. The difference in the number of fatigue cycles between two consecutive data points is ΔN. Dividing Δa by ΔN is the crack growth rate per cycle. This is the simplest way to obtain da/dN data points from a set of a versus N data. More refined methods are available (Ref 5.43).

Stress-intensity factors corresponding to each increment of crack growth are computed using the average crack length of each pair of consecutive data points. Alternatively, one may prefer to draw a smooth curve through the a versus N data points first, then determine the slope for each selected point on the curve. ASTM Standard E 647 includes a Fortran program for data reduction using the seven-point incremental polynomial method. The slope of a given point on the smooth curve is the da/dN for that point. The K_{max} and K_{min} values corresponding to each of these crack lengths also can be determined.

As explained below, da/dN is a function of K_{max} and K_{min}. With this in mind, we let $\Delta K = K_{max} - K_{min}$ and $R = K_{min}/K_{max}$. Then the relation between da/dN and K is expressed as:

$$\frac{da}{dN} = f(\Delta K, R) \qquad \text{(Eq 5.33)}$$

or

$$\frac{da}{dN} = f(K_{max}, R) \qquad \text{(Eq 5.33a)}$$

For zero-to-tension loading (i.e., $R = 0$), these two equations are identical. Therefore, da/dN can be plotted as a function of ΔK, or K_{max}. However, it has become a standard practice to use ΔK as the independent variable for data presentation. Figure 5.41 shows a comparison of typical da/dN versus ΔK data (plotted in a log-log scale) for metals, intermetallics, ceramics, and composites.

To put da/dN and ΔK in an equation, Paris (Ref 5.40, 5.41) described the fatigue crack growth rate data by a power law equation:

$$\frac{da}{dN} = C(\Delta K)^n \qquad \text{(Eq 5.34)}$$

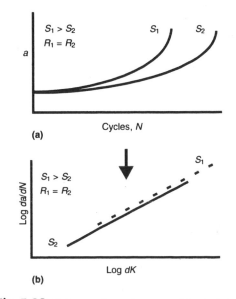

Fig. 5.38 Fatigue crack growth rate correlation technique

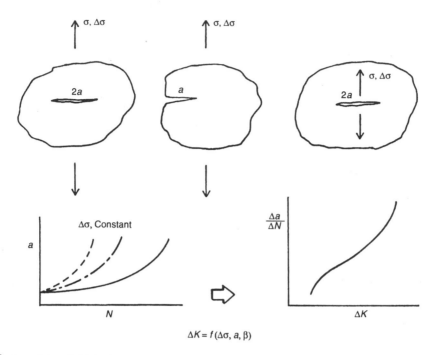

Fig. 5.39 Expanded crack driving force concept

where n is the slope and C is the coefficient at the intercept of a log-log plot. In other words, there is a linear relationship (in log-log scale) between *da/dN* and ΔK. Since then, many equations have been suggested by various investigators. This diversification in crack growth rate description is deemed necessary because each individual investigator faces a different situation that is unique to the material and application associated with the product of a particular industry. More than that, there is an intrinsic element in the crack growth rate data that surfaces as more data are generated. In general, a *da/dN* curve appears to have three regions: a slow-growing region (the so-called threshold), a linear region (the middle section of the curve), and a terminal region (toward the end of the curve where ΔK approaches K_C). A smooth connection of all three regions forms a sigmoidal-shape curve representative of the entire *da/dN* versus ΔK curve for a given *R* (Fig. 5.42). Although some alloys may not exhibit a clearly defined threshold, or the transition from the linear region to final fracture may be rather abrupt (i.e., the entire *da/dN* curve is apparently linear), the existence of these regions has been well recognized.

The physical significance of the slow-growing and the terminal regions is that there are two obvious limits of ΔK in a *da/dN* curve. The lower limit (the threshold value, ΔK_{th}) implies that cracks will not grow if $\Delta K < \Delta K_{th}$, that is,

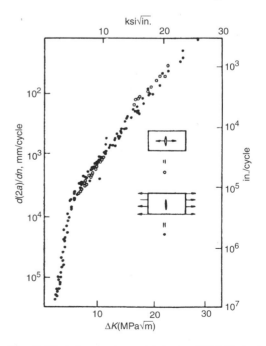

Fig. 5.40 Fatigue crack growth behavior of 7075-T6 aluminum under remote and crack-line loading conditions. Source: Ref 5.41

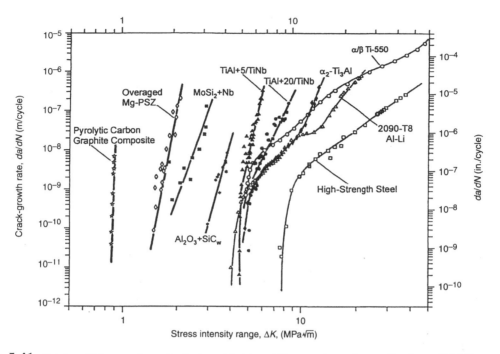

Fig. 5.41 Variation of fatigue crack growth rates for metals, intermetallics, ceramics, and composites. Source: Ref 5.45

$da/dN \to 0$ when $(\Delta K - \Delta K_{th}) \to 0$. On the other hand, if ΔK becomes too high, which implies that K_{max} exceeds the fracture toughness K_C, static failure will follow immediately. This is equivalent to ΔK exceeding $(1 - R)K_C$, that is, $da/dN \to \infty$ when $[(1 - R)K_C - \Delta K] \to 0$. Consequently, Eq 5.34 should be modified to have a general form:

$$\frac{da}{dN} = C(\Delta K)^n \frac{\Delta K - \Delta K_{th}}{(1 - R)K_C - \Delta K} \quad \text{(Eq 5.35)}$$

This equation gives a sigmoidal relation on a log-log plot with two vertical asymptotes. The material constants C and n are the intercept and slope of the line that fit through the linear region

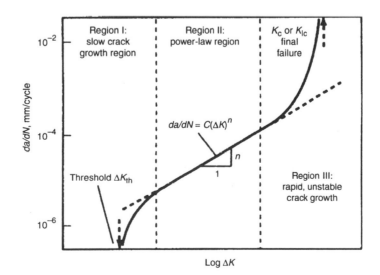

Fig. 5.42 Typical shape of a fatigue crack growth rate curve for a given R value

of the *da/dN* curve. Some refined equations and the reasons behind the refinements are presented in the following section.

5.3.3 Effect of Mean Stress on Constant-Amplitude Fatigue Crack Growth

Now, define mean stress as the average of the maximum and minimum stresses of a given fatigue cycle. When two tests are conducted at different mean stress levels (for example, the tests have the same ΔS but different R values), their cyclic crack growth rates will not be the same. Examining the fractograph in Fig. 5.36, evidently the striation band corresponding to the cycle of lower mean stress is narrower than the striation band for the cycle that has a higher mean stress level. That is, the fatigue crack growth rate $\Delta a/\Delta N$ for the former is slower than the latter. This phenomenon is called the mean stress effect, or stress ratio (R-ratio) effect.

A typical *da/dN* versus ΔK data set of the 2024-T351 aluminum alloy is shown in Fig. 5.43. This data set contains six R-ratios ranging from -2 to 0.7; center-cracked specimens were used. Using Eq 5.35, one can only plot a *da/dN* curve for a given R-ratio. Eventually a series of equations is needed to fully describe the *da/dN* behavior for a wide range of R-ratios. This means that a different pair of C and n constants is needed for each equation (for a particular R-ratio). This type of data presentation is called correlating the data *individually*. Keep in mind that the $(1 - R)$ term in Eq 5.35 is solely for providing the termination point of the *da/dN* curve. It does not form any connection between *da/dN* at different R-ratios. Therefore, Eq 5.35 can only be used for correlating the data individually (i.e., with one R-ratio at a time).

Thus, two problems arise in crack growth rate data description. One of them is the need for establishing a relationship between all the R-ratios in obtaining a mathematical function that can collapse all the *da/dN* data points into a common scale. In so doing, only one pair of C and n constants is needed for the entire data set. This approach is called correlating the data *collectively*. Another problem is the need to select

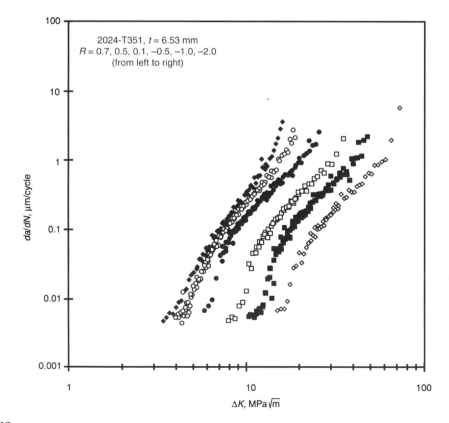

Fig. 5.43 Fatigue crack growth rate data for 2024-T351 aluminum with six *R*-ratios

a curve-fitting equation capable of fitting through all three regions of the *da/dN* data. Among the many crack growth rate equations published in the literature, some dealt with the mathematical formulation of a sigmoidal curve, or normalizing the *R*-ratio effect, or both. Some even divided the *da/dN* curve into multiple segments in an attempt to obtain a closer fit between experimental data and a set of equations. Discussed below are a few techniques that are commonly used today.

Effective ΔK as a Function of Crack Closure. It appears that the actual ΔK acting on a stress cycle does not have the same magnitude as the applied ΔK. In each load cycle the full excursion of the loads between the valley and the peak does not fully contribute to crack growth. Instead, only a portion of the load magnitude effectively contributes to crack extension. Its corresponding ΔK is called the effective ΔK. The differences between the applied and the effective ΔK vary among different *R*-ratios. In order to account for the mean stress effect on *da/dN*, we need to develop a method to estimate the effective level of an applied *K* for a given *R*-ratio. Following an extensive survey of the existing methods published in the literature, it seems that the sound approach would be to develop a relationship between *effective ΔK* and *R* based on the concept of crack closure.

The concept of crack closure (Ref 5.46) considers that material at the crack tip is plastically deformed during fatigue crack propagation. During unloading of a stress cycle, some contact between the crack surfaces will occur due to the constraint of surrounding elastic material. The crack surfaces will not open immediately at the start of the following cycle; instead, they remain closed for some time during the cycle. The hypothesis goes further and states that crack propagation takes place only at the rising portion of a stress cycle, and that the crack is unable to propagate while it remains closed.

The notion that a fatigue crack does not extend during downloading has gained support from an unparalleled investigation. Barsom (Ref 5.47) has come to the same conclusion by running tests on various forms of loading patterns: sinusoidal, triangular, square, skewed-square, positive-sawtooth, and negative-sawtooth. Although the objective of his work was to investigate the corrosion fatigue crack propagation behavior of 12Ni-5Cr-3Mo steel, it led to the assumption that the net effect of closure is to reduce the apparent ΔK to some effective level ΔK_e ($= K_{max} - K_{op}$). In other words, K_{op} is the minimum stress intensity level of that cycle required to reopen the crack, and its magnitude is usually higher than the applied K_{min} (Fig. 5.44). In Ref 5.46, Elber defines the effective stress-intensity range as:

$$\Delta K_e = U \cdot \Delta K \qquad \text{(Eq 5.36)}$$

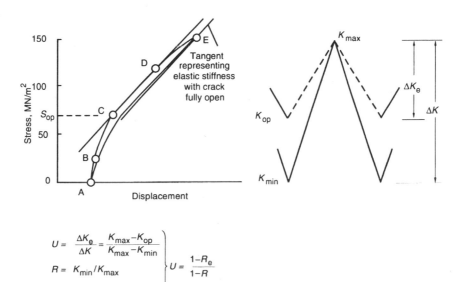

$$U = \frac{\Delta K_e}{\Delta K} = \frac{K_{max} - K_{op}}{K_{max} - K_{min}} \left. \begin{array}{c} \\ \\ \\ \end{array} \right\} U = \frac{1 - R_e}{1 - R}$$
$$R = K_{min}/K_{max}$$
$$R_e = K_{op}/K_{max}$$

Fig. 5.44 Schematic illustration of crack closure

where

$$U = \frac{K_{max} - K_{op}}{K_{max} - K_{min}} \quad \text{(Eq 5.37)}$$

Using a set of crack growth rate data for 2024-T3 aluminum, which contains R values between -0.1 and 0.7, Elber showed that U was a linear function of stress ratios. Since then, numerous investigators have developed various techniques to measure the crack opening load, or conducted studies of the physical mechanism of crack closure.

The experimental aspects of K_{op} measurements (or S_{op}, the stress counterpart of K_{op}) are quite complicated. Experimentalists often disagree among themselves on issues regarding fundamental techniques as well as the definitions for *crack opening load* and *closure load*. Generally, closure load is the load level where the crack starts to close during unloading; crack opening load is the load level where the crack starts to reopen during reloading. Inconsistencies in measured K_{op} values are frequently found in the literature. References 5.48 to 5.50 point out that K_{op} is dependent on material thickness, R-ratio, K_{max} level, crack length, and environment (including frequency and temperature). Test procedure and the method for interpreting crack opening load from the load trace record of a test also are important variables. It has been reported that for a given crack length, K_{op} changes from the surfaces toward the midthickness of the test coupon. The value of K_{op} is also dependent on the smoothness or flatness of the crack faces, which is often associated with small crack sizes or low stress intensity levels, as well as the formation of shear lips at longer crack lengths.

Therefore, it would be impossible (and impractical) to develop a normalizing function for ΔK_e that correlates engineering crack growth rate data from experimental crack closure data. The challenge to a fracture mechanics analyst is to select an analytical function for U as a function of R and then pair it with a crack growth rate equation to obtain correlations with reasonable degrees of accuracy and consistency. It is desirable to have a U-function that can fit to data sets that cover a wide range of R-ratios, extending from an approximate unity to an extreme negative value. Criteria for setting up such a U-function are discussed below.

Examining Eq 5.36 and 5.37, it is clear that for a given R-ratio the value of U should be between 0 and 1, that is, $0 < U \le 1$. Physically speaking, $U = 1$ implies that the crack is fully opened (i.e., $K_{op} = K_{min}$), whereas $U = 0$ would mean that the crack is fully closed (i.e., $K_{op} = K_{max}$). For a given stress cycle, one may define the effective R-ratio as:

$$R_e = K_{op}/K_{max} \quad \text{(Eq 5.38)}$$

where $K_{op} \ge K_{min}$. We know that $R = K_{min}/K_{max}$; therefore, $R_e \ge R$. In addition, the finite element analysis results of Newman (Ref 5.51) have shown that R_e should decrease with R. Also consider that R_e can be a negative value, because K_{op} can be a negative value as long as it is above (or equal to) K_{min}. Therefore, K_{op} (or R_e) should not be set to zero for compression-tension stress cycles.

Because both Eq 5.37 and 5.38 contain a K_{op} term, there is a unique relationship between U and R_e. Dividing both the denominator and the numerator of Eq 5.37 by K_{max} results in:

$$U = (1 - R_e)/(1 - R) \quad \text{(Eq 5.39)}$$

Therefore, $R \le R_e < 1$, because $0 < U \le 1$.

Comparison of Two U-Functions and Correlation with Test Data. To obtain a U-function that covers a full range of R-ratios, Liu (Ref 5.52) postulated that U might reach an asymptotic limit U_o when R approaches $-\infty$. His expression for U is denoted here as:

$$U = U_o + (1 - U_o) \cdot \beta^{(1-R)} \quad \text{(Eq 5.40)}$$

where β and U_o are empirical constants whose values depend on material, thickness, and environment. This equation satisfies the above-stated criteria and leads to $U \to U_o$ for $R \to -\infty$, and $U \to 1$ for $R \to 1$. By definition, the value of U_o is greater than zero. By trial and error, a pair of β and U_o values can be determined for a given data set.

When implementing Eq 5.40 to compress a set of *da/dN* data, it is recommended that U_o be taken as a small value somewhere in the vicinity of 0.1. It is conceivable that U is in the neighborhood of 0.5 for $R = 0$. Therefore, an initial estimate of β can be easily obtained by using Eq 5.40, with $U = 0.5$ and $R = 0$. Visual assessment to judge the fairness of fit can be made by displaying the normalized data points on the monitor screen of a personal computer. A final pair of U_o and β that provides the best fit can be easily reached by making a few iterations.

To demonstrate the usefulness of this equation, a data set for 2024-T351 aluminum alloy (Fig. 5.43) has been normalized by using $\beta = 0.45$ and $U_o = 0.12$. This data set contains six R-ratios ranging from -2 to 0.7; center-cracked specimens were used. For purposes of clarity, the normalized data points have been arbitrarily divided into two groups, randomly mixed with positive and negative R, and are presented in Fig. 5.45 and 5.46. It can be seen that Eq 5.40 has effectively compressed the entire data set of six R-ratios into a common scale. All data points displayed in both figures are fitted by a common da/dN curve (having the same empirical constants for curve fitting). The method used to plot the curve will be discussed in the next section. For now, it is enough to demonstrate that the entire data set (covering R-ratios ranging from -2 to 0.7) has been normalized by a single pair of empirical constants ($\beta = 0.45$ and $U_o = 0.12$).

An alternate approach has been taken by Newman (Ref 5.53). His approach to this problem is to derive an equation for R_e and then use Eq 5.39 to obtain U. The Newman equation for R_e is defined as:

$$R_e = A_o + A_1R + A_2R^2 + A_3R^3 \qquad \text{(Eq 5.41a)}$$

for $R \geq 0$, and

$$R_e = A_o + A_1R \qquad \text{(Eq 5.41b)}$$

for $-1 \leq R < 0$. The polynomial coefficients are defined as:

$$A_o = (0.825 - 0.34\alpha + 0.05\alpha^2)$$
$$\cdot \{\cos[(\pi/2) \cdot (S_{max}/S_o)]\}^{1/\alpha} \qquad \text{(Eq 5.42a)}$$

$$A_1 = (0.415 - 0.071\alpha) \cdot S_{max}/S_o \qquad \text{(Eq 5.42b)}$$

$$A_2 = 1 - A_o - A_1 - A_3 \qquad \text{(Eq 5.42c)}$$

$$A_3 = 2A_o + A_1 - 1 \qquad \text{(Eq 5.42d)}$$

where α is a material constant, its value varying from 1 for plane stress to 3 for plane strain; S_{max} is the peak stress of a stress cycle. Newman

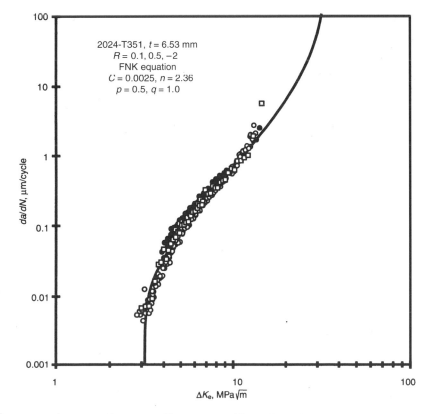

Fig. 5.45 Compressed 2024-T351 data points with $U_0 = 0.12$ and $\beta = 0.45$, part 1

states that S_o is the material flow stress; however, no clear definition for its quantity was given. In practice, Forman et al. (Ref 5.27) treated both α and S_{max}/S_o as fitting constants. By letting $\alpha = 1.0$ and $S_{max}/S_o = 0.23$, the 2024-T351 data were successfully compressed, with the results presented in Fig. 5.47 and 5.48. Again, curve fitting is discussed below. Examining the results shown in Fig. 5.45 to 5.48, it appears that any data set containing multiple R-ratios can be compressed into a common scale by means of any reputable function of ΔK_e.

To validate a given set of empirical constants against the criteria for U and R_e, simply compute U and R_e using the empirical constants in question and display the results on a graph. To illustrate this with the 2024-T351 data set, the graphical solution of Eq 5.40 (with $\beta = 0.45$ and $U_o = 0.12$) is plotted as the solid line in Fig. 5.49. The dotted line is the solution for the Newman U with $\alpha = 1.0$ and $S_{max}/S_o = 0.23$. Their corresponding R_e values are plotted in Fig. 5.50. The abilities of these empirical constants to meet the criteria are validated by the trends displayed in both figures. Despite the differences in these two approaches, there is a very close agreement between their U and R_e values. Although Newman has set a limit of application for his R_e at $R \geq -1$, good correlation with test data in a range outside this limit (down to $R = -2$) has been obtained. In fact, in the NASA/FLAGRO computer code, Forman et al. (Ref 5.27) did lower the limit for Eq 5.41(b) to -2.

5.3.4 Fatigue Crack Growth Rate Equation

The second element in a crack growth rate equation is the mathematical function that describes the characteristics of the da/dN data over all three regions. The plot of the equation should display a close fit to the already normalized data points. As mentioned earlier, the da/dN versus ΔK (or ΔK_e) data usually form a sigmoidal curve. Some material may have a longer linear segment than others, and some material may not have an identifiable threshold value.

Recognizing a crack growth rate equation is merely a convenient tool needed for making computerized life estimates. Therefore, there is

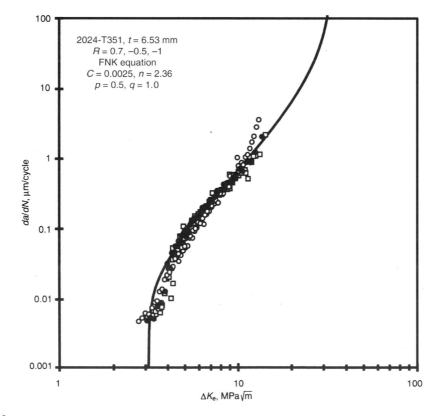

Fig. 5.46 Compressed 2024-T351 data points with $U_0 = 0.12$ and $\beta = 0.45$, part 2

no point in reviewing or evaluating all the equations currently available. To plot a sigmoidal curve through a normalized (da/dN versus ΔK_e) data set, simply use an equation that possesses the same format as Eq 5.35.

FNK Equation. Among the many sigmoidal equations available in the literature, the FNK (Forman, Newman, and de Konig) equation is considered the most suitable for the present purpose. The equation is used in the crack growth prediction program NASA/FLAGRO (Ref 5.27). Similar to Eq 5.35, the equation is written as:

$$\frac{da}{dN} = C\left(\frac{1-f}{1-R}\Delta K\right)^n$$
$$\cdot \frac{[1 - (\Delta K_{th}/\Delta K)]^p}{[1 - (\Delta K/(1-R)K_C)]^q} \quad \text{(Eq 5.43)}$$

where f is Newman's R_e given in Eq 5.41(a) and (b), and ΔK_{th} is the threshold value of ΔK for a given R. Since $U = (1-f)/(1-R)$, and $\Delta K_e = U \cdot \Delta K$, Eq 5.43 can be rewritten as:

$$\frac{da}{dN} = C(\Delta K_e)^n$$
$$\cdot \frac{[1 - (U \cdot \Delta K_{th}/\Delta K_e)]^p}{[1 - (\Delta K_e/U \cdot (1-R)K_C)]^q} \quad \text{(Eq 5.43a)}$$

Thus, this equation consists of two major elements: the ΔK_e term for handling the R-ratio effect, and the p and q exponents for shaping the da/dN curve. The following paragraphs demonstrate that any ΔK_e function can be placed in this equation to achieve the same results.

Plotting a da/dN Curve. To make use of Eq 5.43(a), start by using an effective ΔK function to normalize all the da/dN data points in a given data set. The chosen effective ΔK function can be based on Liu's U or any appropriate U. Once this process is completed—that is, after all the constants and coefficients associated with ΔK_e are determined and the normalized data points are satisfactorily displayed in a log-log plot— ΔK_e can be regarded as an independent variable in Eq 5.43(a).

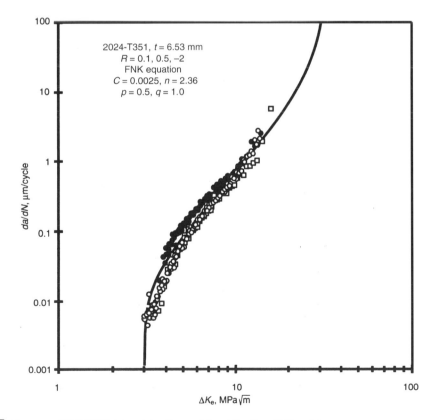

Fig. 5.47 Compressed 2024-T351 data points with $\alpha = 1.0$ and $S_{max}/S_0 = 0.23$, part 1

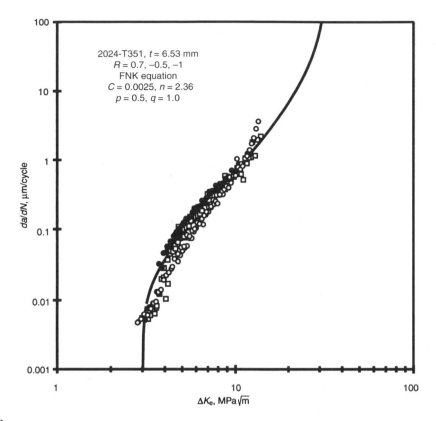

Fig. 5.48 Compressed 2024-T351 data points with $\alpha = 1.0$ and $S_{max}/S_0 = 0.23$, part 2

To make use of Eq 5.43(a) and treat ΔK_e as an independent variable, the easiest way is to let $R = 0$; Eq 5.43(a) then becomes:

$$\frac{da}{dN} = C(\Delta K_e)^n \cdot \frac{[1 - (\overline{U} \cdot \Delta K_o/\Delta K_e)]^p}{[1 - (\Delta K_e/\overline{U} \cdot K_C)]^q}$$

(Eq 5.43b)

where ΔK_o is ΔK_{th} at $R = 0$ and $\overline{U}$ is U at $R = 0$. The quantities $\overline{U} \cdot \Delta K_o$ and $\overline{U} \cdot K_C$ can be regarded as the effective values of ΔK_o and K_C, respectively. When ΔK_e is a function of Liu's U, $\overline{U} = [U_o + (1 - U_o) \cdot \beta]$. When ΔK_e is a function of Newman's U, $\overline{U} = 1 - A_o$, and A_o is given by Eq 5.42(a).

Now we can use Eq 5.43(b) to plot those curves previously shown in Fig. 5.45 to 5.48. First, look at the data set and determine the values for ΔK_o and K_C: $\Delta K_o = 6.04$ MPa$\sqrt{m}$,

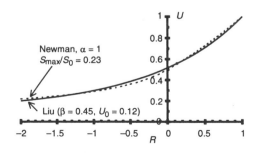

Fig. 5.49 Effective stress-intensity range ratio as a function of R

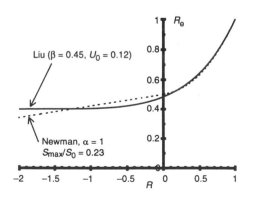

Fig. 5.50 Effective R-ratio as a function of R

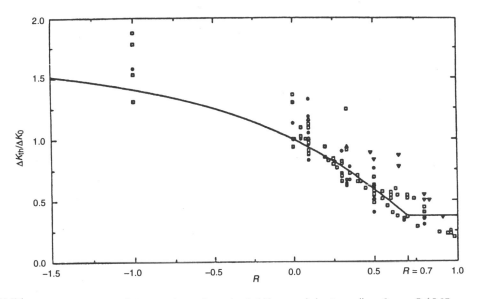

Fig. 5.51 Effect of stress ratio on fatigue crack growth rate threshold for several aluminum alloys. Source: Ref 5.27

$K_C = 65.93$ MPa$\sqrt{m}$. Then select several values for the empirical constants p, q, C, and n. After several iterations, we determine that p = 0.5, q = 1.0, C = 0.0025, and n = 2.36. The curves in all four figures (Fig. 5.45 to 5.48) are plotted by using these coefficients.

At this point, it is important to note that not all the effective ΔK and crack growth rate coefficients used in this exercise will provide a perfect fit for this batch of material, and certainly they are not intended to be used as material allowables. They are used here to demonstrate how to work with these crack growth rate equations. Recommended material constants, including ΔK_{th}, K_C, α, S_{max}/S_o, p, q, C, and n, for many engineering alloys are listed in Ref 5.27.

In the above examples, we plotted the crack growth rate curve with ΔK_e as an independent variable. It is equally possible to use the same FNK equation to plot curves that fit through the original (noncompressed) data points for each individual R-ratio. This involves using many different ΔK_{th} and K_C values, because both ΔK threshold and the termination point of the da/dN curve are functions of the R-ratio. Naturally, the da/dN data set in hand should contain all these data. The termination points are supposedly taken care of by the $(1 - R) \cdot K_C$ term in Eq 5.43, and it seems satisfactory. Empirically determined ΔK_{th} versus R relationships for a number of engineering alloys are available. Figures 5.51 to 5.53 are compilations of the data for alu-

minum alloys, titanium alloys, and steels, respectively.

Speaking of ΔK_{th}, one might have noticed that in Fig. 5.43 there is a "kink" in the data for each R-ratio near $da/dN = 2 \times 10^{-7}$ in./cycle. To investigate whether this is a real phenomenon or merely an experimental scatter, a supplemental test was conducted to generate more data points in this region. The test was conducted at $R = 0.5$, and the results are presented in Fig. 5.54. A literature survey uncovered more data of this type for the 2024-T3

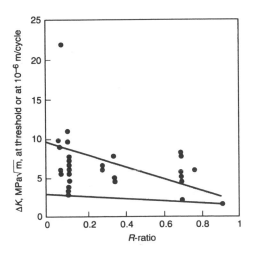

Fig. 5.52 Effect of stress ratio on fatigue crack growth rate threshold for titanium alloys. Source: Ref 5.54

alloy (Ref 5.56, 5.57). A large quantity of data for 7000 series aluminum alloys that also show this behavior is available (Ref 5.58). These results seem to confirm that there is a kink in the vicinity of 2×10^{-7} in./cycle, at least for aluminum. They also show that, depending on the size of the kink, fitting a single sigmoidal curve to a double sigmoidal data set may not be easy. The real ΔK_{th} values for a given material also become questionable.

5.3.5 Factors Affecting the Material Constant-Amplitude da/dN Properties

Unlike tensile yield and ultimate strengths, fatigue crack growth behavior does not have consistent material characteristics. Fatigue crack growth is influenced by many known or unknown factors. Some known factors are discussed below. As a result, a certain amount of

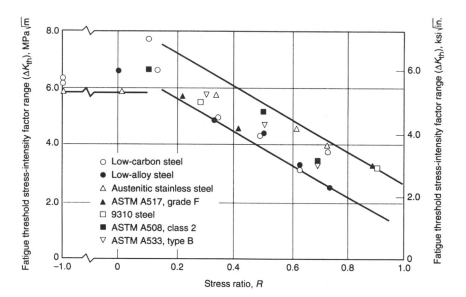

Fig. 5.53 Effect of stress ratio on fatigue crack growth rate threshold for several low- to medium-strength steels. Source: Ref 5.55

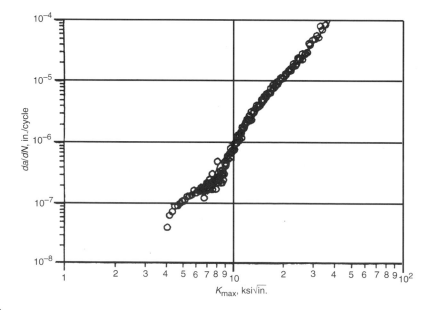

Fig. 5.54 Fatigue crack growth rates for a 2024-T3 plate at $R = 0.5$. Courtesy of J.H. FitzGerald

scatter exists. Therefore, the fairness of the predicted crack growth lives should be judged based on circumstantial factors relevant to in-service conditions.

Before discussing other crack growth phenomena, we will briefly examine some factors that are likely to contribute to scatter in material baseline da/dN data. The discussion will not cover the whole spectrum of variables, but will help to shed light on the sources of statistical variations in material da/dN data.

Metallurgical and processing variables pertain to the material under consideration. The crack propagation characteristics for a particular alloy differ in plate, extrusions, and forgings. Forgings may exhibit a rather large anisotropy, which may have to be considered in the growth of surface flaws and corner cracks, which grow simultaneously in two perpendicular directions. Stretched plates normally exhibit a faster crack growth rate, primarily due to increased tensile yield strength, which causes a smaller plastic zone under load. Consequently, there will be less crack closure (i.e., crack opening starts at a lower applied load level), which results in a larger effective ΔK. The higher tensile yield strength that results from prestraining also causes a smaller shear lip (flatter fracture surface), which further enhances crack propagation (Ref 5.59, 5.60).

For a given product form, the grain orientation effects on fatigue crack growth rates usually are ranked in the following order with respect to applied loads: Short transverse has the fastest rate; long transverse is ranked medium; longitudinal exhibits the slowest rate. Alloys of nominally the same composition but produced by different manufacturers may have largely different crack growth rate properties. Variations in crack growth rates may occur between different batches of the same alloy produced by the same manufacturer. Processing variables, including heat treatment and residual stress, may significantly affect the fatigue crack growth rate of a given material (Ref 5.61). Extensive coverage of the effect of metallurgical and mechanical processing variables on fatigue crack growth is found in Volume 19 of the *ASM Handbook* and other reference sources.

Geometric Variables. Thickness has a small but systematic effect on crack growth rate behavior. Its effect is primarily related to material processing variables and the stress state at the crack tip (i.e., plane stress or plane strain). There is a perception that crack growth rates are higher in thicker plates, because the fracture surface of a thicker plate is rather flat and lacks shear lips. From a design/production standpoint, an alloy is very seldom used for both thin and thick sections in a given product. Material selection is based on meeting the strength, stiffness, and other requirements for intended service. For example, when an alloy is selected for the fuselage of an aircraft, it is unlikely that the same alloy will be used for the wing, and vice versa. Therefore, da/dN data for a full range of thicknesses is seldom available.

Tests conducted at the Naval Research Laboratory showed that da/dN is independent of thickness (Ref 5.62). The material was A516-60 steel ($F_{ty} = 304$ MPa, or 44 ksi) in thicknesses of 6.4, 12.7, 25.4, and 50.8 mm (0.25, 0.5, 1, and 2 in.). A 6.4 mm (0.25 in.) thick plate is not thick as compared to 25.4 or 50.8 mm (1 or 2 in.), but hardly thin by any standard. We do not know whether the da/dN for the thin sheet is the same as those for the thick sections. The chance is that thin sheet is not even available in this alloy. So, one need be concerned only about the effect of thickness within the thickness range under consideration, which is probably minimal.

Other geometric variables can be handled by using appropriate stress-intensity expressions. However, it has been known for some time that fatigue crack growth rates are not the same among different types of specimens. There is lack of documentation that explicitly addresses the issue of which specimen type produces faster (or slower) crack growth rates. In Ref 5.63, Rhodes and Radon report that such discrepancy has been found in 2024-T3 and 7010 (DTD5120) aluminum alloys. The British Aerospace in-house data that they used show that da/dN in a center-cracked specimen can be as much as two times faster than in a compact specimen. They attributed the source of discrepancy to the nonsingular stress parallel to the crack. This stress is tensile in a compact specimen but compressive in the center-cracked specimen, causing different crack-tip buckling behavior.

5.3.6 Effect of Load History

Structural components are often subjected to variable-amplitude loading. The load history during the lifetime of a given structural component consists of various load patterns as a result of in-service applications. The sequence of high, low, and compression ($S_{min} < 0$) loads in a load history is one of the key elements controling crack growth behavior. Figure 5.55 shows several typical loading patterns that are often used for developing experimental crack growth data as a means to study crack growth behavior.

A simple illustration of the load sequence effect on crack growth is schematically presented in Fig. 5.56. The crack growth histories shown here are supposedly developed from identical panels. Both tests start at the same arbitrary crack length. But one of the panels is subjected to a series of higher constant-amplitude cyclic stresses than the other panel. The crack propagation curves for these two panels are shown as solid lines. Now consider a third panel where the stress level at the beginning of the test equals the lower of the two stress levels. Then the stress is increased to the higher stress level when the crack reaches point A. From that point on, the rate of fatigue crack propagation will increase to the expected rate for the higher stress, as indicated by the dashed line in Fig. 5.56(a).

If the sequence of loading is reversed—that is, if the stress level is changed from the higher to the lower stress at point B in Fig. 5.56(b)—the new rate of crack propagation is not what would be expected for the lower stress for that crack length. Actually, the crack remains dor-

mant for a period of time. However, after a number of load cycles (when the crack reaches point C), the crack will resume its normal crack growth behavior as if the change in load level had never occurred during the test.

For a given material, the amount of time that crack propagation remains inactive depends on the level of the higher stress and the difference between the two stress levels. It can be said that the action of normal crack propagation behavior for the lower loads is deferred by the presence of the high loads. The number of load cycles spent on this deferring action is called the *delay cycles*. The overall phenomenon of this event is called delay or, more commonly, *crack growth retardation*. The high load that causes retardation is referred to as the *overload*.

Actually, when the load level changes from high to low in a test, the crack growth rate is not immediately retarded. A microscopic phenomenon that is not apparent in Fig. 5.56(b) exists. Experiments show that some microscopic amount of acceleration may occur immediately after the overload, followed by crack growth rate reduction (Fig. 5.57). Yet, the crack growth rate does not drop to minimum instantly. It reaches a minimum level until the crack has progressed through a short distance into the load interaction

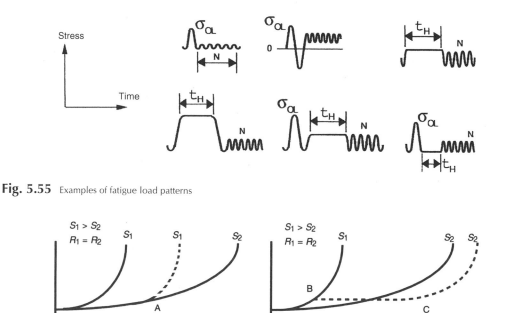

Fig. 5.55 Examples of fatigue load patterns

Fig. 5.56 Effect of applied stress on fatigue crack propagation

zone, approximately one-eighth to one-fourth of the total monotonic plastic zone of the overload (Ref 5.64).

As will be shown later in this chapter, retardation is not the only matter of concern when predicting spectrum crack growth lives. Under certain circumstances the magnitude of retardation might be minimized, but accelerated crack growth rates were also observed.

Basic Concept of Load Interaction. The simple explanation of load sequence effect on crack growth given above has led researchers to believe that some interaction takes place between each pair of consecutive loading steps. The crack-tip plastic zone associated with each loading step is responsible for such interaction and thus the crack growth behavior.

Attempts have been made to develop crack growth modeling techniques to elucidate how the stress-strain field in the plastic zone around the tip of a crack responds to variable-amplitude loads and thereby influences subsequent crack growth. An objective that is common to all the existing load interaction models (or the so-called retardation models) is to predict variable-amplitude crack growth rate behavior by using con-

stant-amplitude crack growth rate data. Some physical background on crack-tip plastic zones and load interactions and their roles in crack growth retardation is reviewed here.

Crack-Tip Plastic Zones. When a crack is subjected to a cyclic stress, due to the high stress concentration at the tip of a sharp crack (even though the applied stress level may have been very low), the bulk of material in front of the crack will undergo plastic yielding. As previously shown in Chapter 4, the size of a crack-tip plastic zone can be expressed as:

$$r_p = \frac{\beta}{\pi} (K_{max}/F_{ty})^2 \qquad \text{(Eq 5.44)}$$

where F_{ty} is the material tensile yield strength, and β accounts for the degree of plastic constraint at the crack tip. For example, $\beta = 1$ for a through-thickness crack in a thin sheet because the stress state at the crack tip is plane stress. The value of β decreases to 0.33 as the stress state changes from plane stress to plane strain.

Equation 5.44 represents the crack-tip plastic zone that is formed during the upward excursion of a load cycle. This plastic zone is usually referred to as the monotonic plastic zone. The rate of crack growth at each load cycle depends on the crack-tip constraint and the magnitude of the stress intensity. After removal of the applied load, or at the valley of each load cycle, there is a residual plastic zone remaining at the crack tip. This is called the reverse plastic zone.

A schematically Figure 5.58 shows the monotonic and reverse plastic zones. The symbols r_p and r_{rp} in this figure refer to monotonic and reverse plastic zone, respectively. The residual stress distribution ahead of the crack (associated with one completed cycle) is also shown. The crossover point, from negative residual stress to

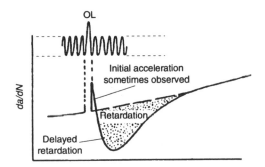

Fig. 5.57 Schematic post-overload crack growth behavior

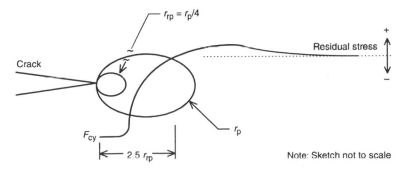

Fig. 5.58 Schematic comparison of the monotonic and cyclic plastic zones

positive residual stress, is theoretically 2.5 times the reverse plastic zone. In Fig. 5.58, the term F_{cy} is the material compression yield strength. Its absolute magnitude is approximately the same as the tensile yield strength of the same material. During unloading, the material immediately in front of the crack tip experiences an excursion load. Its absolute magnitude is equivalent to twice that of F_{ty} (because the load is dropping from F_{ty} to F_{cy}). Following Eq 5.44, the reverse plastic zone can be written as:

$$r_{rp} = \frac{\beta}{\pi} (K_{max}/2F_{ty})^2 = r_p/4 \qquad \text{(Eq 5.45)}$$

Therefore, the size of the reverse plastic zone is one-fourth that of the monotonic plastic zone. The reverse plastic zone is also called the cyclic plastic zone. However, a cyclic plastic zone is sometime defined as:

$$r_{cp} = \frac{\beta}{\pi} (\Delta K/F_{ty})^2 \qquad \text{(Eq 5.46)}$$

because the crack tip experiences an excursion of stresses in between S_{min} and S_{max} (not always in between S_{max} and zero load). Thus, we are dealing with three types of plastic zones: the monotonic zone based on K_{max}, the reverse zone based on K_{max}, and the cyclic zone based on ΔK. Most of the experimental data currently available indicate that crack growth rate behavior is better correlated with the monotonic plastic zone induced by the overload. While some investigators used the full zone (r_p) to make their correlation, others used the plastic zone radius r_y ($= r_p/2$). In any event, it is customarily assumed that load interaction takes place inside an area that is related to the monotonic plastic zone, although its exact size has to be determined. This area is frequently referred to as the overload-affected zone, or the load interaction zone. For the purpose of making life assessment, either approach, if used consistently, would make little difference in the predicted life.

Current crack growth interaction models always incorporate ways to compute crack growth rates inside the load interaction zone. The approaches taken by investigators generally fall into one of these categories: (a) modification of the stress-intensity factors (Ref 5.65), (b) direct computation of the instantaneous crack opening load and residual stress field at the crack tip (Ref 5.66), or (c) empirical adjustment of the crack growth rates (Ref 5.67). We will not discuss any of these models. Instead, we will review some physical background on crack growth retardation behavior by going through a scenario that involves a hypothetical stress profile, depicted in Fig. 5.59.

Mechanical Load Interaction. During crack propagation under variable-amplitude loading,

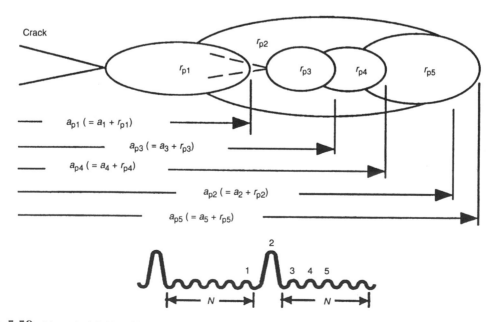

Fig. 5.59 Schematic definition of load interaction zones

or spectrum loading, a crack-tip plastic zone is associated with each stress cycle. Upon unloading, residual stresses remain on the crack plane and interact with the singular stress field of the subsequent stress cycle. The actual driving force for crack growth will be an effective stress field that accounts for the contributions of the applied stress and the residual stresses. In a given loading step, the plastic zone is calculated using Eq 5.44 (or using one-half of this magnitude, if one prefers).

The calculation is based on the current crack length and the load applied to it, and the computed plastic zone is placed in front of that crack length, as shown in Fig. 5.59. In this manner, there will be no load interaction if the elastic-plastic boundary of the current plastic zone exceeds the largest elastic-plastic boundary of the preceding plastic zones. In other words, for a crack to propagate at its baseline constant-amplitude rate, the advancing boundary of the crack-tip plastic zone due to the current tensile loading (the lower loads) must exceed the greatest prior elastic-plastic interface created by a prior higher tensile load. Load interaction exists as long as the plastic zone for the current stress cycle is inside the elastic-plastic boundary remaining from any of the preceding stress cycles.

Consider a stress level S_1, applied to a crack length a_1. This action will create a plastic zone r_{p1} in front of the crack (Fig. 5.59). The elastic-plastic boundary for this stage in the crack growth history will be $a_{p1} (= a_1 + r_{p1})$. Meanwhile, the crack will grow from a_1 to a_2. The stress event that follows, S_2, will create another plastic zone r_{p2} in front of a_2, so that the elastic-plastic boundary will become $a_{p2} (= a_2 + r_{p2})$. Because $a_{p2} > a_{p1}$, no load interaction occurs while the crack is extending from a_2 to a_3.

In the following three steps, both a_{p3} and a_{p4} are smaller than a_{p2}, but a_{p5} finally travels through the previous elastic-plastic boundary (i.e., $a_{p5} > a_{p2}$). Therefore, load interactions exist in the two loading steps that are associated with a_{p3} and a_{p4}. Crack growth rates will be affected by the residual stress field associated with a_{p2} while the crack is propagating from a_3 to a_4, and from a_4 to a_5. By definition, there is no load interaction between a_{p5} and a_{p2}, because $a_{p5} > a_{p2}$. Thus, the growth of a_5 to a_6 (not shown) will be solely due to S_5.

Experimentally Observed Post-Overload Crack Growth Behavior. Several studies have shown that the crack-tip plastic zone shape can play an important role in post-overload crack

growth behavior. Figures 5.60(a) to (d) are sketches typical of experimental results obtained from 6061-T6 aluminum (Ref 5.68, 5.69). In general, the crack-tip plastic zone shape changes with magnitude of the overload (Fig. 5.60e). The solid line in Fig. 5.60(a) to (d) is the experimentally observed boundary of a monotonic overload plastic zone. Within a given overload plastic zone there is a monotonic zone (a dash-dot line) that accompanied fatigue crack growth prior to the overload. The dotted line shows the path followed by the crack during subsequent growth. The messages presented by these diagrams are: (a) As opposed to the analytic solution, experimentally determined plastic zones are irregular and unsymmetrical. (b) The crack accelerates and returns back to steady growth immediately upon crossing the overload boundary nearest the crack tip. (c) The crack changes direction in order to reach the nearest boundary as soon as possible.

On the basis of their experimental observations, Lankford et al. (Ref 5.69) presented a generalized view connecting plastic zone shape (which depends on the magnitude of the overload) and its effect on delay cycles. They stated that a 50% overload causes no retardation because its overloaded zone does not extend directly ahead of the crack (see configuration 1 in Fig. 5.60e). Following a 100% overload, a larger plastic zone appears, which extends a finite distance ahead of the crack (configuration 2) and causes retarded crack growth while the crack transverses the distance to the nearest boundary. A still higher overload closes the "arms" of the overloaded zone, leaving a relatively undisturbed "hole" (configuration 3), which allows the crack a momentary acceleration during the retardation period. Finally, a still higher overload produces a solid zone (configuration 4), whose maximum dimension must be traversed by the crack. A schematic a versus N plot depicting the effect of overload plastic zone shape on delayed behavior is presented in Fig. 5.60(f). The a versus N curves 1, 2, 3, and 4 correspond to the four plastic zone configurations in Fig. 5.60(e); those overloads were 50, 100, 150, and 200%, respectively.

In light of these findings, it is conceivable that the plastic zone shape irregularity will limit the ability of analytical crack growth modeling. Thus, obtaining an accurate life prediction may be difficult. Currently, there is insufficient data to provide a useful input for development of an improved analytical crack growth model. Nonetheless, Lankford et al.'s data certainly help to shed light on one of the possible sources that may affect accuracy in analytical predictions.

Effect of Other Loading Conditions. Crack growth behavior is also affected by underload, hold time, biaxial stress, and load transfer.

Effect of Underload. If a compressive stress cycle, also known as an underload cycle, is inserted in between the overload and the normal load cycles, the extent of retardation will be reduced. It is conceivable that the plastic zone of

the underload counteracts the plastic zone of the overload, the end result perhaps being a smaller net plastic zone that contributes fewer delay cycles.

Effect of Hold Time. At room temperature, cyclic frequency and stress-cycle shape insignificantly affect both constant-amplitude and spectrum crack growth behavior. The magnitude and sequential occurrences of stress cycles are the only key variables that affect room-temperature crack growth behavior. Therefore, an accurate

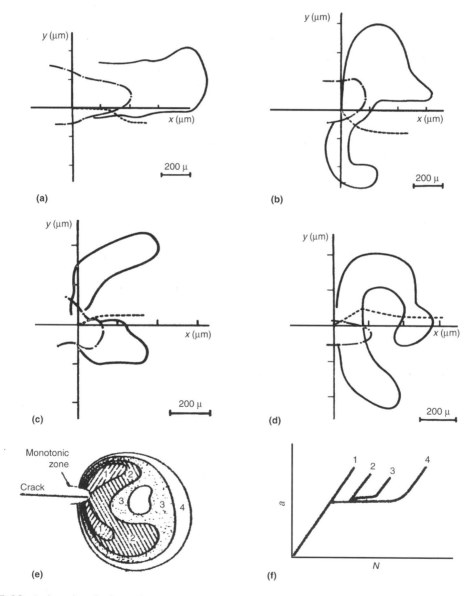

Fig. 5.60 Cyclic and overload (OL) plastic zone boundaries and their relation to post-overload crack growth. (a) 50% OL, $R = 0.4$. (b) 100% OL, $R = 0.4$. (c) 50% OL, $R = 0.05$. (d) 100% OL, $R = 0.05$. $\Delta K = 10.5$ MPa$\sqrt{m}$ for (a) to (d). (e) Overload plastic zone development. (f) Delayed crack growth behavior as a function of plastic zone shape. Source: Ref 5.68, 5.69

representation of the material crack growth rate data as a function of the stress amplitude ratio (i.e., the so-called crack growth law or crack growth rate equation) and a load interaction model for monitoring the load sequence effects on crack growth (commonly called the crack growth retardation/acceleration model) are the only two essential elements in cycle-dependent crack growth life predictive methodology.

Later in this chapter, we will see that fatigue crack growth in a corrosive or high-temperature environment will be significantly affected by the time factor in a given stress cycle. Low frequency or hold time may promote the time-dependent mechanism of fatigue crack growth. Now we will see that applying hold time on a given fatigue load cycle may alter the crack growth retardation characteristics of the subsequent load cycles, even in room-temperature air.

When delay tests are carried out on mill-annealed Ti-6Al-4V in room-temperature air, and hold time is applied at the peak of an overload cycle, the number of delay cycles, N_D, will increase. However, if the hold time is held at K_{min} (either at zero load or at the valley of an underload), it results in a reduced N_D as compared to no hold time (Ref 5.70). The reduction of delay cycles is attributed to relaxation of the crack-tip residual stresses. Validation of this phenomenon is supported by two other experimental studies on aluminum alloy 7075-T651 (Ref 5.71, 5.72). Reference 5.71 reported that residual stress relaxation resulted from conducting the test in vacuum and holding the load at zero (below K_{min}). The tests of Ref 5.72 were conducted in room-temperature air. The test sequences (loading profiles) and results are presented in Fig. 5.61 and Table 5.3, respectively. It should be

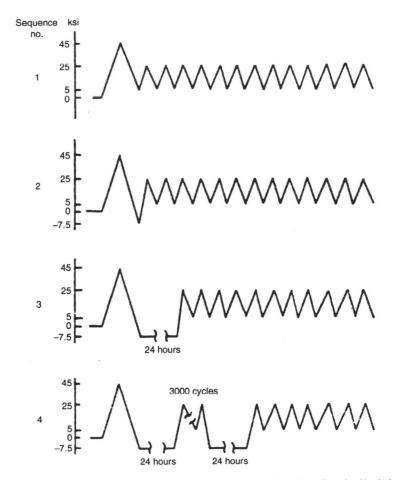

Fig. 5.61 Schematic stress profiles for fatigue crack propagation testing showing the effects of overload/underload with hold time. Source: Ref 5.72

noted that the stress levels labeled in Fig. 5.61 were net section stresses and the specimens used contained a circular hole. Again, it is quite clear that both underload and holding the load at K_{min} caused reduction of delay cycles. In light of these observations—that is, the possibility of erasing the beneficial effect of retardation due to residual stress relaxation—it is conceivable that using a retardation model for crack growth life prediction might result in an overestimated life. More research in this area is warranted.

Biaxial Stress. In the case of constant-amplitude crack growth, there is a general perception that different combinations of biaxial stress ratios will result in different crack growth rates. However, on the basis of a series of well-planned investigations, Liu and Dittmer have found that there is no biaxial stress effect on fatigue crack growth rate behavior (Ref 5.73; see also Chapter 7 of Ref 5.7). That is, the *da/dN* versus ΔK curves for the uniaxial tension test and a biaxial stress test are the same. The key elements in the discrepancies are the stress field and the *K*-factors associated with those nonstandard specimens used in biaxial load testing. Without a group of *K*-factors explicitly representing a specific combination of specimen geometry and loading condition, the *da/dN* versus ΔK behavior in the biaxially loaded specimen will not be correctly displayed. When the *K*-factors are correctly determined, the *da/dN* versus ΔK curves are the same for any biaxial stress ratio, whether the tension stress normal to the crack is combined with a positive or negative stress parallel to the crack.

Out-of-phase biaxial loading is important to the engine/turbine industry because of the nature of the loading exhibited in a turbine disk. In a biaxially loaded fatigue crack growth test, out-of-phase means that there is a time lag between the load waves in the *x* and *y* directions. For example, a 180° out-of-phase condition means that the P_x cycle will not start until the first P_y reaches the peak. The biaxial stress ratio will not be the same (in general) as those originally planned for the test. The resulting biaxial stress ratio will not be constant throughout a test, as it would be in the in-phase loading. It varies from time to time as each component of the sinusoidal waves oscillates across one another at different phase angles.

Therefore, planning and conducting the experiments and interpreting the results for this type of test are not simple tasks. One way to run such a test is to adjust (position) the two sinusoidal waves of the input loads so that the intended nominal biaxial stress ratio will occur once at a selected phase angle in each completed stress cycle. On the basis of a limited scope of investigation, the test results reported in Ref 5.73 showed that there is no effect on fatigue crack growth rate attributed to biaxial loading whether the loads were applied in-phase or out-of-phase. The fatigue crack growth rates are the same for both loading conditions. This conclusion is based on tests where the out-of-phase input loads were adjusted. If the loads had not been adjusted, the results would have been different. Such test data are also available (Ref 5.73).

In the case of variable-amplitude crack growth, Liu's test results (Ref 5.7) show a remarkable effect attributed to biaxial stress ratios. His data show that a crack grows faster at positive biaxial stress states but slower at negative biaxial stress states, as compared to uniaxial tension—the rationale being that crack-tip plastic zones are smaller at positive biaxial stress states but larger at negative biaxial stress states. Similar to variable-amplitude crack growth under uniaxial tension, a larger crack-tip plastic zone will result in a more pronounced crack growth retardation and vice versa.

Load Transfer. Fatigue crack growth analysis of splices, joints, lugs, and so on falls into a category of problems involving load transfer through fasteners from one part to another. This load transfer will tend to change the basic fatigue spectrum in the local area of the load transfer. For example, the lug hole does not recognize the compressive load contained in the fatigue spectrum. Therefore, all the compression loads in the spectrum may be truncated (i.e., the min-

Table 5.3 Fatigue crack propagation test results for overload/underload with hold time

Sequence No.	Specimen life, 1000 cycles	Mean life, 1000 cycles
1	1322.0	910.33
	1009.2	
	399.2	
2	212.2	214.4
	171.0	
	260.0	
3	158.4	126.05
	112.2	
	97.8	
	135.8	
4	118.7	130.5
	124.4	
	148.4	

Source: Ref 5.72

imum load level set to zero). In a spliced joint where the fastener holes are loaded holes, various degrees of load transfer take place at these holes while the structural member is subjected to uniform far-field loading. Therefore, a spectrum modification logic must be incorporated into one's life prediction methodology to realistically account for the structural response to the applied loads.

5.3.7 Issues Concerning Cracks at a Circular Hole

Preexisting Stresses. Geometry-induced residual stress is often a result of load excursions at stress concentration. We have explored this problem with an open hole in Chapter 2. Preyielded holes such as interference-fitted fastener holes or cold-worked holes always attach with a preexisting stress field representing the mechanical work-induced residual stresses. Their effect on crack growth behavior and the applicable analysis methods are discussed in Ref 5.7 and 5.74.

Small Crack. Numerous researchers have discussed the short-crack (or small-crack)* phenomenon. They have found that the Bowie factors are inaccurate or not applicable for very short cracks, and that excessive general yielding in the area near the hole-edge may contribute to this phenomenon. The high loads in a fatigue spectrum or a monotonic increasing load appears to be responsible for the yielding of a hole. It has been postulated that the small crack would have been fully embedded inside a locally deformed area (a general yield zone) that is associated with the circular hole. Both the general yield zone at the hole and the crack-tip plastic zone enlarge as the applied load increases. However, the total effective crack length (the physical crack length plus the crack-tip plastic zone) will always be embedded inside the general yield zone. This results in abnormal fatigue crack growth and fracture behavior, as compared to long crack behavior.

This short-crack behavior is very troublesome and has caused numerous inaccuracies in life predictions. A lengthy discussion and bibliography on the subject of the fatigue growth of small cracks are provided by McClung et al. (Ref 5.75). There is no satisfactory solution at this time, because load history effects have not been adequately characterized. A more sophisticated spectrum crack growth prediction method is also needed.

As always, any approach toward solving the short-crack problem (or any crack problem) will depend on the correctness of the K-solution for the structural configuration under consideration. Any conclusion, or the technology developed from it, will be false without an appropriate K-solution. However, not a single paper has addressed this problem among all the research conducted in the last 20 years. In terms of fracture strength, one attempt has been made to use the R-curve (of Chapter 4) in combination with the J-integral fracture index, which is discussed in Chapter 6. The result is promising (Ref 5.7). More research in this area is needed to develop this method to a quantitative prediction procedure.

To put this matter in perspective, it seems appropriate to quote from the summary of McClung's review paper (Ref 5.75): "Small-crack behavior was first documented in the mid-1970s, extensively investigated in the 1980s, and remains an active research topic. The problem is now well enough understood to facilitate some standardization of concepts, test methods, and analysis techniques, but small-crack technology is not yet routinely applied in industrial practice. At this writing, no general-purpose computer codes for fatigue crack growth (FCG) analysis are available that explicitly address small-crack behavior. Furthermore, several important problems remain unresolved. For example, some small-crack effects appear to be accentuated under variable-amplitude loading, but load history effects have not been adequately characterized. In addition, as noted earlier, it is not yet clear if small cracks exhibit a well-defined threshold or non-propagation condition; if so, how might this be related to the large-crack threshold." More research and development of "small-crack technology" is needed to advance the method of structural life assessment.

5.3.8 Effect of Environment

Corrosion fatigue crack propagation is no different from ordinary fatigue crack growth, except that the test is run in a corrosive environment (e.g., salt water, high-humidity air, etc.). Therefore, it is common practice to conduct tests

---

*There is no clear definition for the size of a short crack. Suggestions include (a) a crack length less than 1 mm (0.04 in.), (b) a crack size smaller than its crack-tip plastic zone radius or the size of the local yield zone if the crack is at the notch, or (c) a crack size smaller than the microstructural unit size (typically the grain size).

and record and implement the *da/dN* data the same way as normally done in room-temperature air. We also can postulate that corrosion fatigue crack propagation is the same as crack growth in a conventional corrosion fatigue test. The only difference between these two tests is that the corrosion fatigue crack propagation test does not involve pitting because a precracked specimen is used.

In terms of microscopic crack propagation path and fracture surface appearance, it is generally agreed that the fracture path in a traditional corrosion fatigue specimen (repeated cyclic loading in a corrosive medium) is typically transgranular. There are exceptions. For example, after several specimens of the same alloy were tested in the same environment (7079-T651 tested in NaCl solution), the fracture paths in these specimens were either transgranular, intergranular, or mixed (Ref 5.76). These disturbing results are probably attributed to variation in the degree of the stress-corrosion cracking component in an individual test. When a specimen exhibits time-dependent behavior (because it is tested at a very low frequency or subjected to a long hold time), its fracture path may have a tendency toward intergranular cracking.

Crack Growth Mechanism. According to McEvily and Wei (Ref 5.77), corrosion fatigue crack propagation exhibits three basic types of crack growth rate behavior (Fig. 5.62). True corrosion fatigue (TCF) describes the behavior when fatigue crack growth rates are enhanced by the presence of an aggressive environment at levels of K below K_{ISCC} (Fig. 5.62a). This be-

havior is characteristic of materials that do not exhibit stress corrosion, i.e., $K_{ISCC} = K_{IC}$. Stress corrosion fatigue (SCF) describes corrosion under cyclic loading that occurs whenever the stress in the cycle is greater than K_{ISCC}. This is characterized by a plateau in crack growth rate (Fig. 5.62b) similar to that observed in stress-corrosion cracking (Fig. 4.20). The most common type of corrosion fatigue behavior, shown in Fig. 5.62(c), is characterized by stress-corrosion fatigue above K_{ISCC}, superimposed on true corrosion fatigue at all stress-intensity levels.

For most of the commonly used alloys, the effect of frequency on constant-amplitude crack growth rate is probably negligible in dry-air environments. A lightly corrosive environment (humid air) gives rise to higher crack growth rates than a dry environment. The cyclic crack growth rate varies significantly at elevated temperature and in chemically aggressive environments. The effects of frequency, waveform, and hold time on constant-amplitude crack growth rate is generally magnified by the presence of a corrosive medium. The faster the testing speed, the slower the fatigue crack growth rate (or the longer the hold time, the faster the growth rate). Although there is no concurrence of opinion as to the reason for the environmental effect, it is certainly due to corrosive action. As a result, the influence of the environment is time and temperature dependent.

Hydrogen embrittlement can be considered a form of corrosion damage. It has been generally established that the principal mechanism of corrosion damage in Ti-6Al-4V alloy is hydrogen

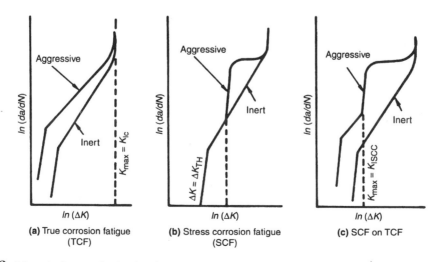

Fig. 5.62 Schematic diagrams showing three basic types of corrosion fatigue crack growth behavior. Source: Ref 5.77

embrittlement. Most engineering alloys, including titanium, owe their corrosion resistance to instantaneous formation of tightly bonded protective oxide layers. If the oxide layer is disturbed, these alloys experience corrosion attack in aqueous, salt, and water-vapor environments. In the case of titanium, once the oxide layer is removed by either mechanical deformation or chemical dissolution, hydrogen can diffuse into the titanium lattice.

As the environment proceeds from normal air to distilled water, increasing amounts of hydrogen are made available through the water molecule dissociation reaction. Apparently, the extra corrosiveness of the aqueous salt environment is due to accelerated breakdown of the oxide film on the crack surface. Lower cyclic load frequency increases the exposure time of the metal to the environment. Since hydrogen embrittlement depends on the rate of hydrogen generation and diffusion, longer exposure of stressed bare-metal surfaces will permit greater concentrations of hydrogen to diffuse into the metal, and thus accelerate crack growth rate.

In studying the environmental fatigue crack growth behavior in titanium, Dawson and Pelloux (Ref 5.78) have classified environments into three groups according to the possible influence of loading frequency. These models are shown schematically in Fig. 5.63. In normally inert environments, such as vacuum, helium, argon, or air, fatigue crack growth rates in titanium exhibit no effect of frequency (Fig. 5.63a). In liquids such as methanol, a "normal" frequency effect is found, in that higher fatigue crack growth rates occur at lower frequencies (Fig. 5.63b). In halide-containing solutions such as salt water, "cyclic stress-corrosion cracking" with a characteristic discontinuity in the da/dN versus ΔK curve is found (Fig. 5.63c). The lower the loading frequency, the lower the ΔK value at which the discontinuity is observed, where the limiting value is such that $K_{max} = K_{ISCC}$. Figure 5.64 illustrates the corrosion fatigue behavior for X-65 line pipe steel tested in salt water with a superimposed cathodic potential. This behavior is somewhat close to type 3 of Fig. 5.63. When the same data are compared to the generalized model of McEvily and Wei (Ref 5.77), they fall somewhere in between types B and C behavior in Fig. 5.62.

Environment-Affected Zone. Diffusion is perhaps the one important mechanism that has not received close attention from fracture mechanics analysts. Penetration of chemical agent into a localized crack-tip region would severely damage the bulk of material near the crack tip. This is particularly true for a loading profile that contains load cycles with long hold time. The damaged material no longer represents the original bulk material and exhibits accelerated crack growth rates as compared to its baseline behavior. This damaged zone is called the environment-affected zone (EAZ) by some investigators. Theoretically, the size of the EAZ can be calculated using a textbook formula. The width of the EAZ (the diffusion distance) is expressed as:

$$r_e = \sqrt{Dt_H} \qquad \text{(Eq 5.47)}$$

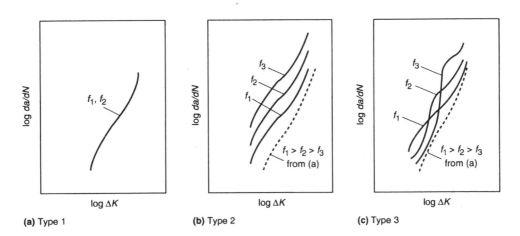

(a) Type 1   **(b) Type 2**   **(c) Type 3**

Fig. 5.63 Schematic influence of loading frequency on fatigue crack growth for three classes of environments. (a) Little or no effect of frequency, as in vacuum, inert gas, or air. (b) Fatigue crack growth increases with decreasing frequency, as in methanol. (c) Cyclic stress-corrosion cracking effect, as in salt water. Source: Ref 5.78

where t_H is hold time. The diffusion coefficient D is a function of the activation energy Q and the gas constant R ($= 8.314$ J/°K), and is given by:

$$D = w^2 \cdot v' \cdot \exp(-Q/RT) \qquad \text{(Eq 5.48)}$$

where T is absolute temperature (in °K), w is jump distance, and v' is jump frequency. The quantity for Q (in J or cal/mole) is sometimes available in the open literature. It is easy to understand that Eq 5.48 is rather academic, not meant for day-to-day use in solving engineering problems.

Experimental testing presently is the only means for determining the EAZ size and the crack growth behavior inside the EAZ. In the case of high-temperature crack growth testing, oxidation due to penetration of hot air into a localized crack-tip region is considered the primary source for creating the damaged zone. A typical example, taken from Ref 5.80, is shown in Fig. 5.65. The data were developed from the Inco 718 alloy, tested at 590 °C (1095 °F). The loading profile consists of one trapezoidal stress cycle (i.e., with hold time) and many triangular (or sinusoidal) stress cycles of the same magnitude, applied before and after the trapezoidal load. The test was conducted such that a constant K level was maintained at each stress cycle. A record of crack lengths versus stress cycles was taken (Fig. 5.65).

This type of data can be reduced to a plot of crack growth rate versus crack length, as shown in Fig. 5.66. The purpose of making this plot is to show the influence of sustained load on crack growth rate, so the crack length immediately following the sustained load cycle is set to 0. It appears that the data points are divided into two parts, each of which can be approximated by a straight line. The first part characterizes the crack growth behavior inside the EAZ, where the crack growth rate is initially higher than normal because the material has been damaged by the hot air. The crack growth rates corresponding to those postsustained load cycles gradually decrease to a level that is otherwise normal for an undamaged material. Thus, the horizontal line implies that the crack had traveled through the EAZ and resumed a stable crack growth behavior. The intersection of these two lines determines the size of the EAZ. Many test data of this type are reported in Ref 5.80, covering a range of combinations of temperature, hold time, and stress-intensity level. Figures 5.65 and 5.66 are typical of the data reported.

There are many ways to formulate an empirical equation for the EAZ. In dealing with crack growth of aluminum in salt water at room temperature, Kim and Manning (Ref 5.81) hypothesized that the EAZ (which is attributed to hydrogen penetration) is in proportion to the crack-tip plastic zone size (which is a function of the applied K), D (the diffusion coefficient of hydrogen in aluminum), and t_H. However, based on high-temperature crack growth rate data of the Inco 718 alloy, Chang (Ref 5.82) concluded that the EAZ (attributed to oxidation) was a function of hold time and temperature, independent of K. Thus, Chang further postulated that EAZs could be fitted by an empirical equation of the form:

$$r_e = \lambda \cdot t_H^n \cdot e^{-Q/RT} \qquad \text{(Eq 5.49)}$$

Here, EAZ is noted as r_e, t_H is the hold time, T is the absolute temperature in °K, Q is the activation energy, R is the gas constant ($= 8.314$ J/°K), and λ and n are empirical constants. Chang's data are plotted in Fig. 5.67. Using Eq 5.49 with $Q = 6 \times 10^4$ cal/mol (251,400 J) for the nickel-base alloy, the empirical constants λ and n can be determined by fitting a regression

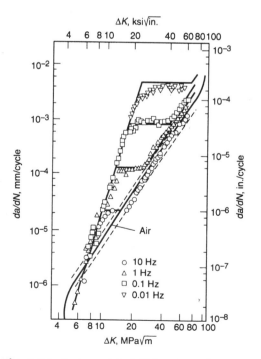

Fig. 5.64 Crack growth in X-65 line pipe steel exposed to air and 3.5% salt water with a superimposed cathodic potential. Source: Ref 5.79

line through all the experimental data points. In this case, $\lambda = 7.55 \times 10^{10}$ mm/s and n = 1.0.

Fatigue Crack Growth at Low Temperature. So far, gathering low-temperature fatigue crack growth rate data and trying to understand exactly what happened is the only type of documentation available in the open literature. At low temperatures the reaction kinetics are slower and the air can contain less water vapor. This may reduce crack growth rates in certain alloys.

Sometimes the effect of temperature on fatigue crack growth in the low-temperature range is very small. However, for some alloys (e.g., steels in particular) that exhibit ductile to brittle transition at low temperature, crack growth rates can be very high when brittleness causes cleavage during crack propagation. Figure 5.68 demonstrates the increase in the crack growth rate exponent with decreasing temperature for a number of iron-base alloys and steels. In any

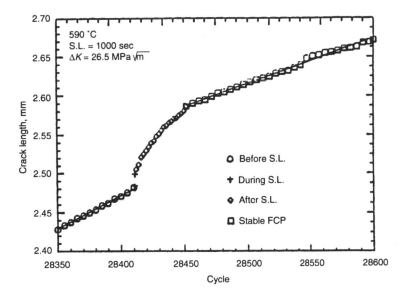

Fig. 5.65 Effect of long hold time on constant-amplitude crack growth of Inco 718 alloy. Source: Ref 5.80

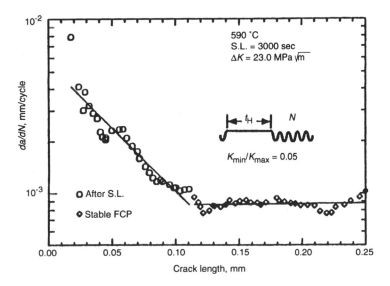

Fig. 5.66 Measured *da/dN* after a sustained load cycle. Source: Ref 5.80

event, using the actual *da/dN* and fracture toughness data for material evaluation and selection and to perform life assessment is the only way to deal with low-temperature crack growth and fracture.

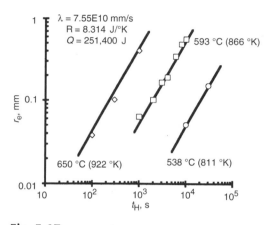

Fig. 5.67 EAZ sizes for Inco 718 alloy. Source: Ref 5.82

Fatigue Crack Growth at High Temperature. In many ways, crack growth at high temperature shares the same platform as crack growth in corrosive environments. Crack growth rates are generally higher in corrosive or high-temperature environments as compared to room-temperature air. However, they may grow under different mechanisms. The EAZ is formed by diffusion of the corrosive medium or hot air into the material at the crack-tip vicinity. For creep ductile materials, such as low-alloy steels and stainless steels, crack growth and fracture behavior may become viscoelastic at high temperatures. A new kind of fracture indice, C_t or C^*, is required for characterizing this type of behavior. These new parameters will be presented in Chapter 6, along with the criteria that divide the elastic and the viscoelastic behaviors. It is also shown that the linear-elastic stress-intensity factor K is adequate for analyzing high-temperature fatigue crack growth and fracture of environment-sensitive and creep-resistant super-

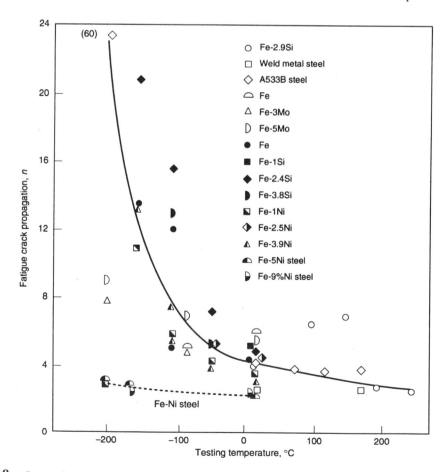

Fig. 5.68 Influence of testing temperature on fatigue crack propagation exponent for iron-base alloys. Source: Ref 5.83

alloys. This section briefly summarizes the factors affecting high-temperature fatigue crack growth in the context of traditional K-factor analysis.

In traditional cyclic crack growth testing at high temperature (i.e., with a sinusoidal or symmetrically triangular waveform at a moderately high frequency), the crack growth rates are functions of ΔK and R. Thus, the phenomenon is similar to those at room temperature, with the following exceptions: (a) For a given R, the value of ΔK_{th} is higher at higher temperature; and (b) for a given R, the terminal ΔK value is higher at higher temperature because K_C is usually higher at a higher temperature (because the material tensile yield strength is lower at a higher temperature).

Figure 5.69 shows a schematic representation of temperature influence on da/dN. The crossover phenomenon of the ΔK_{th} values seems real in high-temperature fatigue crack growth, because it has been experimentally observed in a number of materials (i.e., Inco 718, René 95, titanium alloys, and titanium aluminides) (Ref 5.84). The crossover of the terminal ΔK values happens because K_C usually increases with temperature so that $(1 - R) \cdot K_C$ also increases with temperature. A note of caution is that crack growth rates in the linear range are not always higher at temperatures higher than room temperature, as implied in Fig. 5.69. Depending on frequency and ΔK range, some material (particularly those sensitive to environment) may exhibit slower crack growth rates at intermediate temperatures. Moisture, which might have acted as a corrosive medium, was vaporized by heat. Therefore, the magnitude of the environmental fatigue component that is attributed to moisture would be reduced. A three-dimensional representation of the effects of temperature and frequency on da/dN for an air-cast Cr-Mo-V rotor steel is presented in Fig. 5.70.

In the power-law (Paris equation) crack growth regime, the effects of temperature, stress ratio R, and hold time have been investigated for many high-temperature alloys. Typical behavior and crack growth results for specific alloys are covered elsewhere (e.g., Volume 19 of the *ASM Handbook*). A general comparison of temperature effects on fatigue crack growth of several different high-temperature alloys is shown in Fig. 5.71. Because the reported data are obtained at various ΔK ranges and temperature ranges, the general comparison is based on a constant ΔK, arbitrarily chosen as 30 MPa$\sqrt{m}$

(27 ksi$\sqrt{in.}$). A clear trend of crack growth rate increase with increasing temperature can be seen. At temperatures up to about 50% of the melting point (550 to 600 °C, or 1020 to 1110 °F), the growth rates are relatively insensitive to temperature, but the sensitivity increases rapidly at higher temperatures. The crack growth rates for all the materials at temperatures up to 600 °C (1110 °F), relative to the room-temperature rates, can be estimated by a maximum correlation factor of 5 (2 for ferritic steels).

Cyclic frequency (or duration of a stress cycle, e.g., with hold time) plays an important role in high-temperature crack growth. At high frequency—that is, fast loading rate with short hold time (or no hold time)—the crack growth rate is cycle dependent and can be expressed in terms of da/dN (the amount of crack growth per cycle). At low frequency (or with long hold time), however, the crack growth rate is time dependent; that is, da/dN is in proportion to the total time span of a given cycle. For tests of different cycle times, all crack growth rate data points are collapsed into a single curve, of which da/dt is the dependent variable. A mixed region exists in between the two extremes. The transition from one type of behavior to another depends on material, temperature, frequency, and R (Ref 5.87). For a given material and temperature combination, the transition frequency is a function of R. The frequency range at which the crack growth rates remain time dependent increases as R increases

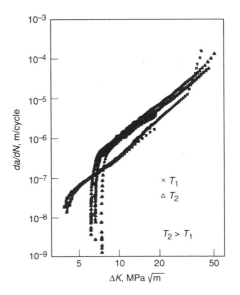

Fig. 5.69 Schematic of temperature effect on fatigue crack threshold and growth rates

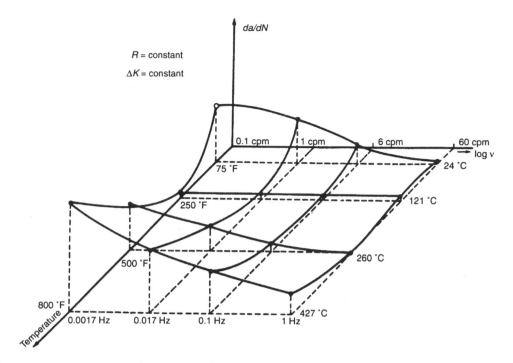

Fig. 5.70 Three-dimensional representation of effects of frequency and temperature on fatigue crack growth rate in air. Source: Ref 5.85

(Ref 5.87). The limiting case is R approaches unity. It is equivalent to crack growth under sustained load, for which the crack growth rates at any frequency will be totally time dependent.

5.3.9 Elements of Life Assessment

A sophisticated crack growth analysis procedure generally contains at least five key elements:

- Operational usage (including stress level)
- Material (environmental) da/dN properties
- Material fracture properties
- A library of stress-intensity factors
- Fracture mechanics methodology integrating all the other elements into a conglomerate tool for life prediction

A pictorial presentation of these elements is shown in Fig. 5.72. Analytical results are sensitive to many input variables, including:

- Operational usage (including stress level, corrosive environment, temperature, and hold time)
- Fracture mechanics methodology
- Structural geometry
- Initial crack length
- Material properties

These variables interact with each other, and it is difficult to separately pinpoint the significance of the sensitivity attributed to each one. However, the sensitivity of most of these variables lies in their effect on the stress-intensity factor. For example, when the initial stress-in-

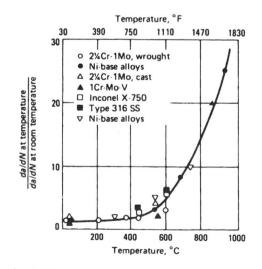

Fig. 5.71 Variation of fatigue crack growth rates as a function of temperature at $\Delta K = 30$ MPa$\sqrt{m}$. Source: Ref 5.86

tensity factor in one case is below or slightly above the threshold value in the *da/dN* curve and the initial stress-intensity factor in another case is relatively higher, the difference in calculated life will be substantial. Therefore, starting from a given initial crack length, a realistic and accurate crack growth history can be predicted, depending on how all the elements in Fig. 5.72 are handled.

This scenario brings up another, but related, area of concern: material selection. Consideration of all the requirements for a design application, such as strength, toughness, corrosion and exfoliation resistance, weight, and so forth, generally narrows a field of candidate materials down to a few choices. Among these finalists, one might have higher fracture toughness but also higher fatigue crack grow rates (in the linear range). Its fatigue crack growth rate threshold could be higher. Sometimes it is difficult to judge this material in comparison to the other candidate materials because of different combinations of fracture toughness, fatigue crack growth rate, and fatigue crack growth rate threshold. Making a decision requires running a life prediction analysis. Many fundamental considerations, concepts, and techniques have been discussed in depth in the preceding sections. The techniques that are required to determine damage accumulation (i.e., do the final integration in Fig. 5.72) are discussed below.

Damage Accumulation in Constant-Amplitude Loading. So far, we have shown that crack length, or the increment of crack growth, is related to number of stress cycles (or flight time) through a driving force K (or ΔK). To calculate a crack growth history, one may choose to determine how much a crack will grow in each applied load cycle or group of load cycles. Or, one may find it more appropriate to calculate the number of load cycles it takes to grow a crack from a prescribed crack length to the next prescribed crack length.

Most crack growth cumulation analysis routines are structured around fatigue crack propagation laws. As discussed earlier, a crack subjected to a constant-amplitude applied differential stress intensity, ΔK, has been observed to propagate at a material-dependent rate, or:

$$\frac{da}{dN} = f(\Delta K, C, n, K_C, \Delta K_o) \qquad \text{(Eq 5.50)}$$

To estimate the crack extension over a discrete number of constant-amplitude load cycles requires integration of Eq 5.50 along a, or:

$$N - N_o = \int_{a_o}^{a_n} \frac{da}{f(\Delta K, \ldots)} \qquad \text{(Eq 5.51)}$$

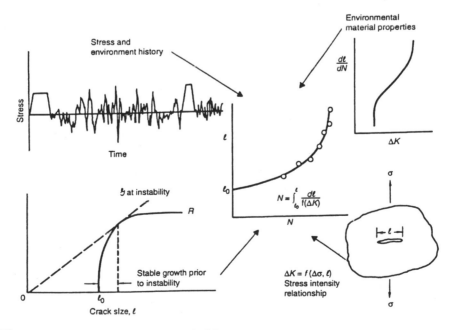

Fig. 5.72 Required elements in a life assessment methodology

Numerical integration of Eq 5.51 is required in most cases due to the complexity of the crack growth rate expression and/or the stress-intensity solution. Two approaches can be taken to this integration: incrementing either crack size or cycles at a chosen fixed rate.

The number of cycles required to grow a crack a discrete distance, Δa, can be calculated by simply evaluating ΔK at the average crack length $(a + \Delta a/2)$ and calculating the average growth rate, da/dN, from Eq 8.1 and solving:

$$\Delta N = \Delta a/(da/dN) = \Delta a/|f(\Delta K, \ldots)|$$

$$\text{(Eq 5.52)}$$

For a crack configuration that experiences large stress-intensity gradients, dK/da, the Δa increment must be small to preserve accuracy. By summing the number of cycles calculated from Eq 5.52, an output of a versus N can be generated and arbitrarily terminated at a specific stress intensity, crack size, or cyclic life, or just grow the crack to failure.

Conversely, an increase in crack size resulting from a discrete number of load cycles can be evaluated by:

$$\Delta a = \Delta N/(da/dN) = \Delta N/|f(\Delta K, \ldots)|$$

$$\text{(Eq 5.53)}$$

The average growth rate for this calculation is slightly more difficult to estimate; however, several integration techniques are available in the literature to provide an accurate evaluation of Eq 5.53.

The third method for calculating damage accumulation is the cycle-by-cycle summation technique. This technique, in conjunction with the deterministic input of the loads spectra, is best suited for incorporation with crack growth interaction models to account for load interaction effects. In general, the procedure consists of the following steps:

1. The initial crack size follows from the damage tolerance assumption as a_1. The stress range in the first cycle is $\Delta\sigma_1$. Then determine ΔK_1.
2. Determine da/dN at ΔK_1 from the tabulated da/dN versus ΔK data (or an equation such as those presented in this chapter). Take into account the appropriate R-ratio at each loading step in the load spectrum.
3. The crack extension Δa_1 in cycle 1 is $\Delta a_1 = (da/dN)_1 \cdot 1$.

4. The new crack length will be $a_2 = a_1 + \Delta a_1$.
5. Repeat steps 1 to 4 for every following cycle, while for the ith cycle replacing a_1 by a_i and a_2 by a_{i+1}.
6. Dealing with a load spectrum that consists of variable-amplitude cycles requires incorporation of a load interaction model. The load interaction model usually involves determining an effective ΔK in each load cycle as crack growth progresses.

Because the stress-intensity factor varies along the periphery of a part-through crack, a two-dimensional crack growth scheme is an essential item in a crack growth computing routine. More importantly, the two-dimensional crack growth computing routine must be able to calculate crack growth increments at each point (e.g., on the surface and at maximum depth) independently. The reason for this is that in some stress level and stress-intensity factor combinations, the stress intensity at one local point on the crack may be lower than ΔK_{th}, but the other point on the same crack may have a stress-intensity value significantly higher than ΔK_{th}. This situation implies that one point on the crack may stop growing while the other point on the same crack is still growing. However, this situation may be only temporary, because the crack shape continuously changes, although the crack is growing in one direction. The stress-intensity level that is originally below ΔK_{th} may become high enough to cause crack extension due to the new flaw geometry or simply due to changes in applied stress level (in the subsequent loading step), or both.

The example given below demonstrates the procedures that can be used for computing one-dimensional fatigue crack growth damage. A scenario goes like this: A flat panel made of alloy A, for which hypothetical da/dN data is shown in Fig. 5.73, is subjected to zero-to-tension $(R = 0)$ constant-amplitude loading. The maximum stress level is 49.6 MPa (7.2 ksi). Predict the number of cycles required to grow a crack with a total length of 12.7 cm (5 in.) to a total length of 22.9 cm (9 in.). In this example we use the actual da/dN curve (i.e., choose da/dN values from the graph instead of using a crack growth rate equation) for each step of the calculations.

The step-by-step procedures are summarized in Table 5.4 for four increments of crack length. To keep the problem simple, the panel is assumed to be an infinite sheet containing a through-thickness crack. Thus, the stress-intensity factor would always be $\sigma\sqrt{\pi a}$. Note that the quantity l in Table 5.4 is the total crack length (i.e., $2a$). The propagation rate over each increment is taken to be the average of the rates at the extremes of the increment. The length of each increment divided by its rate gives the approximate number of cycles required to grow the crack the incremental length. The total number of cycles predicted to grow the crack from 12.7 to 22.9 cm (5 to 9 in.) is the sum of the cycles occurring for all the increments. For this example problem, the total number is 2335 cycles. The computed crack growth history is plotted in Fig. 5.74. If the crack growth history is determined by using a cycle-by-cycle procedure, the predicted life would have been shorter, only 1760 cycles in this case (see Fig. 5.74). Therefore, keeping a small increment is essential for preserving accuracy.

Cycle-Dependent versus Time-Dependent Crack Growth. The integration schemes discussed so far are applicable to constant amplitude, constant frequency, triangular waveform, in room-temperature air. When we introduced the concept of an EAZ, clearly the time factor in a given stress cycle (whether it is associated with hold time or low frequency) that promotes time-dependent crack growth behavior may play a significant role in crack growth in high temperature or a corrosive medium. Although the general layout for life assessment (Fig. 5.72) remains unchanged, the material da/dN properties under different circumstances will change accordingly. New ways to handle the environmental crack growth rate behaviors are required.

Many analytical methods offer a handle on estimating environmental fatigue crack growth rates as functions of frequency and hold time. Some models were originally developed for corrosion fatigue cracking, others for high-temperature applications. Some of them can only handle the frequency effect. It is conceivable that any models that are designed to handle the time effects are equally applicable to crack growth in either type of environment (i.e., corrosion or high temperature). In those regions in

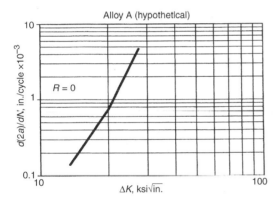

Fig. 5.73 Hypothetical fatigue crack growth rates for example procedure

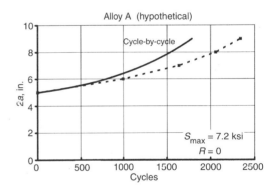

Fig. 5.74 Comparison of the life assessement results obtained from two methods

Table 5.4   Procedures for computing one-dimensional fatigue crack growth damage

l, in.	$\sqrt{\pi l/2}\ \sqrt{\text{in.}}$ (from No. 1)	ΔK, ksi$\sqrt{\text{in.}}$ (7.2 × No. 2)	dl/dN, in./cycle	Average dl/dN(b), in./cycle (from No. 4)	Δl, in. (from No. 1)	ΔN, cycles (No. 6 ÷ No. 5)
No. 1(a)	No. 2(a)	No. 3(a)	No. 4(a)	No. 5(a)	No. 6(a)	No. 7(a)
5	2.80	20.16	0.76×10^{-3}			
				1.005×10^{-3}	1.0	995
6	3.07	22.05	1.25×10^{-3}			
				1.565×10^{-3}	1.0	639
7	3.32	23.85	1.88×10^{-3}			
				2.39×10^{-3}	1.0	419
8	3.54	25.47	2.90×10^{-3}			
				3.55×10^{-3}	1.0	282
9	3.76	27.0	4.20×10^{-3}			
				Total cycles		2335

(a) Numbers identify a given column. (b) Average from two values in column 4.

which both the cycle-dependent and the time-dependent phenomena are present, implementation of a superposition technique may be required. The Wei-Landes superposition model (Ref 5.88) would be the natural choice for this purpose. The Wei-Landes model is theoretically capable of handling loading profiles that contain either the triangular or the trapezoidal stress cycles, or both. Experimental verifications have shown that the Wei-Landes model actually works well with test data for a wide range of frequencies. The time function in the model basically describes the fluctuating loads as a sinusoidal wave. The applicability of this model to stress cycles with hold time has not been tested. In searching for a model better suited for handling the trapezoidal stress cycles, Liu (Ref 5.7, 5.89) modified the Wei-Landes model. Summary descriptions of both models are given here.

Wei-Landes Superposition Model. In this method, the rate of fatigue crack growth in an aggressive environment is considered to be equal to the algebraic sum of the rate in an inert reference environment and that of an environmental component, computed from sustained-load crack growth data obtained in an identical aggressive environment and the load profile represented by $K(t)$. The total crack growth rate in

an aggressive environment ($\Delta a/\Delta N$) can be expressed as (Ref 5.88):

$$\frac{\Delta a}{\Delta N} = \left(\frac{\Delta a}{\Delta N}\right)_r + \int \frac{da}{dt} \cdot K(t) \cdot dt \qquad \text{(Eq 5.54)}$$

where $(\Delta a/\Delta N)_r$ is the crack growth rate in an inert reference environment, usually room-temperature air, at an ordinarily fast cyclic frequency. The term da/dt is the sustained-load crack growth rate in an identical aggressive environment. It can be determined from test data obtained from sustained-load tests. The test data points are usually plotted as da/dt versus K_{max}. The function $K(t)$ represents the load profile as a function of time. The effects of frequency and loading variable are incorporated through $K(t)$. In a constant-amplitude, sinusoidal loading condition, $K(t)$ can be written as:

$$K(t) = \frac{K_{max}}{2}[(1 + R) + (1 - R) \cdot \cos\omega t]$$

$$\text{(Eq 5.55)}$$

This accounts for the effects of frequency, mean load, range of cyclic loads, and the hold time waveform on the sustained-load growth component. The computational procedure is illustrated schematically in Fig. 5.75. An example

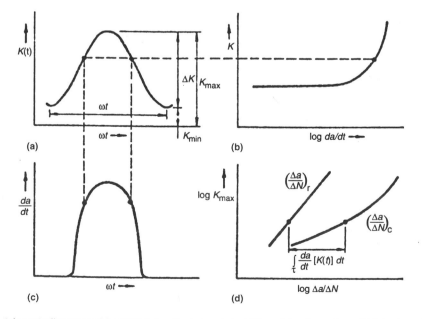

Fig. 5.75 Schematic illustration of the suggested method of analysis. (a) Stress-intensity spectrum. (b) Rate of crack growth under sustained load in an aggressive environment. (c) Environmental contribution to crack growth in fatigue. (d) Integrated effects on environment and K_{max} on fatigue crack growth rate. Source: Ref 5.88

comparing the predictions with actual test data is shown in Fig. 5.76. Except for one test, excellent correlations were obtained. The case that shows a poor correlation was tested at 1 Hz (the fastest among all the tests), and probably can be attributed to the fact that the influence of the sustained load had vanished at this frequency.

In any event, it is thought that this approach allows prediction of corrosion fatigue crack growth rates simply by adding the inert fatigue crack growth rate per cycle and the crack extension due to stress corrosion. Therefore, it eliminates the work needed to conduct fatigue crack growth tests in a corrosive environment.

Most of the test data used by Wei and Landes (Ref 5.88) for developing Eq 5.54 were from constant-amplitude tests. On the basis of the data gathered, the steady-state response of fatigue crack growth to environments may be grouped into three basic types in relation to K_{ISCC}, as previously shown in Fig. 5.62:

- *Type A* behavior is typified by the aluminum-water system. Environmental effects resulted from interaction of fatigue and environmental attack. The integral term in Eq 5.54 vanishes for this type of behavior.
- *Type B* behavior is typified by the hydrogen-steel system. Environmental crack growth is directly related to sustained-load crack growth, with no interaction effects. The second term of Eq 5.54 vanishes for this type of behavior.
- *Type C* behavior is the behavior of most alloy-environment systems. Above K_{ISCC} the behavior approaches that of type B, whereas below K_{ISCC} the behavior tends toward type A, with the associated interaction effects. The transition between the two types of behavior is not always sharply defined. In this case the second term on the right-hand side of Eq 5.54 vanishes when K_{max} is above K_{ISCC}. The third term becomes zero when K_{max} is below K_{ISCC}.

Modified Superposition Model. Liu's modification of the Wei-Landes superposition model (Ref 5.7, 5.89) basically involves the following:

- Considering/identifying which portion of a load cycle contributes to crack growth
- Modifying the *da/dN* and *da/dt* terms in the Wei-Landes model to reflect this result
- Changing the $K(t)$ in the Wei-Landes model to a more generalized $K(t)$ that can describe both trapezoidal and triangular (sinusoidal) waveforms

- Going through all the essential steps and finalizing the updated model

A summary of how to formulate such a collective procedure is given below.

For a stress cycle that is either triangular or trapezoidal waveform, the crack growth rate for that cycle can be regarded as the sum of three parts:

- Uploading (load rising) portion of a cycle
- Hold time
- Downloading (unloading) portion of a cycle

Many experimental tests have shown that the amount of *da* for the unloading part is negligible unless the stress profile is unsymmetric, and the uploading time to the unloading time ratio is significantly small; that is, the unloading time compared to the uploading time is sufficiently long. For simplicity, we will limit our discussion to cases based on two components only, by ignoring the downloading time. However, the uploading term may be cycle dependent, time dependent, or mixed. This term consists of two parts:

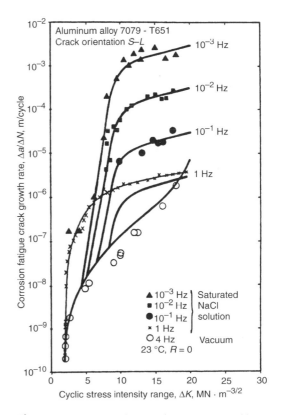

Fig. 5.76 Comparison between theory (Eq 5.54) and fatigue crack growth data of 7079-T651 aluminum alloy tested in vacuum and saturated NaCl solution. Source: Ref 5.76

One part accounts for the cyclic wave contribution and the other for the time contribution. Therefore:

$$\frac{da}{dN} = \left(\frac{da}{dN}\right)_c + \left(\frac{da}{dN}\right)_t + \left(\frac{da}{dN}\right)_H \qquad \text{(Eq 5.56)}$$

The subscripts c and t are the cyclic and time components in the uprising phase of a cycle; the subscript H stands for hold time. The first term on the right-hand side of Eq 5.56 represents the cycle-dependent part of the cycle. It comes from the conventional crack growth rate data at high frequency; that is, it follows those crack growth laws cited in the literature (such as the Paris and FNK equations). In reality, when a stress cycle is totally cycle dependent, the magnitude of the second term on the right-hand side of Eq 5.56 will be negligibly small. On the other hand, when a stress cycle is totally time dependent, the contribution of $(da/dN)_c$ to the total da/dN is negligible; thereby the validity of Eq 5.56 with respect to full frequency range is maintained. Unlike the original Wei-Landes model (Eq 5.54), the crack growth rate of Eq 5.56 has nothing to do with the reference (inert) environment. The test is conducted at temperature.

When a crack growth rate component exhibits time-dependent behavior, it is equivalent to crack growth under a sustained load where the crack growth rate description is defined by da/dt (instead of da/dN) as:

$$\frac{da}{dt} = C \cdot (K_{max})^m \qquad \text{(Eq 5.57)}$$

This quantity is obtained from a sustained-load test. To express the second term on the right-hand side of Eq 5.56 in terms of da/dt, adopt a generalized $K(t)$ function from Ref 5.90. This function can describe K at any given time in a valley-to-peak cycle. That is:

$$K(t) = R \cdot K_{max} + 2K_{max} \cdot (1 - R) \cdot t_r \cdot f \qquad \text{(Eq 5.58)}$$

where t_r is the time required for ascending the load from valley to peak and f is the frequency of the cyclic portion of a given load cycle. For symmetric loading (i.e., $t_r = f/2$), Eq 5.58 gives $K(t) = K_{min}$ at $t_r = 0$, and $K(t) = K_{max}$ at $t_r = \frac{1}{2}f$. The amount of crack extension over a period t_r can be obtained by replacing the K_{max} term of Eq 5.57 by $K(t)$ and integrating:

$$\left(\frac{da}{dN}\right)_t = C \int_0^{t_r} [K(t)]^m \, dt \qquad \text{(Eq 5.59)}$$

For any positive value of m, Eq 5.59 yields:

$$\left(\frac{da}{dN}\right)_t = C \cdot (K_{max})^m \cdot t_r \cdot R_m \qquad \text{(Eq 5.60)}$$

where

$$R_m = (1 - R^{m+1})/[(m + 1) \cdot (1 - R)] \qquad \text{(Eq 5.61)}$$

A parametric plot for R_m as a function of R is shown in Fig. 5.77, which shows that R_m increases as R increases. Therefore, for a given K_{max}, $(da/dN)_t$ increases as R increases in the time-dependent regime. This trend is in reverse to those customarily observed in the high-frequency, cycle-dependent regime.

The third term on the right-hand side of Eq 5.54 is simply equal to da/dt times the time at load. Recognizing that the first term on the right-hand side of Eq 5.60 is actually equal to da/dt, Eq 5.54 can be expressed as:

$$\frac{da}{dN} = \left(\frac{da}{dN}\right)_c + \frac{da}{dt} \cdot t_r \cdot R_m + \frac{da}{dt} \cdot t_H \qquad \text{(Eq 5.62)}$$

where t_H is the hold time.

The applicability of Eq 5.62 is demonstrated in Fig. 5.78. A test data set, extracted from Ref 5.91, was generated from the Inco 718 alloy at 649 °C (1200 °F). The test condition for this data set involves three loading variables: $R = 0.5, f$

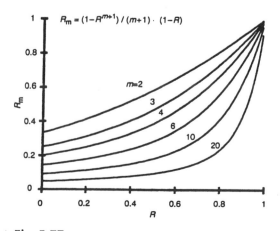

Fig. 5.77 Parametric representation of R_m

$= 0.01$ Hz (i.e., $t_r = 50$ s), $t_H = 50$ s. The stress levels were such that the crack grew from $K_{max} = 20$ to 140 MPa$\sqrt{m}$ ($\Delta K = 10$ to 70 MPa$\sqrt{m}$). The predictions were made by using Eq 5.62 with C $= 2.9678 \times 10^{-11}$ m/s and m $= 2.65$ (the separately determined material da/dt constants). The value for the $(da/dN)_C$ term was set to those experimental data points for $f = 10$ Hz. Evidently, a very good match between test data and prediction was obtained (up to $\Delta K = 35$ MPa$\sqrt{m}$).

In conclusion, the crack growth behavior of a stress cycle having a trapezoidal waveform can be predicted by using the combination of conventional high-frequency da/dN data, sustained-load data (da/dt), and Eq 5.62. For those stress cycles having a triangular waveform, and where the time factor is a function of cyclic frequencies, Eq 5.54 would be an excellent choice. As long as load/environment interactions are absent, the total crack growth rate for a loading block containing both triangular and trapezoidal stress cycles will be:

$$\left(\frac{da}{dN}\right)_{Total} = \sum_i \left(\frac{da}{dN}\right)_i \qquad \text{(Eq 5.63)}$$

where i denotes the ith loading step in the entire group of loads under consideration. The amount of da for each loading step is determined by using Eq 5.54 or 5.62.

The analytical methods described in this chapter are capable of handling constant-amplitude stress cycles involving variations in frequency (with or without hold time in a given cycle). At present, they are inadequate for dealing with loading spectra that contain both trapezoidal and sinusoidal stress cycles with variations in magnitudes. The characterizing elements discussed in this section provide only some insight into the problem that is faced. More research is needed in order to develop a reliable model for predicting spectrum crack growth at high temperature.

REFERENCES

5.1. H. Tada, P.C. Paris, and G.R. Irwin, *Stress Analysis of Cracks Handbook,* 3rd ed., American Society of Mechanical Engineers, 2000

5.2. H. Tada, P.C. Paris, and G.R. Irwin, *Stress Analysis of Cracks Handbook,* Del Research Corp., 1973

5.3. G. Shih, *Handbook of Stress-Intensity Factors,* Institute of Fracture and Solid Mechanics, Lehigh University, 1973

5.4. D.P. Rooke and D.J. Cartwright, *Compendium of Stress Intensity Factors,* Her Majesty's Stationery Office, 1976

5.5. Y. Murakami et al., Ed., *Stress-Intensity Factors Handbook,* Vol 1 and 2, Pergamon Press, 1987

5.6. P.C. Paris and G.C. Sih, Stress Analysis of Cracks, *Symposium on Fracture Toughness Testing and Its Applications,* STP 381, ASTM, 1965, p 30–83

5.7. A.F. Liu, *Structural Life Assessment Methods,* ASM International, 1998

5.8. M. Isida, Effect of Width and Length on Stress Intensity Factors of Internally Cracked Plate Under Various Boundary Conditions, *Int. J. Fract. Mech.,* Vol 7, 1971

5.9. W.F. Brown, Jr. and J.E. Srawley, *Plane Strain Crack Toughness Testing of High Strength Metallic Materials,* STP 410, ASTM, 1966

5.10. O.L. Bowie, Analysis of an Infinite Plate Containing Radial Cracks Originating at the Boundaries of an Internal Circular Hole, *J. Math. Phys.,* Vol 35, 1956, p 60

5.11. J.C. Newman, Jr., "Predicting Failure of Specimens with Either Surface Cracks or Corner Cracks at Holes," Report TN D-8244, NASA, June 1976

5.12. J.C. Newman, Jr., "An Improved Method of Collocation for the Stress Analysis of Cracked Plates with Various Shaped

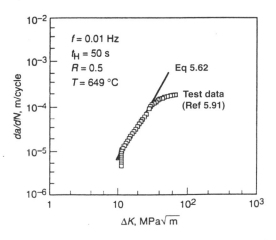

Fig. 5.78 Comparison of actual and predicted crack growth rates for Inco 718

Boundaries," Report TN D-6376, NASA, Aug 1971

5.13. J.C. Newman, Jr., Stress Analysis of Compact Specimens Including the Effects of Pin Loading, *Fracture Analysis,* STP 560, ASTM, 1974, p 105

5.14. Metals Test Methods and Analytical Procedures, *Annual Book of ASTM Standards,* Vol 03.01, *Metals—Mechanical Testing; Elevated and Low Temperature Tests; Metallography,* ASTM

5.15. K. Katherisan, T.M. Hsu, and T.R. Brussat, "Advanced Life Analysis Methods— Crack Growth Analysis Methods for Attachment Lugs," Report AFWAL-TR-84-3080, Vol II, Air Force Flight Dynamics Laboratory, Sept 1984

5.16. A.F. Liu and H.P. Kan, Test and Analysis of Cracked Lugs, *Advances in Research on the Strength and Fracture of Materials,* Vol 3B, *Applications and Non-Metals,* D.M.R. Taplin, Ed., Pergamon Press, 1978, p 657–664

5.17. R.B. Haber and H.M. Koh, "An Eulerian-Lagrangian Crack Extension Method for Mixed-Mode Fracture Problems," Paper 84-0883, AIAA, 1984

5.18. V.E. Saouma and I.J. Zatz, An Automated Finite Element Procedure for Fatigue Crack Propagation Analysis, *Eng. Fract. Mech.,* Vol 20, 1984, p 321–333

5.19. W.T. Kirkby and D.P. Rooke, A Fracture Mechanics Study of the Residual Strength of Pin-Lug Specimens, *Fracture Mechanics in Engineering Practice,* Applied Scientific, 1977, p 339–360

5.20. D.J. Cartwright, Contributed Discussion to Paper No. 8: "Calculation of Stress Intensity Factors for Corner Cracking in a Lug," *Fracture Mechanics Design Methodology,* ARGARD CP221, 1977, p D8-1

5.21. K. Katherisan and T.R. Brussat, "Advanced Life Analysis Methods—Executive Summary and Damage Tolerance Criteria Recommendations for Attachment Lugs," Report AFWAL-TR-84-3080, Vol V, Air Force Flight Dynamics Laboratory, Sept 1984

5.22. G.R. Irwin, Crack-Extension Force for a Part-Through Crack in a Plate, *J. Appl. Mech. (Trans. ASME),* Vol 84 (Ser. E), 1962, p 651–654

5.23. F.W. Smith, The Elastic Analysis of the Part-Circular Surface Flaw Problem by the Alternating Method, *The Surface Crack: Physical Problems and Computational Solutions,* ASME, 1972, p 125–152

5.24. J.C. Newman, Jr. and I.S. Raju, An Empirical Stress-Intensity Factor Equation for the Surface Crack, *Eng. Fract. Mech.,* Vol 15, 1981, p 185–192

5.25. J.C. Newman, Jr. and I.S. Raju, Stress-Intensity Factor Equations for Cracks in Three-Dimensional Finite Bodies, *Fracture Mechanics: 14th Symposium,* Vol I, *Theory and Analysis,* STP 791, ASTM, 1983, p I.238–I.265

5.26. I.S. Raju, S.R. Mettu, and V. Shivakumar, Stress Intensity Factor Solutions for Surface Cracks in Flat Plates Subjected to Nonuniform Stresses, *Fracture Mechanics, Twenty-Fourth Volume,* STP 1207, ASTM, 1994, p 560–580

5.27. R.G. Forman, V. Shivakumar, and J.C. Newman, Jr., "Fatigue Crack Growth Computer Program NASA/FLAGO Version 2.0," Report JSC-22267A, NASA, May 1994

5.28. J.C. Ekvall, T.R. Brussat, A.F. Liu, and M. Creager, "Engineering Criteria and Analysis Methodology for the Appraisal of Potential Fracture Resistant Primary Aircraft Structure," Report AFFDL-TR-72-80, Air Force Flight Dynamics Laboratory, 1972

5.29. A.F. Grandt, Jr., J.A. Harter, and B.J. Heath, The Transition of Part-Through Cracks at Holes into Through-the-Thickness Flaws, *Fracture Mechanics: 15th Symposium,* STP 833, ASTM, 1984, p 7–23

5.30. J.M. Waraniak and A.F. Liu, "Fatigue and Crack Propagation Analysis of Mechanically Fastened Joints," Paper 83-0839, presented at the AIAA/ASME/ASCE/AHS 24th Structures, Structural Dynamics, and Materials Conference (Lake Tahoe, NV), May 2–4 1983; synopsis appears in *J. Aircraft,* Vol 21, 1984, p 225–226

5.31. D.J. Cartwright and D.P. Rooke, Approximate Stress Intensity Factors Compounded from Known Solutions, *Eng. Fract. Mech.* Vol 6, 1974, p 563–571

5.32. M.L. Williams, On the Stress Distribution at the Base of a Stationary Crack, *J. Appl. Mech. (Trans. ASME),* Vol 24 (Ser. E), 1957, p 109–114

5.33. H.M. Westergaad, Bearing Pressures and Cracks, *J. Appl. Mech.,* Vol 6, 1939, p 49–53

5.34. P. Tong, T.H.H. Pian, and S. Lasry, A Hybrid-Element Approach to Crack Problems in Plane Elasticity, *Int. J. Numer. Meth. Eng.,* Vol 7, 1973, p 297–308

5.35. S.R. Murnane, L.F. Voorhees, and O.B. Davenport, Northrop/United States Air Force Application of Failure Predictions to an Operational Aircraft, *Fracture Mechanics,* University Press of Virginia, 1978, p 19–42

5.36. S.K. Chan, I.S. Tuba, and W.K. Wilson, On the Finite Element Method in Linear Fracture Mechanics, *Eng. Fract. Mech.,* Vol 2, 1970, p 1–17

5.37. L. Banks-Sills and D. Sherman, Comparison of Methods for Calculating Stress Intensity Factors with Quarter-Point Elements, *Int. J. Fract.,* Vol 32, 1986, p 127–140

5.38. C.B. Buchalet and W.H. Bamfort, Stress Intensity Factor Solutions for Continuous Surface Flaws in Reactor Pressure Vessels, *Mechanics of Crack Growth,* STP 590, ASTM, 1976, p 385–402

5.39. J.C. McMillan and R.M.N. Pelloux, Fatigue Crack Propagation Under Programmed and Random Loads, *Fatigue Crack Propagation,* STP 415, ASTM, 1966

5.40. P.C. Paris, M.P. Gomez, and W.E. Anderson, A Rational Analytic Theory of Fatigue, *The Trend in Engineering,* Vol 13, University of Washington, 1961, p 9–14

5.41. P.C. Paris, The Fracture Mechanics Approach to Fatigue, *Fatigue: An Interdisciplinary Approach,* Syracuse University Press, 1964, p 107–132

5.42. J. Schijve, *Four Lectures on Fatigue Crack Growth,* Delft University of Technology, Oct 1977

5.43. A. Saxena and C.L. Muhlstein, Fatigue Crack Growth Testing, *ASM Handbook,* Vol 19, *Mechanical Testing and Evaluation,* ASM International, 2000, p 740–757

5.44. A. Saxena and C.L. Muhlstein, Fatigue Crack Growth Testing, *ASM Handbook,* Vol 19, *Fatigue and Fracture,* ASM International, 1996, p 168–181

5.45. R.O. Ritchie and R.H. Dauskardt, Cyclic Fatigue of Ceramics: A Fracture Mechanics Approach to Subcritical Crack Growth and Life Prediction, *J. Ceram. Soc.,* Vol 99, 1991, p 1047–1062

5.46. W. Elber, The Significance of Fatigue Crack Closure, *Damage Tolerance in Aircraft Structures,* STP 486, ASTM, 1970, p 37–45

5.47. J.M. Barsom, "Effect of Cyclic-Stress Form on Corrosion-Fatigue Crack Propagation Below K_{ISCC} in a High-Yield-Strength Steel," International Conference on Corrosion Fatigue, University of Connecticut, June 14–18, 1971

5.48. K.D. Unangst, T.T. Shih, and R.P. Wei, Crack Closure in 2219-T851 Aluminum Alloy, *Eng. Fract. Mech.,* Vol 9, 1977, p 725–734

5.49. P.C. Paris and L. Hermann, Twenty Years of Reflection on Questions Involving Fatigue Crack Growth, Part II: Some Observation of Crack Closure, *Fatigue Thresholds,* Vol. I, Engineering Materials Advisory Services, 1982, p 11–32

5.50. S. Sunesh and R.O. Ritchiee, On the Influence of Environment on the Load Ratio Dependence of Fatigue Thresholds in Pressure Vessel Steel, *Eng. Fract. Mech.,* Vol 18, 1983, p 785–800

5.51. J.C. Newman, Jr., A Finite Element Analysis of Fatigue Crack Growth, *Mechanics of Crack Growth,* STP 590, ASTM, 1976, p 281–301

5.52. A.F. Liu, Application of Effective Stress Intensity Factors to Crack Growth Rate Description, *J. Aircraft,* Vol 23, 1986, p 333–339

5.53. J.C. Newman, Jr., A Crack Opening Stress Equation for Fatigue Crack Growth, *Int. J. Fract.,* Vol 24, 1984, p R131–R135

5.54. J.K. Gregory, Fatigue Crack Growth of Titanium Alloys, *ASM Handbook,* Vol 19, *Fatigue and Fracture,* ASM International, 1996, p 847

5.55. S.T. Rolfe and J.M. Barsom, *Fracture and Fatigue Control in Structures: Application of Fracture Mechanics,* Prentice-Hall, 1977

5.56. E.P. Phillips, The Influence of Crack Closure on Fatigue Crack Growth Thresholds in 2024-T3 Aluminum Alloy, *Mechanics of Fatigue Crack Closure,* STP 982, ASTM, 1988, p 515

5.57. T.L. Mackey, Fatigue Crack Propagation Rate at Low ΔK of Two Aluminum Sheet Alloys, 2024-T3 and 7075-T6, *Eng. Fract. Mech.,* Vol 11, 1979, p 753–761

5.58. P.E. Bretz, R.J. Bucci, R.C. Malcom, and A.K. Vasudevan, Constant-Amplitude Fatigue Crack Growth Behavior of 7XXX Aluminum Alloys, *Fracture Mechanics:*

14th Symposium, Vol II, *Testing and Applications,* STP 791, ASTM, 1983, p II.67–II.86

5.59. J. Schijve, The Effect of Pre-Strain on Fatigue Crack Growth and Crack Closure, *Eng. Fract. Mech.,* Vol 8, 1976, p 575–581

5.60. T.S. Kang and H.W. Liu, The Effect of Pre-Stress Cycles on Fatigue Crack Growth: An Analysis of Crack Growth Mechanism, *Eng. Fract. Mech.,* Vol 6, 1974, p 631–638

5.61. R.J. Bucci, Effect of Residual Stress on Fatigue Crack Growth Rate Measurement, *Fracture Mechanics: 13th Conference,* STP 743, ASTM, 1981, p 28–47

5.62. A.M. Sullivan and T.W. Crooker, The Effect of Specimen Thickness Upon the Fatigue Crack Growth of A516-60 Pressure Vessel Steel, *J. Pressure Vessel Technol. (Trans. ASME),* Vol 99 (Ser. J), 1977, p 248–252

5.63. D. Rhodes and J.C. Radon, Effect of Some Secondary Test Variables on Fatigue Crack Growth, *Fracture Mechanics: 14th Symposium,* Vol II, *Testing and Applications,* STP 791, ASTM, 1983, p II.33–II.46

5.64. E.F.J. Von Euw, R.W. Hertzberg, and R. Roberts, Delay Effects in Fatigue Crack Propagation, *Stress Analysis and Growth of Cracks,* STP 513, ASTM, 1972, p 230–259

5.65. J.P. Gallagher and H.D. Stalnaker, Predicting Flight by Flight Fatigue Crack Growth Rates, *J. Aircraft,* Vol 12, 1975, p 699–705

5.66. H.D. Dill and C.R. Saff, *Fatigue Crack Growth under Spectrum Loads,* STP 595, ASTM, 1976, p 306–319

5.67. O.E. Wheeler, *J. Basic Eng. (Trans. ASME),* Vol 94 (Ser. D), 1972, p 181–186

5.68. J. Lankford, Jr. and D.L. Davidson, Fatigue Crack Tip Plasticity Associated with Overloads and Subsequent Cycling, *J. Eng. Mater. Technol. (Trans. ASME),* Vol 33 (Ser. H), 1976, p 17–23

5.69. J. Lankford, Jr., D.L. Davidson, and T.S. Cook, *Cyclic Stress-Strain and Plastic Deformation Aspects of Fatigue Crack Growth,* STP 637, ASTM, 1977, p 36–55

5.70. O. Jonas and R.P. Wei, An Exploratory Study of Delay in Fatigue-Crack Growth, *Int. J. Fract. Mech.,* Vol 7, 1971, p 116–118

5.71. W.P. Slagle, D. Mahulikar, and H.L. Marcus, Effect of Hold Times on Crack Retardation in Aluminum Alloys, *Eng. Fract. Mech.,* Vol 13, 1980, p 889–895

5.72. D. Simpkins, R.L. Neulieb, and D.J. Golden, Load-Time Dependent Relaxation of Residual Stresses, *J. Aircraft,* Vol 9, 1972, p 867–868

5.73. A.F. Liu and D.F. Dittmer, "Effect of Multiaxial Loading on Crack Growth," Report AFFDL-TR-78-175 (in 3 volumes), Air Force Flight Dynamics Laboratory, Dec 1978

5.74. D.V. Nelson, Effects of Residual Stress on Fatigue Crack Propagation, *Residual Stress Effects in Fatigue,* STP 776, ASTM, 1982, p 172–194

5.75. R.C. McClung, K.S. Chan, S.J. Hudak, Jr., and D.L. Davidson, Behavior of Small Fatigue Cracks, *ASM Handbook,* Vol 19, *Fatigue and Fracture,* ASM International, 1996, p 153–158

5.76. M.O. Speidel, Stress Corrosion and Corrosion Fatigue Crack Growth in Aluminum Alloys, *Stress Corrosion Research,* H. Arup and R.N. Parkins, Ed., Sijthoff & Noordhoff, 1979, p 117–175

5.77. A.J. McEvily, Jr. and R.P. Wei, *Corrosion Fatigue: Chemistry, Mechanics and Microstructure,* National Association of Corrosion Engineers, 1972, p 381–395

5.78. D.B. Dawson and R.M.N. Pelloux, Corrosion Fatigue Crack Growth Rates of Titanium Alloys Exposed in Aqueous Environments, *Metall. Trans.,* Vol 5, 1974, p 723–731

5.79. O. Vosikovsky, Fatigue-Crack Growth in an X-65 Line-Pipe Steel at Low Cyclic Frequencies in Aqueous Environments, *J. Eng. Mater. Technol. (Trans. ASME),* Vol 97 (Ser. H), 1975, p 298–305

5.80. K.M. Chang, Elevated Temperature Fatigue Crack Propagation after Sustained Loading, *Effects of Load and Thermal Histories on Mechanical Behavior of Materials,* TMS/AIME, 1987, p 13–26

5.81. Y.H. Kim and S.D. Manning, *Fracture Mechanics: 14th Symposium,* Vol I, *Theory and Analysis,* STP 791, ASTM, 1983, p I.446–I.462

5.82. K.M. Chang, in *Material Research Symposium Proceedings,* Vol 125, Material Reseach Society, 1988, p 243–252

5.83. W.W. Gerberich and N.R. Moody, A Review of Fatigue Fracture Topology Effects on Threshold and Growth Mechanisms, *Fatigue Mechanisms,* STP 675, ASTM, 1979, p 292–341

5.84. J.E. Allison and J.C. Williams, *Scr. Metall.,* Vol 19, 1985, p 773–778

5.85. T.T. Shih and G.A. Clarke, Effect of Temperature and Frequency on the Fatigue Crack Growth Rate Properties of a 1950 Vintage CrMoV Rotor Material, *Fracture Mechanics,* STP 677, ASTM, 1979, p 125–143

5.86. R. Viswanathan, *Damage Mechanisms and Life Assessment of High-Temperature Components,* ASM International, 1989

5.87. T. Nicholas and N.E. Ashbaugh, Fatigue Crack Growth at High Load Ratios in the Time-Dependent Regime, *Fracture Mechanics: 19th Symposium,* STP 969, ASTM, 1988, p 800–817

5.88. R.P. Wei and J.D. Landes, Correlation Between Sustained-Load and Fatigue Crack Growth in High-Strength Steels, *Material Research and Standards,* TMRSA, Vol 9, ASTM, 1969, p 25–28

5.89. A.F. Liu, Assessment of a Time Dependent Damage Accumulation Model for Crack Growth at High Temperature, *Proceedings 1994,* Vol 3, 19th Congress of the International Council of the Aeronautical Sciences, 1994, p 2625–2635

5.90. T. Nicholas, T. Weerasooriya, and N.E. Ashbaugh, A Model for Creep/Fatigue Interactions in Alloy 718, *Fracture Mechanics: 16th Symposium,* STP 868, ASTM, 1985, p 167–180

5.91. G.K. Haritos, T. Nicholas, and G.O. Painter, Evaluation of Crack Growth Models for Elevated-Temperature Fatigue, *Fracture Mechanics: 18th Symposium,* STP 945, ASTM, 1988, p 206–220

SELECTED REFERENCES

- D. Broek, Concepts of Fracture Control and Damage Tolerance Analysis, *ASM Handbook,* Vol 19, *Fatigue and Fracture,* ASM International, 1996, p 410–419

- D. Broek, The Practice of Damage Tolerance Analysis, *ASM Handbook,* Vol 19, *Fatigue and Fracture,* ASM International, 1996, p 420–426

- D. Broek, Residual Strength of Metal Structures, *ASM Handbook,* Vol 19, *Fatigue and Fracture,* ASM International, 1996, p 427–433

- J. Schijve, "The Accumulation of Fatigue Damage in Aircraft Materials and Structures," AGARDDograph AGARD-AG-157, NATO, Jan 1972

CHAPTER 6

Nonlinear Fracture Mechanics

THE CONCEPT OF LINEAR-ELASTIC FRACTURE MECHANICS (LEFM) assumes that the stress-intensity factor K is a valid fracture index and that the material behaves in a "brittle" manner. As a rule of thumb, the applicability of Eq 4.2(b) is limited to a crack-tip plastic zone radius (r^* or r_y) smaller than one-tenth of the current crack length, for maintaining elasticity. The linear-elastic stress-intensity factor approach to fracture analysis is not applicable to situations where nonlinear behavior is encountered in the vicinity of the crack tip. Large-scale yielding at the crack tip and time-dependent crack growth behavior, such as stress relaxation due to creep, belong to the category of nonlinear behavior.

This chapter introduces two new fracture indices: the J-integral for handling the problem of large-scale yielding, and the C^*-integral for handling creep-fatigue crack growth. An extension of Chapter 1, which presents elastic stress analysis of cracks, this chapter takes the same subject to the plastic range, leading to J and C^*. The available solutions for J and C^* are included here as engineering tools for analysis tasks.

6.1 Elastic-Plastic Fracture Mechanics

Large-scale yielding is not limited to the conventional definition of crack-tip plastic zone size. Geometry-induced plasticity often plays an important role in the crack growth behavior and residual strength of a structural member. The local area in the immediate vicinity of a stress raiser, or a cutout, is known to be susceptible to stress concentration. Taking a circular hole as an example, the local tangential stress at the hole edge is at least three times the applied far-field stress. Because the material cannot forever follow Hooke's law under monotonic increasing load, in reality it follows the stress-strain rela-

tionship of the tensile stress-strain curve. Therefore, the magnified stress at the hole edge eventually causes gross-scale yielding around the hole. Depending on the applied stress level, the yielded zone adjacent to the hole can be very large, to the extent that a very short crack would be totally embedded inside this zone. Although this type of fracture behavior has long been recognized, a simple engineering solution to the problem does not exist. Direct application of current linear fracture mechanics technology to predict the residual strength of this configuration is inappropriate. So far, the J-integral is used as an alternate fracture index to characterize fracture behavior involving large-scale yielding at the crack tip. On the basis of a limited scope of investigation, Shows et al. have shown that the J-integral can also be used to determine the fracture strength of the yielded hole configuration (Ref 6.1).

For a stationary crack, the HRR crack-tip stress field, named after Hutchinson, Rice, and Rosengren, can be used in place of Eq 1.77 as a generalized crack-tip stress distribution (Ref 6.2, 6.3). For mode 1 loading, the HRR field is defined as:

$$\sigma_{ij} = \sigma_0 \tilde{\sigma}_{ij}(\theta,n)\left[\frac{EJ}{\sigma_0^2 I_n r}\right]^{1/(n+1)} \quad \text{(Eq 6.1)}$$

or

$$\sigma_{ij} = \sigma_0 \tilde{\sigma}_{ij}(\theta,n)\left[\frac{J}{\alpha\sigma_0\varepsilon_0 I_n r}\right]^{1/(n+1)} \quad \text{(Eq 6.1a)}$$

where J is the J-integral (in. · lb/in.^{2}), θ and r are the polar coordinates centered at the crack tip, $\tilde{\sigma}_{ij}(\theta,n)$ is a dimensionless function that is a function of θ and n, I_n is a dimensionless constant that is a function of n, and σ_0 is yield stress, which relates to yield strain ε_0. In the pure power stress-strain law, the uniaxial plastic strain ε is related to the uniaxial stress σ by:

$$\frac{\varepsilon}{\varepsilon_0} = \alpha\left(\frac{\sigma}{\sigma_0}\right)^n \tag{Eq 6.2}$$

where α is a material constant and n is the strain-hardening exponent. The numerical values of $\tilde{\sigma}_{ij}(\theta,n)$ and I_n are given in Ref 6.4. For the convenience of the reader, the values for the dimensionless constant I_n are tabulated in Table 6.1, for it will be used in several other places in this chapter.

For an elastic body (i.e., small-scale yielding) under plane stress, with $n = 1$, $I_n = 6.28$ (i.e., 2π), and $J = K^2/E$, Eq 6.1 reduces to the familiar K-field distribution:

$$\sigma_{ij} = \tilde{\sigma}_{ij}(\theta)\left[\frac{K^2}{2\pi r}\right]^{1/2} \tag{Eq 6.3}$$

By comparison, it is clear that $\tilde{\sigma}_{ij}(\theta)$ is the same as those θ-functions in Eq 1.77. In terms of fatigue crack growth and fracture, J replaces K and J_C replaces K_C in an appropriate circumstance.

The following paragraphs define the J-integral. A procedure for doing the numerical integration is presented along with an example problem. Following that, the fully plastic solutions for center-cracked and single-edge-cracked plates and compact specimens are presented.

6.1.1 The J-Integral

The J-integral can be regarded as a change of potential energy of the body with an increment of crack extension. The general expression for J, as defined by Rice (Ref 6.5), is given by:

$$J = \int_{\Gamma}\left(W dy - \vec{T} \cdot \cup \frac{\partial\vec{u}}{\partial x}\, ds\right) \tag{Eq 6.4}$$

Table 6.1 Dimensionless constant I_n associated with the HRR singularity

n	Plane strain	Plane stress
2	5.94	4.22
3	5.51	3.86
4	5.22	3.59
5	5.02	3.41
6	4.88	3.27
7	4.77	3.17
8	4.68	3.09
9	4.6	3.03
10	4.54	2.98
12	4.44	2.9
15	4.33	2.82
20	4.21	2.74
30	4.08	2.66
50	3.95	2.59
100	3.84	2.54

Source: Ref 6.4

where $\vec{T}$ is the traction vector, acting on a segment ds of contour Γ, $\vec{u}$ is a displacement vector (the displacement on an element along arc s), and Γ is any contour in the X-Y plane that encircles the crack tip. Γ is taken in a counterclockwise direction, starting from one crack face, ending on the opposite face, and closing the crack tip (Fig. 6.1). W is the strain energy density, given by:

$$W \int_0^{\varepsilon_{ij}} \sigma_{ij} \cdot d(\varepsilon_{ij}) \tag{Eq 6.5}$$

That is,

$$W = \int\left[\begin{array}{l}\sigma_x + d\varepsilon_x + \tau_{xy} + d\gamma_{xy} + \tau_{xz} + d\gamma_{xz} \\ + \sigma_y + d\varepsilon_y + \tau_{yz} + d\gamma_{yz} + \sigma_z + d\varepsilon_z\end{array}\right] \tag{Eq 6.5a}$$

For generalized plane stress, Eq 6.5(a) is reduced to:

$$W = \int[\sigma_x + d\varepsilon_x + \tau_{xy} + d\gamma_{xy} + \sigma_y + d\varepsilon_y] \tag{Eq 6.5b}$$

The total magnitude of J consists of two parts: an elastic part and a fully plastic part. For small-scale yielding or linear-elastic material behavior, J is equivalent to Irwin's strain-energy release rate $\mathcal{G}$. Therefore, the portion for "elastic-J" is directly related to K, and the plastic part vanishes. However, unlike $\mathcal{G}$ or K, J can be used as a generalized fracture parameter for small-or large-scale yielding. In fact, J is used as an all-purpose fracture index in some literature. Cir-

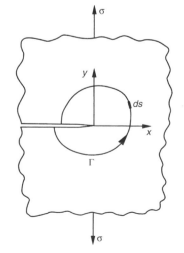

Fig. 6.1 Definition of the J-integral

cumstantially, the *J* reported could have been a *J* with large-scale yielding, or just an elastic-*J*. The latter is actually a value of K^2 divided by *E*, or by $E/(1 - \nu^2)$, depending on the crack-tip constraint (i.e., plane stress or plane strain).

6.1.2 General Procedure for Determining J

The finite element method can be used to determine the stress and strain for points on the path of the selected contour Γ. The results are then substituted into Eq 6.4 and 6.5 for integration. In order to carry out the integration indicated in Eq 6.4 and 6.5, a relationship between stress and strain that realistically describes the plastically deforming material is needed. It has been shown in Ref 6.6 that Eq 6.5(b) can be reduced to:

$$W = \frac{1}{2E} \{\sigma_x + \sigma_y\}^2 + \left(\frac{1 + \nu}{E}\right)\{\tau_{xy}^2 - \sigma_x \cdot \sigma_y\}$$
$$+ \int_0^{\bar{\varepsilon}_p} \bar{\sigma} \cdot (d\bar{\varepsilon}_p) \qquad \text{(Eq 6.6)}$$

where $\bar{\sigma}$ and $\bar{\varepsilon}_p$ are equivalent stress and equivalent plastic strain, respectively, and:

$$\bar{\sigma} = \{\sigma_x^2 - \sigma_x\sigma_y + \sigma_y^2 + 3\tau_{xy}^2\}^{1/2} \qquad \text{(Eq 6.7)}$$

where σ_x, σ_y, and τ_{xy} are determined using a finite element computer code. To execute the finite element code, a stress-strain curve is entered into each element in the finite element model (or at least into those elements along the selected contour Γ). The stress-strain curve can be represented by an empirical equation, or the *X-Y* values of the curve are entered physically. For segments of Γ on which $\bar{\sigma}$ is greater than the material proportional limit and *dy* is nonzero, the value of $\bar{\varepsilon}_p$ that corresponds to a $\bar{\sigma}$ value is obtained from the stress-strain curve. Both $\bar{\sigma}$ and $\bar{\varepsilon}_p$ are then substituted into Eq 6.6 to compute the plastic component of *J*. Physically, the integral in Eq 6.6 may be interpreted as the area under the plastic stress-strain curve. Therefore, the plastic part of *J* is the integral in Eq 6.6, and the elastic part of *J* includes all the terms in Eq 6.4, excluding the integral term in Eq 6.6.

For very ductile materials, considerable care must be taken in evaluating the stress $\bar{\sigma}$, since a small error in $\bar{\sigma}$ will lead to large errors in estimating $\bar{\varepsilon}_p$ and thus may contribute large errors

to the *J*-integral. In studying Eq 6.6 it is clear that *W* will have a unique value only if unloading is prohibited at every point in the structure. Monotonic loading conditions prevail throughout a cracked body under steadily increasing applied loads, provided that crack extension does not occur. Because in any calculation of *J* the crack length is held constant, *W* will be unique and a valid *J*-integral is obtained.

The integration procedure can be demonstrated by running through a numerical exercise of an example problem. The problem, which has been presented in Ref 6.1, deals with a 2024-T351 aluminum rectangular plate specimen that measured 6.35 mm (0.25 in.) thick, 152.4 mm (6 in.) wide, and 457.2 mm (18 in.) long. A 12.7 mm (0.5 in.) diameter hole was located in middle of the specimen. A through-thickness radial crack, 1.45 mm (0.057 in.) long, was at the bore of the hole.

This analysis used the NASTRAN structural analysis computer code. A finite element model of the specimen was constructed, using constant-strain triangles and quadrilaterals, with a very fine mesh in the neighborhood of the crack. Figure 6.2 shows the portion of the finite element mesh adjacent to the hole and the crack. The actual material stress-strain curve shown in Fig. 2.65 was input to each element in the model using the piecewise linear format.

As shown in Fig. 6.2, one-half of a selected contour Γ starts from grid point no. 7 and ends at grid point no. 3. Therefore, one-half of the *J* value is equal to:

$$\frac{J}{2} = \int_7^{49}$$
$$\cdot \left[\begin{matrix} \dfrac{1}{2E}(\sigma_x + \sigma_y)^2 + \dfrac{1 + \nu}{E}(\tau_{xy}^2 - \sigma_x \cdot \sigma_y) \\ -\left\{\sigma_x\left(\dfrac{\partial u}{\partial x}\right) + \tau_{xy}\left(\dfrac{\partial v}{\partial x}\right)\right\} + \int_0^{\bar{\varepsilon}_p} \bar{\sigma} \cdot (d\bar{\varepsilon}_p) \end{matrix} \right] dy$$
$$- \int_{49}^{45}\left[\tau_{xy}\left(\frac{\partial u}{\partial x}\right) + \sigma_y\left(\frac{\partial v}{\partial x}\right)\right]dx$$
$$+ \int_{45}^{3}\left[\begin{matrix} \dfrac{1}{2E}(\sigma_x + \sigma_y)^2 + \dfrac{1 + \nu}{E}(\tau_{xy}^2 - \sigma_x \cdot \sigma_y) \\ -\left\{\sigma_x\left(\dfrac{\partial u}{\partial x}\right) + \tau_{xy}\left(\dfrac{\partial v}{\partial x}\right)\right\} + \int_0^{\bar{\varepsilon}_p} \bar{\sigma} \cdot (d\bar{\varepsilon}_p) \end{matrix} \right] dy$$

$$\text{(Eq 6.8)}$$

The terms associated with $\partial u/\partial x$ or $\partial v/\partial x$ come from the second term of Eq 6.4. The rest belong to Eq 6.6. The double integral:

$$\int\int\int_0^{\bar{\varepsilon}_p} \bar{\sigma} \cdot (d\bar{\varepsilon}_p)dy \qquad \text{(Eq 6.9)}$$

is the "fully plastic-J." The NASTRAN output and the calculated J value for this configuration (under an applied far-field stress of 245 MPa, or 35.5 ksi) are presented in Table 6.2.

Rice (Ref 6.5) also noted that J is path independent. That is, the values of J would be identical whether the contour Γ is taken far away or very close to the crack tip. In addition, there is no restriction as to what specific course Γ should follow. To confirm the path independence of J, another contour farther away from the crack tip (i.e., far from the plastic zone and primary elastic) was also evaluated. Although the analysis result for the second path is not shown here, the values of J calculated from the two paths were found to agree within 1%. Therefore, path independence had been verified.

6.1.3 Handbook Solutions for the Fully Plastic J_p

As mentioned earlier, the total J is a superimposition of an elastic part and a fully plastic part:

$$J = J_e + J_p \qquad \text{(Eq 6.10)}$$

For mode 1, the term J_e is directly related to K_1 by way of:

$$J_e = K_1^2/E' \qquad \text{(Eq 6.11)}$$

where $E' = E$ for plane stress and $E' = E/(1 - v^2)$ for plane strain. The existing handbook

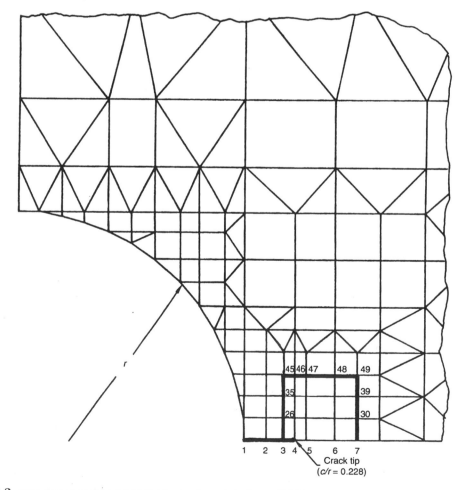

Fig. 6.2 Finite element mesh near a crack emerging from a hole. Source: Ref 6.1

solution for K can be used for a structural configuration under consideration.

Close-form solutions for J_p are available. Solutions for many laboratory specimen geometries and a number of common structural configurations are given in Ref 6.7 and 6.8. Three of these solutions are presented here. The specimen configurations are the center-cracked plate, the single edge-cracked plate, and the compact specimen. Integration of Eq 6.4 is required for those cases where a solution for K (for J_e) or J_p is not available.

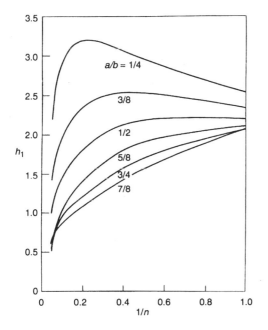

Fig. 6.3 h_1 versus $1/n$ for a center-cracked panel in tension, plane stress. Source: Ref 6.7

Center-Crack Panel. For a center through-thickness crack subjected to far-field uniform tension (e.g., the M(T) specimen configuration, according to ASTM terminology), having a total crack length $2a$ and width W, the fully plastic solution for J is:

$$J_p = \alpha\sigma_0\varepsilon_0 a \cdot (1 - 2a/W) \cdot h_1 \cdot \left(\frac{P}{P_0}\right)^{n+1}$$

(Eq 6.12)

where n and α are material constants defined by the pure stress-strain law of Eq 6.2, σ_0 is the yield stress, ε_0 is the yield strain, and h_1 is a function of $2a/W$ and n; its value is given in Fig. 6.3 (for plane stress) and Fig. 6.4 (for plane strain). In these figures, the dimension b is equal to one-half of the plate width. P is the applied load per unit thickness, and:

$$P_0 = \psi \cdot (W - 2a) \cdot \sigma_0$$

(Eq 6.13)

where $\psi = 1$ for plane stress, and $\psi = 2/\sqrt{3}$ for plane strain.

Single-Edge Crack under Uniform Tension. For a single-edge-cracked plate subjected to far-field uniform tension, having a crack length a and width W, the fully plastic solution for J is:

$$J_p = \alpha\sigma_0\varepsilon_0 \cdot (W - a) \cdot (a/W) \cdot h_1 \cdot \left(\frac{P}{P_0}\right)^{n+1}$$

(Eq 6.14)

where h_1 is a function of a/W and n; its value is given in Fig. 6.5 (for plane stress) and Fig. 6.6 (for plane strain). In these figures, the dimension b is the total width. P is the applied load per unit

Table 6.2   Calculation of J for a 1.45 mm crack at a 12.7 mm diameter hole under 245 MPa applied stress

Grid points	One-half of the J path		NASTRAN results					Calculated results			
	dx, mm	dy, mm	σ_x, MPa	σ_y, MPa	τ_{xy}, MPa	$\frac{\delta u}{\delta r} \times 10^3$	$\frac{\delta v}{\delta r} \times 10^3$	$\bar{\sigma}$, MPa	$\bar{\varepsilon}_p$	$\frac{1}{2}J_e$, MN/m	$\frac{1}{2}J_p$, MN/m
7–30	0	0.635	200.2	439.2	12.5	−0.470	−1.34	381.5	0.01200	4.943	11.230
30–39	0	0.635	151.6	427.3	5.0	−0.450	−3.26	375.3	0.00870	5.277	5.796
34–49	0	0.635	91.3	408.0	−11.7	−0.990	−4.16	371.4	0.00630	4.716	2.053
49–48	0.635	0	65.8	396.3	−25.8	−1.360	−5.02	370.6	0.00608	8.558	0
48–47	−0.762	0	42.0	378.9	−51.2	−1.740	−6.44	370.6	0.00605	12.355	0
47–46	−0.330	0	18.8	342.7	−93.3	−1.950	−8.32	370.7	0.00610	6.077	0
46–45	−0.305	0	15.9	288.3	−134.7	−1.420	−8.67	365.0	0.00528	4.850	0
45–35	0	−0.635	32.8	276.6	−140.9	−0.997	−9.60	357.9	0.00500	2.032	0.121
35–26	0	−0.635	85.4	270.2	−160.6	−0.734	−15.00	366.9	0.00540	6.246	−0.483
26–3	0	−0.635	139.7	268.3	−172.4	+0.625	−27.00	378.4	0.01060	16.369	−8.573
									$\Sigma = 71.423$		9.902
										Total $J = 162.650$	

Source: Ref 6.1

thickness, and:

$$P_0 = \psi \cdot \eta \cdot (W - a) \cdot \sigma_0 \qquad \text{(Eq 6.15)}$$

where $\psi = 1.072$ for plane stress, $\psi = 1.455$ for plane strain, and:

$$\eta = \left[1 + \left(\frac{a}{W - a}\right)^2\right]^{1/2} - \frac{a}{W - a} \qquad \text{(Eq 6.16)}$$

Compact Specimens. The fully plastic J solution for the ASTM standard compact specimen is:

$$J_\mathrm{p} = \alpha \sigma_0 \varepsilon_0 \cdot (W - a) \cdot h_1 \cdot \left(\frac{P}{P_0}\right)^{n+1} \qquad \text{(Eq 6.17)}$$

where h_1 is a function of a/W and n; its value is given in Fig. 6.7 (for plane stress) and Fig. 6.8 (for plane strain). In these figures, the dimension b is the same as W in Fig. 5.7. P is the applied load per unit thickness, and:

$$P_0 = \psi \cdot \eta \cdot (W - a) \cdot \sigma_0 \qquad \text{(Eq 6.18)}$$

where $\psi = 1.072$ for plane stress, $\psi = 1.455$ for plane strain, and:

$$\begin{aligned}
\eta &= \left[\left(\frac{2a}{W - a}\right)^2 + 2\left(\frac{2a}{W - a}\right) + 2\right]^{1/2} \\
&\quad - \left(\frac{2a}{W - a} + 1\right)
\end{aligned} \qquad \text{(Eq 6.19)}$$

6.2 Time-Dependent Fracture Mechanics

Current fracture mechanics theory treats cyclic crack growth as a linear-elastic phenomenon. The residual strength of a test coupon, or a structural component, is frequently computed based on linear-elastic fracture indices. Elastic-plastic or fully plastic analysis such as the J-integral approach is used when large-scale yielding occurs. For crack growth at high temperature, under constant load, the material near the crack tip may undergo viscoelastic deformation, or creep. The crack-tip stress field, which takes the basic form of Eq 1.77, will transform itself once more to adjust to the situation. The process is simply a modification of

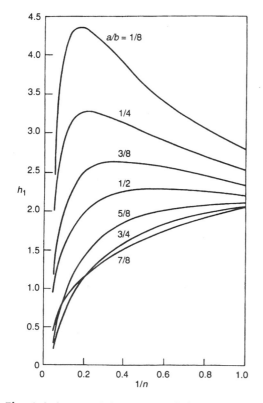

Fig. 6.4 h_1 versus $1/n$ for a center-cracked panel in tension, plane strain. Source: Ref 6.7

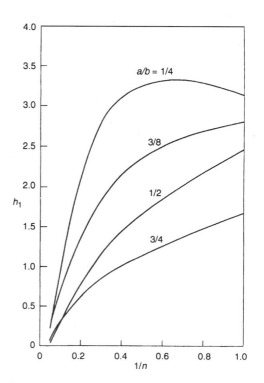

Fig. 6.5 h_1 versus $1/n$ for a single-edge-cracked panel in tension, plane stress. Source: Ref 6.7

the elastic-plastic stress field (Eq 6.1a) where the strain and displacement ε and u are replaced by their rates, $\dot{\varepsilon}$ and $\dot{u}$.

For a material under the steady-state creep range—that is, where the material follows a creep law in the form:

$$\dot{\varepsilon} = \alpha\dot{\varepsilon}_0\left(\frac{\sigma}{\sigma_0}\right)^n \qquad \text{(Eq 6.20)}$$

—the constant α is usually absorbed in the constant $\dot{\varepsilon}_0$. Therefore, the creep law (i.e., Eq 6.20) frequently is also written as:

$$\dot{\varepsilon} = \dot{\varepsilon}_0\left(\frac{\sigma}{\sigma_0}\right)^n \qquad \text{(Eq 6.20a)}$$

For a material that is primarily influenced by second-stage creep, the creep strain rate is:

$$\dot{\varepsilon} = A \cdot \sigma^n \qquad \text{(Eq 6.21)}$$

The steady-state creep crack growth parameter $C*$ is analogous to the J-integral in the fully plastic condition. Therefore, the close-form solution for $C*$ takes the same form as that for J. An expression for $C*$ is derived by substituting the material creep strain-rate coefficients of Eq 6.21 for the material stress-strain coefficients of Eq 6.2. Comparing Eq 6.20 and 6.21, one has:

$$A = \alpha\dot{\varepsilon}_0\sigma_0^{-n} \qquad \text{(Eq 6.22)}$$

Substituting Eq 6.22 into Eq 6.1(a) for α and ε_0, the HRR field for elastic-plastic deformation can be expressed for creep as:

$$\sigma_{ij} = \tilde{\sigma}_{ij}(\theta,n)\left[\frac{C*}{AI_n r}\right]^{1/(n+1)} \qquad \text{(Eq 6.23)}$$

where the physical unit for $C*$ is (MPa · m)/h or (in. · lb/in.^{2})/h.

6.2.1 The C*-Integral

Similar to J, $C*$ is path independent and has been defined by Landes and Begley (Ref 6.9) as:

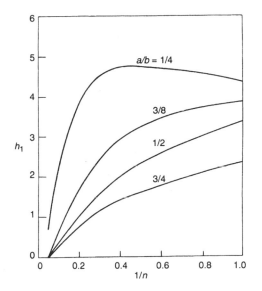

Fig. 6.6 h_1 versus $1/n$ for a single-edge-cracked panel in tension, plane strain. Source: Ref 6.7

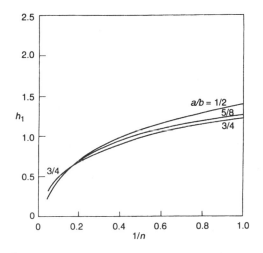

Fig. 6.7 h_1 versus $1/n$ for a compact specimen, plane stress. Source: Ref 6.7

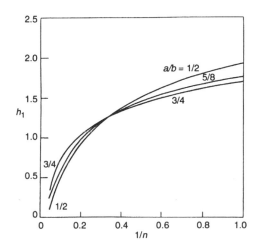

Fig. 6.8 h_1 versus $1/n$ for a compact specimen, plane strain. Source: Ref 6.7

$$C^* = \int_\Gamma W^* \cdot dy - T_i \left(\frac{\partial \dot{u}_i}{\partial x} \right) ds \qquad \text{(Eq 6.24)}$$

where

$$W^* = \int_0^{\dot{\varepsilon}_{min}} \sigma_{ij} d\dot{\varepsilon}_{ij} \qquad \text{(Eq 6.25)}$$

T is the traction vector, Γ is the line contour taken from the lower crack surface in a counter-clockwise direction to the upper surface, and W^* is the strain energy rate density associated with the point stress σ_{ij} and strain rate $\dot{\varepsilon}_{ij}$.

Because the steady-state creep crack growth parameter C^* is analogous to the J-integral in the fully plastic condition, the close-form solution for C^* takes the same form as that for J_p. In other words, an expression for C^* can be derived by adapting an appropriate J_p solution for a geometry under consideration. The expression for C^* is the same as for J_p, except that the material stress-strain coefficients in the J_p equation are replaced by the material creep strain-rate coefficients. Taking the center-cracked panel for example, the fully plastic J (i.e., J_p) is given by Eq 6.12. After making proper substitutions of the creep coefficients, Eq 6.12 becomes (Ref 6.10):

$$C^* = \alpha \dot{\varepsilon}_0 \sigma_0 a \cdot (1 - 2a/W) \cdot h_1 \cdot \left(\frac{P}{P_0} \right)^{n+1} \qquad \text{(Eq 6.26a)}$$

or

$$C^* = A \cdot a \cdot (1 - 2a/W)^{-n} \cdot h_1 \cdot \left(\frac{P}{W} \right)^{n+1} \qquad \text{(Eq 6.26b)}$$

for plane stress, and

$$C^* = A \cdot a \cdot (1 - 2a/W)^{-n} \cdot h_1 \cdot \left(\frac{\sqrt{3}P}{2W} \right)^{n+1} \qquad \text{(Eq 6.26c)}$$

for plane strain. Here, A is the coefficient for the second-stage creep as defined by Eq 6.22. Its value can be determined by experimental testing. The physical unit for C^* is (MPa · m)/h or (in. · lb/in.^{2})/h. For the compact specimen:

$$C^* = \alpha \dot{\varepsilon}_0 \sigma_0 \cdot (W - a) \cdot h_1 \cdot \left(\frac{M}{M_0} \right)^{n+1} \qquad \text{(Eq 6.27a)}$$

or

$$C^* = A \cdot (W - a)^{-n} \cdot h_1 \cdot \left(\frac{P(W + a)/2}{0.2679(W - a)^2} \right)^{n+1} \qquad \text{(Eq 6.27b)}$$

for plane stress, and

$$C^* = A \cdot (W - a)^{-n} \cdot h_1 \cdot \left(\frac{P(W + a)/2}{0.364(W - a)^2} \right)^{n+1} \qquad \text{(Eq 6.27c)}$$

for plane strain.

6.2.2 Steady-State Creep versus Small-Scale Creep

For small-scale creep, Eq 6.23 can be written as:

$$\sigma_{ij} = \tilde{\sigma}_{ij}(\theta, n) \left[\frac{C(t)}{A I_n r} \right]^{1/(n+1)} \qquad \text{(Eq 6.28)}$$

Equation 6.28 is called the RR field, named after Riedel and Rice (Ref 6.11). For small-scale secondary creep (SSC) conditions, $C(t)$ can be approximated by:

$$C(t)_{SSC} = \frac{(1 - v^2)K^2}{(n + 1)Et} \qquad \text{(Eq 6.29)}$$

for plane strain; for plane stress the factor $(1 - v^2)$ must be omitted. At time zero, i.e., $t = 0$, $C(t)$ can be considered as a crack growth driving force similar to K, or elastic-J; it is path dependent. As the extensive creep conditions become prevalent, the value of $C(t)$ changes toward C^*. Under-steady state creep, $C(t)$ becomes time-independent, and its value approaches C^*:

$$C(t \to \infty) = C^* \qquad \text{(Eq 6.30)}$$

The characteristic transition time t_1 (between small-scale creep and steady-state creep) is estimated by setting the value of $C(t)$ of Eq 6.29 equal to C^*; that is, for plane strain:

$$t_1 = \frac{(1 - v^2)K^2}{(n + 1)EC^*} \qquad \text{(Eq 6.31)}$$

Again, the factor $(1 - v^2)$ in Eq 6.31 should be deleted for the plane-stress condition. The characteristic time can be used to decide whether or not the test data set in question is correlated with K. In other words, K can be used as a controlling parameter if $t \ll t_1$. Otherwise, C^* should be used for the steady-state creep. In the transient regime (i.e., between small-scale and extensive creep conditions), an interpolation equation for analytically estimating $C(t)$ is given by (Ref 6.12):

$$C(t) \approx C^* \left[\frac{t_1}{t} + 1 \right] \tag{Eq 6.32}$$

or

$$C(t) \approx \frac{(1 - v^2)K^2}{(n + 1)Et} + C^* \tag{Eq 6.33}$$

6.2.3 Characterizing Parameters

Depending on material, temperature, time, and environment, crack growth at high temperature can exhibit various degrees of creep deformation at the crack-tip region. Consider that creep resistance of a material is characterized by the coefficient and exponent of the relationship of minimum creep rate to stress:

$$\dot{\varepsilon} = A \cdot \sigma^n \tag{Eq 6.34}$$

The degree of creep deformation can be described as small scale-creep, transient creep, or steady-state creep. Under small-scale creep, the creep zone is small in comparison to the crack length, and the size of the body and its growth are constrained by the surrounding elastic material. Under steady-state creep the creep zone engulfs the entire untracked ligament of the material (e.g., in between the crack tip and the free edge of the test coupon). The transient creep condition represents an intermediate level of creep deformation.

Theoretically, these three levels of creep deformation are analogous to the three levels of crack-tip plasticity in the subcreep temperature regime: the small-scale yielding level in which the load-carrying behavior of the specimen (or component) is linear, the fully plastic level, and the intermediate (elastic-plastic) level. Therefore, under small-scale creep the driving force for crack growth will be K. Under steady-state creep, crack growth occurs essentially by creep. The fracture index will be the energy-rate line integral C^*, which is considered to be the counterpart of J_p for a fully plastic body. The only difference between J_p and C^* is that the plat-

form for J_p is plastic strain, and the platform for C^* is the creep strain rate.

Generally speaking, the controlling parameter depends on the level of creep deformation, which is indicated by the size of a creep zone under load (Ref 6.11). The size of creep zone r_{cr} at the tip of a stationary crack is a function of K (elastic stress-intensity factor), E (material elastic modulus), A and n (the minimum creep strain-rate coefficients), and most importantly, t (time at load). The estimated creep zone size will ultimately provide a measure of the degree of creep deformation, thereby hinting at the appropriate fracture mechanics index for a given case under consideration. However, there are no clear boundaries, or criteria, to divide the r_{cr} into small-scale creep and transient creep categories. The Saxena parameter C_t (to be discussed later), which is embedded with a term for the size of a creep zone, provides a universal characterization for the entire range of creep deformation.

6.2.4 Crack-Tip Deformation Mechanisms at High Temperature

At high temperatures, materials around the crack tip can undergo three types of deformation modes: by plastic yielding, by creep, and by environmental attack. A deformation zone is associated with each type of deformation—the plastic zone, the creep zone, and the environment-affected zone (EAZ)—all of which are time dependent. Crack growth is controlled by K at time zero, and perhaps when $t \ll t_1$. Afterward, crack growth is controlled by K inside the plastic or the environment-affected zone, but by C_t or C^* inside the creep zone. It has been suggested (Ref 6.13) that two or all three types of crack-tip deformation zones can coexist at a given time. Crack growth behavior will be dictated by the most prevailing mechanism at that time.

Plastic Zone. Because crack-tip plasticity is a function of mechanical load and material tensile yield strength, we can simply use the material tensile yield strength at temperature for F_{ty}. For short time stress cycles, the equation for r_p as a function of K and F_{ty} is given by Eq 5.44. This concept can be extended to determine the size of r_p for a load cycle with hold times at the peak. That is, r_p can be calculated by using an effective yield strength that is a function of temperature and time. Such an effective yield strength can be obtained from a creep-rupture test that accounts for the combined effect of tem-

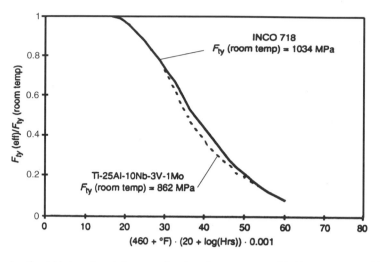

Fig. 6.9 Effective tensile yield strength variations as a function of temperature and hold time represented by the Larson-Miller parameter

perature and time. The notion is that, given time, the dwell time in a trapezoidal stress cycle promotes creep deformation at the crack tip. The F_{ty} term in Eq 5.44 may be considered as an effective yield stress required to achieve a 0.2% strain in a creep test. The ratio of effective yield stress to room-temperature tensile yield strength can be plotted against the Larson-Miller parameter, which takes into account the combined effect of temperature and time at load. Figure 6.9 shows such a relationship for two high-temperature alloys: the Inco 718 alloy and a super-$\alpha 2$ titanium aluminide forging (Ti-25Al-10Nb-3V-1Mo at.%, which is equivalent to Ti-14A1-20Nb-3V-2Mo in wt%).

Creep Zone. Similar to r_p (the crack-tip plastic zone), which is a result of crack-tip plastic yielding, a creep zone r_{cr} will form at the crack tip during creep deformation. Much like the HRR field is the crack-tip stress distribution associated with r_p, the RR field is the crack-tip stress distribution associated with r_{cr}. In small-scale creep, the creep zone is defined as the region around the crack tip where the creep strains exceed the elastic strains. For a stationary crack, the creep zone size is given by Ref 6.11 as:

$$r_{cr}(\theta,t) = \frac{K^2}{2\pi} (EAt)^{2/(n-1)} \cdot F_{cr}(\theta,n) \qquad \text{(Eq 6.35)}$$

where

$$F_{cr}(\theta,n) = \tilde{F}_{cr}(\theta,n)\left[\frac{(n+1)I_n}{2\pi\eta}\right]^{2/(n-1)} \qquad \text{(Eq 6.36)}$$

where $\eta = (1 - v^2)$ for plane strain ($= 1$ for plane stress). The term $\tilde{F}_{cr}(\theta,n)$ is roughly maximum at $\theta = 90°$ for plane strain and at $\theta = 0°$ for plane stress. The values of $\tilde{F}_{cr}(\theta,n)$ are given in Ref 6.11 (equal to 0.2 to 0.5 for plane strain, and 0.5 to 0.6 for plane stress, depending on n). For convenience, Fig. 6.10 shows a plot for Eq 6.36, that is, a plot of $F_{cr}(\theta,n)$ versus n (with $v = 0.3$).

All of the equations for $C(t)$ and r_{cr} have been formulated assuming the following conditions: (a) the crack is stationary; (b) the applied load (either sustained or cyclic) has a step loading profile (i.e., the load rising time is very short); and (c) the material at the crack-tip region is dominated by the second-stage creep deformation mechanism. For discussion of a moving crack, where the crack-tip stress field changes with time during crack propagation, readers are

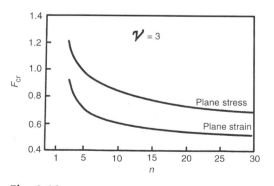

Fig. 6.10 Creep-zone-size dimensionless coefficient F_{cr}

referred to Ref 6.14 and 6.15. For a ramp-type waveform (where it takes time for the load to rise), the magnitude of $C(t)$ is increased by a factor of $(1 + 2n)/(n + 1)$, and the creep zone size thereby changes accordingly (Ref 6.16, 6.17). If the degree of primary creep is significant, a whole new set of analytical solutions are required (Ref 6.18, 6.19).

Environment-Affected Zone. The general concept and empirical-based methods for determining the size of an EAZ have been presented in Chapter 5, where the discussion centered on environment-sensitive materials. A word of caution is that something out-of-the-ordinary can happen. An alloy that is normally susceptible to environment may become susceptible to creep at higher temperatures. When an alloy is somewhat susceptible to creep even though crack extension is still controlled by K, crack growth behavior inside an EAZ may not be the same as shown in Fig. 5.65 and 5.66. The crack growth rates for those load cycles following a sustained load would have decreased rather than increased. An actual example of this phenomenon is shown in Fig. 6.11. In this case, the IN 100 alloy was tested at 732 °C (1350 °F) (Ref 6.20) with a loading profile very much the same as shown in Fig. 5.66. Because the stress levels for the trapezoidal and the sinusoidal stress cycles were the same ($K_{max} = 38.5$ MPa$\sqrt{m}$, or 35

ksi$\sqrt{in.}$), crack growth retardation was not supposed to happen. According to the concept of the EAZ, the post-sustained load crack growth should have behaved as shown in Fig. 5.65. However, the crack growth behavior of this alloy after the sustained load was very much the same as that resulting from a high/low load sequence profile. That is, the post-sustained load crack growth rates were retarded instead of accelerated. The mechanism for such behavior is not fully understood, because this is the only data set of this type ever reported in the literature. Results of a previous survey have shown that the creep crack growth behavior of the IN 100 alloy is rather peculiar (Ref 6.21, 6.22). Elastic-K is not the only crack-growth-controlling parameter for this alloy. Besides K, two other fracture indices, C^* and σ_{net}, also correlate with the sustained-load fracture test results. Thus, one possible explanation for this phenomenon is that the crack-tip deformation zone here resembles a creep zone in which stress relaxation is prominent.

Hypothetical Case History. From a material application standpoint, engineering materials generally behave in the manner of creep ductile or creep brittle. Typical creep ductile materials include low-alloy steels and stainless steels. Creep brittle materials include high-temperature superalloys, high-temperature aluminum alloys,

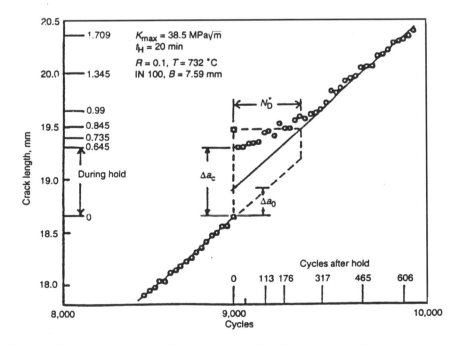

Fig. 6.11 Effect of hold time on constant-amplitude crack growth of IN 100 alloy. Source: Ref 6.20

titanium alloys, intermetallics, and ceramics. There is a general consensus that creep-resistant materials are creep brittle, and perhaps environment sensitive. The creep zone sizes for these materials are smaller than in creep ductile materials. The linear-elastic index K may, in most cases, adequately characterize the high-temperature crack growth behavior in these alloys. The K-based high-temperature fatigue crack growth methodology is presented in Chapter 5.

From a crack-tip deformation mechanism standpoint, it seems that depending on material and service application, the coexistence of r_p and EAZ has brought about conflicting results of the delay behavior in general. In the case of crack growth at high temperature, we have learned that there is a third type of crack-tip deformation zone: the creep zone. The existence of the creep zone may further confuse the subject.

According to the hypothesis stated earlier, two out of three crack-tip deformation models can coexist in a material/temperature system—that is, r_p and r_{cr}, or r_p and EAZ. It may even be possible that all three types of crack-tip deformation zones coexist at a given time. Each of them has a unique way of controlling the crack growth rates of the current and subsequent cycles. The plastic zone is the source for altering the crack-tip stress field during crack propagation. The size of a creep zone indicates the relative degree of creep deformation, and is closely associated with changes in crack-growth-controlling parameters with time. If an alloy is sensitive to environment, an EAZ will exist at the crack tip. Consider, for example, Inco 718 alloy. The sizes for all three crack-tip deformation zones at 593 °C (1110 °F) and at 650 °C (1200

°F), corresponding to an applied K level of 25 MPa$\sqrt{m}$, or 22.7 ksi$\sqrt{in}$.), have been computed using Eq 5.44, 6.35 and 5.49. Using Fig. 6.9, the effective F_{ty} values for Inco 718 are 900 MPa (13 ksi) at 593 °C (1110 °F) and 848 MPa (123 ksi) at 650 °C (1200 °F). The EAZ sizes are directly obtained from Fig. 5.68. The coefficients in Eq 6.35 for the creep zone at 593 °C (1110 °F) are $A = 1.05e - 58$ (MPa)$^{-n/h}$, $n = 19.0$, $E = 1.70e + 05$ MPa. The coefficients in Eq 6.35 for the creep zone at 650 °C (1200 °F) are $A = 1.83e - 47$ (MPa)$^{-n/h}$, $n = 15.8$, $E = 1.66e + 05$ MPa. The results are plotted in Fig. 6.12 and 6.13. It is conceivable that at a lower temperature (i.e., 593 °C, or 1110 °F), and up to the first 2 h of sustained loading, crack growth is primarily controlled by crack-tip plasticity. After that, crack growth will be controlled by the EAZ mechanism. If the temperature had instead been 650 °C (1200 °F), the crack-tip plastic zone would dominate the crack growth behavior for the first 15 min. Then the crack growth mechanism would switch to EAZ.

Equation 6.35 implies that the degree of creep deformation is a function of stress level and time at load. For a stress cycle with a relatively short hold time, the creep zone size will be relatively small (or nonexistent, in the case of no hold time). Equation 6.35 also implies that for a given pair of K and t, r_{cr} is a function of the material properties E, A, and n. The creep zone r_{cr} is in proportion to $t^{2/(n-1)}$ even for a stationary crack (when the crack is not moving but accumulates damage inside the creep zone). The rate of creep zone expansion, $\dot{r}_{cr}$ (i.e., the derivative of r_{cr} with respect to t), is in proportion to $t^{-(n-3)/(n-1)}$. For $n < 3$, $\dot{r}_{cr}$ increases with

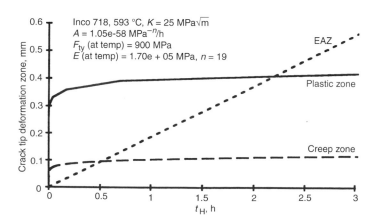

Fig. 6.12 Crack-tip deformation zone sizes for Inco 718 at 593 °C (1110 °F)

time. Thus, steady-state creep is not possible; the crack is dominated by the K field. For $n > 3$, $\dot{r}_{cr}$ decreases with time so that steady-state creep is possible (because $\dot{r}_{cr}$ has to approach zero to reach the time-independent state). For the Inco 718 cases shown in Fig. 6.12 and 6.13, the slope of the creep zone versus time curve decreases with time because $n > 3$. As shown, the creep zone slope eventually approaches zero. This means that C^* may act as a controlling parameter for crack growth. However, as also shown in Fig. 6.12 and 6.13, the plastic zone is significantly larger than its creep zone counterpart at any given time. Therefore, it can be said that crack growth of this material/temperature system is primarily dominated by r_p and EAZ. The driving force for crack extension will be K, not C_t or C^*.

It is clear that choosing a characterizing parameter for creep crack growth is not straightforward. It is always safe to make a correlation check with test data and let the data dictate the outcome. According to the results of a survey reported in Ref 6.22, creep crack growth behavior for most high-temperature superalloys (including Inco 718) is primarily controlled by K.

6.2.5 Creep Crack Growth

As we've seen, the characterizing parameters during creep crack growth are K for small-scale creep and C^* for steady-state creep. Saxena claims that the value of $C(t)$ in the transient regime can only be estimated through the use of Eq 6.32 and 6.33, and that $C(t)$ is not an experimentally measurable quantity. He then proposes a new parameter C_t and states that C_t can cover the entire time spectrum as needed (Ref 6.23).

To perform a remaining-life assessment of a component under creep crack growth conditions, two principal ingredients are needed: (a) an appropriate expression for relating the driving force K, C^*, or Saxena's C_t to the nominal stress, crack size, material constants, and component geometry; and (b) a correlation between this driving force and the crack growth rate in the material, which has been established on the basis of prior data or by laboratory testing of samples from the structural/machine component. Once these two elements are available, they can be combined to derive the crack size as a function of time. The general methodology for doing this is illustrated below, assuming C_t to be the driving crack-tip parameter.

In general, C_t and C^* are determined based on graphical interpretation of the laboratory test record. Load-line deflection rate and specimen geometry are the key elements in the calibration of a generalized expression for calculating the resulting C_t or C^* of a test. Reference 6.24 provides a detailed account of this procedure. However, this type of analysis presents considerable difficulties to fracture mechanics analysts when performing structural crack growth life predictions.

An analytical expression for calculating C_t is given by Saxena (Ref 6.18):

$$C_t = (C_t)_{ssc} + C^* \qquad \text{(Eq 6.37)}$$

where the first term denotes the contribution from small-scale creep, and the second term denotes the contribution from steady-state, large-scale creep. The first term is time variant, whereas the second term is time invariant. In the limit of $t \to 0$, approaching small-scale creep

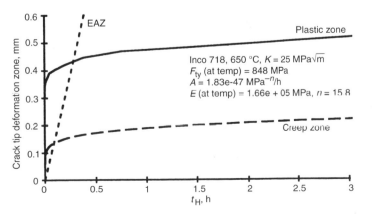

Fig. 6.13 Crack-tip deformation zone sizes for Inco 718 at 650 °C (1200 °F)

conditions, the first term dominates, implying that K is the controlling parameter in crack growth, with time also explicitly entering the relationship. In the limit $t \to \infty$, the first term becomes zero and C_t becomes identical with C^*.

The term $(C_t)_{SCC}$ in Eq 6.37, for the contribution of small-scale creep, has been defined by Saxena (Ref 6.18) as:

$$(C_t)_{SSC} = \frac{4\alpha\chi\tilde{F}_{cr}(\theta,n) \cdot \eta}{E(n-1)} \cdot \frac{K^4}{W} \frac{F'}{F}$$
$$\cdot (EA)^{2/(n-1)} \cdot (t)^{-(n-3)/(n-1)} \qquad \text{(Eq 6.38)}$$

where $\eta = (1 - \nu^2)$ for plane strain (1 for plane stress) and χ is a scaling factor that is approximately equal to 1/3 (for plane strain) as determined by finite element analysis (Ref 6.18). F is a K-calibration factor, a function of a/W. It is given by $KB\sqrt{W}/P$, where K is the elastic stress intensity factor, a function of a/W, geometry, and loading condition. F' is the derivative of F, that is $dF/d(a/W)$. The term α is given by:

$$\alpha = \frac{1}{2\pi} \left[\frac{(n+1)I_n}{2\pi\eta} \right]^{2/(n-1)} \qquad \text{(Eq 6.39)}$$

The material properties A and n can be obtained from creep tests. On the basis of another finite element analysis (Ref 6.18), the term $\chi\tilde{F}_{cr}(\theta,n)$ in Eq 6.38 was found to have a value of approximately 1/7.5 (for plane strain). Therefore, Eq 6.38 can be reduced to:

$$(C_t)_{SSC} = \frac{4\alpha\beta(1 - \nu^2)}{E(n-1)} \cdot \frac{K^4}{W} \cdot \frac{F'}{F}$$
$$\cdot (EA)^{2/(n-1)} \cdot (t)^{-(n-3)/(n-1)} \qquad \text{(Eq 6.40a)}$$

where $\beta \cong 1/7.5$. Alternatively, taking advantage of Fig. 6.10, after proper substitution of Eq

6.36 and 6.39 into Eq 6.38, Eq 6.32 can be simplified as:

$$(C_t)_{SSC} = \frac{2\chi\eta}{\pi(n-1)} \cdot \frac{K^4}{W} \cdot \frac{F'}{F}$$
$$\cdot \frac{A^{2/(n-1)}}{(Et)^{(n-3)/(n-1)}} \cdot F_{cr}(\theta,n) \qquad \text{(Eq 6.40b)}$$

where $\chi \cong 1/3$ (for plane strain). Either Eq 6.40(a) or (b), whichever is convenient, can be used as the first term of Eq 6.37.

Equations 6.37 to 6.40(b) can be used to estimate C_t from an applied load (stress) and from a knowledge of the elastic and creep behavior of the material, the K-calibration expression, and the C^* expression for the geometry of interest. The K expression can be found in handbooks. The J_p expressions (which are required to derive C^*) are not as abundantly available for different geometries. References 6.7 and 7.8 are two of the few sources that provide engineering J_p solutions. At present, this is viewed as a limitation of the technology. More detailed descriptions of the derivations of the C_t and C^* expressions, and the manner of obtaining some of the constants and calculating their values, are given in Ref 6.24.

Once C_t is known, it can be correlated to the crack growth rate through the constants b and m in the following relation:

$$\dot{a} = bC_t^m \qquad \text{(Eq 6.41)}$$

Values of b and $\dot{m}$ for all the materials analyzed by Saxena et al. are listed in Table 6.3. It also has been shown that m should be approximately equal to $n/(n+1)$, where n is the creep rate exponent.

Crack growth calculations are performed with the current values of a and the corresponding

Table 6.3 Creep crack growth constants b and m for various ferritic steels

Material	b				m	
	Upper scatter line		Mean		Upper scatter	Mean
	BU(a)	SI(b)	BU(a)	SI(b)		
All base metal	0.094	0.0373	0.022	0.00874	0.805	0.805
2¼Cr-1Mo weld metal	0.131	0.102	0.017	0.0133	0.674	0.674
1¼Cr-½Mo weld metal	(c)	(c)	(c)	(c)	(c)	(c)
2¼Cr-1Mo and 1¼Cr-½Mo heat-affected-zone/ fusion-line material	0.163	0.0692	0.073	0.031	0.792	0.792

(a) BU = British units: da/dt in in./h; C_t in in. $\cdot$ lb/in. $\cdot$ h $\times 10^3$. (b) SI = Système International units: da/dt in mm/h; C_t in kJ/m^2 $\cdot$ h. (c) Insufficient data; creep crack growth rate behavior comparable to that of base metal. Source: Saxena, Han, and Banerji, "Creep Crack Growth Behavior in Power Plant Boiler and Steam Pipe Steels," EPRI Project 2253-10, published in Ref 6.25

values da/dt to determine the time increment required for increasing the crack size by a small amount Δa, that is, $\Delta t = \Delta a/\dot{a}$. This provides new values of a, t, and C_t, and the process is then repeated. When the value of a reaches the critical size a_c as defined by K_{IC}, J_{IC}, or any other appropriate failure parameter, failure is deemed to have occurred.

In conclusion, creep crack growth can be characterized in terms of fracture mechanics crack growth by the steady-state parameter C^* (in the large-scale crack growth regime) and the transient parameter C_t. Under small-scale creep, C_t is designated as $(C_t)_{SSC}$ and is theoretically in proportion to K (the crack-tip stress intensity factor). It becomes C^* when the amount of creep deformation approaches steady-state creep condition. Therefore, C_t can be used for the entire range of creep deformation. Because K is a more convenient parameter and all the crack growth analysis methodologies have been developed based on K, it is better to use K in lieu of C_t when the situation permits.

REFERENCES

6.1. D. Shows, A.F. Liu, and J.H. FitzGerald, Application of Resistance Curves to Crack at a Hole, *Fracture Mechanics: 14th Symposium—Volume II: Testing and Applications*, STP 791, ASTM, 1983, p II.87–II.100

6.2. J.W. Hutchinson, Singular Behavior at the End of a Tensile Crack in a Hardening Material, *J. Mech. Phys. Solids*, Vol 16, 1968, p 13–31

6.3. J.R. Rice and G.F. Rosengren, Plane Strain Deformation Near a Crack Tip in a Power-Law Hardening Material, *J. Mech. Phys. Solids*, Vol 16, 1968, p 1–12

6.4. C.F. Shih, "Table of Hutchinson-Rice-Rosengren Singular Field Quantities," Report No. MRL E-147, Brown University, June 1983

6.5. J.R. Rice, A Path Independent Integral and the Approximate Analysis of Strain Concentration by Cracks and Notches, *J. Appl. Mech. (Trans. ASME) E*, Vol 35, 1968, p 379–386

6.6. M.M. Ratwani and D.P. Wilhem, "Development and Evaluation of Methods of Plane Stress Fracture Analysis, Part II, Vol I, A Technique for Predicting Residual Strength of Structure," Report AFFDL-

TR-73-42, Air Force Flight Dynamics Laboratory, Aug 1977

6.7. V. Kumar, M.D. German, and C.F. Shih, "An Engineering Approach for Elastic-Plastic Fracture Analysis," Report NP-1931, Electric Power Research Institute, July 1981

6.8. A. Zahoor, "Ductile Fracture Handbook, Vol 1, Circumferential Throughwall Cracks," Report NP-6301-D, Electric Power Research Institute, 1989

6.9. J.D. Landes and J.A. Begley, A Fracture Mechanics Approach to Creep Crack Growth, *Mechanics of Crack Growth*, STP 590, ASTM, 1976, p 128–148

6.10. A. Saxena, Evaluation of C* for the Characterization of Creep-Crack-Growth Behavior in 304 Stainless Steel, *Fracture Mechanics, 12th Conference*, STP 700, ASTM, 1980, p 131–151

6.11. H. Riedel and J.R. Rice, Tensile Cracks in Creeping Solids, *Fracture Mechanics, 12th Conference*, STP 700, ASTM, 1980, p 112–130

6.12. R. Ehlers and H. Riedel, A Finite Element Analysis of Creep Deformation in a Specimen Containing a Macroscopic Crack, *Advances in Fracture Research*, Vol 2, Pergamon Press, 1981, p 691–698

6.13. A. Saxena and J.L. Bassani, Time-Dependent Fatigue Crack Growth Behavior at Elevated Temperature, *Fracture: Interaction of Microstructure, Mechanisms and Mechanics*, TMS/AIME, 1984, p 357–383

6.14. C.Y. Hui and H. Riedel, The Asymptotic Stress and Strain Field Near the Tip of a Growing Crack Under Creep Conditions, *Int. J. Fract.*, Vol 17, 1981, p 409–425

6.15. D.E. Hawk and J.L. Bassani, Transient Crack Growth Under Creep Conditions, *J. Mech. Phys. Solids*, Vol 34, 1986, p 191–212

6.16. H. Riedel, Crack-Tip Stress Fields and Crack Growth Under Creep-Fatigue Conditions, *Elastic-Plastic Fracture: Second Symposium, Vol I, Inelastic Crack Analysis*, STP 803, ASTM, 1983, p I.505–I.520

6.17. A. Saxena, Limits of Linear Elastic Fracture Mechanics in the Characterization of High-Temperature Fatigue Crack Growth, *Basic Questions in Fatigue*, Vol II, STP 924, ASTM, 1988, p 27–40

6.18. A. Saxena, Mechanics and Mechanism of Creep Crack Growth, *Fracture Mechan-*

ics: Microstructure and Micromechanisms, ASM International, 1989, p 283–334

6.19. H. Riedel, Creep Crack Growth, *Flow and Fracture at Elevated Temperatures*, American Society for Metals, 1983, p 149–177

6.20. J.M. Larsen and T. Nicholas, Load Sequence Crack Growth Transients in a Superalloy at Elevated Temperature, *Fracture Mechanics, 14th Symposium, Vol II: Testing and Applications*, STP 791, ASTM, 1983, p II.536–II.552

6.21. A.F. Liu, *Structural Life Assessment Methods*, ASM International, 1998

6.22. A.F. Liu, "Element of Fracture Mechanics in Elevated Temperature Crack Growth," Paper No. 90-0928, *Proceedings AIAA/ASME/ASCE/AHS/ASC 31st Structures, Structural Dynamics and Materials Conference,* Part 2, 1990, p 981–994

6.23. A. Saxena, Creep Crack Growth under Non-Steady State Conditions, *Fracture Mechanics: Vol 17*, STP 905, ASTM, 1986, p 185–201

6.24. R. Norris, P.S. Crover, B.C. Hamilton, and A. Saxena, Elevated-Temperature Crack Growth, *ASM Handbook,* Vol 19, *Fatigue and Fracture*, ASM International, 1996, p 507–519

6.25. R. Viswanathan, *Damage Mechanisms and Life Assessment of High-Temperature Components*, ASM International, 1989

SELECTED REFERENCES

• J.W. Hutchinson, *A Course on Nonlinear Fracture Mechanics*, Technical University of Denmark, 1979

• A. Saxena, *Nonlinear Fracture Mechanics for Engineers* CRC Press, 1998

CHAPTER 7

Mechanical Behavior of Nonmetallic Materials

IN PREVIOUS CHAPTERS, discussion has focused on metals—their mechanical behavior and the analytical methods applicable to them. The structural mechanics and fracture mechanics tools used for metals are believed to be equally applicable to nonmetals, as long as the substance is homogeneous and isotropic. This chapter briefly discusses the essential features associated with ceramics, glasses, and polymers. Special attention is paid to the characteristics of these materials as compared to metals.

7.1 Ceramics and Glasses

Most ceramics are composed of oxides, carbides, nitrides, or borides. Advanced ceramics used for engineering applications are in a polycrystalline form. They consist of individual grains, which are actually single crystals. The grains are bonded to one another during processing at elevated temperature. Silicon carbide (SiC), silicon nitride (Si_3N_4), alumina (Al_2O_3), and silicon aluminum oxynitride (Sialon) belong to the advanced ceramics category. Intentionally or otherwise, the silicon nitrides and Sialons are incorporated with glassy grain-boundary phases that are present in different compositions and amounts. Thus, a set of materials can be tailored for different temperature ranges and different applications. Inorganic glasses are considered a subset of ceramics. The difference between the two is that ceramics have a periodic crystal structure, whereas glasses possess a short-range order structure.

What distinguishes metallics and nonmetallics are their bonding forces. Metallic elements release electrons to develop metallic bonding, whereas nonmetallics (ceramics and polymers) share electrons to develop covalent bonds. Exceptions are the ceramic compounds such as refractory carbides and nitrides, which have combinations of metallic and covalent bonds.

Ceramics share many commonalities with metals. Both are crystalline, homogeneous, and isotropic. Ceramics are used in a variety of engineering applications that utilize their hardness, wear resistance, refractoriness, and high compression strength. They are seldom used in tensile-loaded structures, because they are brittle (with minimal ductility) and very sensitive to stress raisers. (However, ceramics may be toughened with a more ductile second phase to create a composite or aggregate.) This nil ductility and high sensitivity to stress raisers make mechanical testing of ceramics using conventional test methods very difficult. Applicable ASTM standards are listed in Table 7.1.

7.1.1 Fracture

The brittle nature of ceramics causes great difficulty in detecting flaws or precracking fracture toughness test specimens. Flaws in ceramics are not easily detectable by ordinary nondestructive evaluation (NDE) methods. Fractographic studies conducted in the 1970s showed that critical flaw sizes in ceramics range from ≈1 μm to ≈500 μm (typically 10 to 50 μm). Today's new, toughened ceramics are somewhat improved, but their critical flaw sizes are still very small. These earlier fractographic studies revealed the characteristics of flaw extension in ceramics. The fracture surface of a ceramic material, according to Freiman and Wiederhorn (Ref 7.1), consists of four distinct regions: the *smooth mirror region* that separates the *original crack* (the flaw origin) and the *mist region,* which is a transitional area preceding the *rough/hackle region.*

A quantitative relationship exists between the stress at fracture and the distance from the flaw to each of the boundaries. Through the model below, we can estimate the critical flaw size and/or the fracture toughness of the material.

Consider a very small flaw originating on a ceramic surface as shown in Fig. 7.1, which schematically depicts a fracture surface typically found in fractographs of ceramics and glasses.* The mirror region is flat, smooth, and brittle in nature. The mist region is somewhat less smooth. The rough hackle region is easily recognized by the outward divergent lines running along the crack propagation direction. This region is associated with a large amount of strain energy absorption and thus is ductile in nature. The smooth mirror region is unique. For a wide variety of ceramics, the ratio of the mirror size to the flaw size is a constant; its value is 13. Because the mirror radius r_M is measurable, the flaw dimensions a and b can be estimated. The initial and critical flaw sizes often are the same in ceramics; therefore, the critical stress intensity factor can be calculated by modeling the crack as an elliptical surface flaw. The stress intensity expression in this case would be (Ref 7.1):

*Polymers that fracture in a brittle manner exhibit a similar appearance.

$$K = \sqrt{1.2\pi\left(\frac{\sqrt{a \cdot b}}{r_M}\right)} \cdot \frac{A_M}{\Phi} \qquad \text{(Eq 7.1)}$$

and

$$A_M = S \cdot r_M^{1/2} \qquad \text{(Eq 7.2)}$$

where Φ is a crack shape parameter, the complete elliptical integral of the second kind, given as Eq 5.23(a) or 5.23(c); K becomes K_{IC} when $a = a_{cr}$ and $b = b_{cr}$. Equation 7.1 can be used for calculating K_{IC}, or to estimate the critical flaw size when a separately determined K_{IC}

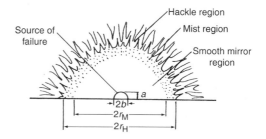

Fig. 7.1 Schematic of fracture surface features observed on many ceramics. The dimensions a and $2b$ denote the minor and major axes of the flaw dimensions, r_M denotes the beginning of the mist region, and r_H denotes the beginning of the hackle region. Source: Ref 7.1

Table 7.1   Standards for mechanical testing of ceramics

Standard No.	Title
ASTM	
C 1121	Standard Test Method for Flexural Strength of Advanced Ceramics at Elevated Temperatures
C 1145	Standard Definition of Terms Relating to Advanced Ceramics
C 1161	Standard Test Method for Flexural Strength of Advanced Ceramics at Ambient Temperature
C 1239	Standard Practice for Reporting Uniaxial Strength Data and Estimating Weibull Distribution Parameters for Advanced Ceramics
C 1273	Standard Test Method for Tensile Strength of Monolithic Advanced Ceramics at Ambient Temperatures
C 1286	Standard System for Classification of Advanced Ceramics
C 1322	Standard Practice for Fractography and Characterization of Fracture Origins in Advanced Ceramics
C 1361	Standard Practice for Constant-Amplitude, Axial Tension-Tension Cyclic Fatigue of Advanced Ceramics at Ambient Temperature
C 1368	Standard Test Method for Determination of Slow Crack Growth Parameters of Advanced Ceramics by Constant-Stress Rate Flexural Testing at Ambient Temperature
C 1421	Standard Test Methods for Determination of Fracture Toughness of Advanced Ceramics at Ambient Temperature
Japanese Standards Association (JIS)	
R 1607	Testing Methods for the Fracture Toughness of High Performance Ceramics
R 1621	Testing Method for Bending Fatigue of Fine Ceramics
R 1632	Test Methods for Static Bending Fatigue of Fine Ceramics
British Standards Institution	
DD ENV 843-3	Advanced Technical Ceramics—Monolithic Ceramics—Mechanical Properties at Room Temperature, Part 3: Determination of Subcritical Crack Growth Parameters from Constant Stressing Rate Flexural Strength Tests
International Organization for Standardization (ISO)	
DIS15732	Fine Ceramics (Advanced Ceramics, Advanced Technical Ceramics)—Test Method for Fracture Toughness of Monolithic Ceramics at Room Temperature by Single Edge Pre-Cracked Beam (SEPB) Method

value is available. Due to the brittleness of ceramics, ASTM E 399 test methods are not applicable to fracture and crack growth testing of them. Instead, a number of specially prepared standards have been developed (Table 7.1).

Typical K_{IC} values for some common ceramics are available. They are listed along with other mechanical properties in Table 7.2. In the case of mixed modes 1 and 2, a survey of K_{IC} and K_{IIC} test data for Al_2O_3, Si_3N_4, ZrO_2, and glass has been conducted by Munz and Fett (Ref 7.3) (Fig. 7.2). The empirical constants ξ and ζ in Eq 4.19 have also been determined: $\xi = 1$ and $\zeta = 2$.* It is also shown that $\xi = \zeta = 1.5$ can fit the data equally well.

---

*These values are given in Ref 7.3, based on a report by H.A. Richard (in German).

7.1.2 Fatigue and Subcritical Crack Growth

In some literature, fatigue behavior in ceramics (and also in polymers) is described as cyclic fatigue (or dynamic fatigue) and static fatigue (or environmentally induced fatigue). In this book, these terms are equivalent to the conventional meaning of fatigue (S-N or da/dN) and crack growth under sustained load (da/dt), respectively. S-N fatigue data for ceramics are available, but very rare. Stress hysteresis is nonexistent due to lack of plasticity. The term "cyclic fatigue" in ceramics basically refers to fatigue crack propagation, da/dN. The general method of fatigue crack growth testing of ceramics can be very similar to that given in ASTM Standard E 647 for measuring fatigue crack growth rate in metallics, except that precracking

---

Table 7.2 Typical properties of selected engineering ceramics

Material	Crystal structure	Theoretical density, g/cm³	Poisson's ratio	Transverse rupture strength, MPa (ksi)	Fracture toughness, MPa√m (ksi√in.)	Young's modulus, GPa (10^6 psi)
Glass-ceramics	Variable	2.4–5.9	0.24	70–350 (10–51)	2.4 (2.2)	83–138 (12–20)
Pyrex glass	Amorphous	2.52	0.2	69 (10)	0.75 (0.7)	70 (10)
TiO_2	Rutile tetragonal	4.25	0.28	69–103 (10–15)	2.5 (2.3)	283 (41)
	Anatase tetragonal	3.84	...	...	...	...
	Brookite orthorhombic	4.17	...	...	...	...
Al_2O_3	Hexagonal	3.97	0.26	76–1034 (40–150)	2.7–4.2 (2.5–3.8)	380 (55)
Cr_2O_3	Hexagonal	5.21	...	>262 (>238)	3.9 (3.5)	>103 (>15)
Mullite	Orthorhombic	2.8	0.25	185 (27)	2.2 (2.0)	145 (21)
Partially stabilized ZrO_2	Cubic, monoclinic, tetragonal	5.70–5.75	0.23	600–700 (87–102)	8–9 (7.3–8.2)(a) 6–6.5 (5.5–5.9)(b) 5 (4.6)(c)	205 (30)
Fully stabilized ZrO_2	Cubic	5.56–6.1	0.23–0.32	245 (36)	2.8 (2.5)	97–207 (14–30)
Plasma-sprayed ZrO_2	Cubic, monoclinic, tetragonal	5.6–5.7	0.25	6–80 (0.9–12)	1.3–3.2 (1.2–2.9)	48 (7)(d)
CeO_2	Cubic	7.28	0.27–0.31	...	...	172 (25)
TiB_2	Hexagonal	4.5–4.54	0.09–0.13	700–1000 (102–145)	6–8 (5.5–7.3)	514–574 (75–83)
TiC	Cubic	4.92	0.19	241–276 (35–40)	...	430 (62)
TaC	Cubic	14.4–14.5	0.24	97–290 (14–42)	...	285 (41)
Cr_3C_2	Orthorhombic	6.70	...	49 (7.1)	...	373 (54)
Cemented carbides	Variable	5.8–15.2	0.2–0.29	758–3275 (110–475)	5–18 (4.6–16.4)	96–654 (57–95)
SiC	α, hexagonal	3.21	0.19	96–520 (14–75)(e) 250 (36)(f) 230–825 (33–120)(g) 398–743 (58–108)(h)	4.8 (4.4)(e) 2.6–5.0 (2.4–4.6)(f) 4.8–6.1 (4.4–5.6)(g) 4.1–5.0 (3.7–4.6)(h)	207–483 (30–70)
	β, cubic	3.21	...	...	...	...
SiC (CVD)	β, cubic	3.21	0.16	1034–1380 (150–200)(j) 2060–2400 (300–350)(k)	5–7 (4.6–6.4)	415–441 (60–64)
Si_3N_4	α, hexagonal	3.18	0.24	414–650 (60–94)(m) 700–1000 (100–145)(n) 250–345 (36–50)(p)	5.3 (4.8)(m) 4.1–6.0 (3.7–5.5)(n) 3.6 (3.3)(p)	304 (44)
	β, hexagonal	3.19	...	...	...	...
TiN	Cubic	5.43–5.44	...	...	...	251 (36)

(a) At 293 K (20 °C, 70 °F). (b) At 723 K (450 °C, 840 °F). (c) At 1073 K (800 °C, 1470 °F). (d) 21 GPa (3 × 10^6 psi) at 1373 K (1100 °C, 2010 °F). (e) Sintered, at 300 K (27 °C, 80 °F). (f) Sintered, at 1273 K (1000 °C, 1830 °F). (g) Hot pressed, at 300 K (27 °C, 80 °F). (h) Hot pressed, at 1273 K (1000 °C, 1830 °F). (j) At 300 K (27 °C, 80 °F). (k) At 1473 K (1200 °C, 2190 °F). (m) Sintered. (n) Hot pressed. (p) Reaction bonded. Source: Ref 7.2

is extremely difficult. Again, test methods for *da/dN* and *da/dt* in ceramics are different than for metals (see Table 7.1). Typical *da/dN* data for ceramics are shown in Fig. 7.3 and 7.4. It is interesting to note that these data exhibit very much the same effect and behavior as in metals, whether tested in room-temperature air or in water. That is, the Mg-PSZ data in Fig. 7.3 show that its fatigue crack growth rates are faster in moist air and much faster in water as compared to in inert nitrogen, resembling typical corrosion fatigue behavior in metals.

Temperature, moisture, water, and water vapor always cause stress-corrosion cracking (SCC)-type crack growth and failure in ceramics and glasses. Impurities such as silicate glass in the grain boundaries and interfaces, compounded with nil crack-tip plasticity, are responsible for SCC failure in ceramics. Frequently in the literature, *da/dt* behavior in glasses and ceramics is described by the type C behavior depicted in Fig. 4.20(b). In that model, the stage 1 crack growth occurs with rapid acceleration and long linear *K*-dependent behavior. It is then followed by *K*-independent stage 2 behavior before accelerating again. Such a model is partially supported by experimental test data, as shown by Fig. 7.5(b) and curve B in Fig. 7.5(d). However, the other *da/dt* curves in Fig. 7.5 seem to agree better with the other models shown in Fig. 4.20(a).

Cyclic slip is the basis for fatigue in metals; however, the strong bonds in highly brittle solids restrict the dislocation movement (slip) for fatigue initiation. Thus, there has been a perception that ceramics are free of true cyclic fatigue effects, and that the cyclic crack growth rates in ceramics or glasses are simply manifestations of environmentally assisted crack growth under

sustained tensile loading. That means *da/dN* and *da/dt* are interchangeable (after converting the time-based *da/dt* to the cycle-based *da/dN*, and vice versa). Many curve-fitting equations were developed to fit the *da/dt* data and used for converting to *da/dN*. Conclusive evidence of true cyclic crack growth behavior in glasses and ceramics has surfaced in recent studies, as shown in Fig. 7.3 and 7.4. Experimental test results for Al_2O_3-33SiC have shown that for the same material, and at the same *K* level, cyclically loaded tests produce much slower crack growth rates compared to sustained-load testing (Ref 7.6). However, it has also been shown that crack growth behavior in mid-toughness MgO-PSZ under 55% relative humidity air or in distilled water is just the opposite: *da/dN* is faster than *da/dt* (Ref 7.7). These discrepancies can be attributed, in part, to ignoring the stress ratio and frequency effects in cyclic crack growth testing. Nevertheless, the practice of using converted crack growth rates result in unreliable fatigue life estimates for ceramics.

7.2 Polymers

Polymers generally are noncrystalline. Thus, they constitute a material class of their own, separate from metals and ceramics. They are composed of covalently bonded long chain mol-

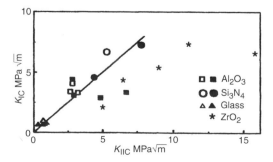

Fig. 7.2 Interrelation between K_{IC} and K_{IIC} from literature data. □■, Al_2O_3; ○●, Si_3N_4; △▲, glass; *, ZrO_2. Open symbols, specimens with slot; solid symbols, precrack and chevron notch. Straight line, $K_{IC} = K_{IIC}$. Source: Ref 7.3

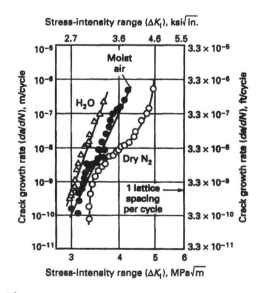

Fig. 7.3 Cyclic fatigue crack growth rates in low-toughness Mg-PSZ, showing acceleration due to moisture or water. Source: After Ref 7.4

ecules (macromolecules), consisting of a series of small repeating molecular units (monomers). Most common polymers have carbon (organic material) backbones, although polymers with inorganic backbones (e.g., silicates and silicones) are possible. Polymeric materials exhibit strong covalent bonds within each chain. However, individual chains are frequently linked via secondary bonds (e.g., the van der Waals attractive forces). Under applied stresses, polymer chains slide over one another. Failure is the result of chain separation rather than breakage of interchain bonds.

Certain polymers cannot be crystallized at all, but others can exhibit limited crystallinity, which results in higher density and increased stiffness and strength. Crystalline or semicrystalline polymers can be obtained by slow cooling from the molten state. For example, atactic polystyrene cannot be crystallized, whereas polyethylene terephthalate (PET) is amorphous when cooled rapidly from the melt but can be crystallized by slow cooling or annealing at an elevated temperature.

A large number of polymers are available, many with complex chemical structures and long names. We can separate them into three distinct groups on the basis of their physical properties: thermosets, thermoplastics, and elastomers (rubbers). The first two are considered as glassy polymers.

Thermosets are rigid, highly cross-linked polymers that degrade rather than melt by heat as thermoplastics do. They are used at temperatures below the glass transition temperature. Thermoplastics have a linear structure (i.e., not cross-linked). They usually contain various stabilizers, fillers, or toughening agents; therefore, a wide range of properties are possible. Due to their toughness, thermoplastics also are used as the matrix for fiber-reinforced composites. In contrast, rubbers are lightly cross-linked polymers that consist of long flexible chain-like molecules interconnected at various points by cross-links to form a loose molecular network. The best-known characteristics of rubber are its capability to undergo large deformations and, when the stress is released, return to its original shape. In elastomers, the amount of recoverable elastic strain can be 500% or more.

Discussion here centers on engineering plastics, which consist of thermoset and thermoplastic resins. Thermosetting resins are usually brittle compared to thermoplastic resins, which are

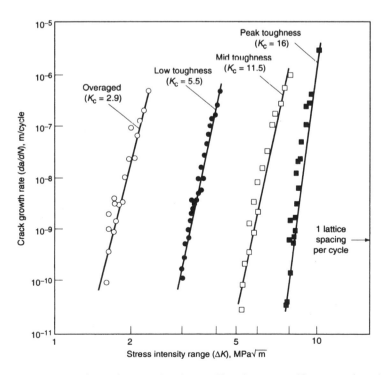

Fig. 7.4 Cyclic fatigue crack growth rates for MgO-PSZ subeutectoid aged to a range of fracture toughness levels. Data obtained from C(T) specimens in room-temperature air, 50 Hz, $R = 0.1$. Source: Ref 7.4

tougher, more temperature sensitive, and much higher in cost. Common thermoset and thermoplastic resins include:

- *Thermosetting resins:* Epoxy (EP), bismaleimide (BMI), polyimide (PI), cyanate ester, polyester, vinyl ester, and polyurethane. Among these, polyester, polyimide, and polyurethane are also available as thermo-

plastics. The applicable service temperature ranges for epoxy and toughened epoxy resins are about 120 °C (250 °F) and 80 to 105 °C (180 to 220 °F), respectively.

- *Thermoplastic resins:* Polycarbonate (PC), polyethylene (PE), polyvinyl chloride (PVC), polyvinyl acetate (PVA), polystyrene (PS), polyetheretherketone (PEEK), polymethylmethacrylate (PMMA), polypheny-

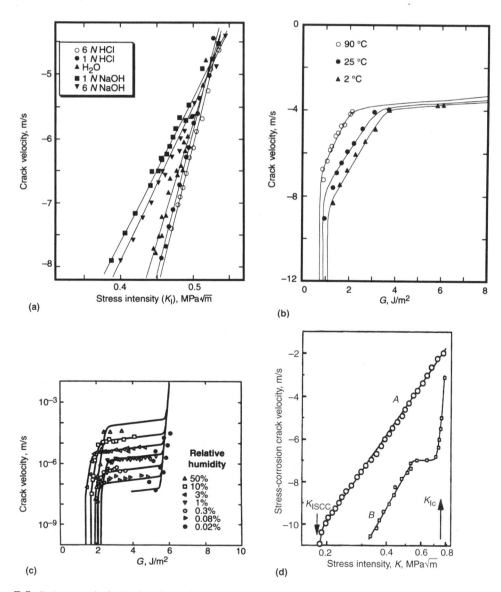

Fig. 7.5 Environmental subcritical crack growth in glasses. (a) Crack velocity as a function of environment and pH for vitreous silica glass. Source: Ref 7.5. (b) Soda-lime glass tested at different temperatures. Source: Ref 7.5. (c) Crack velocity curves for sapphire in moist air (25 °C, or 75 °F) at different relative humidity, double-cantilever-beam specimens. Source: Ref 7.5. (d) Soda-lime glass tested at room temperature (25 °C, or 75 °F). Curve A: tested in water. Curve B: tested in toluene. Note: The term K_{ISCC} in (d) actually appears as K_{Ith} in the source, which means threshold and is a common practice in the ceramics engineering community. The alteration was made to be in line with the models in Fig. 4.20. Source: Ref 7.3

Table 7.3   Typical room-temperature properties of selected thermoplastics

Thermoplastic	Tensile strength		Elongation, %	Tensile modulus		Compression strength		Glass transition temperature	
	MPa	ksi		GPa	psi × 10^6	MPa	ksi	°C	°F
ABS	35–45	5.1–6.5	15–60	1.7–2.2	0.25–0.32	25–50	3.6–7.2	. . .	. . .
PMMA	50–70	7.2–10.1	2–10	3	0.43	80–115	11.6–16.7	105–110	220–230
PS	35–60	5.1–8.7	1–4	3–4	0.43–0.58	80–110	11.6–15.9	100–105	210–220
PVC, rigid	40–60	5.8–8.7	5	2.4–2.7	0.35–0.39	60	8.7	75–105	165–220
PC	62	9	110	2.3	0.33	86	12.5	150	300
Nylon 6	81	11.7	5–50	2.76	0.40	90	13	56–70	133–160
Nylon 6/6	79	11.4	60	2.83	0.41	34	4.9	57–80	135–175

Source: Ref 7.8, 7.9

lene sulfide (PPS), polyetherimide (PEI), polysulfone (PSU), polypropylene (PP), polyacrylonitrile (PAN), polyethylene tere-phthalate (PET), polymethylpentene, acrylonitrile butadiene styrene (ABS), polytera-fluoroethylene (PTFE), polyamide (PA) (i.e., nylons), and polyamideimide (PAI). The last resin, PAI, is originally molded as a thermoplastic, but is postcured in the final composite to produce partial thermosetting characteristics and thus improved subsequent temperature resistance. Typical properties for selected thermoplastics are listed in Table 7.3.

7.2.1 Viscoelastic Behavior

Many polymers exhibit viscoelastic behavior. When viscoelastic polymers are stressed, there is an immediate elastic response. This is followed by viscous flow, which decreases with time until a steady state is reached (Fig. 7.6). If the material is then unloaded, the elastic strain is recovered, followed by time-dependent (delayed) recovery. Some permanent strain, denoted as permanent recovery in Fig. 7.6, remains. Comparison of Fig. 7.6 with Fig. 2.81(b) shows that the recovery phenomenon in polymers is very much the same as in creep of metals.

Because viscoelastic behavior is time dependent, data based on short-term tests may misrepresent the tested polymer in a design application that involves long-term loading. The magnitude of the time dependence of polymers is very temperature dependent. At temperatures well below glass transition temperature, plastics exhibit a high modulus and are only weakly viscoelastic. At these temperatures, test data based on a time-independent analysis will probably be adequate. Making use of the "time-temperature superposition principle," one might derive long-term de-

sign data from a short-term test. This principle states that the mechanical response at long times at some particular temperature is equivalent to the mechanical response at short times but at some higher temperature. By determining shift factors, it is possible to determine which temperature to use in obtaining long-term data from short-term tests. Discussion of the theories and test procedures are given in Ref 7.10 and 7.11. Other accelerated testing methods also are used by engineers (Ref 7.12).

7.2.2 Crazing

The fracture process in thermoplastics often is dominated by crazing. In other words, crack initiation in thermoplastics is preceded by formation of crazes. Whether or not thermosets will craze is inconclusive. It is customarily assumed that thermosets do not craze; its failure mode is all by shear yielding. Geometrically, a craze is a planar defect consisting of bundles of stretched molecules. Crazing begins with microvoid formation under the action of the hydrostatic tension component of the stress tensor. A craze is easily visible, especially when viewed at the correct angle with the aid of a directed light source. Crazing is a mode of plastic deformation rather

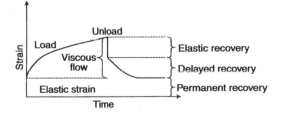

Fig. 7.6 Schematic representation of viscoelastic behavior of a polymer. Loading produces an immediate elastic strain followed by viscous flow. Unloading produces an immediate elastic recovery followed by additional recovery over a period of time.

than mechanical cracking. However, deformation during crazing is constrained laterally; that is, it resists Poisson contraction. Thus, its formation is accompanied by an increase in speci-

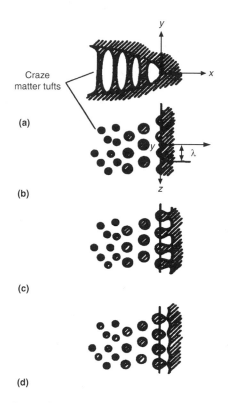

Fig. 7.7 Schematic of craze formation. (a) Outline of a craze tip and the upper and lower surfaces (side view). (b) Cross section in the craze plane (top view) across craze matter tufts (fibrils). (c) and (d) Advance of the craze front by a completed period of interface convolution. Source: Ref 7.13

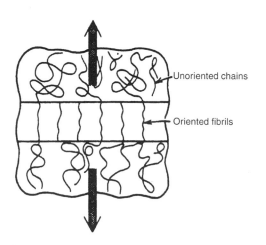

Fig. 7.8 Schematic showing oriented/lengthened fibrils in between the upper and lower surfaces of a craze

men volume. The upper and lower surfaces of what appears to be a crack in a polymer is not a crack, but a craze. These surfaces are interconnected by systems of discrete fibrils in the craze interior, as shown schematically in Fig. 7.7. Crack surfaces cannot support a load, but craze surfaces can (Ref 7.14). The density of the material in the craze zone ranges from 40 to 60% of that of the matrix material (Ref 7.15). For a given polymer and temperature, there is a stress level for craze initiation. A test procedure for determining the stress-to-craze value is specified in ASTM Standard F 484.

After crazing has initiated, the voids increase in size and elongate along the direction of the maximum principal stress. Craze growth occurs by extension of the craze tip into uncrazed material, drawing in new material from the craze flanks. At the same time, the craze thickens by lengthening and disentanglement of the fibrils (Fig. 7.7 and 7.8). As the craze faces separate while the fibrils increase in length, the fibrils fail at a critical strain and a true crack is thus formed. Figure 7.9 schematically illustrates this process.

7.2.3 Static Strength

Engineering plastics are not as strong as metals. Due to their lower density, however, structural plastics possess higher specific strengths than those of metals. Table 7.4 compares mechanical properties for several common engineering materials. A full listing of ASTM and ISO standards for mechanical testing of plastics is provided in Volume 8 of the *ASM Handbook, Mechanical Testing and Evaluation*.

Whether a plastic will behave as brittle or ductile depends on the test temperature or application temperature. A general rule of thumb is that a polymer will be brittle at temperatures well below its own glass transition temperature (T_g); otherwise, it will be ductile. This generalization is based on the fact that the elastic modulus of a plastic abruptly decreases at temperatures near its glass transition temperature.

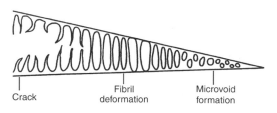

Fig. 7.9 Schematic crack formation from a craze. Source: Ref 7.16

Figure 7.10 shows how the modulus of an amorphous polymer changes as temperature increases or decreases through this critical region. At temperatures below T_g, this polymer behaves like a glassy material, with a relatively high modulus. As shown, the same polymer has lower modulus values at temperatures above T_g, and these values continue to decrease as temperature increases. Its state of substance also changes with increasing temperature—from glassy to leathery, then to rubbery, and finally like a very viscous liquid. This temperature dependence phenomenon is generally the same for all types of polymers. The exception is semicrystalline polymers, which will remain in the rubbery state over an extended temperature range. Their transition temperature (from rubbery to viscous) is higher than for the amorphous polymers. The cross-linked polymers will not become viscous.

As always, there are exceptions to the rule. The glass transition temperatures for PS and PMMA are 102 and 107 °C (215 and 225 °F), respectively. According to the glass transition temperature rule, these polymers should be brittle at room temperature—and indeed they are. In contrast, PC is ductile at room temperature, even though its glass transition temperature is 150 °C (300 °F). That means that the rule of glass transition temperature does not always work. One possible explanation is that crazing may contribute to brittle behavior. Comparing the minimum crazing stress with the material yield stress of the polycarbonate, Petrie (Ref 7.17) believed that the lower of the two would determine the fracture behavior of the material. Minimum crazing stress is the stress level required for crazing to start. That is, if the minimum craze stress is lower than the tensile yield stress, the polymer will fail in a brittle manner. In other words, if the applied stress reaches the

minimum craze stress first, the polymer will be brittle. If the applied stress reaches the yield stress first, as the PC does, the polymer will be ductile. Petrie further states that the minimum crazing stress for PMMA must be lower than the tensile yield stress at ambient temperature, and therefore brittle behavior prevails.*

Many polymers fail in a brittle manner at ambient and low temperatures. Examples of commercial plastics that normally fracture in a brittle manner are PS, PMMA, and thermosetting resins such as epoxy and polyester. On the contrary, PC, PE, PP, PA, PVC, and PET exhibit ductile fracture behavior. Just like metals, decreasing strain rate and/or increasing temperature will cause a normally brittle material to become ductile (Fig. 7.11). Thus, it is the combination of temperature, strain rate, and, of course, the material itself. Figure 7.12(a) shows a group of en-

---

*This theory has been substantiated by statements in Williams' book (Ref 7.18) that the craze stress is indeed lower than the yield stress in PMMA and PS.

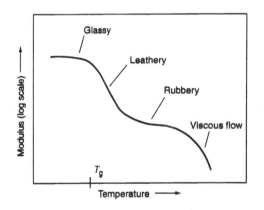

Fig. 7.10 Schematic modulus versus temperature for a typical amorphous polymer

---

Table 7.4   Range of mechanical properties for common engineering materials

Material	Elastic modulus		Tensile strength		Maximum strength/density		Elongation at break, %
	GPa	10^6 psi	MPa	ksi	(km/s)^2	(kft/s)^2	
Ductile steel	200	30	350–800	50–120	0.1	1	0.2–0.5
Cast aluminum alloys	65–72	9–10	130–300	19–45	0.1	1	0.01–0.14
Polymers	0.1–21	0.02–30	5–190	0.7–28	0.05	0.5	0–0.8
Glasses	40–140	6–20	10–140	1.5–21	0.05	0.5	0
Copper alloys	100–117	15–18	300–1400	45–200	0.17	1.8	0.02–0.65
Moldable glass-filled polymers	11–17	1.6–2.5	55–440	8–64	0.2	2	0.003–0.015
Graphite-epoxy	200	30	1000	150	0.65	1.3	0–0.02

Source: Ref 7.8, p 26–48

gineering stress-strain curves (for a plastic) at a constant strain rate, but at different temperatures. At low temperatures (e.g., represented by curve 1), the material is completely brittle with an approximately linear relationship between stress and strain. At higher temperatures (e.g., represented by curve 2), a yield point is observed and the load falls before failure. Sometimes, necking is also observed. However, the breaking strain is still quite low, typically 10 to 20% according to Ward and Hadley (Ref 7.10). At still higher temperatures, approaching the glass transition temperature, the neck grows in a stable manner in some thermoplastics, with material being fed into the neck region from the thicker adjacent regions. This type of deformation in polymers is called "drawing" or "cold drawing." The strains in this case are generally very large, and can reach 1000% (see curve 3). Finally, at temperatures above the glass transition temperature, the polymer (either thermoset or thermoplastic) undergoes a quasi-rubber-like behavior; the deformations are homogeneous and the final strain is very large (see curve 4). Actual test data for the PMMA showing the stress-strain behavior at different temperatures are given in Fig. 7.12(b). Comparing Fig. 7.12(a) with Fig. 7.12(b), it is evident that the tests conducted at 4 to 30 °C (39 to 86 °F) exhibit type 1 behavior. The test conducted at 50 °C (122 °F) exhibits type 2 behavior. The test conducted at 60 °C (140 °F) could have behaved like type 2, and probably not quite like type 3, because the test temperature is still much below the material's glass transition temperature, which is 107 °C (225 °F).

Comparing the plastic deformation behavior in plastics with metals, both materials neck at temperatures at which both are ductile. The difference is that plastics can sustain much greater strains before fracture. Plastic deformations in

metals are evidenced by slips, which take place along crystallographic planes. However, there is no crystallographic plane in plastics. Even for crystalline polymers, which are inevitably semicrystalline, the plastic deformation mechanism is not clearly identified with some kind of slip system. Therefore, the shear bands formed in polymers are generally more diffused or delocalized than those in metals. The potential for extensive elongation after necking is a consequence of the long-chain nature of polymers. Chains in the necked area can uncoil and align. If the necking can maintain long enough, the chains will become completely aligned. At this point, any added elongation will begin to stretch the bonds in the polymer chain. This stretching will lead to failure (Ref 7.20).

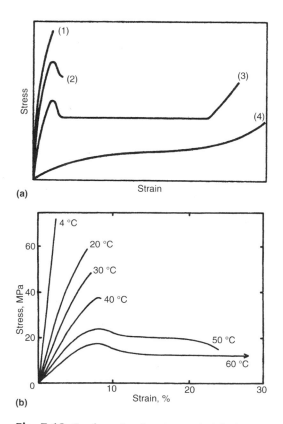

Fig. 7.12 Tensile engineering stress-strain behavior as a function of temperature. (a) Schematic stress-strain curves for a typical polymer tested at four temperatures while the strain rate is kept constant. 1, Low temperature, brittle behavior; 2, intermediate temperature, somewhat ductile; 3, higher temperature (approaching glass transition), necking and cold drawing; 4, above glass transition temperature, homogeneous deformation (quasi-rubber-like behavior). Source: Ref 7.10. (b) Variation of the stress-strain behavior of PMMA with test temperature. Source: Ref 7.19

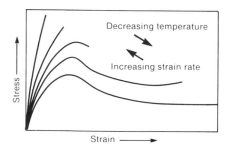

Fig. 7.11 Schematic tensile stress-strain curves for a plastic, showing the effect of strain rate and temperature

7.2.4 Fatigue

Traditional fatigue analysis methods such as those used for metals can be applied to polymers. Polymers, however, involve a significant complication in terms of hysteresis heating and subsequent softening. The temperature of a polymer body will rise during fatigue cycling. Therefore, polymers are highly susceptible to frequency effects: The higher the loading frequency, the lower the fatigue endurance limit. For a given stress level, fatigue life (number of cycles to failure) decreases with increase in cyclic frequency, due to greater heat buildup at higher frequencies. The size of a structural piece also contributes to the extent of thermal heating. Taking two smooth round bars of the same length, for example, the one with a larger diameter will undergo a greater temperature increase because its outer surface area to volume ratio is smaller. That is, the larger piece heats up more, but has a relatively smaller outer surface to release the heat. For the same reason, a flat specimen that is thicker tends to generate more heat, which leads to a greater temperature rise under a given set of test conditions.

Higher stress levels also cause more heat buildup in plastics. Figure 7.13 demonstrates the effect of stress level at a given load frequency. A conventional S-N curve can be plotted for this set of test data by connecting the fracture points, marked F, and the fatigue threshold will be the stress level marked U. In fatigue testing of plastics, mechanical factors (applied stress, R-ratio, notch geometries) and thermal factors both contribute to the outcome of a test.

Thermohysteresis is why most (if not all) polymers exhibit cyclic strain softening in their cyclic stress-strain curves (Fig. 7.14). Metals exhibit either cyclic strain hardening or softening. However, all the polymers shown in Fig. 7.14 exhibit cyclic strain softening regardless of their chemical structure (i.e., semicrystalline, amorphous, or two phase). Note that the yield points that were developed in monotonic tension were eliminated during cyclic stabilization.

Moreover, polymer fatigue behavior is generally sensitive to environment (including test temperature), molecular weight, molecular density, and aging. Additionally, for polymer crystals, fatigue resistance increases as crystallinity increases (Fig. 7.15). S-N curves that do not account for these effects should not be used exclusively without looking at test conditions. In the case of metals, some exhibit an endurance limit

below which no fatigue failure will occur. On the basis of experimental observation, if not derived from theory, we know that lattice structures of metals are the source for endurance limits, or the lack thereof. Like metals, polymers may or may not exhibit a traditional fatigue endurance limit (Fig. 7.16); however, the reason is not known. For those data shown in Fig. 7.16, only nylon and PET do not exhibit an endurance limit.

7.2.5 Fatigue Crack Propagation

Polymers display $\Delta a/\Delta N$ versus ΔK behavior similar to that in metals. They also display the same frequency effect as in metals; that is, the higher the loading frequency, the lower the fatigue crack growth rate (Fig. 7.17). This behavior is directly opposite the S-N fatigue behavior discussed earlier. One possible reason is that temperature buildup in this case is concentrated in a local area: the crack tip. The material surrounding the crack tip and the free surfaces of the crack may help to dissipate heat from the crack tip. This rationale is in some way supported by the data in Fig. 7.13, which shows that temperature in each specimen leveled off at the final stage of the test. It could mean that heating was discontinued after crack initiation, or at least when the crack became longer. So, for a fatigue crack growth rate test where a fatigue crack has been inserted into the specimen right from the beginning, temperature buildup during the test may be minimal. Therefore, although some heating does occur at the crack tip, an excessive temperature rise is prevented; fatigue crack

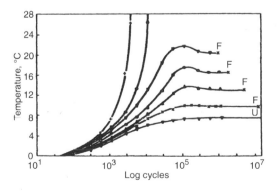

Fig. 7.13 Temperature increase resulting from uniaxial cycling at 5 Hz with load control, sine wave, and zero mean stress in polyacetal. U, unbroken; F, fracture. The applied stress levels are, counterclockwise starting from U: 15, 16, 17.4, 19.7, 21.6, 22.4, and 27.8 MPa (2.2, 2.3, 2.5, 2.9, 3.1, 3.2, and 4 ksi). Source: Ref 7.21

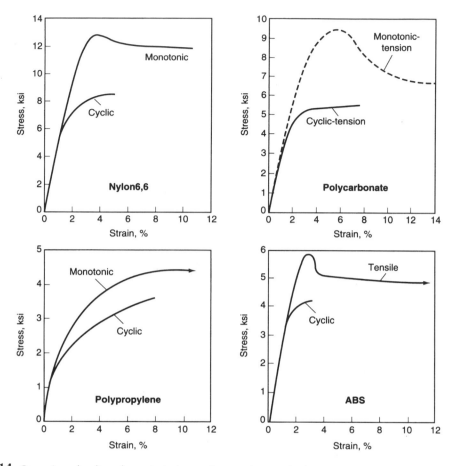

Fig. 7.14 Comparison of cyclic and monotonic stress-strain curves for several polymers at 298 K. Source: Ref 7.22

propagation is primarily a mechanical and time-dependent process. We also observe in Fig. 7.17 that PC is much less sensitive to mechanical load frequencies than PS. An explanation has been offered by Hertzberg and Manson (Ref 7.25),

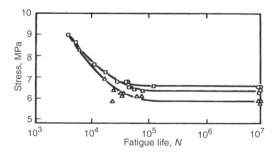

Fig. 7.15 Fatigue life in PTFE with increasing crystallinity. △, low crystallinity, air quenched; □, medium crystallinity, 33.3 °C/h cooling; ○, high crystallinity, 5.6 °C/h cooling. Test frequency, 30 Hz. *R*-ratio not identified. Source: Ref 7.23

who say that polymers that are prone to crazing are more susceptible to cyclic frequency effect; PMMA and PS are examples.

The effect of mean stress on fatigue crack growth rate is another matter. Some polymers exhibit a trend similar to that in metals; that is, the higher the *R*-ratio, the higher the fatigue crack growth rate. Some other polymers exhibit just the opposite behavior. In Ref 7.26, Pruitt examined a vast amount of test data from a number of sources. The results, which identify and classify the polymers examined into two categories, are listed in Table 7.5.

The appearance of fatigue striations in polymers is similar to that in metals, with one minor exception. Electron fractographs for PMMA and other polymers show that each striation contains a fine linear structure oriented normal to the striation line itself (Fig. 7.18). Such lines may reflect evidence of excessive material tearing during the striation

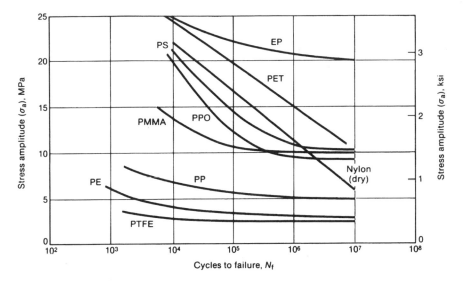

Fig. 7.16 Typical S-N curves for several commodity plastics at 30 Hz. Note: R-ratio is missing in this figure. Source: Ref 7.24

formation process (Ref 7.25). Overall, there appears to be no change in morphology from one side of the striation to the other. In theory, as in metals, a striation has a one-to-one cor-relation with an applied stress cycle. That is, for a given stress cycle, the microscopic stri-ation spacing and the macroscopically mea-sured da/dN should have the same magnitude.

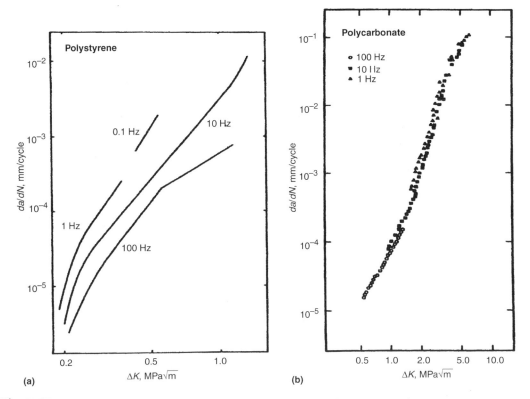

Fig. 7.17 Effect of cyclic frequency on fatigue crack growth rates in (a) PS and (b) PC. Source: Ref 7.25

Figure 7.19 shows comparisons for striations in three types of polymers, thus substantiating that striations can be found on the fracture surfaces of polymers. Under certain test conditions, however, fatigue cracks in some polymers progress in a discontinuous manner. The striationlike parallel markings in Fig. 7.20 are called discontinuous growth bands. Each of these bands is associated with a large number of loading cycles—as many as several hundred. Some polymers exhibit only discontinuous growth bands on the fracture surface (e.g., PVC); some have both (e.g., PC and PS). Examples of discontinuous bands for six different polymers are shown in Fig. 7.21.

For fatigue crack propagation in polymers, the discontinuous growth band is analogous to the crack-tip plastic zone in metals. It is more like a Dugdale zone. The mechanism of the discon-

Table 7.5 Effect of increasing mean stress on polymer fatigue crack propagation

Increasing da/dN with increasing mean stress

High-density polyethylene
Nylon
High-molecular-weight PMMA
Polystyrene
Epoxy
Polyethylene copolymer

Decreasing da/dN with increasing mean stress

Low-density polyethylene
Polyvinyl chloride
Low-molecular-weight PMMA
Rubber-toughened PMMA
High-impact polystyrene
Acrylonitrile-butadiene-styrene
Polycarbonate
Toughened polycarbonate copolyester

Source: Ref 7.26

tinuous growth band is described by a model proposed by Hertzberg and Manson (Ref 7.25). It simply states that the craze at the tip will grow with time (during fatigue cycling). Meanwhile, the fatigue crack front does not move. As mentioned earlier, when some critical condition is satisfied (e.g., the craze grows to a critical size

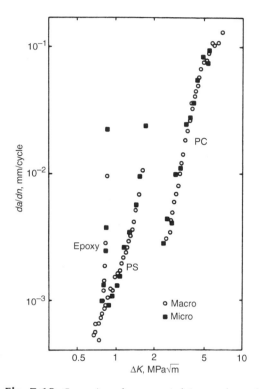

Fig. 7.19 Comparison of macroscopic fatigue crack growth rates and striation spacing measurements in epoxy, PC, and PS. ■, micro (measured from electron fractograph); ○, macro (physical measurement). Source: Ref 7.25

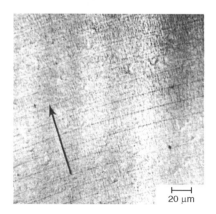

Fig. 7.18 Fatigue striations in PMMA. Arrow indicates crack growth direction. Source: Ref 7.27

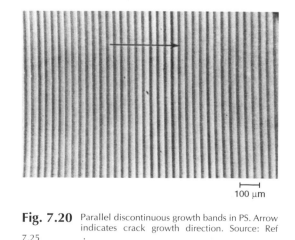

Fig. 7.20 Parallel discontinuous growth bands in PS. Arrow indicates crack growth direction. Source: Ref 7.25

with many broken fibrils), the craze becomes a crack and joins with the existing crack. Put another way, the existing crack will jump through the entire broken craze and arrest at the tip of the craze. This can be considered a new crack length. The period that the old crack sat waiting for the craze to grow accounts for the accumulated load cycles.

The series of photographs in Fig. 7.22(a) demonstrate this discontinuous cracking pro-

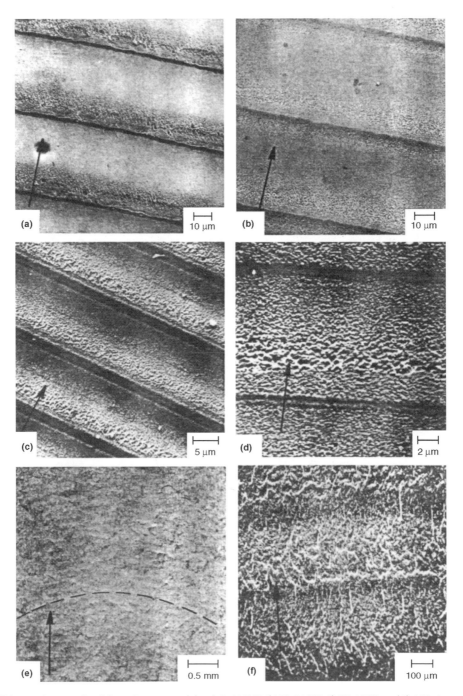

Fig. 7.21 SEM fractographs of discontinuous growth bands in (a) PVC, (b) PS, (c) PSF, (d) PC, (e) PA, and (f) ABS. Arrows indicate crack growth direction. Source: Ref 7.27

cess, with the arrows in each photograph pointing to the crack tip and craze tip locations. The first four photographs show that while the craze

tip advances during fatigue cycling, the crack tip stays at its original position. The last photograph shows that the crack tip moved into the old craze

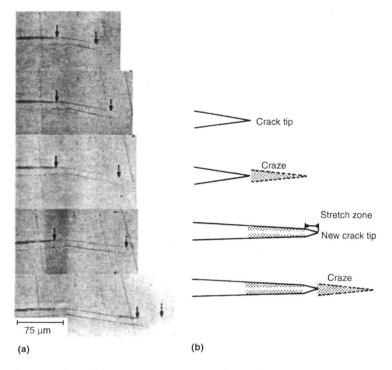

(a)   (b)

Fig. 7.22 Discontinuous crack growth process. (a) Composite microphotograph of PVC showing positions of crack tip (arrows on the left) and craze tip (arrows on the right) at given cyclic intervals. (b) Model of discontinuous cracking process. Source: Ref 7.25

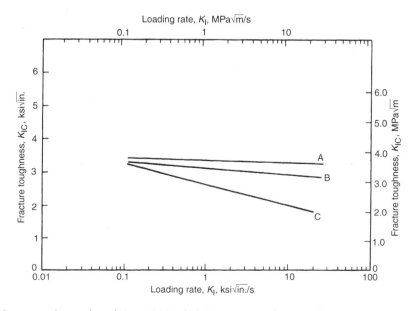

Fig. 7.23 Fracture toughness values of 13 mm (0.5 in.) thick PC specimens as functions of environment and loading rate. The lines A, B, and C denote ambient, dry, and wet, respectively. Note: Ambient temperature values are higher than those given in Table 7.6. Source: Ref 7.29

tip location and a new craze tip appeared. A model of the sequence involving continuous craze growth and discontinuous crack growth is shown in Fig. 7.22(b).

7.2.6 Impact Strength and Fracture Toughness

Fracture toughness of polymers is generally low. The K_{IC} value for a standard epoxy is as low as 0.454 MPa$\sqrt{m}$ (0.413 ksi$\sqrt{in.}$) for the Hercules 3501-6 resin. The K_{IC} values for thermoplastics are much higher. ICI 450G (PEEK resin) has a K_{IC} value greater than 6.6 MPa$\sqrt{m}$ (6 ksi$\sqrt{in.}$) (Ref 7.28). Polycarbonate falls in between; its values as functions of moisture and loading rate are shown in Fig. 7.23. More K_{IC} values for common engineering plastics at room temperature are given in Table 7.6. As mentioned before, ductility of plastics generally increases with temperature. Therefore, fracture toughness and impact strength also increase with temperature. On the basis of impact strength, some common plastics are rated for their brittleness versus temperature and are presented in Table 7.7.

Ductile-to-brittle transition behavior in thermoplastics is very much the same as in metals. Figures 7.24 and 7.25 show some S-shaped curves that resemble those seen in metals. Figure 7.24 shows the K_Q values for a rubber-toughened thermoplastic as a function test temperature. The K_Q values were developed using compact specimens. The exact material was not identified; nonetheless, a ductile-to-brittle transition behavior is evident. The data in Fig. 7.25, developed from Izod impact tests, show that impact strength for PS increases at 100 °C (212 °F), which is the material's glass transition temperature. According to the glass transition temperature rule discussed earlier, PS is brittle at temperatures below its glass transition temperature. The second material in Fig. 7.25, high-impact polystyrene (HIPS), has two transition temperatures. The HIPS consists of a second phase of individual rubber particles embedded in a PS matrix phase. At temperatures below the T_g of

Table 7.6 Typical K_{IC} and Izod impact strength values for engineering plastics

	K_{IC}		Izod impact strength	
Material	MPa$\sqrt{m}$	ksi$\sqrt{in.}$	J/m	ft · lbf/in.
PMMA	0.7–1.6	0.6–1.5	. . .	. . .
PS	0.7–1.1	0.6–1.0	13.3–21.3	0.25–0.4
High-impact PS	1.0–2.0	0.9–1.8	. . .	. . .
PE	1.0–6.0	0.9–5.5	. . .	. . .
ABS	2.0	1.8	213.6	4.0
PC	2.2	2.0	747.6	14
PVC	2.0–4.0	1.8–3.6	. . .	. . .
PI	. . .	. . .	0.9	0.02
PP	3.0–4.5	2.7–4.1	. . .	. . .
PET	5.0	4.5	. . .	. . .
PSU	. . .	. . .	69.4	1.3
PEI	. . .	. . .	53.4	1.0
PEEK	. . .	. . .	96.1	1.8
PAI	. . .	. . .	133.5	2.5
PTFE	. . .	. . .	107–125	2.0–2.4
Nylon 6/6	. . .	. . .	53.3–160	1.0–3.0
Nylon 6	. . .	. . .	53.3–160	1.0–3.0
Nylon 6/12	. . .	. . .	53.3–74.6	1.0–1.4
Nylon 1	. . .	. . .	96	1.8
Epoxy	0.6	0.5	. . .	. . .
Polyester	0.6	0.5	. . .	. . .

Source: Ref 7.30, 7.31

Table 7.7 Fracture behavior of plastics as a function of temperature

	Temperature, °C (°F)							
Plastics	−20 (−4)	−10 (14)	0 (32)	10 (50)	20 (68)	30 (85)	40 (105)	50 (120)
Polystyrene	A	A	A	A	A	A	A	A
Polymethyl methacrylate	A	A	A	A	A	A	A	A
Glass-filled nylon (dry)	A	A	A	A	A	A	A	B
Polypropylene	A	A	A	A	B	B	B	B
Polyethylene terephthalate	B	B	B	B	B	B	B	B
Acetal	B	B	B	B	B	B	B	B
Nylon (dry)	B	B	B	B	B	B	B	B
Polysulfone	B	B	B	B	B	B	B	B
High-density polyethylene	B	B	B	B	B	B	B	B
Rigid polyvinyl chloride	B	B	B	B	B	B	C	C
Polyphenylene oxide	B	B	B	B	B	B	C	C
Acrylonitrile-butadiene-styrene	B	B	B	B	B	B	C	C
Polycarbonate	B	B	B	B	C	C	C	C
Nylon (wet)	B	B	B	C	C	C	C	C
Polytetrafluoroethylene	B	C	C	C	C	C	C	C
Low-density polyethylene	C	C	C	C	C	C	C	C

A, brittle even when unnotched; B, brittle, in the presence of a notch; C, tough. Source: Ref 7.32

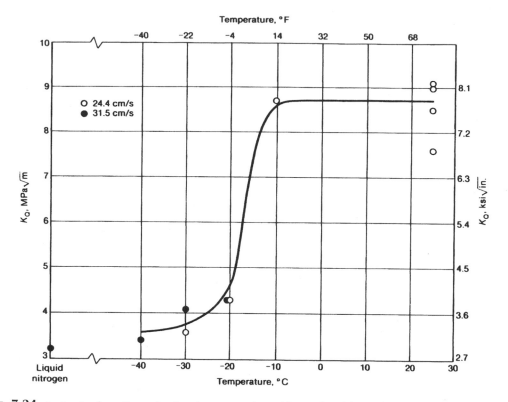

Fig. 7.24 Fracture toughness K_Q as a function of temperature for a rubber-toughened thermoplastic material. Source: Ref 7.32

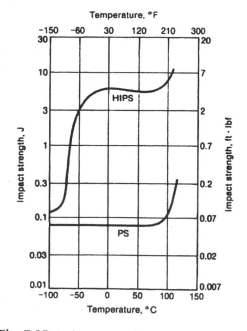

Fig. 7.25 Izod impact strength as a function of temperature for PS and HIPS thermoplastics. Source: Ref 7.33

rubber, the impact strength is very low. At temperatures above the T_g of rubber, the rubber phase helps to increase the impact strength by absorbing some energy during fracture. An additional increase is observed at 100 °C (210 °F), which is the T_g for the base material. Reference 7.33 does not further elaborate on this phenomenon. We can only speculate that at temperatures below 100 °C (between −70 to 100 °C, or −95 to 210 °F), the material is not completely brittle.

Puncture testing is another often used indicator of impact resistance for plastics. Figure 7.26 compares the manner of fracture in PC. Two quick tests indicate that PC is ductile at room temperature but brittle at −90 °C (−130 °F).

Mixed mode 1 and mode 2 fracture testing of polymers is almost nonexistent in the literature. Limited pure mode 2 test data (listed in Table 4.11) show that the K_{2C} to K_{1C} ratio is 0.89 for PMMA (despite coming from two sources and tested using two different specimen configurations).

(a)   (b)

Fig. 7.26 Comparison of failed PC disk from puncture tests. (a) Room temperature. (b) −90 °C (−130 °F). Source: Ref 7.32

7.2.7 Stress Rupture

Stress rupture behavior in plastics is significantly different from that in metals. A stress versus time curve for plastics (at a given temperature) is schematically shown in Fig. 7.27. In region I, the plastic undergoes continuous plastic deformation under a constant load. During that time, the specimen's cross-sectional area gradually reduces. Final fracture of a specimen occurs when the specimen can no longer support the applied load, and thus is ductile. At some lower stress levels and after a long hold time at load, the specimen's surface shows sign of damage and perhaps cracking. When that happens, it takes little time to break the piece (region II), or the part breaks almost immediately (region III). The transitions that take place are indicated by the kinks in the graph. It is almost stress level independent in region III. The switching is quite abrupt. In the last two regions, the fracture behavior is brittle. The surface damage may be very mild, or the crack depth may be very shallow, but that is all it takes to change the fracture mode (Ref 7.34). Quoting the work of Gedde and Ifwarson, Miller (Ref 7.35) has stated that region II may not exist in some materials. The test data of Gedde and Ifwarson (on cross-linked PE) showed that PE of normal cross-linked density exhibits fracture behavior in all three regions. In contrast, region II is missing in PE of low cross-linked density. In the sample test data (for an unidentified PE) shown in Fig. 7.28, the tests apparently stopped somewhere in region II. This figure also shows that the PE plastic underwent creep and stress rupture at room temperature.

It is interesting to note that the ductile-to-brittle transition behavior is not limited to stress rupture testing of plastics. For the same reason—that is, exposure to high temperature (or corrosive medium, or radiation) for a long time to accumulate surface damage—the plastic will become brittle in a short-term tension test. Whether or not the fracture strength of a damaged specimen degrades, the reduction in elongation is significant (Ref 7.34).

7.3 Fractography

Cleavage fracture can also occur in ceramics, inorganic glasses, and in some polymeric materials. As in metals, the fracture surface of these

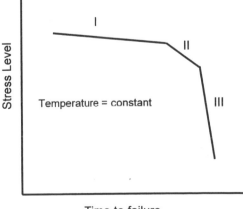

Fig. 7.27 Schematic stress rupture behavior of plastics

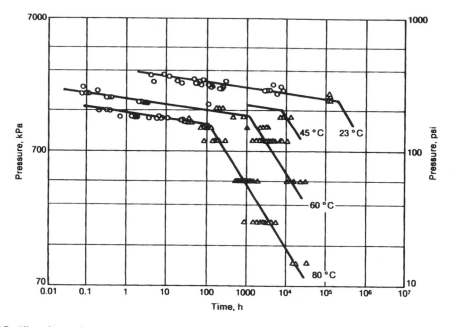

Fig. 7.28 Effect of internal pressure on time-to-failure of PE gas pipe at various temperatures, showing ductile (region I) and brittle (region II) behaviors. Source: Ref 7.36

materials is flat (at low magnification). Cleavage steps (river patterns) may be visible at the macroscale in organic glasses and brittle polymers, but are visible only at the microscale in metallic materials (Ref 7.37). For the soda lime glass shown in Fig. 7.29, cleavage fracture corresponds to the flat fracture surface marked as the "mirror region," which is visible at the macro-

scale. It is brittle, initiates from the fracture origin, and then turns into a transitional region called "mist." The hackle region is associated with a large amount of strain energy absorption and thus is ductile in nature.

Polymers exhibit basically the same fracture appearance as shown in Fig. 7.29 for soda lime glass. Figure 7.30 shows a typical fracture sur-

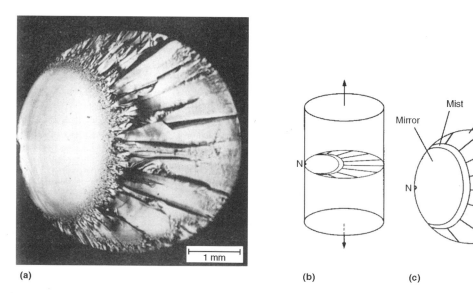

Fig. 7.29 Cleavage fracture in soda lime glass. Crack progresses from left to right. (a) Fracture surface shows the initiation region (featureless mirror region), surrounded by the mist and hackle marks. (b) Geometry of a tensile test bar showing position of fracture surface normal to tensile axis. (c) Arrangement of mirror, mist, and hackle regions on fracture surface. Source: Ref 7.37

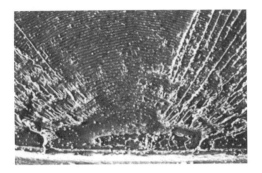

Fig. 7.30 Fracture surface of a PC specimen after Izod impact showing the mirror, mist, and hackle regions, along with Wallner lines that spread over the mist and hackle regions. Magnification: 41×. Source: Ref 7.16

face of a PC specimen after impact. For polymers, the mirror region is the result of nucleation and growth of crazes. The mist region, commonly found in glass, is also observed in some glassy polymers, such as PMMA, PS, and PC. This region is typically flat, smooth, and featureless, except for a slight change in surface texture resembling a fine mist. In polymers, the mist regions are not necessarily confined to the vicinity of the fracture origin. The hackle lines are indications of the final stage of the fracture process. The divergent nature of these lines is advantageous in locating the crack origin. The Wallner lines, which are absent from the fracture surfaces of ceramics and glasses, are not fatigue striations, nor are they true crack front markings. They form when reflected stress waves intersect a propagating crack. The curved crack front appearance of these markings is another useful feature that helps to locate the fracture origin, which is on the concave side of these markings.

REFERENCES

7.1. S.W. Freiman and S.M. Wiederhorn, Fracture Mechanics Applied to Ceramics, *Fracture Mechanics,* University Press of Virginia, 1978, p 299–316

7.2. R.L. Lehman, Overview of Ceramic Design and Process Engineering, *Engineered Materials Handbook,* Vol 4, *Ceramics and Glasses,* ASM International, 1991, p 30

7.3. D. Munz and T. Fett, *Ceramics: Mechanical Properties, Failure Behavior, Materials Selection,* Springer-Verlag, 1999

7.4. R.H. Dauskardt, D.B. Marshall, and R.O. Ritchie, *J. Am. Ceram. Soc.,* Vol 73 (No. 4), 1990, p 893–903

7.5. Y. Katz, N. Tymiak, and W.W. Gerberich, Evaluation of Environmentally Assisted Crack Growth, *ASM Handbook,* Vol 8, *Mechanical Testing and Evaluation,* ASM International, 2000, p 612–648

7.6. L.X. Han and S. Suresh, High Temperature Failure on an Al_2O_3-SiC Composite under Cyclic Loads: Mechanisms of Fatigue Crack-Tip Damage, *J. Am. Ceram. Soc.,* Vol 72, 1989, p 1233–1238

7.7. R.H. Dauskardt, W. Yu, and R.O. Ritchie, Fatigue Crack Propagation in Transformation-Toughened Zirconia Ceramic, *J. Am. Ceram. Soc.,* Vol 70, 1987, p C248–C252

7.8. Mechanical Testing of Polymers and Ceramics, and H.A. Kuhn, Overview of Mechanical Properties and Testing for Design, *ASM Handbook,* Vol 8, *Mechanical Testing and Evaluation,* ASM International, 2000, p 26–48, 49–69

7.9. Guide to Engineering Materials (GEM 2002), *Adv. Mater. Process.,* Vol 159 (No. 12), 2001

7.10. I.M. Ward and D.W. Hadley, *An Introduction to the Mechanical Properties of Solid Polymers,* John Wiley & Sons, 1993

7.11. T. Osswald, *Polymer Processing Fundamentals,* Hanser/Gardner Publications, 1998, p 19–43

7.12. H.F. Brinson, Accelerated Life Prediction, *Engineered Materials Handbook,* Vol 2, *Engineering Plastics,* ASM International, 1985, p 787–795

7.13. A.S. Argon, J.G. Hannoosh, and M.M. Salama, Initiation and Growth of Crazes in Glassy Polymers, *Fracture 1977,* Vol 1, ICF4, Waterloo, 1977, p 445–470

7.14. J.A. Sauer, I. Martin, and C.C. Hsiao, *J. Appl. Phys.,* Vol 20, 1949, p 507

7.15. R.P. Kambour, Structure and Properties of Crazes in Polycarbonate and Other Glassy Polymers, *Polymer,* Vol 5, 1964, p 143–155

7.16. Fracture of Plastics, *ASM Handbook,* Vol 11, *Failure Analysis and Prevention,* ASM International, 2002, p 650–661

7.17. S.P. Petric, Crazing and Fracture, *Engineered Materials Handbook,* Vol 2, *Engineering Plastics,* ASM International, 1985, p 734–740

7.18. J.G. Williams, *Fracture Mechanics of Polymers,* Ellis Horwood, 1984

7.19. E.H. Andrews, *Fracture in Polymers,* Oliver & Boyd, 1968

7.20. C.B. Bucknall, *Toughened Plastics,* Applied Science, 1977

7.21. R.J. Crawford and P.P. Benham, *Polymer,* Vol 16, 1975, p 908

7.22. P. Beardmore and S. Rabinowitz, *Treatise on Material Science and Technology,* Vol 6, 1975, p 267

7.23. M.N. Riddle, G.P. Koo, and J.L. O'Toole, *Polym. Eng. Sci.,* Vol 6, 1966, p 363

7.24. M.N. Riddle, *Plast. Eng.,* Vol 30, 1974, p 71

7.25. R.W. Hertzberg and J.A. Manson, *Fatigue of Engineering Plastics,* Academic Press, 1980

7.26. L. Pruitt, Fatigue Testing and Behavior of Plastics, *ASM Handbook,* Vol 8, *Mechanical Testing and Evaluation,* ASM International, 2000, p 758–767

7.27. M.D. Skibo, R.W. Hertzberg, J.A. Manson, and S. Kim, *J. Mater. Sci.,* Vol 12, 1977, p 531

7.28. J.W. Coltman and S.M. Arndt, "Evaluation of Elastomeric Matrix Materials for Use in Aircraft Primary Structures," Report TR-87437, Simula Inc., July 1987

7.29. S.A. Sutton, J. Tirosh, R.W. Thomas, P.W. Mast, and I. Wolock, The Effect of Loading Rate, Temperature and Moisture on the Fracture Toughness of Polycarbonate, *Proceedings of the 27th National SAMPE Symposium,* San Diego, 1982, p 1003–1021

7.30. J.M. Margolis, Thermoplastic Resins, *Engineered Materials Handbook,* Vol 2, *Engineering Plastics,* ASM International, 1988, p 618–625

7.31. S.P. Petrie, Crazing and Fracture, *Engineered Materials Handbook,* Vol 2, *Engineering Plastics,* ASM International, 1988, p 734–740

7.32. R. Nimmer, Impact Loading, *Engineered Materials Handbook,* Vol 2, *Engineering Plastics,* ASM International, 1988, p 679–700

7.33. L.R. Pinckney, Phase-Separated Glasses and Glass-Ceramics, *Engineered Materials Handbook,* Vol 2, *Engineering Plastics,* ASM International, 1988, p 433–438

7.34. D.E. Duvall, Effect of Environment on the Performance of Plastics, *ASM Handbook,* Vol 11, *Failure Analysis and Prevention,* ASM International, 2002, p 796–799

7.35. E. Miller, *Introduction to Plastics and Composites,* Marcel Dekker, 1996

7.36. C.G. Bragaw, in *8th Plastic Fuel Pipe Symposium,* American Gas Association, 1983, p 40

7.37. D. Hull, *Fractography,* Cambridge University Press, 1999

CHAPTER 8

Mechanics of Fiber-Reinforced Composites

A COMPOSITE MATERIAL is a macroscopic combination of two or more distinct materials with a recognizable interface between them. The constituents retain their identities in the composite; that is, they do not dissolve or otherwise merge completely into each other, although they act in concert. They typically have a fiber or particle phase that is stiffer and stronger than the continuous matrix phase. Therefore, a given composite system is generally stiffer and stronger than the baseline/unreinforced matrix material. Because of the mechanical properties of the fibers, most composites maintain an elastic behavior in their stress-strain relationship.

Many types of composite systems have been developed. Detailed descriptions of the fiber and the matrix materials, including their physical and mechanical properties, are given in MIL-HDBK-17, Volume 21 of the, *ASM Handbook*, and other handbooks and reference books. This chapter focuses on fiber/matrix combinations currently used in industrial applications. The reinforcing fibers may be continuous or discontinuous, as shown in Fig. 8.1(a) and (b). The latter can be in the form of short fibers, whiskers, or particulates. Longitudinal fibers are used in both epoxy and metal-matrix composites. Whiskers and particulates are more often found in metal-matrix composites. The main advantage of discontinuously reinforced composites is that they can be fabricated using processing techniques similar to those commonly used for unreinforced-matrix materials, which makes them more cost effective. In addition, discontinuously reinforced composites have relatively more isotropic properties than continuously reinforced composites, due to the lower aspect ratio and more random orientation of the reinforcements.

Among the composite systems discussed in the next section, the widely used polymer-matrix composites (PMCs) have received the most attention. Compared to other types of composites, the PMC has been thoroughly studied both theoretically and experimentally. Analytical methods have been fully developed and documented, and numerous test data for this material are available in the literature. In this chapter, therefore, discussion is limited to those PMCs known as filamentary composites, which are made up of long continuous fibers (generally graphite or boron) embedded in a matrix of epoxy or thermoplastic. The stress and strain systems, failure mechanisms, and fatigue and fracture of these

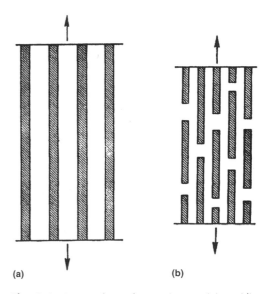

(a)   (b)

Fig. 8.1 Common forms of composites containing unidirectional high-strength/high-modulus fibers embedded in a softer matrix. (a) Straight, continuous fibers. (b) Discontinuous or chopped fibers

composites are covered here. Discussions will be kept in generic terms equally applicable to other composite systems whose laminae are reinforced with continuous fibers.

Advantages and Disadvantages. Filamentary composites offer many advantages compared to conventional materials such as metals. Their light weight provides a tremendous advantage in terms of strength-to-weight ratio. In terms of fracture mechanics, a broken fiber will limit the crack size to the fiber diameter, which is very small. Therefore, the chance of reaching the critical crack size is slim, even when a bundle of nearby fibers are cracked. An unbroken fiber adjacent to a cracked matrix is analogous to a skin-stringer structure, where the stringer (attached to the skin panel) helps to suppress crack opening of the skin.

However, PMCs are highly susceptible to moisture and temperature. Their mechanical properties consequently decrease, to various degrees, in such environments. The so-called hot-wet condition is most detrimental. This chapter touches only on mechanical properties and failure mechanisms, without considering environmental factors.

The ability to tailor a laminate, by adding plies in the directions most needed and leaving them out elsewhere, greatly contributes to the efficiency of advanced composite structures. Other than the ply properties themselves, the orientation and percentage of plies are the major design variables affecting the biaxial and shear strength of a laminate. Varying the combination of ply orientations and fiber volume affords a virtually limitless range of orthotropic properties for almost any conceivable application. Interlaminar strengths, however, are quite low, because they depend solely on the matrix and the quality of the bond between the matrix and the fibers.

carbon-carbon composites, or CCCs). The matrix material for the CCC is actually a polymer (a thermosetting resin), but is formed with extra carbonizing and densifying processing steps. The densifying process fills the open volume of the preform with a dense, well-bonded carbon-graphite matrix.

A composite laminate consists of a stack of individual plies that are selectively oriented and proportioned by the designer according to the directional strength or stiffness requirements of the structure. Figure 8.2 shows examples of layups. The laminates are built up from the matrix and fiber raw materials, and most commonly are fabricated by the layup of a tape containing both the fibers and epoxy matrix materials. The fibers have extremely small diameters: 0.14 and 0.2 mm (0.0056 and 0.008 in.) are the two common sizes for boron; graphite and glass fibers have much smaller diameters of 0.08 to 0.1 mm (0.003 to 0.004 in.). The amount of fiber material is normally specified by giving the fiber volume ratio, which is the ratio of the fiber volume to the total volume. For example, a 60% fiber volume is written as 60 V/O, or $V_f = 60$. Ply material comes in various forms; the most common is the 76 to 1219 mm (3 to 48 in.) wide unidirectional tape, with a thickness slightly greater than the fiber diameter. The fibers are placed parallel to one another, and closely spaced. The prepreg (preimpregnated) tapes can be stored in a refrigerator for future use. In making a composite laminate, the prepreg tapes are stacked together and cured in an autoclave under controlled time, temperature, and pressure.

Both thermoset and thermoplastic resins are used as the matrix for fiber-reinforced PMCs. The latter is tougher and highly resistant to delamination. Polymer-matrix composites are normally reinforced with continuous fibers of glass,

8.1 Types of Composites

There are three major classes of composites:

- Organic-matrix composites (OMCs)
- Metal-matrix composites (MMCs)
- Ceramic-matrix composites (CMCs)

8.1.1 Organic-Matrix Composites

Within the class of OMCs are PMCs and carbon-matrix composites (commonly known as

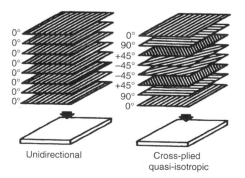

Fig. 8.2 Unidirectional versus quasi-isotropic layups

carbon, boron, aramid, or ceramic. The preforms may be a unidirectional prepreg tape or a multidirectional fabric (e.g., woven, weaved, braided, or knitted). In general, prepreg tapes made of thermoplastic resins have much longer shelf lives. The fiber-reinforced composite laminates also can be used to make sandwiches reinforced with lightweight honeycomb core or corrugated foil. The following is a short list of factors affecting PMC properties and allowables:

- *Volume fraction* (percent fiber to total volume): Generally in the range of 45 to 65%
- *Temperature:* High cure temperatures generally are indicative of higher operating temperatures; the higher the cure temperature, the higher the cost of materials and processing. Therefore, it is important to know the application before setting the cure temperature for cost savings. Temperature differentials in processing and in service will cause internal thermal stresses.
- *Moisture absorption:* The matrix and the reinforcement respond differently to moisture. Service operating temperature and moisture detrimentally affect PMC properties.
- *Layup and stacking sequence*
- *Manufacturing process control*
- *Testing and data reduction procedures:* These may not resemble the real situation.
- *Scaling effect:* Failure load prediction using material coupon test data is not compatible with actual failure load in full-scale built-up structures.

8.1.2 Metal-Matrix Composites

The matrix material for MMCs can be aluminum, titanium, or intermetallics. Reinforcements include:

- *Continuous fibers* made of aluminum oxide, silicon carbide, boron, or carbon
- *Discontinuous fibers,* mainly in the form of particulates; whiskers, chopped fibers, and platelets also are used
- *Particulate reinforcements,* such as SiC, B_4C, Al_2O_3 (for casting applications), and TiC (for high-temperature applications)

8.1.3 Ceramic-Matrix Composites

Ceramic-matrix composites are designed to offer improved damage tolerance or increased toughness through the addition of second-phase reinforcements in the form of fibers, particulates, or whiskers. Among these, the continuous fiber has the highest potential for improving tensile stress-strain behavior and damage tolerance. Significantly improved tensile strains are obtained via the combination of SiC fibers with a SiC matrix. The engineering strain in this CMC system is approximately 3%, compared to less than 0.2% for the monolithic ceramic. Since the tensile strength in this CMC system is kept approximately the same as in the monolithic ceramic, significantly improved fracture toughness also results (Ref 8.1).

A typical CMC is the carbon-fiber-reinforced silicon carbide (C/SiC) used for turbine disks in rocket engines. Particulate reinforcement materials include ZrO_2 and SiC. The ZrO_2 reinforcement, when added in the correct amount and particle size, makes the material more flaw tolerant. The particulates act to deflect cracks from the main propagation path and can absorb energy by a transformation mechanism as well.

Whisker-toughened ceramics can be used for a variety of applications, such as heat exchangers, tool bits, and other aerospace components. Carbon-carbon, which is not the same as the CCC subset of polymer-matrix composites, is another category of CMCs. These composites can withstand temperatures up to 600 °C (1110 °F); they must be protected by coatings or surface sealants when subjected to oxidizing atmospheres. Carbon-carbon composites are used primarily in the military, space, and aircraft industries. Applications include rocket engine nozzles and the tiles in the wing leading edge of the space shuttle orbiter.

8.2 Coding System

In composites literature, a code notation is used to designate the stacking sequence of the laminate plies and the angular direction of the fibers in each ply with respect to the laminate axis. This coding system, while primarily used for fiber-reinforced PMCs, is also used in other composite systems when applicable. Figure 8.3 shows the numbering system for identifying the stacking sequence of plies in a laminate. The ply numbers start from 1 for the top ply and end as the *k*th ply at the bottom. The *z*-directional coordinates are also identified.

In each lamina, there exist two orthogonal planes of symmetry that affect the material prop-

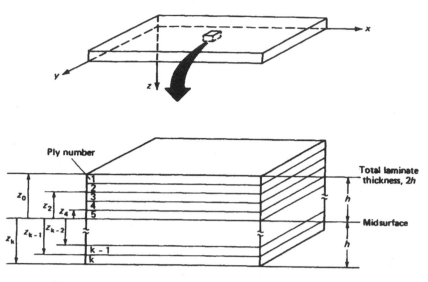

Fig. 8.3 Laminate numbering system

erties in the lamina plane. One of these planes is parallel to the fibers and the other is transverse; they are named the "1" and "2" axes, respectively (Fig. 8.4), and together are referred to as the material axis system. Each lamina in a laminate can be arbitrarily oriented with respect to a reference axis system, commonly represented by the x-y axes. Therefore, the angle representation of a lamina actually refers to the angle between axis 1 of the lamina and axis x of the laminate, which is normally designated as the direction of the applied load. When the reference axes are taken to coincide with the material axes, it is called the on-axis representation (Fig. 8.5). Otherwise, it is called the off-axis representation. For example, the lamina may be rotated at an angle θ with respect to the x-y axes,

where the x-axis is the direction of loading (Fig. 8.6).

Figure 8.2 shows an eight-ply laminate that consists of two 0° plies, two 90° plies, and four 45° plies, in the order of $[0/90/45/-45/-45/45/90/0]_T$. The coding sequence starts at the most positive z lamina, which is the top layer of the laminate. The subscript T indicates that the total laminate is shown. For laminates with midplane symmetry, such as the current example, we can simplify the code as $[0/90/45/-45]_S$. The subscript S is used to indicate that only half the laminate is shown. If the symmetric laminate has an odd number of laminae, the code denoting the center lamina is overlined (i.e., has an overbar marked on top of the orientation), indicating that half of the laminate lies on either side of that lamina. For instance, if we insert a 90° ply to the midplane of the current example, the code becomes $[0/90/45/-45/\overline{90}]_S$. Furthermore, the

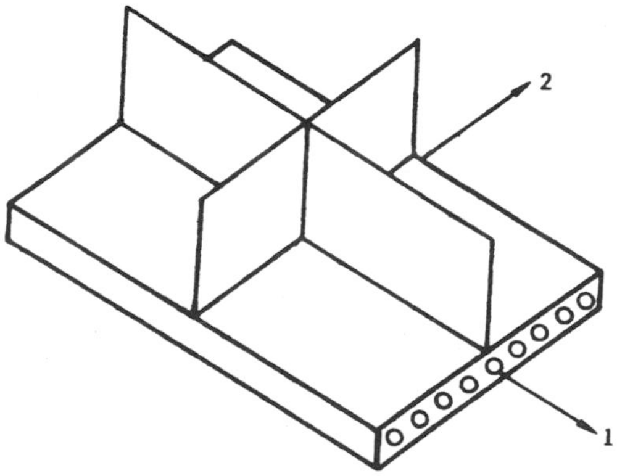

Fig. 8.4 Planes of material symmetry for a lamina

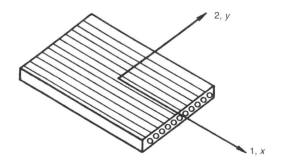

Fig. 8.5 On-axis representation of a lamina

notation for the 45° plies can be simplified as [0/90/±45]$_S$, or [0/90/±45/$\overline{90}$]$_S$, whichever the case might be. In the literature, the subscript T (for total) is sometimes omitted. However, the subscript S is a must for the symmetric laminate.

Numerical subscripts can be used to indicate repeated stacking of plies of the same orientation. For example, if we insert a 90° ply in between the top two plies of the laminate shown Fig. 8.2, and also add another 0° ply to the bottom of the same laminate, the code for this stacking layup is then written as [0/90$_2$/45/−45/−45/45/90/0$_2$]$_T$ or [0/90$_2$/45/−45$_2$/45/90/0$_2$]$_T$. If the ply sequence is unimportant and the code is used for total ply count only, these codes can be condensed to [0$_3$/90$_3$/±45$_2$]. However, this coding method (for total ply counting) can be easily confused with the rest of the codes unless the intent is clearly noted.

To illustrate how to take full advantage of the numerical subscript system, consider again the example given in Fig. 8.2: [0/90/±45]$_S$. If the layups on each side about the midplane are repeated—that is, the laminate is stacked in such order as [0/90/45/−45/0/90/45/−45/−45/45/90/0/−45/45/90/0]$_T$, we can write them as [0/90/±45]$_{2S}$, or [(0/90/±45)$_2$]$_S$.

In addition to the stacking sequence, a laminate is identified by the general content of the ply orientations. For example, the laminate [(0/90/±45)$_2$]$_S$ has 16 plies: four 0° plies, eight 45° plies, and four 90° plies. This laminate is designated as 25/50/25, meaning that there are 25% 0° ply, 50% 45° ply, and 25% 90° ply. A 36/55/9 layup can have a stacking sequence of [±45/0$_2$/±45/0$_2$/±45/90]$_S$, [±45/0$_2$/±45/0/±45/90/0]$_S$, or some other combination.

8.3 Stresses and Strains in Composite Laminates

The mechanical properties of composites can be tailored to meet specific engineering requirements by selecting a particular reinforcement and varying the amount added to the matrix. Increasing the reinforcement volume in a composite system generally increases elastic modulus, ultimate strength, and yield strength, but reduces ductility, fracture toughness, thermal expansion, and, in some cases, the density of the composite system. This section presents the stress-strain relations for a unidirectional lamina and the constitutive equations for the laminate.

8.3.1 Lamina Stress-Strain Relations

On-Axis Stress-Strain Relations. Consider the coordinate system of the lamina shown in Fig. 8.5. The lamina is subjected to generalized in-plane loads. Because the lamina is very thin, a plane-stress condition can be assumed. This assumption simplifies the analysis by reducing the number of nonzero stress components to three (the two normal stresses in directions perpendicular to the thickness direction and the in-plane shear stress). Consequently, only the three corresponding strains are of importance. That is, when the lamina is subjected to in-plane loads in both x and y directions as well as in-plane shear, we can derive the on-axis stress-strain relations for each individual case and add them together.

Uniaxial Longitudinal Loading. A normal load applied in the 1-direction (parallel to the fibers) causes strain in the 1 and 2 directions. There is no shear strain because of symmetry. Therefore:

$$\varepsilon_1 = \sigma_1/E_1 \qquad \text{(Eq 8.1a)}$$

$$\varepsilon_2 = -\nu_{12} \cdot \sigma_1/E_1 = -\nu_{12} \cdot \varepsilon_1 \qquad \text{(Eq 8.1b)}$$

Here E_1 is the longitudinal Young's modulus and ν_{12} is the longitudinal Poisson's ratio; their values are determined by the "rule of mixtures." That is:

$$E_1 = V_m \cdot E_m + V_f \cdot E_f \qquad \text{(Eq 8.1c)}$$

or

$$E_1 = (1 - V_f) \cdot E_m + V_f \cdot E_f \qquad \text{(Eq 8.1d)}$$

and

$$\nu_{12} = V_m \cdot \nu_m + V_f \cdot \nu_f \qquad \text{(Eq 8.1e)}$$

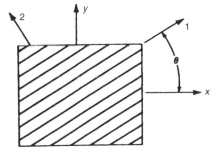

Fig. 8.6 Coordinate systems: 1, 2, principal material coordinates; x, y, laminate or arbitrary coordinates

Figure 8.7 shows the rule of mixtures prediction of composite stiffness for various ratios of fiber to matrix stiffness. Both MMCs and epoxy-matrix composites show good correlation with this analysis.

Uniaxial Transverse Loading. A normal load applied in the 2-direction (perpendicular to the fibers) causes strains only in the 1 and 2 directions. The stress-strain relations for this case can be written as:

$$\varepsilon_2 = \sigma_2/E_2 \qquad \text{(Eq 8.2a)}$$

$$\varepsilon_1 = -\nu_{21} \cdot \sigma_2/E_2 = -\nu_{21} \cdot \varepsilon_2 \qquad \text{(Eq 8.2b)}$$

where E_2 is the Young's modulus in the 2-direction. On the basis of the rule of mixtures, we have:

$$E_2 = \frac{E_f \cdot E_m}{V_m \cdot E_f + V_f \cdot E_m} \qquad \text{(Eq 8.2c)}$$

or

$$\frac{E_2}{E_m} = \frac{1}{V_m + V_f \cdot (E_m/E_f)} \qquad \text{(Eq 8.2d)}$$

The transverse Poisson's ratio ν_{21} can be determined by:

$$\nu_{12}/E_1 = \nu_{21}/E_2 \qquad \text{(Eq 8.3)}$$

Thus, the total number of independent elastic constants is reduced by one.

Longitudinal Shear Loading. When a shear stress is applied in the 1–2 plane, the only non-zero strain of interest is the shear strain ε_{12}. It can be written as:

$$\varepsilon_{12} = \sigma_{12}/G_{12} \qquad \text{(Eq 8.4a)}$$

where G_{12}, by definition, is the longitudinal shear modulus. Again, its value is obtained by using the rule of mixtures. That is:

$$G_{12} = \frac{G_f \cdot G_m}{V_m \cdot G_f + V_f \cdot G_m} \qquad \text{(Eq 8.4b)}$$

or

$$\frac{G_{12}}{G_m} = \frac{1}{V_m + V_f \cdot (G_m/G_f)} \qquad \text{(Eq 8.4c)}$$

By applying the principle of superposition, the contribution of each stress component to the strain component can be summed. The resulting equations are:

$$\varepsilon_1 = \sigma_1/E_1 - \nu_{21} \cdot \sigma_2/E_2 \qquad \text{(Eq 8.5a)}$$

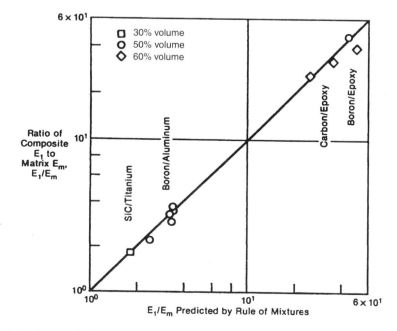

Fig. 8.7 Correlation of composite stiffness by the rule of mixtures. Source: Ref 8.2

$$\varepsilon_2 = -v_{21} \cdot \sigma_1/E_1 + \sigma_2/E_2 \qquad \text{(Eq 8.5b)}$$

$$\varepsilon_{12} = \sigma_{12}/G_{12} \qquad \text{(Eq 8.5c)}$$

Through Eq 8.3, the total number of independent elastic constants is reduced by one. Thus, this set of equations can be written in matrix form as:

$$\begin{Bmatrix} \varepsilon_1 \\ \varepsilon_2 \\ \varepsilon_{12} \end{Bmatrix} = \begin{bmatrix} S_{11} & S_{12} & 0 \\ S_{12} & S_{22} & 0 \\ 0 & 0 & S_{66} \end{bmatrix} \begin{Bmatrix} \sigma_1 \\ \sigma_2 \\ \sigma_{12} \end{Bmatrix} = [S] \begin{Bmatrix} \sigma_1 \\ \sigma_2 \\ \sigma_{12} \end{Bmatrix}$$
$$\text{(Eq 8.6)}$$

where **S** is called the lamina compliance matrix, and

$$S_{11} = 1/E_1 \qquad \text{(Eq 8.7a)}$$

$$S_{22} = 1/E_2 \qquad \text{(Eq 8.7b)}$$

$$S_{12} = -v_{12}/E_1 \qquad \text{(Eq 8.7c)}$$

$$S_{66} = 1/G_{12} \qquad \text{(Eq 8.7d)}$$

$$S_{16} = S_{26} = 0 \qquad \text{(Eq 8.7e)}$$

The 66 subscript is used commonly rather than 33 to avoid confusion with the "3" or thickness direction.

The inverse of the **S** matrix is called stiffness matrix, and is designated by **Q**. That is,

$$[Q] = [S]^{-1} \qquad \text{(Eq 8.8)}$$

and

$$\begin{Bmatrix} \sigma_1 \\ \sigma_2 \\ \sigma_{12} \end{Bmatrix} = \begin{bmatrix} Q_{11} & Q_{12} & 0 \\ Q_{12} & Q_{22} & 0 \\ 0 & 0 & Q_{66} \end{bmatrix} \begin{Bmatrix} \varepsilon_1 \\ \varepsilon_2 \\ \varepsilon_{12} \end{Bmatrix} \qquad \text{(Eq 8.9)}$$

with

$$Q_{11} = \frac{E_1}{1 - (v_{12}^2 E_2/E_1)} \qquad \text{(Eq 8.10a)}$$

$$Q_{22} = \frac{E_2}{1 - (v_{12}^2 E_2/E_1)} \qquad \text{(Eq 8.10b)}$$

$$Q_{12} = \frac{v_{12}E_2}{1 - (v_{12}^2 E_2/E_1)} \qquad \text{(Eq 8.10c)}$$

$$Q_{66} = G_{12} \qquad \text{(Eq 8.10d)}$$

$$Q_{16} = G_{26} = 0 \qquad \text{(Eq 8.10e)}$$

Off-Axis Stress-Strain Relations. Normally the principal axes of the lamina do not coincide with the reference x and y axes for the laminate. In that case, the constitutive relationship must be transformed through an angle θ to the laminate axes x and y. Figures 8.6 and 8.8 show the two coordinate systems for transformation of stresses. In these figures, θ is positive for a counterclockwise rotation. Because we usually know

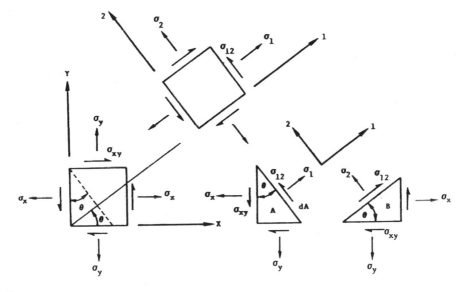

Fig. 8.8 Stress coordinate rotation

the material properties with respect to the 1–2 axes and wish to determine them with respect to the plate axes, the transformation of stresses from the 1-2 system to the x-y system is expressed in matrix form as:

$$\begin{Bmatrix} \sigma_x \\ \sigma_y \\ \sigma_{xy} \end{Bmatrix} = \begin{bmatrix} m^2 & n^2 & -2mn \\ n^2 & m^2 & 2mn \\ mn & -mn & m^2 - n^2 \end{bmatrix} \begin{Bmatrix} \sigma_1 \\ \sigma_2 \\ \sigma_{12} \end{Bmatrix}$$

$$= [\theta] \begin{Bmatrix} \sigma_1 \\ \sigma_2 \\ \sigma_{12} \end{Bmatrix}$$

(Eq 8.11a)

where $m = \cos\theta$ and $n = \sin\theta$. The matrix θ is called the transformation matrix. Similarly, we obtain the transformation matrix for the strains as follows:

$$\begin{Bmatrix} \varepsilon_x \\ \varepsilon_y \\ 2\varepsilon_{xy} \end{Bmatrix} = \begin{bmatrix} m^2 & n^2 & -mn \\ n^2 & m^2 & mn \\ 2mn & -2mn & m^2 - n^2 \end{bmatrix} \begin{Bmatrix} \varepsilon_1 \\ \varepsilon_2 \\ 2\varepsilon_{12} \end{Bmatrix}$$

$$= [\psi] \begin{Bmatrix} \varepsilon_1 \\ \varepsilon_2 \\ 2\varepsilon_{12} \end{Bmatrix}$$

(Eq 8.11b)

The off-axis stiffness matrix is then obtained by substituting Eq 8.11(a) and (b) into Eq 8.9:

$$\begin{Bmatrix} \sigma_x \\ \sigma_y \\ \sigma_{xy} \end{Bmatrix} = \begin{bmatrix} \overline{Q}_{11} & \overline{Q}_{12} & \overline{Q}_{16} \\ \overline{Q}_{12} & \overline{Q}_{22} & \overline{Q}_{26} \\ \overline{Q}_{16} & \overline{Q}_{26} & \overline{Q}_{66} \end{bmatrix} \begin{Bmatrix} \varepsilon_x \\ \varepsilon_y \\ \varepsilon_{xy} \end{Bmatrix}$$

(Eq 8.12)

where

$$\overline{Q}_{11} = Q_{11}m^4 + 2(Q_{12} + 2Q_{66})m^2n^2 + Q_{22}n^4$$
(Eq 8.13a)

$$\overline{Q}_{22} = Q_{11}n^4 + 2(Q_{12} + 2Q_{66})m^2n^2 + Q_{22}m^4$$
(Eq 8.13b)

$$\overline{Q}_{12} = (Q_{11} + Q_{22} - 4Q_{66})m^2n^2 + Q_{12}(m^4 + n^4)$$
(Eq 8.13c)

$$\overline{Q}_{66} = (Q_{11} + Q_{22} - 2Q_{12} - 2Q_{66})m^2n^2 + Q_{66}(m^4 + n^4)$$
(Eq 8.13d)

$$\overline{Q}_{16} = (Q_{11} - Q_{12} - 2Q_{66})m^3n + (Q_{12} - Q_{22} + 2Q_{66})mn^3$$
(Eq 8.13e)

$$\overline{Q}_{26} = (Q_{11} - Q_{12} - 2Q_{66})mn^3 + (Q_{12} - Q_{22} + 2Q_{66})m^3n$$
(Eq 8.13f)

$$\overline{Q}_{21} = \overline{Q}_{12}$$
(Eq 8.13g)

$$\overline{Q}_{61} = \overline{Q}_{16}$$
(Eq 8.13h)

$$\overline{Q}_{62} = \overline{Q}_{26}$$
(Eq 8.13i)

In the material axes system 1-2, the lamina is said to be "specially" orthotropic; that is, Q_{16} and Q_{26} equal zero. In the general x-y system the lamina is said to be "generally" orthotropic, and the $\overline{Q}$ matrix has all the nonzero terms as indicated in Eq 8.12. However, the six $\overline{Q}_{ij}$ terms function as only four independent constants. The terms $\overline{Q}_{16}$ and $\overline{Q}_{26}$ are merely linear combinations of the first four Qs.

Figure 8.9 provides an example showing the variation of the $\overline{Q}_{ij}$ constants with θ for a lamina. The numerical values were computed for a boron epoxy, where the engineering constants are $E_{11} = 30.0 \times 10^6$ psi, $E_{22} = 3.0 \times 10^6$ psi, $G_{12} = 1.0 \times 10^6$ psi, and $v_{12} = 0.3$. Using Eq 8.10, the four values in the **Q** matrix have been computed. Their values are $Q_{11} = 30.3 \times 10^6$ psi, $Q_{22} = 3.03 \times 10^6$ psi, $Q_{12} = 0.91 \times 10^6$ psi, and $Q_{66} = 1.0 \times 10^6$ psi. Finally, Fig. 8.9 shows that the stiffness constants $\overline{Q}_{ij}$, computed using Eq 8.13, change significantly with the orientation angle.

The stiffness matrix of Eq 8.12 can be inverted to find the corresponding compliance matrix $\overline{S}_{ij}$ for an orthotropic plate. Thus,

$$[\overline{S}] = [\overline{Q}]^{-1}$$
(Eq 8.14)

and

$$\begin{Bmatrix} \varepsilon_x \\ \varepsilon_y \\ 2\varepsilon_{xy} \end{Bmatrix} = \begin{bmatrix} \overline{S}_{11} & \overline{S}_{12} & \overline{S}_{16} \\ \overline{S}_{12} & \overline{S}_{22} & \overline{S}_{26} \\ \overline{S}_{16} & \overline{S}_{26} & \overline{S}_{66} \end{bmatrix} \begin{Bmatrix} \sigma_x \\ \sigma_y \\ \sigma_{xy} \end{Bmatrix}$$

(Eq 8.15)

Now the elastic constants for the plate, which uses the arbitrary x-y coordinate system, can be obtained via Eq 8.12 or 8.15. Thus,

$$E_x = \frac{1}{\overline{S}_{11}} \left(or = \overline{Q}_{11} - \frac{\overline{Q}_{12}^2}{\overline{Q}_{22}} \right)$$
(Eq 8.16a)

$$E_y = \frac{1}{\overline{S}_{22}} \left(or = \overline{Q}_{22} - \frac{\overline{Q}_{12}^2}{\overline{Q}_{11}} \right)$$
(Eq 8.16b)

$$G_{xy} = \frac{1}{\overline{S}_{66}} (or = \overline{Q}_{66})$$
(Eq 8.16c)

$$v_{xy} = -\frac{\overline{S}_{12}}{\overline{S}_{11}} \left(or = \frac{\overline{Q}_{12}}{\overline{Q}_{22}} \right)$$
(Eq 8.16d)

8.3.2 Laminate Stress-Strain Relations

Given the location and orientation of all the plies, it is possible to compute the macroscopic elastic properties of a laminate from the elastic properties of its constituent plies. In general, the laminate exhibits coupling of extension and bending. This means that applied extension causes warping (bending) or, conversely, applied moment causes in-plane forces. In this chapter, discussion is limited to those laminates that possess midplane symmetry. Midplane symmetry means that for every ply above the laminate midplane, there is a ply having equal properties and orientated an equal distance below the midplane. The problem of bending-moment coupling does not exist in the midplane symmetric laminates. For further simplification, applied moments are not considered and thus moment-constitutive relations are not required. With these limitations, the average in-plane stresses for the laminate $\bar{\sigma}_x$, $\bar{\sigma}_y$, $\bar{\sigma}_{xy}$ are presented in the form of a matrix as:*

$$\begin{Bmatrix} \sigma_x \\ \sigma_y \\ \sigma_{xy} \end{Bmatrix} = \frac{1}{B} \begin{bmatrix} \bar{A}_{11} & \bar{A}_{12} & \bar{A}_{16} \\ \bar{A}_{12} & \bar{A}_{22} & \bar{A}_{26} \\ \bar{A}_{16} & \bar{A}_{26} & \bar{A}_{66} \end{bmatrix} \begin{Bmatrix} \varepsilon_x \\ \varepsilon_y \\ \varepsilon_{xy} \end{Bmatrix} \qquad \text{(Eq 8.17)}$$

where

$$A_{ij} = \sum_{i-1}^{k} (\bar{Q}_{ij})_i (z_i - z_{i-1}) \qquad \text{(Eq 8.18a)}$$

or

$$\bar{A}_{ij} = \sum_{i-1}^{k} (\bar{Q}_{ij})_i (z_i - z_{i-1})/B \qquad \text{(Eq 8.18b)}$$

Here $(Z_i - Z_{i-1})$ is the thickness of the ith ply, and B is the total thickness of the laminate. The bar denoting average has been dropped and the stresses, when applied to a laminate, are understood as average stresses. As before, the compliance matrix [a] can be found by taking the inverse of the stiffness matrix:

$$[a] = [C]^{-1} \qquad \text{(Eq 8.19)}$$

*Derivations and detailed discussions of the macromechanics analysis methods can be found in any of the textbooks on composites (e.g., Ref 8.3 to 8.8).

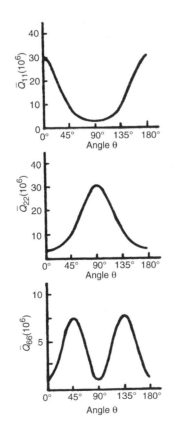

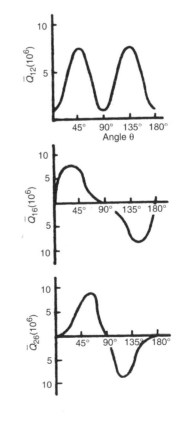

Fig. 8.9 Off-axis stiffness constants as functions of rotation angle θ. Source: Ref 8.3

where

$$[C] = \frac{1}{B} [A] \qquad \text{(Eq 8.20)}$$

so that

$$\begin{Bmatrix} \varepsilon_x \\ \varepsilon_y \\ \varepsilon_{xy} \end{Bmatrix} = \begin{bmatrix} \bar{a}_{11} & \bar{a}_{12} & \bar{a}_{16} \\ \bar{a}_{12} & \bar{a}_{22} & \bar{a}_{26} \\ \bar{a}_{16} & \bar{a}_{26} & \bar{a}_{66} \end{bmatrix} \begin{Bmatrix} \sigma_x \\ \sigma_y \\ \sigma_{xy} \end{Bmatrix} \qquad \text{(Eq 8.21)}$$

The A_{16}, A_{26}, a_{16}, and a_{26} terms in Eq 8.17 and 8.21 are called coupling terms because they couple the in-plane normal and shear behavior: An in-plane normal stress gives rise to an in-plane shear strain and vice versa. A laminate is called an orthotropic laminate when the values of these terms equal zero. This can be achieved by having only 0° and 90° plies (called a cross-ply laminate) or having a $+\theta°$ ply for every $-\theta°$ ply (called a balanced laminate). In this case, the elements a_{ij} of the compliance matrix are defined in terms of the laminate's engineering constants in a manner identical to Eq 8.7, which are derived for a single orthotropic lamina.

8.4 Failure Mechanisms of Composites

8.4.1 Static Strength

The static strength of a composite lamina takes the same formulation as Eq 8.3. However, the term for the matrix strength requires some modification. As depicted in Fig. 8.10, the fiber possesses a much higher modulus than the matrix. In addition, the fiber usually has higher fracture strength than the matrix, but a much lower maximum strain at fracture. It is reasonable to assume that the lamina fails at the maximum strain of the fiber. Consequently, the fracture strain of the composite lamina is limited to the fracture strain of the fiber. At that point, the stress in the matrix will have reached a level that is only a fraction of the matrix strength. Therefore, the rule of mixtures for the tensile strength of the composite lamina becomes:

$$\sigma_{max}^T = \sigma_{max}^f \cdot V_f + (\sigma^m)_{f\varepsilon=max} \cdot (1 - V_f) \qquad \text{(Eq 8.22)}$$

Here the superscript T stands for tension; m and V denote matrix and volume, respectively. Fiber is designated by f, as a subscript or a superscript.

In the case of compression, failure of the composite depends on the buckling strength of the fiber. The final expression is (Ref 8.5):

$$\sigma_{max}^C = 2 \left\{ V_f + (1 - V_f) \frac{E_m}{E_f} \right\} \sqrt{\frac{V_f E_m E_f}{3(1 - V_f)}} \qquad \text{(Eq 8.23)}$$

Here the superscript C stands for compression; the subscripts m and f denote matrix and fiber, respectively; V and E are volume and Young's modulus, respectively.

Tension Failure of a Laminate. A laminate contains a stack of on-axis and off-axis plies. Obviously, the 90° ply will fail first and the 0° ply will fail last. There will be a sequence of ply failures at different loads, culminating in ultimate laminate failure when all plies have failed. Thus, the ultimate load-carrying capacity of the laminate may be significantly higher than the failure load of the first or subsequent plies. In the analysis, the stiffness matrix for the laminate must be modified after each ply failure to reflect the effects of those failures. Computing laminate strength is therefore a very tedious process.

The following numerical example clearly demonstrates the procedure. The scenario considered by Halpin (Ref 8.3) involves a four-ply glass/epoxy laminate: [0/±45/90]. Halpin also conducted experimental tests; a comparison of predicted and measured results is shown in Fig. 8.11. Generally, the modulus of elasticity for each ply is computed. The maximum strain criterion is used to determine ply failure. The stresses corresponding to those strain levels are computed. The ply stiffness for the remaining plies (after each occurrence of ply failure) is computed by setting the stiffness of the failed ply to zero. The same routine is repeated to de-

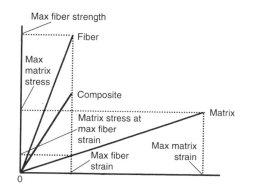

Fig. 8.10 Schematic composite stress-strain curve

termine failure of the next ply, and so on. The final failure stress is taken to be the sum of the incremental stresses. The computational scheme is as follows:

- Engineering properties of the lamina (given):

$$V_f = 0.5$$
$$E_{11} = 5.6 \times 10^6 \text{ psi}$$
$$E_{22} = 1.2 \times 10^6 \text{ psi}$$
$$G_{12} = 0.6 \times 10^6 \text{ psi}$$
$$\nu_{12} = 0.6$$
$$\nu_{21} = 0.6 \times 1.2 \times 10^6/5.6 \times 10^6 = 0.0557$$
$$\varepsilon_1 = 0.0275$$
$$-\varepsilon_1 = 0.0158$$
$$\varepsilon_2 = 0.0038$$
$$-\varepsilon_2 = 0.0142$$
$$\varepsilon_6 = 0.028$$
$$-\varepsilon_6 = 0.028$$

The ε_i terms are strain allowables of the lamina. Here, ε_1 is the ultimate tensile strain in the fiber direction; ε_2 is the ultimate tensile strain transverse to the fiber. The minus sign preceding the strain symbol means compression. The $\pm\varepsilon_6$ terms are the positive and negative shear strain allowables.

- On-axis plane stress moduli (for the 0° ply, computed using Eq 8.10); in this case, the symbol quantity $\overline{Q}_{ij}$ is identical to Q_{ij}:

$$Q_{11} = 5.69 \times 10^6 \text{ psi}$$
$$Q_{22} = 1.22 \times 10^6 \text{ psi}$$
$$Q_{12} = 0.317 \times 10^6 \text{ psi}$$
$$Q_{66} = 0.6 \times 10^6 \text{ psi}$$
$$Q_{16} = 0$$
$$Q_{26} = 0$$

- Off-axis plane stress moduli (for the 45° plies, computed using Eq 8.13):

$$\overline{Q}_{11} = 2.48 \times 10^6 \text{ psi}$$
$$\overline{Q}_{22} = 1.28 \times 10^6 \text{ psi}$$
$$\overline{Q}_{12} = 0.317 \times 10^6 \text{ psi}$$
$$\overline{Q}_{66} = 1.57 \times 10^6 \text{ psi}$$
$$\overline{Q}_{16} = \pm 1.12 \times 10^6 \text{ psi}$$
$$\overline{Q}_{26} = \pm 1.12 \times 10^6 \text{ psi}$$

- Off-axis plane stress moduli (for the 90° ply, computed using Eq 8.13):

$$\overline{Q}_{11} = 1.22 \times 10^6 \text{ psi}$$
$$\overline{Q}_{22} = 5.69 \times 10^6 \text{ psi}$$
$$\overline{Q}_{12} = 0.317 \times 10^6 \text{ psi}$$
$$\overline{Q}_{66} = 0.6 \times 10^6 \text{ psi}$$
$$\overline{Q}_{16} = 0$$
$$\overline{Q}_{26} = 0$$

- The stiffness of the laminate is obtained by summing the plane stress moduli through the thickness in proportion to the percentage of the thickness the ith ply occupying the n-ply laminate, and is done by using Eq 8.18(b). By treating each ply as one-fourth the thickness of the laminate (i.e., taking each ply as one out of a total of four), we have:

$$\overline{A}_{ij} = (\overline{A}_{ij}^0 + \overline{A}_{ij}^{45} + \overline{A}_{ij}^{-45} + \overline{A}_{ij}^{90})/4$$

(Eq 8.24a)

for the initial laminate (i.e., all four plies intact). After the 90° ply fails, Eq 8.24(a) becomes:

$$\overline{A}_{ij} = (\overline{A}_{ij}^0 + \overline{A}_{ij}^{45} + \overline{A}_{ij}^{-45})/4$$

(Eq 8.24b)

Then, after failure of the $\pm 45°$ plies:

$$\overline{A}_{ij} = \overline{A}_{ij}^0/4$$

(Eq 8.24c)

The results derived from these equations are as follows:

		Initial laminate	After 90° ply failed	After 45° plies failed
$\overline{A}_{11}$	=	2.96×10^6 psi	2.66×10^6 psi	1.42×10^6 psi
$\overline{A}_{22}$	=	2.96×10^6 psi	1.54×10^6 psi	0.302×10^6 psi
$\overline{A}_{12}$	=	0.8×10^6 psi	0.721×10^6 psi	0.0792×10^6 psi
$\overline{A}_{66}$	=	1.08×10^6 psi	0.931×10^6 psi	0.149×10^6 psi

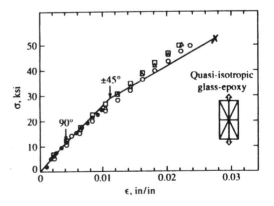

Fig. 8.11 Comparison of predicted and measured stress-strain response of [0/±45/90] Scotchply 1002 glass-epoxy laminate. Note that the curve has two "knees," at which the 90° and 45° plies fail. Source: Ref 8.3

- The overall engineering properties of the laminate are obtained from the laminate stiffness. They take the same form of expression as those in Eq 8.16. That is:

$$\overline{E}_x = \overline{A}_{11} - \frac{\overline{A}_{12}^2}{\overline{A}_{22}} \qquad \text{(Eq 8.25a)}$$

$$\overline{E}_y = \overline{A}_{22} - \frac{\overline{A}_{12}^2}{\overline{A}_{11}} \qquad \text{(Eq 8.25b)}$$

$$\overline{G}_{xy} = \overline{A}_{66} \qquad \text{(Eq 8.25c)}$$

$$\overline{\nu}_{xy} = \frac{\overline{A}_{12}}{\overline{A}_{22}} \qquad \text{(Eq 8.25d)}$$

We now obtain these elastic constants for the laminate:

		Initial laminate	After 90° ply failed	After 45° plies failed
$\overline{E}_x$	=	2.74×10^6 psi	2.32×10^6 psi	1.40×10^6 psi
$\overline{E}_y$	=	2.74×10^6 psi	1.512×10^6 psi	0.297×10^6 psi
$\overline{G}_{xy}$	=	1.08×10^6 psi	0.931×10^6 psi	0.1492×10^6 psi
$\overline{\nu}_{xy}$	=	0.27	0.468	0.262

- Taking the maximum strain as the ply failure criterion, we let each ply be loaded to a strain level equal to the strain allowable of that given lamina. In other words, we say that the 90° ply (which fails first) will fail at a strain level equal to ε_2. The 0° ply (which fails last) will fail at a strain level equal to ε_1. The maximum strain at failure of the 45° plies is determined by transformation of one of the on-axis strain allowables to the 45° angle. That is:

$$\overline{\varepsilon}_x = \overline{\varepsilon}_2/(\sin^2 \theta - \overline{\nu}_{xy} \cos^2 \theta) \qquad \text{(Eq 8.26)}$$

Therefore, the maximum strain at failure of each of these plies will be:

		90° ply	45° ply	0° ply
ε_x	=	0.0038	0.0141	0.0275

Next, we compute the incremental strain in between each ply failure, and the stress corresponding to each incremental strain:

		90° ply	±45° ply	0° ply
$\Delta\varepsilon_x$	=	0.0038	0.0103	0.0134
$\overline{E}_x$	=	2.74×10^6 psi	2.32×10^6 psi	1.40×10^6 psi
$\Delta\sigma_x$	=	10,412 psi	23,896 psi	18,760 psi

- The final fracture stress will be the sum of the three incremental failure stresses:

$$\sum \Delta\sigma_x = (10{,}412 + 23{,}896 + 18{,}760)$$
$$= 53{,}068 \text{ psi}$$

Static Strength of Notched Laminates. As stated in Chapters 1 and 2, stress concentration in a metal part may be reduced from its initial value to lower values as the applied load goes beyond the material's elastic limit. Clearly, the failure mechanism for composites will not be the same as those for metals, because elasticity is maintained in most (if not all) composite materials. A consensus has been reached: It is believed that static strength reduction due to the presence of an open hole is much more severe in composites. Figure 8.10 schematically illustrates such behavior. Several failure criteria for composite laminates with holes have been proposed. The two models proposed by Whitney and Nuismer (Ref 8.9) are by far the simplest to use and are described below.

Similar to metals, the stress distribution at the circular hole in an orthotropic plate also displays such a shape, as shown in Fig. 8.12. This curve can be approximated by a mathematical function (Ref 8.2, 8.9). The one given by Whitney and Nuismer (Ref 8.9) will be used because it correlates with the failure models that are described here. That is:

$$\sigma_\theta = \frac{S}{2}\left[2 + \frac{r^2}{X^2} + 3\frac{r^4}{X^4} - (K_{t\infty} - 3)\left(5\frac{r^6}{X^6} - 7\frac{r^8}{X^8}\right)\right] \qquad \text{(Eq 8.27)}$$

where S is the applied far-field stress, r is hole radius, and X is the distance along the x-axis starting from the center of the hole. Here $K_{t\infty}$ is

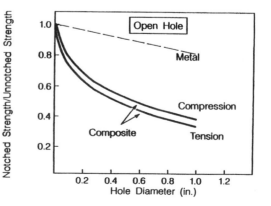

Fig. 8.12 Schematic notch sensitivity comparison of graphite/epoxy and aluminum to open holes

the stress-concentration factor at the edge of the hole, given by Lekhnitskii (Ref 8.10). That is:

$$K_{t\infty} = 1 + \sqrt{\frac{\overline{E}_x}{\overline{G}_{xy}} - 2\overline{v}_{xy} + 2(\overline{E}_x/\overline{E}_y)^{1/2}}$$

(Eq 8.28)

The parameters $\overline{E}_x$, $\overline{E}_y$, $\overline{G}_{xy}$, $\overline{v}_{xy}$ are the elastic constants of the laminate. The bar over the moduli denotes the effective elastic modulus of the orthotropic laminate, and the subscript x denotes the axis parallel to the applied load. Note that Eq 8.27 is for a circular hole inside a sheet of infinite width. For a finite-width sheet, the term $K_{t\infty}$ is replaced by K_t, whose value is:

$$K_t = K_{t\infty} \cdot \frac{2 + (1 - 2r/W)^3}{3(1 - 2r/W)}$$

(Eq 8.29)

Point Stress Criterion. The concept behind the models of Whitney and Nuismer is called "characteristic dimension." The first of the two models is referred to as the "point stress criterion." It assumes failure of a notched laminate to occur when at some fixed distance, d_0, ahead of the hole, σ_θ first reaches the tensile strength of the unnotched laminate of the same material (Fig. 8.13). That is, when:

$$\sigma_\theta(x,0)|_{x=r+d_0} = \sigma_0$$

(Eq 8.30)

where σ_0 is the tensile strength of the unnotched laminate. This means that S in Eq 8.27 will be the applied stress; its value at fracture of the notched plate satisfies Eq 8.30. In other words, S becomes the fracture stress of the notched laminate and σ_θ becomes the fracture strength of the unnotched laminate. We now designate the ratio of these two terms as σ_N^∞/σ_0. From Eq 8.27, we have:

$$\frac{\sigma_N^\infty}{\sigma_0} = \frac{2}{2 + \xi_1^2 + 3\xi_1^4 - (K_{t\infty} - 3)(5\xi_1^6 - 7\xi_1^8)}$$

(Eq 8.31)

where

$$\xi_1 = r/(r + d_0)$$

(Eq 8.32)

For a given composite material (which came with a set of elastic constants), there exist the lower and upper limits of the σ_N^∞/σ_0 ratios. For very large holes, $\xi_1 = \to 1$, Eq 8.31 gives $\sigma_N^\infty/\sigma_0 = 1/K_{t\infty}$. Thus, the classical stress concentra-

tion result is recovered. On the other hand, for hole sizes that are very small, $\xi_1 \to 0$, the ratio $\sigma_N^\infty/\sigma_0 \to 1$, as would be expected.

A sizable test program (116 tests in all) was carried out by Nuismer and Whitney (Ref 8.11) to evaluate this model (and the model discussed in the next paragraph). The specimens were made of T300/5208 graphite epoxy or Scotchply 1002 glass/epoxy. Each of these materials had a stacking sequence of either $[0/\pm45/90]_{2S}$ or $[0/90]_{4S}$. Two of their graphs comparing test data and predictions are shown in Fig. 8.14 and 8.15.

Average Stress Criterion. The second of the two models, referred to as the "average stress criterion," assumes that failure occurs when over some fixed distance, a_0, ahead of the hole, the average value of σ_θ first reaches the tensile strength of the laminate. That is, when:

$$\frac{1}{a_0} \int_r^{r+a_0} \sigma_\theta(x,0)dx = \sigma_0$$

(Eq 8.33)

Again, using this criterion with Eq 8.31 gives the ratio of the notched to unnotched strength as:

$$\frac{\sigma_N^\infty}{\sigma_0} = \frac{2(1 - \xi_2)}{2 - \xi_2^2 - \xi_2^4 + (K_{t\infty} - 3)(\xi_2^6 - \xi_2^8)}$$

(Eq 8.34)

where

$$\xi_2 = r/(r + a_0)$$

(Eq 8.35)

The expected σ_N^∞/σ_0 limits are again recovered for very small and very large holes. In practice,

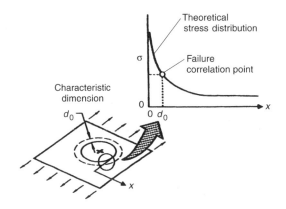

Fig. 8.13 Characteristic dimension d_0 and distribution of stresses at a circular hole

the quantities d_0 and a_0 are determined experimentally from strength reduction data. It has been hypothesized that these characteristic dimensions are material properties, independent of the hole dimension. The correlations shown in Fig. 8.14 and 8.15 seem to indicate that both d_0 and a_0 also are independent of material and lay-up configuration. After looking through all the data generated from the entire experimental program, Nuismer and Whitney concluded that neither of these models actually is superior to the other, and that the correlations for both are considered good. On the basis of their data, it seems that these two models are related and that:

$$a_0 \cong 4d_0 \qquad \text{(Eq 8.36)}$$

The fracture test data of Nuismer and Whitney show that residual strengths of the notched specimens depend on hole diameter. This is in qualitative agreement with the fracture test data developed by Waddoups et al. (Ref 8.12), which also exhibit dependence on hole diameter. In addition to circular holes, it has been shown that these models are equally applicable to cracks (Ref 8.11) and loaded holes (Ref 8.13). In Ref 8.11, the same d_0 and a_0 values were used for circular holes and cracks. In Ref 8.13, however, the pin-loaded specimens were made of the AS/

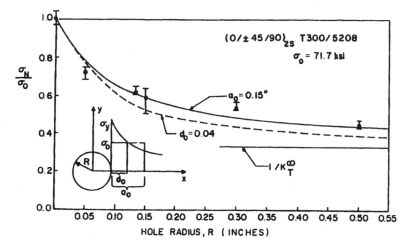

Fig. 8.14 Comparison of predicted and experimental failure stresses for circular holes in $[0/\pm45/90]_{2S}$ T300/5208 graphite/epoxy laminates. Source: Ref 8.11

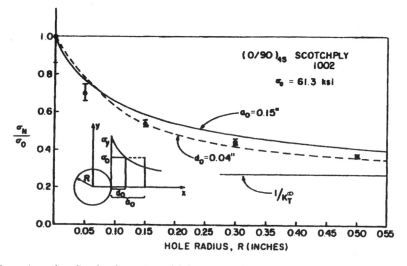

Fig. 8.15 Comparison of predicted and experimental failure stresses for circular holes in $[0/90]_{4S}$ Scotchply 1002 glass/epoxy laminates. Source: Ref 8.11

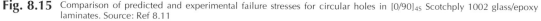

3501-6 graphite/epoxy (50/40/10) laminate; d_0 was set to 0.5 mm (0.02 in.).

8.4.2 Fatigue Behavior

The fatigue failure mechanism in composite laminates is quite different from that in metals. Fatigue damage in composites usually is extensive, occurring throughout the entire volume of the test specimen. That is in contrast to the localized, single-crack phenomenon normally observed in metals. The four basic failure mechanisms in composites are layer cracking, delam-

ination, fiber breakage, and fiber-matrix interfacial debonding. Any combination of these can cause fatigue damage that results in reduced strength and stiffness. Both the type and degree of damage vary widely, depending on material properties, lamination and stacking sequence, and type of fatigue loading. Therefore, the S-N behavior of a given composite laminate strongly depends on the constituent material properties, as well as its layup and stacking sequence. Most advanced fibers are very insensitive to fatigue, and the resulting composites show good fatigue resistance.

The direction of load relative to the fiber orientation of each individual lamina is also important. Figure 8.16 illustrates this effect on a graphite composite under tension-tension. Laminates with fiber-dominant failure modes exhibit better fatigue resistance than laminates with matrix-dominant failure modes (Ref 8.15). The slope of the S-N curve for the 0° ply laminate is relatively flat because of the fatigue insensitivity of the graphite fiber. Consequently, the slope of the S-N curve tends to become more negative as the content of the 0° ply decreases, because the angle plies are subjected to matrix cracking failures.

A different trend is observed when fatigue tests are conducted at negative R-ratios. Comparing the S-N curves in Fig. 8.16 and 8.17, the slopes of the multidirectional ply laminates for the negative R-ratios are less negative compared to those for the

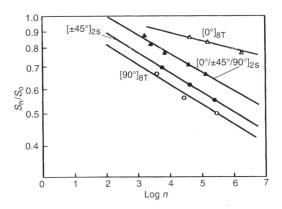

Fig. 8.16 S-N plot showing the effect of fiber orientation on fatigue performance of AS4/3502 graphite/epoxy laminates. The ordinate represents the fatigue strength ratio, which is the ratio of fatigue stress to static strength. Source: Ref 8.14

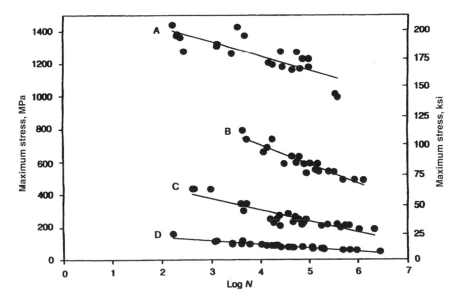

Fig. 8.17 S-N curves for T300/934 graphite/epoxy laminates, $R = -1$. A, unidirectional $[0]_{16}$; B, cross-ply $[0/90]_{4S}$; C, quasi-isotropic $[0/\pm45/90]_{2S}$; D, angle-ply $[\pm45]_{4S}$. Source: Ref 8.16

positive R-ratios. However, greater reduction in fatigue strength is seen in the cases of negative R-ratios. Another phenomenon worth noting is that the fatigue strength of multidirectional laminates is closely related to their tensile strengths. When the ratio of laminate fatigue strength to unidirectional fatigue strength is plotted against the ratio of laminate static strength to unidirectional static strength, a nearly one-to-one correlation is obtained (Fig. 8.18).

Unlike static strength, tension fatigue is not a threat to notched composite laminates. The test

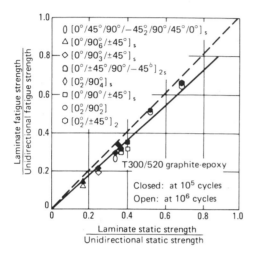

Fig. 8.18 Fatigue and static strength data for the T300/520 graphite/epoxy showing a one-to-one correlation between the ratio of laminate fatigue strength to unidirectional fatigue strength and the ratio of laminate static strength to unidirectional static strength. Source: Ref 8.17

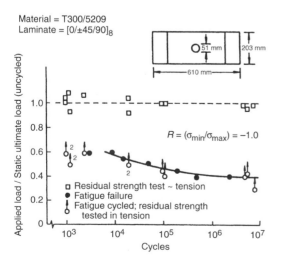

Fig. 8.19 S-N curve for T300/5209 graphite/epoxy laminates and residual strength of specimens after fatigue cycling. Source: Ref 8.18

program conducted by Walter et al. (Ref 8.18) revealed that tension fatigue strength is usually very high, only slightly lower than static strength. For $R = 0.05$ and maximum stress equal to 90% of the composite's ultimate tensile strength, the tests can run for 10^6 cycles without specimen failure. However, typical S-N fatigue behavior shows up in those tests having $R = -1$ (complete reverse) or $R = 10$ (compression-compression). A set of fatigue data generated at $R = -1$ is shown in Fig. 8.19. Thus, the compressive stress component of a fatigue spectrum is the dominating parameter in determining the fatigue life of composites with holes.

Figure 8.19 also shows that there will be no further degradation in static strength after prolonged fatigue cycling. Again, this observation agrees with the set of test data developed by Waddoups et al. (Ref 8.12), who showed that the residual strength of notched, fatigued laminates exceeded the static strength of those not fatigued.

8.4.3 Delamination

Three common types of cracks occur in composites: intralaminar, interlaminar, and translaminar (Fig. 8.20). The interlaminar (or interply) cracking mode has received the most attention. It is caused by out-of-plane loads that usually occur at structural edges (Fig. 8.21). The pri-

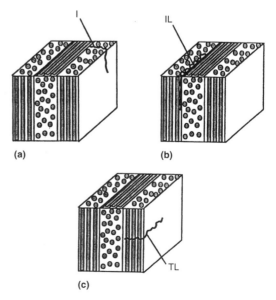

Fig. 8.20 Schematic illustrations of intralaminar (a), interlaminar (b), and translaminar (c) fracture paths in continuous-fiber/polymer composite laminates

mary reason for this type of delamination is that the assumption of plane stress used in laminate analysis is not 100% correct. Detailed stress analysis results have revealed that three-dimensional stress does exist, even for in-plane loading. Other than free-edge delamination (the first diagram in Fig. 8.21), eccentricities in structural load paths may include out-of-plane loads; discontinuities in the structure can also create out-of-plane loads. The rest of the diagrams in Fig. 8.21 include a curved edge such as a hole boundary, a drop-off of the interior laminate plies to taper thickness, a bonded joint, and a bolted joint. Low-velocity impacts also cause delamination, usually in the subsurface layers.

Free-edge delamination, which is a separation of the individual plies, normally occurs at the interface between the 90° ply and a ply of different orientation. Delamination causes stiffness loss, local strain concentration, and local instability, any of which can lead to structural failure. Stress analysis for out-of-plane interlaminar and interplane shear stress distributions across the laminate is quite complicated. The techniques are documented in the literature (Ref 8.19). Figures 8.22 and 8.23 show an example involving free-edge delamination of an 11-ply composite laminate under longitudinal in-plane load. The magnitude of the out-of-plane stress increases rapidly near the plate edge. This stress distribution corresponds to an interlaminar stress at the $-30/90$ interface, nearest to the midplane of the laminate, where σ_z at the edge is the highest. Though the actual magnitude of σ_z is not very high, it would be high enough to cause interply cracking. Once it does, rapid fracture follows

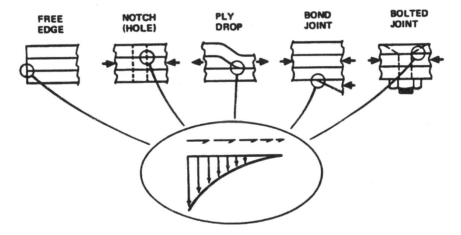

Fig. 8.21 Sources of out-of-plane loads in local areas in a structure

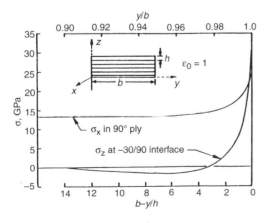

Fig. 8.22 Schematic cross-the-width stress distributions of σ_x and σ_z in a $[\pm 30/\pm 30/90/\overline{90}]_s$ laminate. Tensile load is applied along the x-axis.

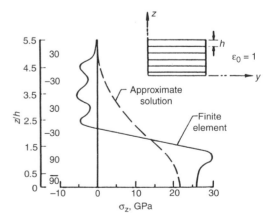

Fig. 8.23 Schematic through-thickness stress distributions of σ_z at the edge of a $[\pm 30/\pm 30/90/\overline{90}]_s$ laminate. Tensile load is applied along the x-axis.

because the interlaminar fracture toughness is extremely low for most resins. The example shown in Fig. 8.24 is typical of the T300/5208 class of graphite/epoxy. The figure plots critical stress-intensity factor K_{IC} as a function of the percentage of 0° plies in a balanced, symmetric laminate containing 0° and ±45° plies. The plot is self-explanatory, but the point is that in the interlaminar mode, toughness is always at the minimum; its value is almost nil.

8.5 Fracture Mechanics for Fibrous Composites

8.5.1 Stress-Intensity Factors

The crack-tip stresses in an infinite anisotropic plate exhibit crack-tip stress singularity similar to those in the isotropic plate (Ref 8.21). That is:

$$\begin{Bmatrix} \sigma_y \\ \sigma_x \\ \tau_{xy} \end{Bmatrix} = \frac{K_1}{\sqrt{2\pi r}} \{\mu_1, \mu_2, \theta\} \qquad \text{(Eq 8.37)}$$

Here μ_1 and μ_2 are the roots of the characteristic equation that contains the compliance coefficients for the plate; its stress-strain relationship has been presented as Eq 8.21. Omitting the details, it can be said that crack-tip stress distributions in isotropic and anisotropic plates are

analogous, as are the stress-intensity factors associated with them. The notion is that the general concept of stress-intensity factor is equally applicable for predicting fracture of isotropic, anisotropic, and filamentary composites (Ref 8.22, 8.23). However, the effect of finite width on anisotropic plates is not the same as for isotropic plates. A few examples follow.

Table 8.1 presents some of the data developed by Snyder and Cruse (Ref 8.24) for a center-cracked panel. The lamina properties are $E_{11} = 21,000$ ksi, $E_{22} = 1700$ ksi, $G_{12} = 1400$ ksi, and $\nu_{12} = 0.21$, which are representative of graphite/epoxy. The 0° fibers are oriented parallel to the loading direction, perpendicular to the crack. The stress-intensity factors for the composites are compared with their counterparts for the isotropic material. Also listed in Table

Table 8.1 Finite-width correction factors for center-cracked plates ($L/W = 3$)

Material	φ for 2a/W			H for 2a/W		
	0.4	0.6	0.8	0.4	0.6	0.8
Isotropic	1.102	1.302	1.642			
[0/±45/90]$_S$	1.107	1.296	1.781	1.004	0.995	1.085
[0]$_S$	1.115	1.295	1.755	1.012	0.995	1.069
[±30]$_S$	1.140	1.371	1.943	1.034	1.053	1.183
[±45]$_S$	1.157	1.406	2.029	1.050	1.080	1.235
[±60]$_S$	1.135	1.362	1.849	1.030	1.046	1.261
[90]$_S$	1.090	1.252	1.660	0.989	0.962	1.011

Note: φ is the width correction factor, defined by Eq 5.6; $H = $ φ (anisotropic)/φ (isotropic); graphite/epoxy ($E_{11} = 21,000$ ksi, $E_{22} = 1700$ ksi, $G_{12} = 1400$ ksi, $\nu_{12} = 0.21$). Source: Ref 8.24

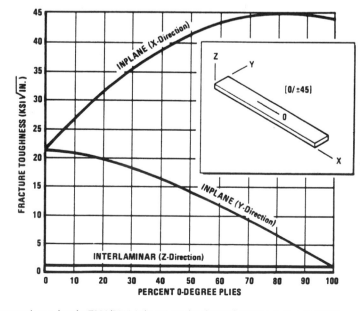

Fig. 8.24 Fracture toughness data for T300/5208 [0/±45] graphite/epoxy laminate. Source: Ref 8.20

8.1 is the anisotropy factor H, which is the ratio of the φ-factors for the two materials. Thus, the deviation of H from unity is a direct measure of the effect of material anisotropy on the stress-intensity factor. For an infinite sheet (i.e., $2a/W = 0$), the stress-intensity factors for all materials should be the same. Therefore, no comparisons are made here. Table 8.1 shows comparisons for three intermediate $2a/W$ ratios. The results for the quasi-isotropic and the unidirectional laminates agree well with the isotropic material, except for the long crack. Deviations from the norm become significantly high for laminates with angled plies, particularly for the ±45° laminates.

Table 8.2 represents another set of stress-intensity factors for a center-cracked panel. These data were developed by Konish (Ref 8.25). The material was T300/5208 graphite/epoxy; $E_{11} = 20,500$ ksi, $E_{22} = 1370$ ksi, $G_{12} = 752$ ksi, and $v_{12} = 0.31$. In this table, however, only the H factors are listed. The data in both Tables 8.1 and 8.2 show that the deviation in stress-intensity factor is highest for long cracks, thereby suggesting that the effect of anisotropy is strongest when the crack tip is near a free boundary.

Regarding cracks that emanate from a hole, comparisons between Bowie factors for isotropic and anisotropic materials are shown in Fig. 8.25, which includes results for the boron/epoxy and the graphite/epoxy. The Bowie factors are generally higher in the anisotropic materials, but merge with the values for the isotropic material at longer crack lengths. The exceptionally high Bowie factors shown for the 0.25 mm (0.01 in.) crack can probably be attributed to the so-called short-crack phenomenon mentioned earlier in Chapter 5.

8.5.2 Application to Delamination

In the case of metals, residual strength is a function of the critical crack size and is controlled by the fracture toughness (or critical stress-intensity factor) of the material. Unlike metals, delamination is the primary source of composite failure. Determining the axial strain at delamination onset is perhaps the most important aspect of applying fracture mechanics technology to composites. Fracture mechanics is useful in relating axial strain with crack extension force. Development of such technology has been carried out by T.K. O'Brien and his team since the 1980s (Ref 8.27–8.33). Highlights of the application of fracture mechanics to delamination are touched on below.

The interlaminar fracture toughness $\mathcal{G}_C$ of a composite laminate is the critical value of the strain energy release rate $\mathcal{G}$ required to grow a delamination. For composite laminates, $\mathcal{G}$ is related to the laminate thickness, t, the remote strain, ε, and the difference of the elastic modulus before and after delamination, but independent of the lamination size. That is:

$$\mathcal{G} = \frac{\varepsilon^2 t}{2}(E_{LAM} - E^*)$$

(Eq 8.38a)

Here, t is the total thickness of the laminate, ε is the axial strain, E_{LAM} is the original elastic

Table 8.2 Anisotropy factor in center-cracked tension angle-ply specimens of T300/5208 graphite/epoxy

Material	H for $2a/W$						
	0.1	0.2	0.3	0.4	0.5	0.6	0.7
$[0]_S$	1.000	0.997	0.991	0.984	0.975	0.964	0.952
$[\pm 10]_S$	1.001	1.000	0.997	0.994	0.990	0.985	0.979
$[\pm 15]_S$	1.002	1.002	1.003	1.003	1.004	1.004	1.005
$[\pm 20]_S$	1.003	1.005	1.010	1.016	1.023	1.030	1.037
$[\pm 25]_S$	1.004	1.010	1.020	1.033	1.047	1.061	1.074
$[\pm 30]_S$	1.005	1.016	1.032	1.052	1.073	1.094	1.113
$[\pm 35]_S$	1.007	1.021	1.043	1.070	1.098	1.125	1.145
$[\pm 45]_S$	1.007	1.025	1.053	1.087	1.120	1.152	1.164
$[\pm 55]_S$	1.002	1.016	1.037	1.062	1.087	1.107	1.149
$[\pm 60]_S$	0.999	1.008	1.023	1.040	1.058	1.080	1.115
$[\pm 65]_S$	0.996	1.002	1.010	1.020	1.033	1.050	1.073
$[\pm 70]_S$	0.995	0.997	1.001	1.006	1.012	1.021	1.034
$[\pm 75]_S$	0.995	0.995	0.995	0.995	0.996	0.997	1.002
$[\pm 80]_S$	0.996	0.995	0.992	0.988	0.984	0.980	0.976
$[90]_S$	0.998	0.995	0.990	0.983	0.974	0.964	0.952

Source: Ref 8.25

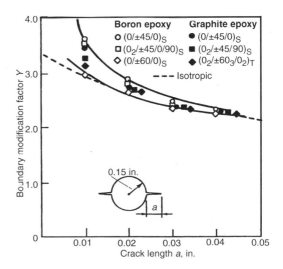

Fig. 8.25 Comparison of Bowie factors for anisotropic and isotropic materials. Source: Ref 8.26

modulus of the laminate, and E^* is the elastic modulus of the laminate after complete delamination along one or more interfaces. The values of both Es are computed using laminated plate theory and rule of mixtures. The term $(E_{LAM} - E^*)$ can be regarded as stiffness loss due to delamination. At the onset of delamination, the required axial strain is related to the fracture toughness of the laminate as:

$$\mathcal{G}_C = \frac{\varepsilon_C^2 t}{2}(E_{LAM} - E^*) \quad \text{(Eq 8.38b)}$$

or

$$\varepsilon_C = \sqrt{\frac{2G_C}{t(E_{LAM} - E^*)}} \quad \text{(Eq 8.38c)}$$

The modulus of the edge-delaminated laminate E^* is the sum of the modulus for each sublaminate. For the layup configuration in Fig. 8.22, i.e., $[\pm 30/\pm 30/90/\overline{90}]_S$, there will be three sublaminates: one for [90/90/90] and two for [$\pm 30/\pm 30$], because delamination takes place in between the two interfaces at $(-30/90)$. The E^* for this case is:

$$E^* = \frac{8E_{(\pm 30/\pm 30)_s} + 2nE_{90}}{8 + 2n} \quad \text{(Eq 8.39)}$$

Here, $n = 1.5$, standing for one and a half 90° plies on the laminate top and bottom.

It has been recognized that matrix cracking may occur in the 90° plies. Local delamination can be the result of growing the matrix crack(s) across the entire section of the 90° plies (Fig. 8.26). A similar equation has been derived to account for this situation:

$$\varepsilon_C = \frac{1}{E_{LAM}t} \sqrt{\frac{2mG_C}{\left(\frac{1}{E_{LD}t_{LD}} - \frac{1}{E_{LAM}t}\right)}} \quad \text{(Eq 8.40)}$$

where m is the number of delaminations growing from a matrix crack, E_{LD} is the modulus of a locally delaminated cross section, and t_{LD} is the thickness of a locally delaminated cross section. To be clear, if local delamination ever occurs in an 11-ply laminate, E_{LD} and t_{LD} are simply the modulus and thickness, respectively, of the remaining eight 30° plies. The value for m is simply 2; that is, the final delamination still takes place at the $-30/90$ interfaces, no matter how many 90° plies are in the middle.

O'Brien has investigated these models (Eq 8.38c and 8.40) using the test data of Crossman and Wang (Ref 8.34). The correlations are shown in Fig. 8.27. The test specimens were made of graphite/epoxy laminates, with a stacking sequence of $[\pm 25/90_n]_S$. The value for n is

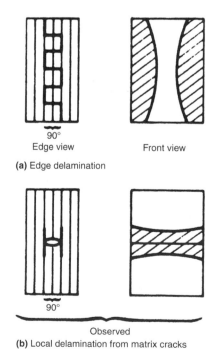

90°
Edge view   Front view

(a) Edge delamination

90°

Observed

(b) Local delamination from matrix cracks

Fig. 8.26 Two types of delamination in unnotched laminates. Only two 90° plies and four angle plies ($n = 4$, $t = 6$) are shown in these models. Source: Ref 8.28

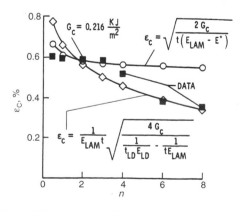

Fig. 8.27 Prediction of delamination onset in $-25/90$ interfaces of $[25/-25/90_n]_S$ laminates. Note that the G_C value was separately determined by O'Brien (for the same material and stacking arrangement). Source: Analysis, Ref 8.28; test data points, Ref 8.34

½, 1, 2, 3, 4, 6, and 8; that is, there will be one to 16 plies of the 90° laminae, sandwiched by two plies of the 45° laminae on each side. The lamina properties are $E_{11} = 19,500$ ksi, $E_{22} = 1480$ ksi, $G_{12} = 800$ ksi, and $v_{12} = 0.30$. The analytical/experimental correlations presented in Fig. 8.27 show that Eq 3.38(c) agrees well with test data up to $n = 4$; Eq 8.40 correlates well with test data for $n \geq 4$. It can be seen that matrix cracking caused delamination to occur at lower ε_C values, compared to specimens without local delamination. The reasons for this are discussed in Ref 8.28, but we will not elaborate here. Figure 8.27 also shows that the two equations of the O'Brien model cross each other at $n = 2$ instead of 3, or 4 (as indicated by the test data). O'Brien attributed this to the missing $\mathcal{G}_2$ term in these equations, because both tension and shear stresses are present at the interface of delamination. Nevertheless, the predicted ε_C values at the transitional region would be conservative.

Mode 1 is traditionally viewed as the only driving force for edge delamination. However, on the basis of published data, K_{2C} is always higher than K_{1C} in PMCs, whether the resin is brittle or toughened (Table 8.3). In the mixed-mode condition, the interaction envelope used in Chapter 4 is adopted. That is:

$$\left(\frac{K_1}{K_{1C}}\right)^{\xi} + \left(\frac{K_2}{K_{2C}}\right)^{\zeta} = 1 \qquad \text{(Eq 8.41)}$$

where ξ and ζ are empirical constants. Theoretically, $\xi = \zeta = 2$; its counterpart expressed in terms of $\mathcal{G}_1$ and $\mathcal{G}_2$ is:

$$\frac{\mathcal{G}_1}{\mathcal{G}_{1C}} + \frac{\mathcal{G}_2}{\mathcal{G}_{2C}} = 1 \qquad \text{(Eq 8.42)}$$

We know that K and $\mathcal{G}$ are interrelated, in a general form of Eq 4.1(c) and 4.1(d). However, the precise relationship between K and $\mathcal{G}$ for composites is not known. The analytical model will be complicated, because many directional E and v values are involved in a laminate that is associated with a unique stacking sequence. Some solutions for a generally anisotropic material, modeled as an orthotropic system, relating K_1 to $\mathcal{G}_1$ and K_2 to $\mathcal{G}_2$ are given in Ref 8.39. In Ref 8.22, Sih et al. compared the computed and the experimentally determined Young's moduli for the unidirectional graphite/epoxy and glass/epoxy. They also used another model that treats the entire composite body as an isotropic material and compared the computed Young's moduli with the same test data. It was found that the isotropic model (Eq 4.1f), which uses equivalent elastic constants for the entire composite body, provided the best correlation with specimens containing a small amount of glass fibers. The orthotropic model (Ref 8.39) provided good correlation with glass/epoxy having a fiber volume over 50%. Both models show poor correlation with graphite/epoxy test results.

Table 8.3 lists fracture toughness values of various composites by G_C or K_C, as provided in the literature. Mixed-mode interaction curves were constructed using data extracted from the same references listed in the table, and using the original units (G or K). The graphs are presented in Fig. 8.28 and 8.29. Mixed correlations are shown—some reasonably good, and some not so good. The contribution of K_2 to delamination and growth is not clear. Numerous tests reported in the literature have shown that delamination

Table 8.3 Mode 2 and mode 1 fracture toughness ratios of PMC materials(a)

Material	K_{2C}/K_{1C}	G_{2C}/G_{1C}	Ref
Scotchply 1022 glass/epoxy(b)	3.9	...	8.35
ERL 2256/0820 E-glass(b)	1.3(c)	...	8.36
Graphite/epoxy(b)	5.2	...	8.37
IM7/E7T1-2 graphite/epoxy(d)	...	10.7	8.32
C6000/H205 graphite/epoxy(d)	...	2.2(e)	8.31
Graphite/epoxy with F185 toughened resin(d)	...	1.4(e)	8.31
T300/BP907 graphite/epoxy(d)	...	6.7	8.29
T300/5028 graphite/epoxy(d)	...	6.6	8.29
Glass-textile/epoxy	...	4.9	8.38

(a) Except for data from Ref 8.37 and 8.38, the listed fracture toughness ratios are approximate (interpreted from graphs given in the literature). (b) Unidirectional laminates. (c) Extrapolated K_{2C} value used. (d) Quasi-isotropic laminates. (e) Extrapolated G_{2C} value used

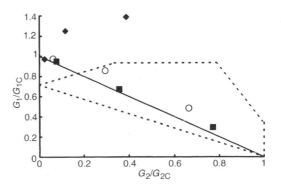

Fig. 8.28 Comparison of G_1 and G_2 interaction behavior for three quasi-isotropic PMC materials. ■, graphite/epoxy with F185 toughened resin; ○, C6000/H205 graphite/epoxy; ◆, IM7/E7T1-2 graphite/epoxy; dotted line enclosure shows scatter of 27 data points of glass-textile/epoxy; solid line represents theoretical value. Source: Ref 8.29, 8.31, 8.32, 8.38

growth is primarily controlled by K_1, though the magnitude of K_2 is actually greater than K_1 in that system. The crack growth rate (da/dN) is actually higher in mode 2 (as compared to mode 1 for a given K-value). As cited by O'Brien et al. and others, the exact role of mode 2 in delamination onset and growth is yet to be determined.

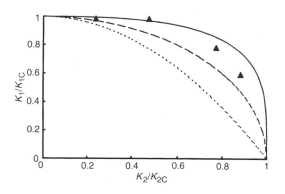

Fig. 8.29 Comparison of K_1 and K_2 interaction behavior for three unidiretional PMC materials. Dotted line, Scotchply 1002 glass/epoxy ($\xi = 1$, $\zeta = 2$); dashed line, theoretical value ($\xi = \zeta = 2$); solid line, graphite/epoxy ($\xi = 2$, $\zeta = 3$); ▲, ERL 2256/0820 E-glass. Source: Ref 8.35–8.37

8.6 Damage Tolerance of Composites

The major concern in the area of damage tolerance in composites is damage resulting from impact, and fatigue and fracture after impact. The type of damage resulting from impact on composites depends on the energy level of the impact. As schematically shown in Fig. 8.30, confined, through-thickness damage typically results from high-energy ballistic penetration. Lower-energy impact does not produce penetration, but leaves behind a large delamination zone inside the laminate. Low-energy impact damage, also known as low-velocity impact damage, is a major problem in structural applications of composites. This type of damage is subsurface, invisible most the time, and leaves a huge interior area of delaminated and cracked laminae. Typically, the damaged area is conical in shape, expanding toward the back face of the laminate. Figure 8.31 shows two photographs of the cross sections of graphite/epoxy and graphite/PEEK laminates. The damage closely resembles the sketches in Fig. 8.30. Both top surfaces after impact are quite smooth, barely visible, and would show no sign of excessive interior delamination

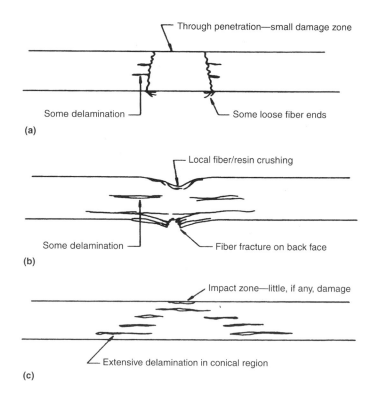

(a)

(b)

(c)

Fig. 8.30 Schematic failure modes of impact damage in composites. (a) High-energy impact damage. (b) Medium-energy impact damage. (c) Low-energy impact damage

if not sectioned or inspected by x-ray or other nondestructive inspection means. This type of damage can easily happen in a number of circumstances—for instance, a shop worker dropping a toolbox, or during on-site repair or walk-around inspection.

Different approaches are used to handle each type of impact damage. A hole resulting from ballistic penetration is treated as a crack, and thus fracture mechanics is applied. A number of Air Force reports written by J.G. Avery and his team in the 1970s and 1980s gave detailed accounts in this subject area. In the case of low-velocity impact, analysis involves modeling the damage area and determining the shape and size of the impact damage. A number of computer programs have been constructed or are under development. This is straightforward solid mechanics analysis, without fracture mechanics, and so far only moderate success has been achieved. Experimental work has emphasized nondestructive measurement of the damaged area, as well as determination of the effects of impact level on static and fatigue strengths, and static strength after fatigue cycling. Development of improved structural design concepts is

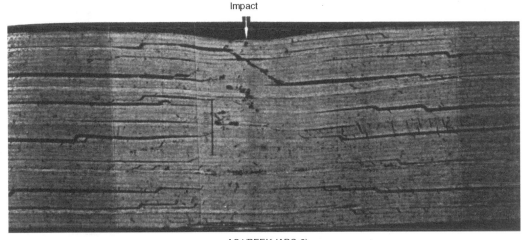

Nominal ply thickness
0.0056 in. (0.0142 cm)

AS4/PEEK (APC-2)
500 in.-lb (56.5J)

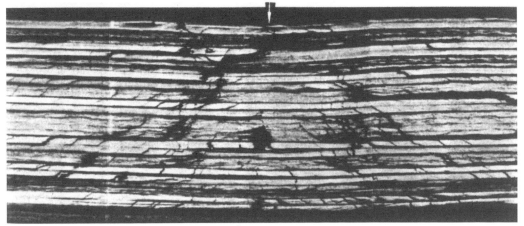

Nominal ply thickness
0.0073 in. (0.0185 cm)

AS6/2220-3
350 in.-lb (39.5J)

Fig. 8.31 Through-thickness damage comparison of impact in thermoplastic (top) and thermoset (bottom) laminates. Note the smooth surfaces after impact and the excessive delamination in both cases, and the large conical-shape internal damage in the thermoset laminate. Source: Ref 8.40

also high on the agenda. Some test results extracted from an Air Force report (Ref 8.40) are discussed in the following paragraphs.

At the laboratory test level, impact damage is produced by dropping a steel impactor onto the composite plate at a right angle. The impactor has a 13 mm (0.5 in.) diameter hemispherical tip and weights 4.5 kg (10 lb), per NASA specification (Ref 8.41).* For those data reported in Ref 8.40, a 9 kg (20 lb) impactor was used. A sketch of the test setup is shown in Fig. 8.32. The drop height is adjusted to obtain a desirable

velocity/energy level. Table 8.4 shows the static strength of virgin AS4/PEEK laminate with two types of layup configurations. These laminates were also subjected to two levels of low-velocity impact. Figures 8.33 and 8.34 show the damage area revealed by C-scan and the measured dent depth as a function of impact energy level. The

---

*The Air Force requirement for certification of military aircraft structures specifies use of a 25 mm (1 in.) hemispherical impactor. A detailed description of the Air Force damage tolerance criteria along with the probable damage mechanisms is given in Ref 8.42.

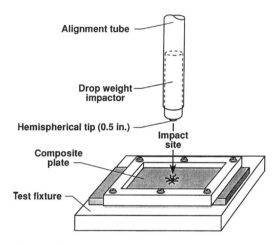

Fig. 8.32 Specimen impact apparatus

Table 8.4 Static compression strengths for AS4/PEEK (APC-2)

Comparison of virgin material post-impact, and post-impact followed by 1 million cycles of fatigue testing at $R = 10$

	Impact level		Maximum cyclic strain after	Failure stress	
Layup(a)	J	in. · lbf	impact	MPa	ksi
40/50/10	0	0	None	385/441	55.8/63.9
	28.2	250	None	286/306	41.4/44.4
	28.2	250	0.0031	329/362	47.7/52.5
	28.2	250	0.0036	368/387	53.3/56.1
	56.5	500	None	273/336	39.6/48.7
10/80/10	0	0	None	307/315	44.5/45.7
	28.2	250	None	295/314	42.7/45.5
	56.5	500	None	225/238	32.6/34.5

(a) All laminates are 48 ply; fiber orientations are 0°, ±45°, and 90°. The notation system indicates the percentage of these plies; for example, 40/50/10 means 40% 0° ply, 50% 45° ply, and 10% 90° ply. The stacking sequences are listed in Ref 8.40.

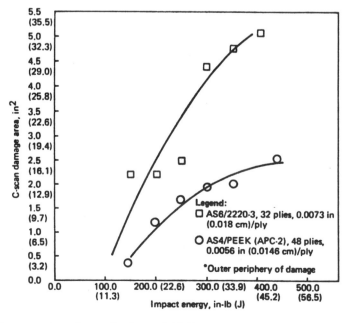

Fig. 8.33 Effect of impact energy on damage size. Source: Ref 8.40

extent of damage in the thermoplastic panels is considerably less than on the thermoset panels.

After impact, compression fatigue tests ($R = 10$) were conducted. Laminates that did not fail after 1 million cycles were tested again for static strength. Table 8.4 indicates that for two types of laminates tested, the 40/50/10 layup lost nearly 30% of its static strength after impact. However, the 10/80/10 layup was not affected by impact at the lower energy level (28.2 J, or 250 in. · lbf). Though the quantity of data is limited, it is evident that impact-damaged thermoplastic laminates are insensitive to fatigue loading. Post-impact compression strength and compression fatigue data for both the thermoplastic and the thermoset laminates are plotted in Fig. 8.35 and 8.36, respectively. Again, the 10/80/10 layup of the thermoplastic exhibits superior fatigue strength over the 40/50/10 layup of the same material. Judging from the graphs in Fig. 8.33 to 8.36, and the photographs in Fig. 8.31, it is clear that the thermoplastic had much better impact resistance and post-impact strength.

The impression that layup and stacking sequence can affect the post-impact compression strength of a composite laminate is also supported by test data. The data in Fig. 8.37 were obtained from a composite material made of 8551-7 toughened epoxy resin reinforced with IM7 fiber. It is evident that the layup and stacking sequence did play a big role in post-impact strength. Unfortunately, it is impossible to develop a trend, or a rationale, from these data to explain the behavior.

8.7 Some Practical Issues

Stress analysis of composite laminates is complicated, particularly for out-of-plane loading. The performance of a composite laminate depends on its fiber and matrix constituents, as well as fiber fraction, orientation, and stacking sequence. The layup of a composite laminate can be classified in four general categories:

- Unidirectional
- Cross-ply

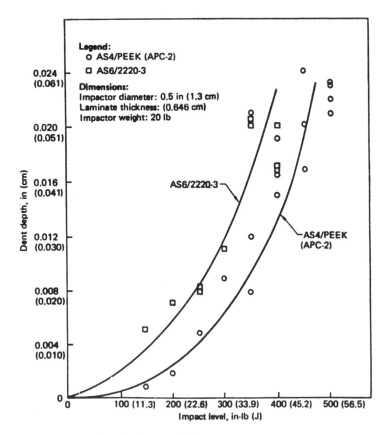

Fig. 8.34 Effect of impact energy on dent depth. Source: Ref 8.40

- Angled-ply
- Quasi-isotropic, which contains all the above

Cutouts in composites may not affect fatigue, but do affect static strength; they cause more static strength reduction than in metals. However, a hole in a specimen subjected to tension-tension poses no problem in terms of fatigue strength. This is exactly the opposite of metals. Delamination is a problem in every respect, whether induced by a third dimensional stress or by impact of a foreign object. We will review some of the methods being used to fight delamination. Scaling effect will also be discussed, along with the "building-block" approach for ensuring the performance of the full-scale structure.

8.7.1 Methods for Improving Out-of-Plane Delamination Resistance

Structural efficiency of laminated composites can be significantly degraded by delaminations. Therefore, an effective means must be attained to improve delamination resistance and mini-

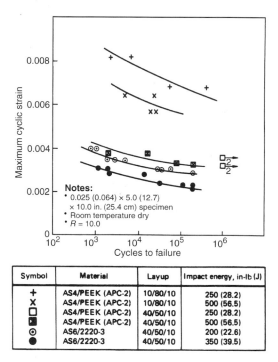

Symbol	Material	Layup	Impact energy, in.-lb (J)
+	AS4/PEEK (APC-2)	10/80/10	250 (28.2)
X	AS4/PEEK (APC-2)	10/80/10	500 (56.5)
□	AS4/PEEK (APC-2)	40/50/10	250 (28.2)
⊠	AS4/PEEK (APC-2)	40/50/10	500 (56.5)
⊙	AS6/2220-3	40/50/10	200 (22.6)
●	AS6/2220-3	40/50/10	350 (39.5)

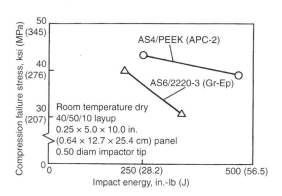

Fig. 8.35 Comparison of residual compression strength after impact for AS4/PEEK (APC-2) versus AS6/2220-3 graphite/epoxy panels. Source: Ref 8.40

Fig. 8.36 Comparison of compression fatigue response of AS4/PEEK (APC-2) with AS6/2220-3 grahite/epoxy after low-velocity impact. Source: Ref 8.40

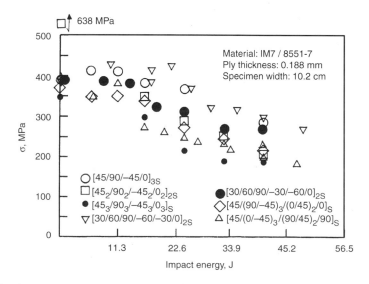

Fig. 8.37 Test data showing post-impact compression strength as a function of laminate stacking sequence. Source: Ref 8.43

mize delamination growth in a laminated composite system. High-strain fibers and high-toughness epoxy resins can be used to improve design strain levels. Another approach has been to incorporate through-thickness reinforcement to withstand out-of-plane loads using the potentially economical textile processes of stitching (Ref 8.44–8.49), weaving (Ref 8.50), or braiding (Ref 8.51, 8.52). Some three-dimensional configurations of this sort are shown in Fig. 8.38.

Following a brief discussion of material selection, we will present the results of a survey on current practices in three-dimensional stitch-

ing. This case history shows how an effective stitch configuration can be selected.

High-Toughness Resins. Using Eq 8.38(c) or Eq 8.40, the strain required for delamination initiation and growth will be higher for a material with a higher fracture toughness value. Therefore, a matrix that has a high fracture toughness value is an obvious choice to help a structure withstand a higher load before delamination. As shown in Fig. 8.39, standard epoxies have very low fracture toughness, whereas toughened epoxies have improved fracture toughness values. It is also clear that escalated improvements are obtained in the thermoplastics. The two ther-

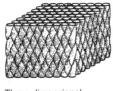
Three-dimensional braiding

Stitched fabric

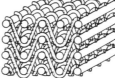
Multiharness woven cloth

Multiaxial multilayer warp knit

Fig. 8.38 Types of three-dimensional textiles

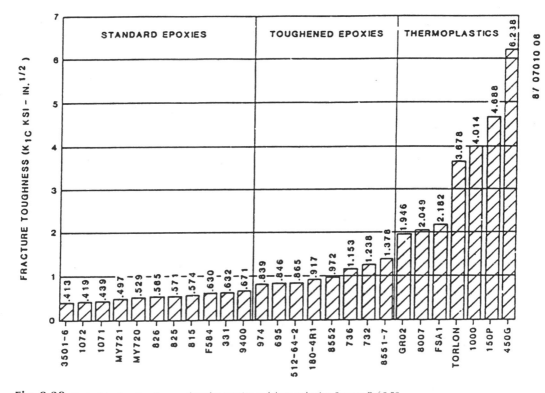

Fig. 8.39 Typical fracture toughness values for epoxies and thermoplastics. Source: Ref 8.53

moplastics (150P and 450G) that possess the highest fracture toughness values are part of the PEEK system.

Stitching has been used to minimize delamination, thereby improving impact resistance and damage tolerance. The following case history starts with a baseline composite system containing high-modulus fiber and standard epoxy. Some commonly used stitch patterns, thread materials, and gages are introduced. Then post-impact compression strengths are determined, the data analyzed, and the likely choices discussed. Note that only English units of measure are given.

Composite Systems Selection. The uniweave fabric made of 3K AS4 carbon fibers (tensile strength = 550 ksi, tensile modulus = 34 msi, ultimate elongation = 1.53%) and conventional 3501-6 epoxy resin was selected as the baseline material. The baseline composite system had 48 plies and was stacked in a $[45/0/-45/90]_{6S}$ sequence as the preform for a symmetric quasi-isotropic laminate. Compression test results (Ref 8.44–8.46) showed that this composite system, made of either uniweave cloth or prepreg tapes, exhibited an 80 ksi compression strength. This value will be used as a reference point in making

comparisons with stitched laminates of the same material, or laminates of other composite systems.

Both IM7 (tensile strength = 683 ksi, tensile modulus = 41 msi, ultimate elongation = 1.6%) and IM6 (tensile strength = 635 ksi, tensile modulus = 40 msi, ultimate elongation = 1.5%) carbon fibers were considered as candidates for mixing with toughened epoxy resins. The IM7/8551-7 tapes were also laid up into a 48-ply quasi-isotropic laminate with a $[45/0/-45/90]_{6S}$ stacking sequence. However, the IM6/1808I laminates had only 40 plies with a $[45/0/-45/90]_{5S}$ stacking sequence. The compression strengths for these materials were, respectively, 12 and 9% higher than the baseline AS4/3501–6 material.

Stitched Graphite/Epoxy Panel. The performance of a stitched graphite/epoxy panel is a function of many variables: the type and density of the stitches, the material and heaviness of the thread, and the method of impregnation. In addition to the conventional chain and lock stitches shown in Fig. 8.40 and 8.41, a variety of stitch patterns have been considered by investigators. Some of those configurations are covered in the listed references. However, most of the data compiled here pertains to modified lock-stitched laminates.

The advantage of using the modified lock stitch, rather than the standard lock stitch, is that the knots formed by the needle and bobbin threads are positioned on the outer surface of the stacked fabrics (instead of in the middle of the layers). This way, carbon fiber damage that might be caused by the stitches is minimized. On the other hand, chain stitching uses a single thread and requires access to only one surface. Figure 8.41 schematically illustrates the through-thickness, modified lock-stitch composite laminate.

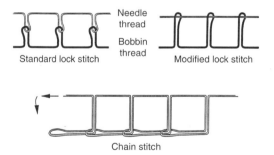

Fig. 8.40 Lock and chain stitch configurations

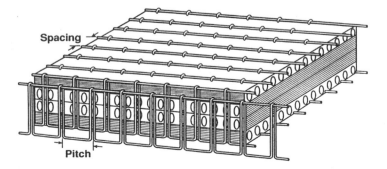

Fig. 8.41 Schematic illustration of modified lock-stitched composite laminate

The density of a given stitch configuration is defined by the stitch spacing and pitch. For example, a notation of ($\frac{1}{4}$ × $\frac{1}{4}$ × 8) indicates that the laminate has stitches in the form of a square array, with $\frac{1}{4}$ in. spacing in both the longitudinal and transverse directions. On each row of stitches, stitching was done at 8 penetrations per inch. If the stitches were made in one direction only, the notation for the same stitch pattern would be ($\frac{1}{4}$ × 8).

Most of the experimental data collected here involves through-thickness ($\frac{1}{8}$ × 8) modified lock stitches (parallel to the loading axis) made with S2 glass thread of various weight (ranging from 2400 to 6000 d). A limited number of test data points for other stitch patterns, or stitches made with Kevlar 29 thread (1000 or 1500 d), are also included.

Compression and Post-Impact Compression Strengths. All the compression and post-impact compression test data published in the literature (Ref 8.43–8.47) were developed using the standard specimens and experimental procedures of

NASA (Ref 8.41). Interpretation of these data is presented below.

Figure 8.42 shows the compression strength (S_C) and the compression after impact strength (S_{CAI}) for three types of composites: the baseline 48-ply quasi-isotropic AS4/3501-6 laminates, with and without stitches, and the IM7/8551-7 toughened composite. A baseline pattern of ($\frac{1}{8}$ × 8) modified lock, with 2400, 3000, 3600, 3678, or 6000 d S2 glass thread, was used. All test data points were normalized by 80 ksi, which was the compression strength (without impact) of the unstitched AS4/3501-6 material. It is evident that the toughened material (i.e., IM7/8551-7, unstitched) had greater S_C and S_{CAI} values than the baseline AS4/3501-6 material. The stitched materials had higher post-impact compression strengths than both the baseline and the toughened materials; the heavier the thread, the higher the residual compression strength. However, the laminates that were stitched with a heavier-gage thread (i.e., heavier than 3000 d) exhibited a reduction in zero-impact compression strength (S_C), probably due to the presence of an excessive amount of damaged carbon fibers. Thus, it is apparent that the 3000 d S2 glass thread provides optimized properties where the S_{CAI} values are reasonably high and an uncompromised zero-impact compression strength is also maintained. The post-impact compression properties for these three materials are represented by the fitted lines shown in Fig. 8.42.

Figure 8.43 shows the compression and post-impact compression strengths for other stitched laminates—that is, the chain and the modified lock of ($\frac{1}{8}$ × $\frac{1}{8}$ × 8) or ($\frac{1}{4}$ × $\frac{1}{4}$ × 8) with S2 glass or Kevlar thread. Data for other toughened composites—the IM6/1808I and some unidentified material (labeled as unpublished Lockheed data)—are also included. The trendlines previously shown in Fig. 8.42, along with data points for the single-directional 3678 and 6000 d S2 glass thread stitched laminates, are replotted in Fig. 8.43 for comparison with the new data.

In a low-velocity impact test, the applied impact energy is customarily denoted by ft · lbf (the total applied energy) or in. · lbf/in. (the applied energy per unit thickness of the composite laminate). For the data compiled here, the stitched panels had the same number of plies but unequal total thicknesses, or vice versa. The consequence of these discrepancies might be significant in terms of the actual applied impact

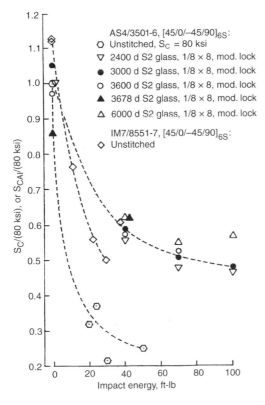

Fig. 8.42 Comparison of post-impact compression strengths for the unstitched and stitched AS4/3501-6, and the toughened composite laminate, part 1

energy level in a laminate. For the purpose of fair comparison, a normalized parameter, in. · lbf/ply (applied impact energy per ply) has been used in Fig. 8.43 as the independent variable. A review of Fig. 8.43 reveals the following:

- For a given impact energy level, the chain stitch pattern has much higher residual compression strength than the modified lock configuration. The S_{CAI} value for the chain stitched laminates (with 3600 d S2 glass thread) is higher than the S_{CAI} value for all the other data points shown in Fig. 8.42 and 8.43. However, its zero-impact compression strength (S_C) was only 8% lower than the unstitched laminate.
- For the Kevlar 29 thread stitched laminates, the S_{CAI} values for the heavier thread (1500 d) are higher than the S_{CAI} values for the lighter thread (1000 d). Although both thread gages exhibit higher post-impact compression strengths than the unstitched materials, a significant loss in the S_C property is also noted.
- Within experimental scatter, the S_C and S_{CAI} values for panels stitched with 1500 d Kevlar

29 thread and those stitched with 3678 d S2 glass thread are approximately the same. This similarity can probably be attributed to the fact that the knot-breaking strengths for both thread materials were approximately the same.

- Within experimental scatter, there are insufficient data points to show a noticeable difference in compression strengths (S_C and S_{CAI}) between the single-directional and the square-array stitch patterns.
- The S_C and S_{CAI} values for the IM7/8551-7 material are slightly higher than those for the IM6/1808I material.

Open Hole Compression Strength. Available open-hole compression test data for the aforementioned composites are plotted in Fig. 8.44 as a function of hole diameter. Although all the tests reported in the source references were conducted in accord with NASA standard specimen configurations and procedures, the specimens consisted of three hole sizes (0.25, 0.5, and 1.0 in. diameter) and two specimen widths (3 and 5 in.). Therefore, a normalized net-section stress ratio (i.e., the failure load divided by the net cross-sectional area divided by 80 ksi) is used for data presentation. Due to limited data, an incomplete, simplified correlation is presented here:

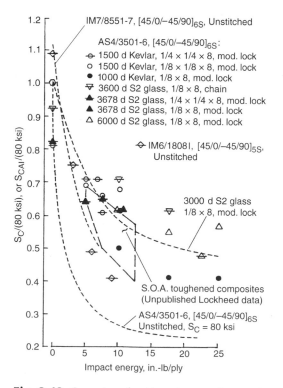

Fig. 8.43 Comparison of post-impact compression strengths for the unstitched and stitched AS4/3501-6, and the toughened composite laminate, part 2

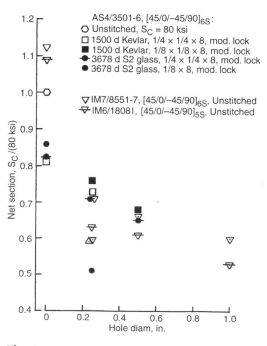

Fig. 8.44 Comparison of open-hole compression strengths for the AS4/3501-6 and the toughened composites

- For the toughened composites, the open-hole compression strength for the IM7/8551-7 material is higher than for the IM6-1808I material.
- The open-hole compression strengths for the unstitched IM7/8551-7 and the stitched AS4/3501-6 composites are compatible, whether the stitches were made using 1550 d Kevlar 29 or 3678 d S2 glass thread. Of the two epoxies, the fracture toughness for the toughened 8551-7 resin is much higher than that of the standard 3501-6 resin (see Fig. 8.39). This means that stitching can improve the S_C value of a material to the level of another material with an initially higher S_C value.

Compression Fatigue Strength. Figure 8.45 shows the effect of stitching on zero-impact compression fatigue strength. These limited data, which include the stitched and unstitched AS4/3501-6 panels, show a clear reduction in fatigue strength attributed to stitching. For $R = 10$, at 10^6 cycles the maximum compression stress level for the stitched laminate was approximately 25% lower than for the unstitched laminate. This is consistent with the same amount of reduction in static strength, as shown in Fig. 8.43.

Summary. The effectiveness of a fiber/resin system, or a stitching configuration, can be measured by its open-hole and post-impact compression strengths. Among the materials and stitching methods considered, modified lock stitching, with (⅛ × 8), 2400 to 3000 d S2 glass thread, offers adequate S_C and S_{CAI} values without loss of zero-impact compression strength in the stitched panel. The square-array stitching configuration—(⅛ × ⅛ × 8)—helps to maintain adequate open-hole strength of the material. Based on knot-breaking strength, it is fair to assume that 1300 to 1400 d Kevlar thread can be substituted for the 3000 d S2 glass thread for stitching. It is also expected that further enhancement of open-hole and post-impact strength can be accomplished by stitching the toughened composite material (i.e., IM7/8551-7). Furthermore, on the basis of limited data, stitching can improve static compression strength but not compression fatigue strength.

8.7.2 Effect of Scaling on Strength Prediction

Scale-up/scale-down may affect predicted strengths in composite structures. For metals, the failure load is proportional to the cross-sectional area of the structural piece. This load-carrying behavior is represented by the solid line in Fig. 8.46. In composites, however, it is impossible to literally predict the failure load of a laminate layup configuration by using the allowable developed from another layup configuration of the same material. The reason for this is best explained by the scenarios considered by Deo and Kan (Ref 8.54). Two of the cases are presented here.

Consider a 24-ply, 1 in. wide baseline laminate subjected to tensile loading along the 0° direction. The laminate stacking sequence is $[\pm 45/0_2/\pm 45/0_2/\pm 45/90/0]_S$, which gives a (42/50/8) distribution of plies. The lamina mechanical properties are $E_1 = 18.7 \times 10^6$ psi, $E_2 = 1.9 \times 10^6$ psi, $G_{12} = 10.85 \times 10^6$ psi, $\nu_{12} = 0.3$, $t = 0.0052$ in., and $\varepsilon_f = 0.011$ in./in. The computed Young's modulus in the loading direction is $E_x = 9.977 \times 10^6$ psi. The strain response of the laminate can be approximated by:

$$\varepsilon = \frac{P}{AE_x} = \frac{P}{nbtE_x} \qquad \text{(Eq 8.43)}$$

where P is the applied load, n is the number of plies, and b is the laminate width. To scale down the laminate and simulate the strain responses, two assumptions were made. First, it was assumed that laminate symmetry is maintained and, second, that laminate orthotropy is maintained throughout the scaling process. These assumptions ensure that Eq 8.43 holds true for all of the scaled-down laminates.

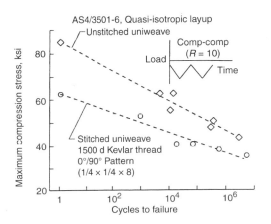

Fig. 8.45 Comparison of compression fatigue strengths for unstitched and stitched graphite/epoxy laminates

Based on the maximum strain criterion, the failure load P_f will be:

$$P_f = n \cdot b \cdot t \cdot E_x \cdot \varepsilon_f \qquad \text{(Eq 8.44)}$$

The failure load calculated for the baseline laminate is 13,681 lb. Unlike metals, for which the modulus E_x is a constant, E_x in a composite laminate is a function of thickness (number of plies) and layup. Scaling in thickness by adding or reducing the number of plies gives rise to a nonlinear relationship between the failure load and cross-sectional area. To illustrate this point, suppose that the laminate width is kept constant; the failure load for a scaled-down laminate would depend on which ply or plies are removed. For example, if we remove two of the 0° plies from the 24-ply baseline laminate, the failure load will be 11,500 lb for the scaled-down laminate. But if we remove two 90° plies instead, the failure load will be 12,900 lb for the new laminate. Of course, we obtain other values of failure load if we remove plies of other orientation, or a combination of any two ply orientations. As the number of plies is reduced still further, the possible failure loads are found to lie in an envelope centered about the linear failure load versus cross-sectional area relationship, as shown in Fig. 8.46.

The second case concerns the scale-up effect that exists in between small test panels and simulated structural components. Figure 8.47 compares the post-impact compression strength of a 5 in. wide coupon to that of built-up panels of similar layup and thickness subjected to the same level of impact energy. The shift in strength data from coupons to structural components is indicative of the scale-up effect. An explanation of this phenomenon was offered by Horton and Whitehead (Ref 8.40), who stated that the overall post-impact strength of a built-up structure is supposedly influenced by the structural arrangement. Most of these panels exhibited a two-stage failure. At initial failure, the damage propagated to the spar fastener lines and was arrested there. The load level at initial failure of the spar panel is equivalent to the failure load of the small coupon (which is a single-stage failure), so that final failure takes place at a higher load.

8.7.3 Building-Block Approach

The type of failure mode in the second example described above is not unique. It is very much the same as those crack-arrest mechanisms observed in metal skin-stringer structures. That is, the fibers take some loads from the matrix,

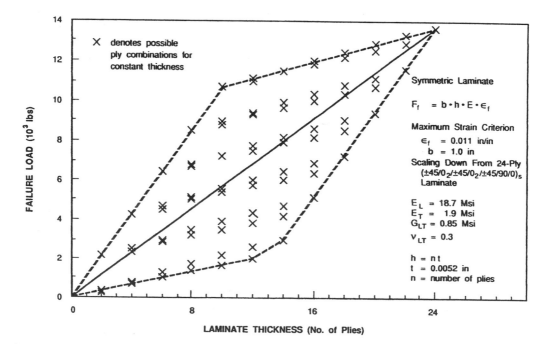

Fig. 8.46 Tensile failure loads as functions of layup and stacking sequence of the composite laminate. Source: Ref 8.54

just as the stringers do to the skin.* The end result is a higher failure load for the given structural system. Nonetheless, the two examples together do point out the need for a design/analysis/testing package that can ensure proper handling of scale-up effects. The "building-block" approach presented here divides the design/analysis process into several stages. Each stage (or level) stands for a certain degree of complexity regarding structural configuration. In each stage, an analytical model(s) is developed from test data of that stage. This model(s) is used to predict the failure mode(s) of the next stage, and is verified by the test results of that stage.

The building-block concept is illustrated by the pyramid in Fig. 8.48; details of the approach are discussed in Ref 8.56 and Volume 1 of MIL-HDBK-17. The building block begins at the lowest level of the pyramid, where the basic material properties are determined. Each type of test contains a number of specimens large enough for estimating the multibatch population scatter. At each succeeding level, progressively more complicated structures are built and tested, and the failure mode and failure load are predicted by analysis based on the lower-level data. When more data are obtained, the structural

analysis models are refined as needed to agree with the test results. As the test structures become more complex, the number of replicates (repeated tests using the same type of specimen and environment) and number of environments are reduced. The culmination of the building-block pyramid is often a single confirming test of a full-scale component or structural assembly.

Upper-level (especially full-scale) tests are not always performed in the worst-case design environment. Data from the lower levels of the building block can be used to establish environment compensation values to be applied to the loads of higher-level room-temperature tests. Similarly, other lower-level building-block tests determine truncation approaches for fatigue spectra and compensation for fatigue scatter at the full-scale level.

The building-block approach can be summarized in the following steps:

- Procure and process materials to a specification that includes controls on constituent, prepreg, and lamina/laminate physical properties.
- Determine multibatch materials properties at the lamina and/or laminate level; calculate the statistical basis values and preliminary design allowables.
- Based on the design/analysis of the structure, select critical areas and design features for subsequent test verification.

---

*The concepts of load transfer are frequently documented in the open literature (e.g., Ref 8.55).

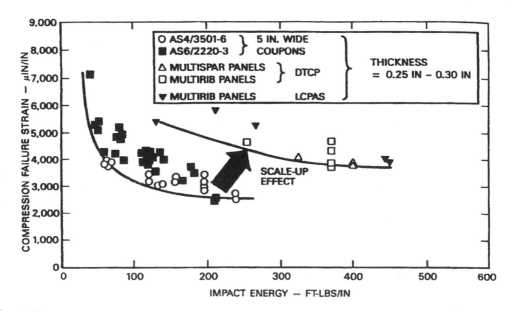

Fig. 8.47 Scale-up effects on post-impact compression strength. Source: Ref 8.54

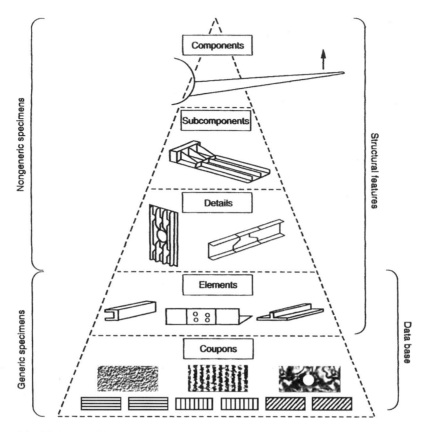

Fig. 8.48 Building-block pyramid for testing of composites. Source: Ref 8.56

- For each design feature, determine the load/environment combination that is expected to produce a given failure mode with the lowest matrix-sensitive failure modes (such as compression, out-of-plane shear, and bond lines) and potential "hot spots"caused by out-of-plane loads or stiffness-tailored designs.
- Design and test a series of specimens, each of which simulates a single selected failure mode and load/environment condition. Compare test results to previous analytical predictions and adjust analytical models or design properties as necessary.
- Conduct increasingly more complicated tests for the purpose of evaluating loading situations; consider the possibility of failure resulting from several potential failure modes. Compare the results with analytical predictions; adjust analytical models as necessary.
- Conduct, as required, full-scale component (or assembly) static and fatigue testing for final validation of internal loads and structural integrity. Compare the results to analysis.

REFERENCES

8.1. P.R. Naslain, Inorganic Matrix Composite Materials for Medium and High Temperature Applications: A Challenge to Material Science, *Development in the Science and Technology of Composite Materials,* A.R. Bunsell et al., Ed., Elsevier Science, 1985, p 34–35

8.2. C.R. Saff, "Durability of Continuous Fiber Reinforced Metal Matrix Composites," Report AFWAL-TR-85–3107, Air Force Flight Dynamics Laboratory, 1986

8.3. J.C. Halpin, *Primer on Composite Materials: Analysis,* Technomic Publishing, 1984

8.4. J.E. Ashton and J.M. Whitney, *Theory of Laminated Plates,* Technomic Publishing, 1970

8.5. R.H. Jones, *Mechanics of Composite Materials,* 2nd ed., Taylor & Francis, 1999

8.6. J.M. Whitney, *Structural Analysis of Laminated Anisotropic Plates,* Technomic Publishing, 1987

8.7. R.F. Gibson, *Principles of Composite Material Mechanics,* McGraw-Hill, 1994

8.8. S.W. Tsai and H.T. Hahn, *Introduction to Composite Materials,* Technomic Publishing, 1980

8.9. J.M. Whitney and R.J. Nuismer, Stress Fracture Criteria for Laminated Composites Containing Stress Concentrations, *J. Compos. Mater.,* Vol 8, 1974, p 253–265

8.10. S.G. Lekhnitskii, *Anisotropic Plates,* transl. from the 2nd Russian ed. by S.W. Tsai and T. Cheron, Gordon & Breach, 1968

8.11. R.J. Nuismer and J.M. Whitney, Uniaxial Failure of Composite Laminates Containing Stress Concentrations, *Fracture Mechanics of Composites,* STP 593, ASTM, 1975, p 117–142

8.12. M.E. Waddoups, J.R. Eisenmann, and B.E. Kaminski, Macroscopic Fracture Mechanics of Advanced Composite Materials, *J. Compos. Technol. Res.,* Vol 5, 1971, p 446–454

8.13. S.P Garbo and R.L. Gallo, "Strength of Laminates with Loaded Holes," Report MCAIR No. 81–013, McDonnell Aircraft Co., presented at the 5th DoD/NASA Conference on Fibrous Composites in Structural Design, New Orleans, Jan 27–29, 1981

8.14. *Engineered Materials Handbook,* Vol 1, *Composites,* ASM International, 1987, p 438

8.15. R. Talreja, Fatigue of Composite Materials: Damage Mechanisms and Fatigue Life Diagrams, *Proc. R. Soc. (London) A,* Vol 378, 1981, p 461–475

8.16. A. Rotem and H.G. Nelson, Residual Strength of Composite Laminates Subjected to Tension-Compressive Fatigue Loading, *J. Compos. Technol. Res.,* Vol 12, 1990, p 76–84

8.17. M. Whitney, "Fatigue Characterization of Composite Materials," Report AFWAL-TR-79–4111, Air Force Materials Laboratory, 1979

8.18. R.W. Walter, R.W. Johnson, R.R. June, and J.E. McCarthy, *Designing for Integrity in Long-Life Composite Aircraft Structures,* STP 636, ASTM, 1977, p 228–247

8.19. N.J. Pagano, Ed., *Interlaminar Response of Composite Materials,* Composite Materials Series 5, Elsevier Science, 1989

8.20. H.J. Konish, J.L. Swedlow, and T.A. Cruse, Fracture Phenomena in Advanced Fiber Composite Materials, *AIAA J.,* Vol 11, 1973, p 40–43

8.21. G.C. Sih, P.C. Paris, and G.R. Irwin, On Cracks in Rectilinearly Anisotropic Bodies, *Int. J. Fract. Mech.,* Vol 1, 1965, p 189–203

8.22. G.C. Sih, P.D. Hilton, R. Badaliance, P.S. Shenberger, and G. Villarreal, Fracture Mechanics for Fibrous Composites, *Analysis of the Test Methods for High Moludus Fibers and Composites,* STP 521, ASTM, 1973, p 98–132

8.23. H.J. Konish, T.A. Cruse, and J.L. Swedlow, Method for Estimating Fracture Strength of Specially Orthotropic Composite Laminates, *Analysis of the Test Methods for High Modulus Fibers and Composites,* STP 521, ASTM, 1973, p 133–142

8.24. M.D. Snyder and T.A. Cruse, "Crack Tip Stress Intensity Factors in Finite Anisotropic Plates," Report AFML-TR-73–209, Air Force Materials Laboratory, 1973

8.25. H.J. Konish, Mode I Stress Intensity Factors for Symmetrically-Cracked Orthotropic Strips, *Fracture Mechanics of Composites,* STP 593, ASTM, 1975, p 99–116

8.26. E.F. Olster and H.A. Woodbury, "Evaluation of Ballistic Damage Resistance and Failure Mechanism of Composite Materials," Report AFML-TR-72–79, Air Force Materials Laboratory, 1972

8.27. T.K. O'Brien, Charaterization of Delamination Onset and Growth in a Composite Laminate, *Damage in Composite Materials,* STP 775, ASTM, 1982, p 140–167

8.28. T.K. O'Brien, Analysis of Local Delaminations and Their Influence on Composite Laminate Behavior, *Delamination and Debonding of Materials,* STP 876, ASTM, 1985, p 282–297

8.29. T.K. O'Brien, N.J. Johnson, I.S. Raju, D.H. Morris, and R.A. Simonds, Comparisons of Various Configurations of the Edge Delamination Test for Interlaminar Fracture Toughness, *Toughened Composites,* STP 937, ASTM, 1987, p 199–221

8.30. T.K. O'Brien, Towards a Damage Tolerance Philosophy for Composite Materials and Structures, *Composite Materials: Testing and Design, (Ninth Volume),* STP 1059, ASTM, 1990, p 7–33

8.31. T.K. O'Brien, Characterizing Delamination Resistance of Toughened Resin Composites, *Tough Composites,* Report CP 2334, NASA, 1984

8.32. T.K. O'Brien, Fracture Mechanics of Composite Delamination, *ASM Handbook,* Vol 21, *Composites,* ASM International, 2001, p 241–245

8.33. R.H. Martin and G.B. Murri, Characterization of Mode I and Mode II Delamination Growth and Thresholds in AS4/PEEK Composites, *Composite Materials: Testing and Design, (Ninth Volume),* STP 1059, ASTM, 1990, p 251–270

8.34. F.W. Crossman and A.S.D. Wang, The Dependence of Transverse Cracking and Delamination on Ply Thickness in Graphite/Epoxy Laminates, *Damage in Composite Materials,* STP 775, ASTM, 1982, p 118–139

8.35. E.M. Wu, Application of Fracture Mechanics to Anisotropic Plates, *J. Appl. Mech. (Trans. ASME), Ser. E,* Vol 34, 1967

8.36. G.R. Irwin, Fracture Mechanics Applied to Adhesive Systems, *Treatise on Adhesives and Adhesion,* R.L. Patrick, Ed., Marcel Dekker, 1966, p 233–267

8.37. S.H. Yoon and C.S. Hong, Interlaminar Fracture Toughness of Graphite/Epoxy Composite Under Mixed-Mode Deformations, *Exp. Mech.,* Sept 1990, p 234–239

8.38. V.V. Bolotin, Delamination in Composite Structures: Its Origin, Buckling, Growth and Stability, *Composites, Part B: Engineering,* Vol 27B, 1996, p 129–145

8.39. G.C. Sih and H. Liebowitz, Mathematical Theories of Brittle Fracture, *Fracture: An Advanced Treatise,* Vol 2, Academic Press, 1968, p 67–190

8.40. R.E. Horton and R.S. Whitehead, "Damage Tolerance of Composites," Report AFWAL-TR-87–3030 (3 vol), Flight Dynamic Laboratory, Air Force Wright Aeronautical Laboratories, 1988

8.41. "Standard Tests for Toughened Resin Composites," rev. ed., NASA Reference Publication 1092, compiled by ACEE Composites Project Office, Langley Research Center, 1983

8.42. M.R. Woodward and R. Stover, Damage Tolerance, *ASM Handbook,* Vol 21, *Composites,* ASM International, 2001, p 295–301

8.43. M.B. Dow and D.L. Smith, "Development of Two Composite Materials Made of

Toughened Epoxy Resin and High Strain Graphite Fiber," Report TP2826, NASA, 1988

8.44. M.B. Dow, D.L. Smith, and S.J. Lubowinski, "An Evaluation of Stitching Concepts for Damage Tolerant Composites," Report CP 3038, NASA, 1989, p 53–73

8.45. M.A. Portanova, C.C. Poe, and J.D. Whitcomb, "Open Hole and Post-Impact Compression Fatigue of Stitched and Unstitched Carbon/Epoxy Composites," Report TM 102676, NASA, June 1990

8.46. R.J. Palmer, M.B. Dow, and D.L. Smith, Development of Stitching Reinforcement for Transport Wing Panels, *First NASA Advanced Composites Technology Conference,* Report CP 3104, Part 2, NASA, 1991, p 621–646

8.47. H.B. Dexter, G.H. Hasko, and R.J. Cano, Characterization of Multiaxial Warp Knit Composites, *First NASA Advanced Composites Technology Conference,* Report CP 3104, Part 2, NASA, 1991, p 589–619

8.48. H.B. Dexter and J.G. Funk, "Impact Resistance and Interlaminar Fracture Toughness of Through-the-Thickness Reinforced Graphite/Epoxy," Paper 86–1020, AIAA, 1986

8.49. B.N. Cox, M.S. Dadkhah, R.V. Inman, M.R. Mitchell, W.L. Morris, and S. Schroeder, Micromechanics of Fatigue in Woven and Stitched Composites, *First NASA Advanced Composites Technology Conference,* Report CP 3104, Part 2, NASA, 1991, p 579–587

8.50. D.L. Smith and H.B. Dexter, "Woven Fabric Composites with Improved Fracture Toughness and Damage Tolerance," Report CP 3038, NASA, 1989, p 75–89

8.51. L.W. Gause and J.M. Alpaer, "Mechanical Properties of Magnaweave Composites," Report NADC-84030–60, Naval Air Development Center, 1983

8.52. L.W. Gause, J.M. Alpaer, and R.H. Dalrympte, "Fatigue Properties of Multidirectional Braided Composites," Report NADC-85022–60, Naval Air Development Center, 1985

8.53. J.W. Coltman and S.M. Arndt, "Evaluation of Elastomeric Matrix Materials for Use in Aircraft Primary Structures," Report TR-87437, Simula Inc., July 1987

8.54. R.B. Deo and H.P. Kan, Effects of Scale in Predicting Global Structural Response, *First NASA Advanced Composites Tech-*

nology Conference, Report CP 3104, Part 2, NASA, 1991, p 761–777

8.55. D. Broek, *Elementary Engineering Fracture Mechanics,* 4th rev. ed., Martinus Nijhoff, 1986

8.56. R.E. Fields, Overview of Testing and Certification, *ASM Handbook,* Vol 21, *Composites,* ASM International, 2001, p 734–740

SELECTED REFERENCES

- M.R. Allen, D.T. Sawdy, S.J. Bradley, and J.G. Avery, "Survivability Characteristics of Composite Compression Structures," Report AFWAL-TR-88-3014, Flight Dynamics Laboratory, Air Force Wright Aeronuatical Laboratories, 1988

- J.G. Avery, "Design Manual for Impact Damage Tolerant Aircraft Structure," Report AG-238, AGARD, Oct 1981

- J.G. Avery and T.R. Porter, *Comparison of the Ballistic Response of Metals and Composites for Military Aircraft Applications,* STP 568, ASTM, 1975, p 3–29

- A.A. Baker, S. Dutton, and D. Kelley, *Composite Materials for Aircraft Structures,* AIAA, 2004

- S.M. Bishop and G. Dorey, The Effect of Damage on the Tensile and Compressive Performance of Carbon Fibre Laminates, *Characterization, Analysis, and Significance of Defects in Composite Materials,* CP-355, AGARD, April 1983

- Composite Materials, MIL-HDBK-17–1E, Vol 1, *Guidelines for Characterization of Structural Materials,* U.S. Department of Defense, 1977

- Composite Materials, MIL-HDBK-17–1E, Vol 2, *Polymer Matrix Composites: Material Properties,* U.S. Department of Defense, 1977

- Composite Materials, MIL-HDBK-17–1E, Vol 3, *Material Usage, Design, and Analysis,* U.S. Department of Defense, 1977

- K. Friedrich, Ed., *Application of Fracture Mechanics to Composite Materials,* Elsevier, 1989

- W.S. Johnson, Fatigue of Fiber-Reinforced Metal-Matrix Composites, *ASM Handbook,* Vol 21, *Composites,* ASM International, 2001, p 914–919

- C.C. Poe, Residual Strength of Composite Aircraft Structures with Damage, *ASM Handbook,* Vol 21, *Composites,* ASM International, 2001, p 920–935

- A.S. Tetelman, Fracture Processes in Fiber Composite Materials, *Composite Materials: Testing and Design,* STP 460, ASTM, 1969, p 473–502

- R.S. Whitehead and R.W. Winslow, "Composite Wing/Fuselage Program," Report AFWAL-TR-88–3098 (4 vol), Air Force Wright Aeronuatical Laboratories, 1989

- G. Ziegler and W. Huttner, Engineering Properties of Carbon-Carbon and Ceramic-Matrix Composites, *Engineered Materials Handbook,* Vol 4, *Ceramics and Glasses,* ASM International, 1991, p 835–844

APPENDIX 1

Lattice Structure and Deformation Mechanisms in Metallic Single Crystals

EXCEPT FOR SPECIALLY PRODUCED ALLOYS that solidify into a single crystal, most metallic materials have a polycrystalline structure comprised of many crystalline grains separated by grain-boundary regions. Each grain contains many crystals, which are composed of atoms, ions, or molecules arranged in a pattern that is periodic in three dimensions. In contrast, the grain boundaries are disruptions between the crystal lattices of individual grains, and these disruptions in the grain-boundary regions provide a source of strengthening by pinning the movement of dislocations. Thus, grain boundaries are typically stronger than individual grains in properly processed polycrystalline plastic/elastic solids at temperatures below about 0.4 T_M (where T_M is the melting point on the Kelvin scale). At temperatures above about 0.4 T_M, grain boundaries tend to become weaker than the crystalline grains (see the section "Microscopic Aspects of Fracture" in Chapter 2).

The crystal structure of grains is important in describing the deformation of polycrystalline metallic materials, because the grain region is most prone to slip in properly processed materials at temperatures below the creep regime. Deformation within a crystal lattice is governed principally by the presence of dislocations, which are line-type (i.e., two-dimensional) defects in crystal lattice. This is shown in a simplified comparison of an ideal crystal lattice (Fig. A1.1) and a lattice with a dislocation (Fig. A1.2). In the case of the ideal crystal, shear displacement would occur by simultaneous relative motion of all the atoms along a slip plane (Fig. A1.1). In this idealized case, calculated stresses on the order of 10 to 30 GPa (1 to 5 ksi × 10^3)

would be required for the onset of slip deformation. In reality, however, metals typically have dislocations in the crystal structure, and slip can occur by the motion of relatively few atoms at a given instance (Fig. A1.2b and c). The presence of dislocations thus allows deformation within a crystalline lattice of metals at relatively lower stresses in the range of 35 kPa (5 psi) for unalloyed metals to 345 MPa (50 ksi) for high-strength structural alloys. The theory of dislocations can provide an understanding of slip deformation in metals below 0.4 T_M, when slip is most likely to occur within grains rather than in the grain boundaries.

Slip from shear stress is the most common deformation mechanism within crystalline lattices of metallic materials, although deformation of crystal lattices can also occur by other shearing processes such as twinning and, in special circumstances, by the migration of vacant lattice sites. This appendix describes the processes of both slip and twinning. However, before discussing these two basic types of deformation mechanisms, it is useful to describe the notation conventions for specifying the various planes and directions in a crystal lattice. More detailed discussions on crystal geometries and their roles in plastic deformation can be found in Ref A1.2 to A1.6.

A1.1   Crystal Lattice Structures

As noted, a crystal is a solid consisting of atoms or molecules arranged in a pattern that is repetitive in three dimensions. The arrangement of the atoms or molecules in the interior of a crystal is called its crystal structure. All crystals are classified as either triclinic, monoclinic, orthorhombic, hexagonal, rhombohedral, tetragonal, or cubic. The unit cell of a crystal is the smallest pattern of arrangement that can be contained in a parallelepiped with the possibility of three distinct edge lengths along the three axes of the crystal cell. The edge lengths (or so-called lattice parameters) are defined by the symbols a, b, and c for each crystal axis (Fig. A1.3). In addition, the crystal axes are not necessarily orthogonal, so that three distinct interaxial angles (α, β, and γ) may also specify the structure of a crystal cell. The interaxial angles are located on the following faces of a cell:

- The α interaxial angle resides on the side (face A), which contains axes of b and c.
- The β interaxial angle resides on the front (face B), which contains axes of c and a.
- The γ interaxial angle resides on the bottom (face C), which contains axes of a and b.

General relationships for edge lengths and interaxial angles are summarized in Table A1.1 for the seven basic crystal systems.

When many unit cells are connected together, they form a three-dimensional space lattice, or Bravais point lattice. A point lattice is defined as the set of points in space located so that each point has identical surroundings. The five basic arrangements (and their lettering notation) for lattice points within a unit cell are:

- *Simple (primitive) lattice* (P) having lattice points solely at cell corners
- *End-centered lattice* (C) with lattice points centered on the two C faces (containing a and b axes) or ends of the cell (also referred to as base centered, e.g., drawings 3 and 5 in Fig. A1.3)
- *Face-centered lattice* (F) having lattice points centered on all faces

- *Inner-centered (body-centered) lattice* (I) having lattice points at the center of the volume of the unit cell
- *Primitive rhombohedral unit cell* (R), which is considered a separate basic arrangement

When the seven crystal systems are considered together with the above five groups of space lattices, the 14 possible combinations of arranging points in space correspond to the 14 Bravais lattices (Fig. A1.3). These 14 combinations form the basis of the Pearson symbols, a system developed by William B. Pearson, which are widely used to identify crystal types (Table A1.2). The Pearson symbol uses a small letter to identify the crystal system and a capital letter to identify the space lattice. To these is added a number equal to the number of atoms in the unit cell conventionally selected for the particular crystal type.

Of these 14 possible lattice structures, the most common types of point lattices in commercial alloys are hexagonal close-packed (hcp), body-centered cubic (bcc), and face-centered cubic (fcc) structures (drawings 10, 13, and 14, respectively, in Fig. A1.3). The packing atoms in the fcc and hcp lattices result in a closest-packed plane, where all adjoining atoms on a plane are in the closest possible proximity to one another. Because the fcc and hcp lattices have closest-packed planes, 74% of the volume of the unit cells are occupied by atoms.* In contrast, only 68% and 52% of cell volumes are packed with atoms in the bcc and simple (primitive) cubic cells, respectively.

Miller Indices of Crystallographic Planes. In 1839, William H. Miller proposed a special notation system for unambiguous specification of plane orientation based on the scalar equation of a plane with intercepts (a, b, c) along the x-, y-, and z-axes (Fig. A1.4). All planes parallel to this plane (except, of course, the plane passing through the origin) can be described by the scalar relation:

$$\frac{x}{a} + \frac{y}{b} + \frac{z}{c} = 1 \tag{Eq A1.1a}$$

or

$$\frac{1}{a} + \frac{1}{b} + \frac{1}{c} = h + k + l \tag{Eq A1.1b}$$

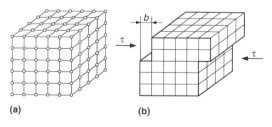
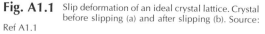

Fig. A1.1 Slip deformation of an ideal crystal lattice. Crystal before slipping (a) and after slipping (b). Source: Ref A1.1

---

*A c/a ratio of $\sqrt{8/3}$ is ideal for hcp packing; actual values of c/a for hcp metals vary from 1.56 to 1.88.

where h, k, and l are the simplest integers that define Miller indices $(h\ k\ l)$ of the plane. Miller indices for planes in a cubic crystal are shown in Fig. A1.5. In a hexagonal system, it is convenient to use a four-axis description (Fig. A1.6) that results in a four-indices notation $(h\ k\ i\ l)$.

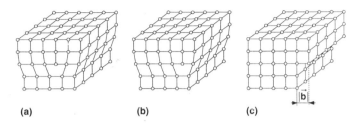

(a) (b) (c)

Fig. A1.2 Deformation in a crystal lattice from slip of line defect (dislocation) from a position in (a) to the edge in (c). The vector **b** is the Burgers vector, which is defined as the unit displacement of a dislocation. Source: Ref A1.1

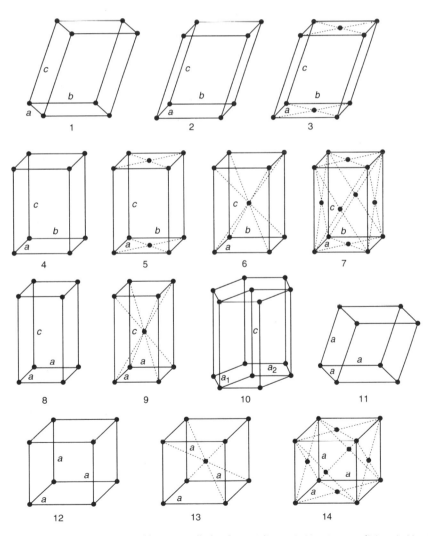

Fig. A1.3 The 14 Bravais lattices illustrated by a unit cell of each: 1, triclinic, primitive; 2, monoclinic, primitive; 3, monoclinic, base centered; 4, orthorhombic, primitive; 5, orthorhombic, base centered; 6, orthorhombic, body centered; 7, orthorhombic, face centered; 8, tetragonal, primitive; 9, tetragonal, body centered; 10, hexagonal, primitive; 11, rhombohedral, primitive; 12, cubic, primitive; 13, cubic, body centered; 14, cubic, face centered. Source: Ref A1.7

This system, referred to as the Miller-Bravais indices for the hexagonal system, helps in the notation of symmetries. Similar indices are used to specify crystallographic directions in crystal systems, as described below for cubic and hexagonal lattices.

Crystallographic Planes and Directions in a Cubic Cell. The relative orientation of atoms in a given crystal lattice can be classified in terms of crystallographic planes and directions along those planes. Crystallographic planes and directions are described using a notation system of Miller indices, as summarized in Fig. A1.7 for a cube. Indices in parentheses define planes, while bracketed indices, such as [100], indicate directions. An overbar is shorthand for a minus sign, indicating a negative Miller coordinate relative to the origin (which is at the center of the cube in Fig. A1.7).

As an example, Fig. A1.8 illustrates planes in a simple (primitive) cubic lattice, consisting of eight atoms occupying each corner of a cubic cell. Three mutually perpendicular axes are arbitrarily placed through one of the corners of the cell. Each of the six crystallographic planes (surfaces) is specified in terms of the length of its intercepts on the three axes, measured from the origin of the coordinate axes. For example, taking point F as the origin, the $ABCD$ plane is designated as the (010) plane because it is parallel to the x- and z-axes (no intercept) and intersects the y-axis at a distance a, which is taken as one interatomic spacing.*

In similar fashion, the plane $EHGF$ is specified as the $(0\bar{1}0)$ plane, where the bar above the numeral one indicates that the intersection is at the opposite direction (because we have conveniently moved the origin from F to C in order to fit the model). There are six crystallographically equivalent planes of the type (100) in this cube, any one of which can have the indices (100), (010), (001), $(\bar{1}00)$, $(0\bar{1}0)$, $(00\bar{1})$, depending on the choice of axes. This group, or family of

---

*As noted in Eq A1.1(b), the indices for the crystallographic planes are actually taken as the reciprocals. That is, an index 1 comes from 1/1, and the index 0 is equivalent to intercepting at infinity ($= 1/\infty$). Whenever appropriate, an index 2 would mean the intercept is at 1/2 atomic spacing (i.e., the reciprocal of 1/2 is equal to 2). See Ref A1.2 or A1.8 for details. The physical unit for the atomic spacing, also called the atomic constant, measured between the centers of two adjacent atoms, is the angstrom unit ($1\text{Å} = 10^{-8}$ cm).

Table A1.1 Relationships of edge lengths and interaxial angles for the seven crystal systems

Crystal system	Edge lengths	Interaxial angles	Examples
Triclinic (anorthic)	$a \neq b \neq c$	$\alpha \neq \beta \neq \gamma \neq 90°$	HgK
Monoclinic	$a \neq b \neq c$	$\alpha = \gamma = 90° \neq \beta$	β-S; $CoSb_2$
Orthorhombic	$a \neq b \neq c$	$\alpha = \beta = \gamma = 90°$	α-S; Ga; Fe_3C (cementite)
Tetragonal	$a = b \neq c$	$\alpha = \beta = \gamma = 90°$	β-Sn (white); TiO_2
Hexagonal	$a = b \neq c$	$\alpha = \beta = 90°$; $\gamma = 120°$	Zn; Cd; NiAs
Rhombohedral(a)	$a = b = c$	$\alpha = \beta = \gamma \neq 90°$	As; Sb; Bi; calcite
Cubic	$a = b = c$	$\alpha = \beta = \gamma = 90°$	Cu; Ag; Au; Fe; NaCl

(a) Rhombohedral crystals (sometimes called trigonal) also can be described by using hexagonal axes (rhombohedral-hexagonal).

Table A1.2 The 14 space (Bravais) lattices and their Pearson symbols

Crystal system	Space lattice	Pearson symbol
Triclinic (anorthic)	Primitive	aP
Monoclinic	Primitive	mP
	Base centered(a)	mC
Orthorhombic	Primitive	oP
	Base centered(a)	oC
	Face centered	oF
	Body centered	oI
Tetragonal	Primitive	tP
	Body centered	tI
Hexagonal	Primitive	hP
Rhombohedral	Primitive	hR
Cubic	Primitive	cP
	Face centered	cF
	Body centered	cI

(a) The face that has a lattice point at its center may be chosen as the c face (the xy plane), denoted by the symbol C, or as the a or b face, denoted by A or B, because the choice of axes is arbitrary and does not alter the actual translations of the lattice.

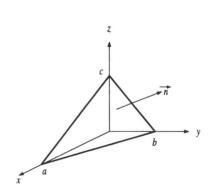

Fig. A1.4 Schematic illustration of intercepts that define the direction of a plane. Source: Ref A1.1

planes, is referred to as a form and is designated as the {100} form. The form {100} comprises the planes outlining the cube. The {110} planes are the most densely packed (but *not* closely packed) planes in bcc crystals.

Crystallographic directions are indicated as bracketed indices [hkl], where the integral digits are regarded as directional indices based on perpdicular projections to the coordinate axes. For example, Fig. A1.9 illustrates indices for

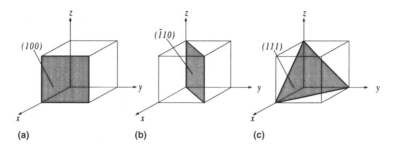

Fig. A1.5 Miller indices of cubic crystal planes. Shading indicates (a) side face, (b) face diagonal plane, and (c) octahedral plane. Source: Ref A1.1

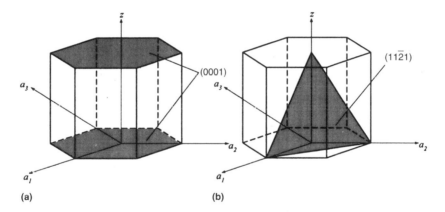

Fig. A1.6 Miller-Bravais indices for various planes in the hexagonal system. (a) Indices (0001). (b) Indices (11$\bar{2}$1). Source: Ref A1.1

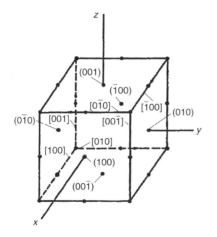

Fig. A1.7 Directional indices and Miller indices of planes in a cube

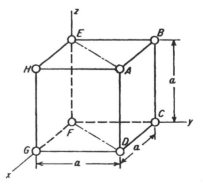

Fig. A1.8 Example of interior planes in a primitive cube structure

various directions. The [110] on the top has the same directional indices as its projection onto the *x-y* plane at the bottom of the cell. A family of directions are noted with indices contained in carets ⟨*xyz*⟩. For example, the ⟨100⟩ directions are the edges of a cubic cell, while ⟨110⟩ indicates the six face diagonals.

Crystallographic Planes and Directions in a Hexagonal Cell. As previously noted in Fig. A1.6, a four-digit (Miller-Bravais) notation is used to define planes in a hexagonal system. Likewise, four-digit indices are used to define directions in a hexagonal system (Fig. A1.10). Three axes connect the center atom on the basal plane to the three outer corners (120° apart), and a vertical axis connects the center atoms of the basal planes. The center atom on one of the basal planes is normally chosen as the coordinate origin. The notation of $\bar{2}$ merely means that the

coordinate origin has been moved to one of the outer corners in order to fit the model, and that there is a two-atom distance between the two corners.

A1.2 Lattice Response under Stress

Under stress, a crystalline lattice may change by the following three mechanisms: slip, cleavage, and mechanical twinning. Cleavage is a brittle process that involves abrupt separation of the lattice under tension, while slip and twinning involve rearrangement (plastic deformation) of the lattice under shear stress. This rearrangement requires some expenditure of energy (work), as in the incremental (progressive, stepwise) displacement of dislocations in a lattice during slip. Slip is the most prevalent mechanism for properly processed alloys below temperatures of $0.4\ T_M$.

Crystals can also deform by other shearing processes such as twinning and, in special circumstances, by the migration of vacant lattice sites. In addition to line defects (dislocations), crystal lattices may include internal surface imperfections such as twins and stacking faults. Like dislocations, surface (planar) imperfections of a crystal lattice occur in conjunction with the plastic deformation of metals, as briefly described in sections A1.3 and A1.4. Planar-type crystal imperfections also may influence the mechanical and metallurgical behavior of metals during deformation. However, twins do not affect mechanical behavior to the same degree as stacking faults (although an important exception is low-temperature deformation of bcc metals). Twins generally play only a minor role in plastic

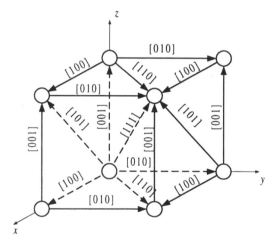

Fig. A1.9 Directional indices in a primitive cube. Source: Ref A1.1

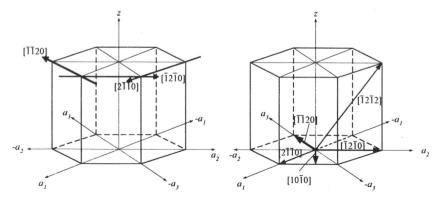

Fig. A1.10 Directional indices in a hexagonal system. Source: Ref A1.1

flow, while stacking faults can affect the work-hardening behavior of metals with close-packed structures (such as fcc metals).

A1.2.1 Slip Deformation

The mechanism of slip (shear strain) in a crystalline lattice is essentially the sliding of blocks of the crystal over one another along definite crystallographic planes, called slip planes. An analogous example is the distortion produced in a deck of cards when pushed from one end (Fig. A1.11). The material in the slip planes also remains crystalline during slip.

As noted, slip generally occurs from the incremental (stepwise) displacement of dislocations, which are one-dimensional (line) imperfections in a crystal lattice (e.g., Fig. A1.2). That is, slip starts from structural irregularities in the crystal, at either one place or a few places in the slip plane, and then spreads outward over the rest of the slip plane at some finite speed. Ob-

viously, macroscopic deformation involves movement of millions of dislocations, but this is possible because the dislocations, which are present in the metal prior to plastic deformation, create other dislocations by a multiplication mechanism during plastic deformation.

The movement (slip) of atoms on one layer over the other layer is a consecutive process, not a simultaneous one. Slip within a crystal structure occurs along specific crystal planes and in specific directions on a slip plane, depending on how the atoms are packed together. Generally, the slip plane is the plane of greatest atomic density, and the slip direction is usually along the line of closely packed atoms within the slip plane. Slip planes are usually either on the closest-packed planes (where adjacent atoms "touch" each other) or along a closely packed plane. These planes tend to have the widest spacing between dislocations, and so dislocations move more easily along these planes.

For a given slip plane, the possible directions of slip are determined by the direction of closest packing of atoms. Slip is more likely in the closest-packed direction, because the Burgers vector (i.e., the unit displacement of dislocation, **b**, in Fig. A1.2) is at a minimum in these directions (thus causing the energy associated with dislocation movement to be at a minimum of $E \sim 1$ Gb^2, where G is the shear modulus and b^2 is the magnitude of the Burgers vector). The values of shortest interatomic distance for some metal crystals are shown in Table A1.3.

Slip occurs when the shear stress exceeds a critical value. The extent of slip in a single crystal depends on the magnitude of the shearing stress produced by external loads, the geometry of the crystal structure, and the orientation of the

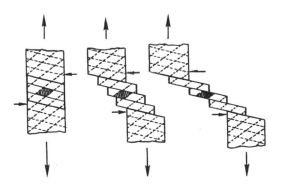

Fig. A1.11 Diagram of crystal lattice after slip (horizontal arrows). Note that each cube is undistorted after slip.

Table A1.3   Room-temperature crystal structure and critical shear stress for metal single crystals

Metal	Crystal structure	Atomic constants at 20 °C a, Å	c, Å	Closest interatomic distance, Å	Purity, %	Critical shear stress(a), g/mm²
Mg	hcp	3.2022	5.1991	3.196	99.996	77
Cd	hcp	2.9727	5.6061	2.979	99.996	58
Ti	hcp	2.953	4.729	2.89	99.99	1400
					99.9	9190
Zn	hcp	2.659	4.936	2.664	99.999	18
Ag	fcc	4.0774	. . .	2.888	99.99	48
					99.97	73
					99.93	131
Cu	fcc	3.608	. . .	2.556	99.999	65
					99.98	94
Ni	fcc	3.517	. . .	2.491	99.8	580
Fe	bcc	2.861	. . .	2.481	99.96	2800
Mo	bcc	3.1403	. . .	2.725	. . .	5000

(a) See Table A1.5 for other reported values of critical stress. Source: Ref A1.2, A1.3

active slip planes with respect to the shearing stresses. Consider a cylindrical crystal under a uniaxial tensile load P. The component of force in the slip direction is $P \cdot \cos \lambda$, where λ is the angle between this direction and the axis of tension (Fig. A1.12). This force is spread over a slope plane of area $A/\cos \phi$, where ϕ is the angle between the normal to the plane and the axis of tension. Thus the resolved shear stress is:

$$\tau_R = \frac{P}{A} \cos \lambda \, \cos \phi \qquad \text{(Eq A1.2)}$$

Table A1.3 gives values of critical shear stress for a number of metals. As shown, the critical shear stress increases as impurity increases. Thus, the critical shear stress for an alloy is usually higher than for pure metals. For example, a single crystal of 50% Au and 50% Ag would have a critical shear stress value nearly 10 times higher than that for the pure gold or silver (Ref A1.2). Also, it must be noted that commercial alloys are, in almost all cases, not single crystals, but rather polycrystalline distributions of individual crystalline grains separated by grain boundaries. The slip directions are randomly oriented among the grains that are randomly attached to each other. Thus, the gross critical shear stress will not be the same as in single crystals. The grain boundaries (in between each grain) further affect the mechanical properties of commercial alloys.

In slip, combinations of slip planes and directions are known as slip systems. Figure A1.13 schematically illustrates the most common slip systems for fcc, bcc, and hcp structures. The fcc and hcp structures are close packed, where atoms (on specific crystallographic planes in the lattice) are in the closest possible proximity to one another. In close-packed structures, slip is always on one of the closest-packed planes and in one of the closest-packed directions on that plane. In contrast, the bcc system is not a close-packed structure; in this case, nearly close-packed planes may serve as slip planes.

According to the von Mises criteria, at least five independent slip systems must be available in polycrystalline materials for the material to be capable of plastic deformation. Thus, the greater the number of independent slip systems, the greater the possibility for plastic deformation. Most technically significant metals and alloys with cubic structures have five or more independent slip systems and thus exhibit substantial plasticity. For example, nickel and copper,

which are fcc metals, each have five independent slip systems and are extremely ductile at ambient temperatures. However, cobalt, which is an hcp metal, has less than five independent slip systems and is brittle at ambient temperatures. Metals with less than five slip systems can exhibit plasticity at ambient temperatures, provided that other deformation modes are available. A good example of this is Zinc, which has only three systems but also exhibits twinning.

When a crystal is loaded, the shear stress increases on all slip systems. However, slip initiates only on the system that first attains the critical resolved shear stress for the material. It is possible to have an orientation such that two slip systems have identical resolved shear stresses, and that these two attain the critical resolved shear stress first. This leads to simultaneous slip on these two slip systems. This type of phenomenon is called "cross-slip."

Slip Systems in an fcc Structure. An fcc structure has 12 slip systems. The {111} planes provide four distinct (nonparallel) slip planes, which each have three $\langle \bar{1}10 \rangle$ slip directions (Fig. A1.14a). Slips occurs on the {111} planes, which have the highest density of packed atoms. Slip occurs in the $\langle \bar{1}10 \rangle$ directions, because these are the shortest distance between two atoms. Thus, the Burgers vector is of type $a/2$ in $\langle 110 \rangle$ directions on {111} planes, where a is the unit cell constant. Lattices with fcc structures maintain the {111}$\langle \bar{1}10 \rangle$ slip system at low temperature, which accounts for the cryogenic application of fcc metals due to their good ductility at low temperatures.

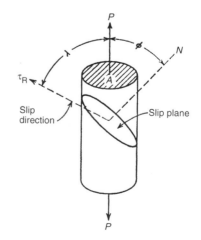

Fig. A1.12 Diagram for calculating resolved shear and resolved normal stresses

Slip Systems in a bcc Structure. The bcc lattice is not a close-packed structure, but the {110} planes are the most closely packed planes in a bcc structure. Thus, the {110} planes are the primary slip plane, because they have the highest atomic density. There are six {110} planes in the bcc structure, and each has two slip direc-

tions in the close-packed direction of the $\langle\bar{1}11\rangle$ type (see Fig. A1.14b).

Slip in a bcc lattice always takes place along a direction that moves one corner atom toward the atom in the center of the cube. This results in 12 slip systems associated with the {110} planes. However, slip planes in bcc lattices are

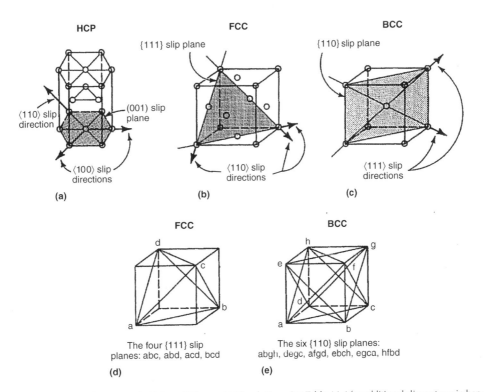

Fig. A1.13 Dominant slip systems in (a) hcp, (b) fcc, and (c) bcc lattices. See Table A1.4 for additional slip systems in hcp and bcc structures. (d) Corners of the four slip (close-packed) planes in an fcc structure. (e) Corners of the six {110} slip planes with the highest atomic density in a bcc structure. Note: The plane and directional indices in the hcp cell of (a) are the three-digit convention instead of the four-digit convention.

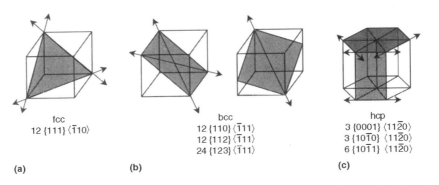

Fig. A1.14 Common slip systems in fcc, bcc, and hcp structures. (a) Slip in an fcc structure is governed solely by the 12 possible combinations of {111}⟨$\bar{1}$10⟩. (b) Slip in a bcc structure is most prevalent in two directions on the six {110} planes with high atomic density, although slip is also possible on other planes with ⟨$\bar{1}$11⟩ directions (see text). (c) Depending on the c/a ratio and the specific system, slip in an hcp system may occur on the basal {0001} planes or, in some cases, along the {10$\bar{1}$0} or {10$\bar{1}$1} planes.

not necessarily restricted to just the {110} planes. Slip can occur on any plane containing a ⟨111⟩ slip direction, such as the {112} and {123} planes. There are 12 unique (orthogonal) {112} planes and 24 unique {123} planes, which each has one possible slip direction. An example is shown in Fig. A1.14(b) (right).

Thus, there are a total of 48 possible slip systems (Table A1.4) in a bcc lattice from the glide of dislocations with a Burgers vector of type 1/2 in the ⟨111⟩ directions on {110}, {112}, and {123} planes. However, these slip planes are not all operative at the same time, as active slip planes depend on the temperature and the specific element. For example, the prevalence of bcc slip planes is noted by Cottrell (Ref A1.4) as follows (where T_M is the melting point on the Kelvin scale):

- {112} planes are preferred below 0.25 T_M.
- {110} planes are preferred from 0.25 to 0.5 T_M.
- {123} planes are preferred at high temperatures (~0.8 T_M).

For example, dislocations in bcc iron are known to move on {110} and also {112} or {123} planes, depending on temperature systems (Ref A1.2, A1.9). The most prevalent slip plane is the most closely packed {110} plane, but additional

slip systems can occur with increased temperature. Nonetheless, the slip direction remains in the ⟨111⟩ direction even when the slip plane changes. This type of behavior is sometimes referred to as "pencil glide" (Fig. A1.15), where the slip surface is regarded as a prismatic cylinder made up of slip planes bounded by the common slip direction (Ref A1.4).

Slip Systems in an hcp Structure. In hcp metals, the number of slip systems and the twinning shear are less than for cubic metals. Depending on the c/a ratio, polycrystalline hcp metals may or may not have the necessary number of slip systems to allow for appreciable plastic deformation. The predominant Burgers vector is 1/3 ⟨11$\bar{2}$0⟩, and the slip planes are usually the basal {0001} planes or {1$\bar{1}$00} prismatic planes (Fig. A1.14c). In magnesium, 1/3 ⟨11$\bar{2}$0⟩ slip on pyramidal {1$\bar{1}$01} planes has been reported (Ref A1.10). When slip is restricted, deformation twinning is common even at strains as small as 0.05.

In ideal hcp packing, the stacked atoms are represented as undistorted spheres, and the ratio of c/a is $\sqrt{8/3}$; actual values for metals vary between 1.56 and 1.88. The ideal hcp structure has only three slip systems in three independent (orthogonal) directions of ⟨11$\bar{2}$0⟩ on the close-packed basal plane. Three slip systems are insufficient to permit polycrystalline plastic deformation, and so hcp polycrystals with slip restricted to the basal plane are not malleable. When c/a is less than the ideal ratio, basal planes become less widely separated and other planes compete with them for slip activity. In these instances, the number of slip systems increases and material ductility is beneficially affected. These nonbasal planes are {10$\bar{1}$0} and {10$\bar{1}$1}, depending on the specific systems and its c/a ratio. Again, ⟨11$\bar{2}$0⟩ is the slip direction (Table A1.4). Table A1.5 lists slip systems in several common hcp metals.

Of the 22 elements with an hcp structure, only zinc and cadmium exhibit a c/a ratio greater than

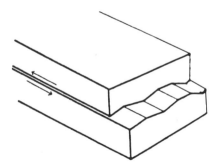

Fig. A1.15 Pencil glide. Slip takes place along different planes in one direction, giving the appearance of a pencil-like surface. Source: Ref A1.4

Table A1.4 Slip systems in fcc, bcc, and hcp crystal lattices

Crystal structure	Slip plane	Slip direction	Number of nonparallel planes	Slip directions per plane	Number of slip systems
Face-centered cubic	{111}	⟨1$\bar{1}$0⟩	4	3	12 = (4 × 3)
Body-centered cubic(a)	{110}	⟨$\bar{1}$11⟩	6	2	12 = (6 × 2)
	{112}	⟨11$\bar{1}$⟩	12	1	12 = (12 × 1)
	{123}	⟨11$\bar{1}$⟩	24	1	24 = (24 × 1)
Hexagonal close-packed(b)	{0001}	⟨11$\bar{2}$0⟩	1	3	3 = (1 × 3)
	{10$\bar{1}$0}	⟨11$\bar{2}$0⟩	3	1	3 = (3 × 1)
	{10$\bar{1}$1}	⟨11$\bar{2}$0⟩	6	1	6 = (6 × 1)

(a) Slip other than {110}⟨$\bar{1}$11⟩ in bcc metals depends on temperature and the system; see text. (b) Slip in hcp depends on c/a ratio; see text.

the ideal close-packed value of 1.633. Metals atoms in most hcp crystals are oblate spheres (squashed at the poles), resulting in c/a ratios lower than the ideal (for hard spheres). In contrast, the basal planes in zinc and cadmium are more widely separated than the theoretical distance for hard spheres. Thus, it is not surprising that the $\{0001\}\langle11\bar{2}0\rangle$ systems are important for zinc and cadmium, and even magnesium, where atoms are scarcely distorted at all. Not only is slip difficult because of the limited number of available slip systems, but cleavage readily occurs along the basal plane. Zinc has a ductile-to-brittle transition temperature (DBTT) just below room temperature, and sharp radii must be avoided to reduce its notch sensitivity. Zinc and cadmium also are reported to slip on $\{11\bar{2}2\}\langle11\bar{2}\bar{3}\rangle$ systems (Ref A1.10).

A1.2.2 Lattice Cleavage

In a manner entirely different from slip, brittle fracture can occur in either single-crystal or polycrystalline materials. On a macroscale, brittle fractures generally appear to have a flat fracture surface normal to the direction of the applied load. There is very little (or nearly no) plastic deformation prior to fracture. On a microscale, brittle fracture occurs by separation of the atoms on either side of a cleavage plane,

Table A1.5 Observed slip systems in metallic materials

Lattice	Material	Slip system	Critical stress, kg/mm^2	Purity, %	Notes
fcc	Ag	$\{111\}\langle1\bar{1}0\rangle$ (dominant)	0.038	99.999	
	Al		0.08	99.994	
	Cu		0.05	99.98	
	Ni		0.33–0.75	99.98	
fcc	Al	$\{100\}\langle\bar{1}10\rangle$ (minor)(a)	...	...	High temperature and high strain rate
		$\{110\}\langle110\rangle$(b)			High temperature
bcc	Fe	For all bcc materials:	2.80	99.96	
	Mo	$\{110\}\langle\bar{1}11\rangle$ (dominant)	7.30	Zone refined	
	Nb	$\{112\}\langle\bar{1}11\rangle$	3.40	Zone refined	
	Ta	$\{123\}\langle\bar{1}11\rangle$	4.20	Zone refined	
hcp	Ti	$\{10\bar{1}0\}\langle2\bar{1}10\rangle$	1.40 (Ti)	99.99 (Ti)	Low c/a
	Zr		0.65–0.7 (Zr)	99.0 (Zr)	
	Rh				
	Hf				
	Be	$\{0001\}\langle2\bar{1}10\rangle$	0.14	Zone refined	Low c/a
	Cd	$\{0001\}\langle2\bar{1}10\rangle$ (dominant)	0.058	99.96	High c/a
	Zn		0.018	99.999	
	Mg		0.04–0.05	99.95	
	Mg	$\{10\bar{1}0\}\langle2\bar{1}10\rangle$	4.00	99.95	High c/a
		$\{11\bar{2}2\}\langle10\bar{1}0\rangle$			
		$\{10\bar{1}1\}\langle11\bar{2}0\rangle$			
	TiZr Hf	$\{10\bar{1}1\}\langle11\bar{2}0\rangle$	...	...	
		$\{0001\}\langle11\bar{2}0\rangle$			
		$\{11\bar{2}2\}$			
	Zn	$\{11\bar{2}2\}\langle11\bar{2}\bar{3}\rangle$	1.05–1.60	99.999	High c/a
Rhombohedral-hexagonal	Bi	$\{111\}\langle10\bar{1}\rangle$	...	...	
	Hg	$\{11\bar{1}\}\langle\bar{1}10\rangle$	...	...	
		$\{11\bar{1}\}\langle011\rangle$			
Diamond cubic	Cd	$\{11\bar{2}2\}\langle11\bar{2}\bar{3}\rangle$	...	...	High c/a
	Te	$\{10\bar{1}0\}\langle11\bar{2}0\rangle$?	...	...	
	Bi	(111) and (11$\bar{1}$) [10$\bar{1}$] and [101]	...	...	
	Ge(c)	$\{111\}$	...	...	
	Si whiskers(d)		28 (torsion test)		
Tetragonal	βU	$\{110\}\langle001\rangle$	...	...	
	βSn	$\{110\}\langle001\rangle$	...	...	
		$\{100\}\langle001\rangle$			
		$\{100\}\langle011\rangle$?			
		$\{10\bar{1}\}\langle101\rangle$			
		$\{121\}\langle101\rangle$			
		Ordered lattices			
B1-NaCl type	NaCl AgCl	$\{110\}\langle1\bar{1}0\rangle\{100\}$	...	...	
B1-CsCl type	LiTl MgTl	$\{110\}\langle100\rangle$	...	...	
	AuZn				
	AgMg	$\{321\}\langle111\rangle$			
	βCuZn	$\{110\}\langle111\rangle$	...	...	

(a) Minor slip system or slip system when load unfavorably oriented for dominant system. (b) Ref A1.11, A1.12. (c) Ref A1.13. (d) Ref A1.13. Source: Ref A1.14

which is unique to a particular crystal lattice structure. The cleavage planes are low-index planes: (100) and (0001), respectively, for bcc and hcp crystals. It is known that bcc lattices also cleave on {110} planes, and that {11$\bar{2}$0} planes also act as cleavage planes for beryllium. Face-centered cubic metals normally do not cleave, because slip systems are prevalent and are unaffected by temperature in fcc systems. However, some fcc metals may cleave only when subjected to severe environmental conditions—for example, austenitic stainless steels and nitrogenated austenitic stainless steels in stress-corrosion cracking conditions (Ref A1.6). The separation is caused by the stress normal to the cleavage plane, which is the resolved normal stress. If the cross-sectional plane labeled the slip plane in Fig. A1.12 is considered to be the cleavage plane, the stress normal to that plane will be:

$$\sigma_R = \frac{P}{A} \cos^2 \phi \qquad \text{(Eq A1.3)}$$

It is believed that there is a competition between slip and cleavage. When the external load on a crystal increases, both the resolved shear stress and the resolved normal stress increase. Which stress component, normal or shear, first reaches a critical value determines the failure mode. As we have frequently pointed out in this book, low temperature and fast loading rate promote brittle fracture, which is particularly relevant to bcc metals. Recently, convincing evidence has shown cleavage fracture in crystalline materials to occur by prior dislocation motion to create cleavage nucleus. Therefore, it is conceivable that a shear stress is also required on one or more slip planes. In the last half of the

20th century, a number of models that refine the definition of cleavage fracture were developed, most of them based on dislocation theory. See the comprehensive reviews in Ref A1.5 and A1.15.

The term "ideal cleavage" or "pure cleavage" refers to totally brittle fracture in metals at the microscopic level. Pure cleavage occurs only under well-defined conditions, primarily when the component is in single-crystal form and has a limited number of slip systems; it is correctly described as "cleavage fracture." In metals, however, it often contains varying fractions of transgranular cleavage and evidence of plastic deformation by slip. When both fracture processes operate simultaneously, especially in the fracture of quenched and tempered steels, the fracture process is called "quasi-cleavage." The dividing line between cleavage and quasi-cleavage is somewhat arbitrary.

A1.3 Mechanical Twinning

Twinning is another deformation mechanism in metals, and is particularly important in materials where slip is restricted. Twinning produces a reorientation of the lattice, resulting in a region that is a mirror image of the parent lattice (Fig. A1.16). Twins are a type of planar (two-dimensional) defect in the crystal lattice. The lattices of the twinned and untwinned parts of the crystal are mirror images of each other by reflection in some simple crystallographic plane. The plane of symmetry between the two portions is called the twin plane.

Like dislocations, surface (planar) imperfections of a crystal lattice also occur in conjunction

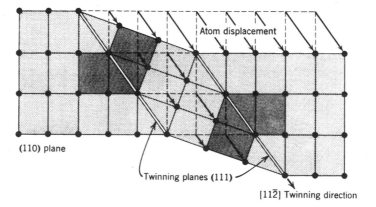

Fig. A1.16 Schematic of twinning as it occurs in an fcc lattice

with the plastic deformation of metals. In addition, planar-type crystal imperfections may influence the mechanical and metallurgical behavior of metals during deformation. However, twins do not affect mechanical behavior to the same degree as stacking faults (an important exception being low-temperature deformation of bcc metals). Twins generally play only a minor role in plastic flow, while stacking faults (discussed in section A1.4) can affect the work-hardening behavior of metals with close-packed structures (such as fcc metals). The actual amount of strain that is gained from twinning is very small—typically less than 5%. The true benefit of twinning is that the lattice inside the twin is frequently reoriented such that the slip systems are more favorably aligned with respect to the applied stress, which can allow some limited plastic deformation to occur.

Twins may be produced by either crystal growth or plastic deformation. The former are referred to as annealing twins, while the latter are termed either mechanical or deformation twins. Mechanical twinning involves shear displacement like slip, but the twinning mechanism is essentially different in that the crystal orientation changes in the deformation region of twinning. Furthermore, it is not known whether there is a critical resolved shear stress for twinning. The shear stress at which twinning occurs is influenced by prior deformation.

Mechanical twinning is not a slow, continuous process like slip. It occurs rather suddenly and happens only once. In slip, the atoms on one side of the slip plane all move an equal distance (Fig. A1.17), and the orientation of crystals above and below the slip plane remains more or less unchanged. In contrast, the atoms during twinning deformation move distances proportional to their distance from the twin plane (Fig. A1.17). Thus, slip involves discrete multiples of the atomic spacing, while twinning involves cooperative movement of atoms, producing a macroscopic shear.

Twinning occurs on a specific crystallographic plane in a definite direction for each crystal structure:

- (111)/[112] for twinning in the fcc lattice
- (112)/[111] for twinning in the bcc lattice
- (101̄2)/[1̄011] for twinning in the hcp lattice

Mechanical twins are produced in bcc and hcp metals under conditions of rapid rate of loading (shock loading) and decreased temperature. Deformation twinning is not commonly seen in the fcc lattice, because of the substantial opportunities for slip. However, there is some evidence in the literature that fcc materials can, in fact, form deformation twins (Ref A1.16, A1.17), but only with considerable difficulty. Annealing twins are more commonly found in fcc crystals.

Alternatively, hcp materials mechanically twin more easily—so easily, in fact, that sometimes mechanical polishing of a metallographic specimen can introduce artifacts. There are considerable differences, however, in the ease of twinning among alloys having an hcp structure. In hcp metals, twinning occurs in the [101̄1] direction along the (101̄2) plane. In zinc and cadmium, which have atoms elongated along the c-axis, twinning occurs by compression along the c-axis. On the other hand, in magnesium and titanium, which have atoms compressed along the c-axis, twinning occurs by tension along the c-axis. Tin, which has a tetragonal structure, twins very easily when it is deformed, giving rise to a squeaking sound referred to as "tin cry."

Deformation twinning also occurs in materials having a bcc lattice, with twinning more likely at low temperature and elevated strain rates. A twinned structure in this lattice can be produced by a shear of $(1/\sqrt{2})$ or $(\sqrt{2})$ in a $\langle 111 \rangle$ direction on a $\{112\}$ plane (Ref A1.4). Strain rates commonly encountered in tensile testing on the order of 10^{-3} to 10^{-2}/min do not typically cause deformation twinning at room temperature, but a hammer blow will cause twinning. When deformation twinning does occur during plastic straining, it is usually activated after prior deformation by slip. That is, single crystals may be oriented to yield by twinning, but polycrystalline material must strain harden sufficiently in order to favor twinning deformation over slip deformation at room temperature.

Role of Twinning in Fracture. Crack formation is generally considered the result of intersecting deformation twins. The appearance of twins before cleavage fracture is a general phe-

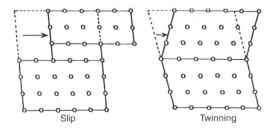

Fig. A1.17 Comparison of slip and twinning deformation

nomenon in the case of both single-crystal and polycrystalline bcc and hcp materials. Twinning often is associated with brittle or ductile-to-brittle transition behaviors in bcc and hcp metals, because only a relatively small fraction of the total volume of a crystal is reoriented by twinning. The twin plane of the annealing twin in the fcc crystal acts like a subgrain boundary and thus provides a secondary barrier for dislocation pileup. It also is similar to a depletion zone, which is susceptible to stress-corrosion cracking.

Twins play a number of roles in fracture. Reference A1.6 provides a comprehensive review.

A1.4   Stacking Faults

Another type of planar defect in a crystalline structure is known as a stacking fault. This type of internal surface defect occurs only in closely packed structures (such as fcc and hcp lattices), and is associated with the stacking sequence of close-packed planes. For example, consider the stacking of planes in both the fcc and hcp systems. In an fcc structure, closest-packed planes can have any {111} orientation, such as the (111) plane shown in Fig. A1.18. When atoms are stacked on top of a close-packed plane in an fcc system, there are two positions of voids between the atoms for stacking the next layer of atoms. Therefore, a second and third layer can follow in a different manner, with the fourth layer being another close-packed plane (Fig. A1.19). The order of distinct crystallographic planes in an fcc system thus has the general stacking sequence of *ABC-ABC-ABC*, and so on.

Now consider the stacking sequence on the closest-packed (0001) plane in an hcp system (Fig. A1.18 and A1.19). In this case, the stacking sequence consists of two distinct layers, *AB-AB-AB*, and so on. The hcp closest-packed plane also has completely the same (equivalent) planes from a crystallographic standpoint as the (111) close-packed planes in an fcc system. That is, the atoms being stacked on a closest-packed plane (the *A* plane in either hcp or fcc) have the alternative of stacking like an hcp or an fcc structure. This equivalence of closest-packed planes means that the stacking of atomic layers in an fcc system may sometimes follow the *AB-AB* sequence rather than an *ABC-ABC* sequence. Atoms do not necessarily have to follow the fcc structure when stacked on a close-packed plane. The atoms being stacked on a closest-packed fcc plane sometimes position themselves as a thin layer of hcp-like material (with an *AB-AB*-type stacking). This "mistake" in stacking and the sequencing of crystallographic planes results in an internal surface defect in the crystal structure that is called a stacking fault.

Stacking faults occur only in close-packed systems, and they may be produced during grain growth or when a partial dislocation moves through a lattice. A full dislocation produces a displacement equivalent to the distance between the lattice points, while a partial dislocation produces a movement that is less than a full distance. If stacking faults can occur easily in a metal, the metal has a low stacking-fault energy. Stacking-fault energy is related to surface energy of the fault and depends on the width of the fault and the repulsive energy of dislocation pairs. Some fcc metal structures have high stack-

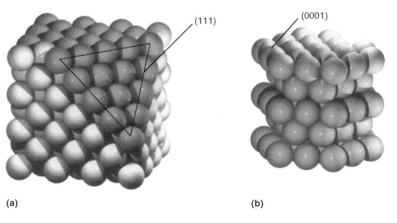

(a)   (b)

Fig. A1.18   Ball models of stacking. (a) An fcc structure with a cross section shown for the (111) close-packed plane. (b) An hcp structure. Source: Ref A1.1

ing-fault energies, while copper has a low stacking-fault energy. Alloying additions of such elements to another metal often significantly reduce the stacking-fault energy. Additions of zinc, aluminum, and silicon to copper, for example, produce this effect.

The thickness of stacking faults is only several atomic diameters in the direction normal to the close-packed planes. Stacking faults in fcc materials generally occur as ribbons (Fig. A1.20). The fault extends normal to the plane over distances that are large compared to an

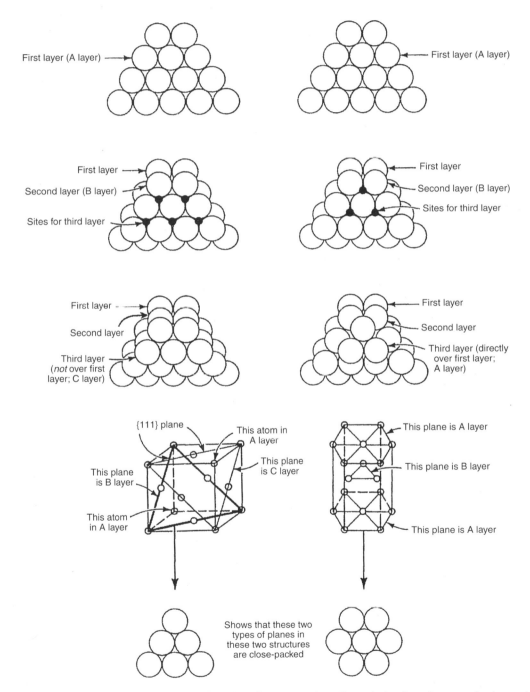

Fig. A1.19 Illustration of the generation of either a fcc or hcp structure depending on the locations of atoms on the close-packed third layer

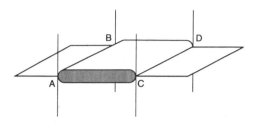

Fig. A1.20 Three-dimensional sketch of a stacking fault in an fcc crystal. The fault is a narrow ribbon, several atomic diameters in thickness, bounded by partial dislocations (the lines *AB* and *CD*). Source: Ref A1.18

atomic size. The ribbon width (the distance between points *A* and *C* or *B* and *D* in Fig. A1.20) is highly variable, ranging in size from the order of one to many atomic diameters. Generally, if the energy of the hcp and fcc allotropic forms of the solid are comparable, the width is large, and vice versa. The boundaries at the edges of the faults (lines *AB* and *CD* in Fig. A1.20) are defined by a special type of dislocation that accommodates the disregistry between the hcp and fcc stacking at the boundaries. Stacking faults play an important role in the work-hardening behavior of some fcc metals and alloys. If their width is large, the material work hardens more than if it is narrow. Stacking-fault energy also influences the occurrence of recovery and recrystallization.

Twin boundaries are somewhat akin to stacking faults, although differences exist between these types of planar defects. The stacking sequence across a twin boundary is *ABCABACBA*; the position of the boundary is denoted by *B*. Note that to either side of this boundary the stacking sequence is typical of fcc. (*ACBACB* represents the same stacking as does *ABCABC*, in that close-packed layers repeat every fourth layer.) At the twin boundary, a layer of *ABA* (hcp stacking) exists. Differences between twins and stacking faults arise from the different positioning of the atoms in the atomic plane twice removed from the respective boundaries. Twins also typically have a width much greater than that of stacking faults (Ref A1.19).

REFERENCES

A1.1. M. Tisza, *Physical Metallurgy for Engineers,* ASM International and Freund Publishing, 2001

A1.2. G.E. Dieter, Jr., *Mechanical Metallurgy,* 3rd ed., McGraw-Hill, 1986

A1.3. A.H. Cottrell, *Theoretical Structural Metallurgy,* St. Martin's Press, 1960

A1.4. A.H. Cottrell, *Dislocations and Plastic Flow in Crystals,* Oxford University Press, 1953

A1.5. W.T. Becker and S. Lampman, Fracture Appearance and Mechanisms of Deformation and Fracture, *ASM Handbook,* Vol 11, *Failure Analysis and Prevention,* ASM International, 2002, p 559–586

A1.6. W.T. Becker and D. McGarry, Mechanisms and Appearances of Ductile and Brittle Fracture in Metals, *ASM Handbook,* Vol 11, *Failure Analysis and Prevention,* ASM International, 2002, p 587–626

A1.7. H. Baker, Structure and Properties of Metals, *Metals Handbook Desk Edition,* 2nd ed., ASM International, 1998, p 85–121

A1.8. A.J. McEvily, *Metal Failures: Mechanisms, Analysis, Prevention,* John Wiley & Sons, 2002

A1.9. D. Hull and D.J. Bacon, *Introduction to Dislocations,* Pergamon Press, 1984

A1.10. Plastic Deformation Structures, *ASM Handbook,* Vol 9, *Metallography and Microstructures,* ASM International, 1985, p 684

A1.11. R.L. Hazif and J.P. Poirier, *Scr. Metall.,* Vol 6, 1972, p 367

A1.12. C.S. Barrett and T.B. Massalski, *Structure of Metals,* 3rd ed., McGraw-Hill, 1966

A1.13. D.C. Drucker and J.J. Gilman, Eds., Gordon & Breach, 1963

A1.14. W.T. Becker, University of Tennessee, private communication

A1.15. W. Soboyejo, *Mechanical Properties of Engineered Materials,* Marcel Dekker, 2003

A1.16. R.W. Cahn, Survey of Recent Progress in the Field of Deformation Twinning, *Deformation Twinning,* R.E. Reed-Hill et al., Ed., Gordon & Breach, 1964, p 1

A1.17. G.Y. Chin, Formation of Deformation Twins in fcc Crystals, *Acta Metall.,* Vol 21, 1973, p 1353

A1.18. T.H. Courtney, *Mechanical Behavior of Materials,* McGraw-Hill, 1990

A1.19. T.H. Courtney, Fundamental Structure-Property Relationships in Engineering Materials, *ASM Handbook,* Vol 20, *Materials Selection and Design,* ASM International, 1997

APPENDIX 2

Close-Form Representation of Tangential Stress Distribution at Circular and Elliptical Holes

CONSIDER THE CASE of an elliptical hole at the center of a very long strip subjected to far-field uniform loading (Fig. A2.1). The strip material can be either isotropic or orthotropic. The local stresses in the vicinity of the hole have been raised above the nominal level due to redistribution of load paths. The problem also involves whether the hole is inside a sheet of infinite width, or finite width, and whether or not a close-form solution is available.

We will tackle this problem by making a list of all the available mathematical solutions. Then we will present a close-form solution adopted from Ref A2.1. A multipurpose equation, it can be used for any combination of hole shape (circular or elliptical), hole size to sheet width ratio, and material (isotropic or orthotropic).

A2.1 Infinite Sheet

For the case of an isotropic sheet, stress-concentration factors at the tip of the major axis of an elliptical hole in an infinite sheet have been given by Inglis (Ref A2.2) as:

$$K_{t\infty} = 2\left(\frac{a}{b}\right) + 1 \qquad \text{(Eq A2.1a)}$$

or

$$K_{t\infty} = 2\left(\frac{a}{\rho}\right)^{1/2} + 1 \qquad \text{(Eq A2.1b)}$$

where $\rho = b^2/a$ by definition. The subscript ∞ implies that the load is applied at infinity and the

stress-concentration factor is actually the ratio of the local stress to the far-field gross area stress. This value reduces to 3 for a circular hole, identical to those given by Kirsch (Ref A2.3). Numerical solutions for the stresses along the X-axis for both cases are also given by Ref A2.2 and A2.3. A close-form representation of the stress distribution of the latter has been given by Timoshenko and Goodier (Ref A2.3) and is presented as Eq 1.74. That is:

$$\sigma_\theta = S\left(1 + \frac{1}{2}\frac{r^2}{X^2} + \frac{3}{2}\frac{r^4}{X^4}\right) \qquad \text{(Eq A2.2)}$$

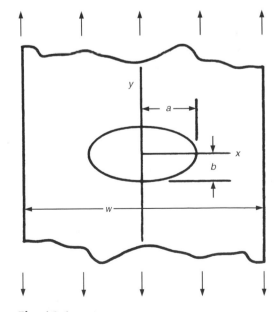

Fig. A2.1 Elliptical hole in a finite-width plate

Here S is the far-field gross area stress, r is hole radius, and X is the distance (along the X-axis) from the center of the circular hole. Note that the σ_θ/S ratio at $X = r$ is designated the stress-concentration factor, $K_{t\infty}$.

As for the orthotropic material, numerical solution has been given by Lekhnitskii (Ref A2.4), including an equation that represents the stress-concentration factor at the edge of the hole:

$$K_{t\infty} = 1 + \frac{a}{b}\sqrt{\frac{E_1}{G_{12}} - 2\nu_{12} + 2(E_1/E_2)^{1/2}}$$

$$\text{(Eq A2.3)}$$

where E_1, E_2, and G_{12} are defined in Chapter 8. This equation reduces to the isotropic plate solution for $E_1 = E_2$.

An extension of Eq A2.2 to include orthotropic plates is given by Nuismer and Whitney in Ref A2.5:

$$\sigma_\theta = \frac{S}{2}\left[2 + \frac{r^2}{X^2} + 3\frac{r^4}{X^4}\right.$$
$$\left. - (K_t - 3)\left(5\frac{r^6}{X^6} - 7\frac{r^8}{X^8}\right)\right] \quad \text{(Eq A2.4)}$$

The value of $K_{t\infty}$ is given by Eq A2.3 with $a = b$. Again, for $E_1 = E_2$, this equation reduces to Eq A2.2.

Note that Eq A2.4 applies only to circular holes in isotropic or orthotropic sheets. Saff's

equation (Ref 2.1), which was proposed later, works the same, except that it also applies to elliptical holes. The Saff equation is described later in this appendix.

A2.2 Finite-Width Sheet

Many published works have dealt with finite-width effects on circular and elliptical holes (Ref A2.5–A2.9). The most familiar is the Howland solution for a circular hole inside a finite-width sheet. A set of numerical values covering a range of hole diameter to plate width ratios has been presented in Fig. 1.48. Nuismer and Whitney (Ref A2.5) justified the use of a simple finite-width correction factor by matching Eq 8.30 to the result of a finite element analysis. The Saff close-form equation mentioned earlier also is designed to work with sheets of finite widths. This equation, along with plots showing comparisons with numerical solutions, is described below.

A2.3 Saff's Close-Form Equation

Recognizing that the local stresses at the hole edge decay from a value of $K_{t\infty}$ to a nominal level at distances away from the hole, the Saff equation uses an assumed mathematical function to represent such stress distribution:

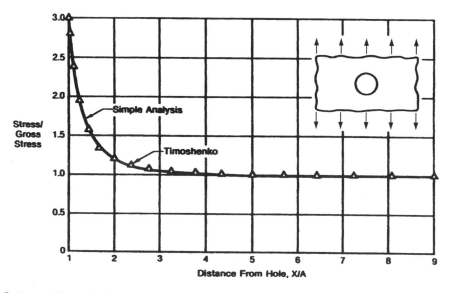

Fig. A2.2 Tangential stress distribution adjacent to a circular hole in an isotropic sheet of infinite width. Comparison of the simple analysis (Eq A2.5b) with Timoshenko's equation (Eq A2.2). Source: Ref A2.1

$$\frac{\sigma_\theta}{S} = \alpha + \beta\left[1 + \frac{X - a}{\rho}\right]^{-\gamma} \quad \text{(Eq A2.5a)}$$

This equation reduces to:

$$\frac{\sigma_\theta}{S} = \alpha + \beta\left[1 + \frac{X}{r}\right]^{-\gamma} \quad \text{(Eq A2.5b)}$$

for a circular hole. These equations apply to holes in a finite-width sheet. They are also suitable for an infinite sheet when proper values are assigned to the variables α, β, and γ, which are defined as:

$$\alpha = K_t - \beta \quad \text{(Eq A2.6)}$$

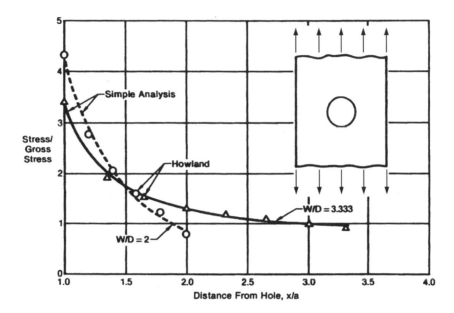

Fig. A2.3 Tangential stress distribution adjacent to a circular hole in an isotropic sheet (hole diameter to width ratio: $D/W = 0.3$, 0.5). Comparison of the simple analysis (Eq A2.5b) with Howland's solution. Source: Ref A2.1

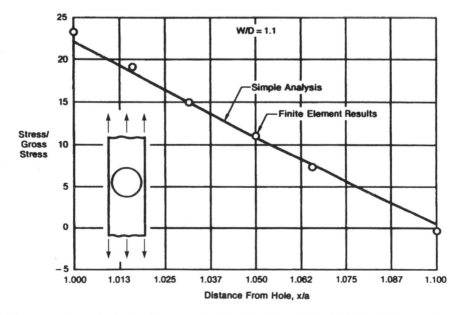

Fig. A2.4 Tangential stress distribution adjacent to a circular hole in an isotropic sheet ($D/W = 0.9091$). Comparison of the simple analysis (Eq A2.5b) and finite element analysis results. Source: Ref A2.1

where K_t is the stress-concentration factor for the hole in a finite-width sheet, on the basis of gross area stress. The gross area stress finite-width K_t is related to the net section area stress K_{tn} by:

$$K_t = K_{tn}/(1 - 2a/W) \qquad \text{(Eq A2.7)}$$

where

$$K_{tn} = 2 + f_1 f_2^2 + f_1 f_2^4 + 0.643 f_3 (1 - f_4^2)$$
$$+ 0.167 f_3 (1 - f_4^4) + 0.109 f_3 f_5 (2a/W)$$
$$\text{(Eq A2.8)}$$

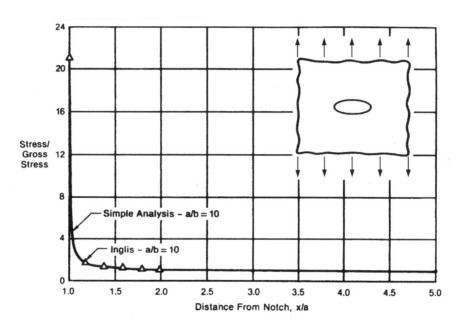

Fig. A2.5 Tangential stress distribution adjacent to an elliptical hole in an isotropic sheet ($D/W = 0$). Comparison of the simple analysis (Eq A2.5a) with Inglis' solution. Source: Ref A2.1

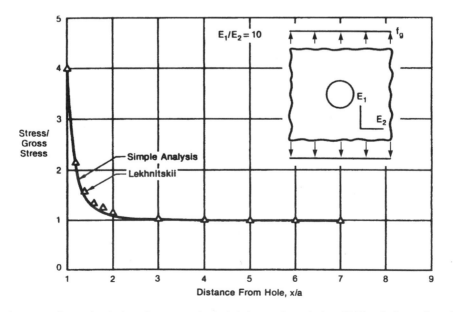

Fig. A2.6 Tangential stress distribution adjacent to a circular hole in an orthotropic sheet ($D/W = 0$). Comparison of the simple analysis (Eq A2.5a) with the Lekhnitskii solution. Source: Ref A2.1

where

$$f_1 = \frac{K_{t\infty}}{2} - 1 \qquad \text{(Eq A2.9a)}$$

$$f_2 = 1 - 2a/W \qquad \text{(Eq A2.9b)}$$

$$f_3 = \frac{a}{b} - 1 \qquad \text{(Eq A2.9c)}$$

$$f_4 = \frac{4a}{W} - 1 \qquad \text{(Eq A2.9d)}$$

$$f_5 = 1 - \left(\frac{2a}{W}\right)^{100} \qquad \text{(Eq A2.9e)}$$

and $K_{t\infty}$ is given by Eq A2.1(a). The second term in Eq A2.6 is equal to:

$$\beta = \beta_1/(\beta_2 \cdot \beta_3 - \beta_4) \qquad \text{(Eq A2.10)}$$

where

$$\beta_1 = 1 - K_t \cdot (1 - 2a/W) \qquad \text{(Eq A2.11a)}$$

$$\beta_2 = 2b^2/[aW \cdot (1 - \gamma)] \qquad \text{(Eq A2.11b)}$$

$$\beta_3 = 1 - K_t \cdot (1 - 2a/W) \qquad \text{(Eq A2.11c)}$$

$$\beta_4 = \left[\frac{aW}{2b^2} + 1 - \left(\frac{a}{b}\right)^2\right]^{1-\gamma} - 1 \qquad \text{(Eq A2.11d)}$$

Finally, the exponential γ in Eq A2.5(a) and (b) equals:

$$\gamma = C \cdot K_t/(K_t - 1) \qquad \text{(Eq A2.12)}$$

where

$$C = \left(\frac{1}{2} + \frac{5}{3}\sqrt{\frac{b}{a}}\right)\sqrt{1 - \frac{2a}{W}} \qquad \text{(Eq A2.13)}$$

A2.4 Comparison between Saff's Close-Form Equation and the Analytical Solutions

Figures A2.2 to A2.7 are graphic plots that present favorable comparisons between the Saff equation and classical solutions for six sample problems:

- Circular hole in an infinite sheet
- Circular hole in finite-width sheets with $2a/W$ ratios of 0.3 and 0.5
- Circular hole in a narrow strip of very high hole diameter to width ratio ($2a/W = 0.9091$)
- Elliptical hole in an isotropic sheet of infinite width
- Circular hole in an orthotropic sheet of infinite width

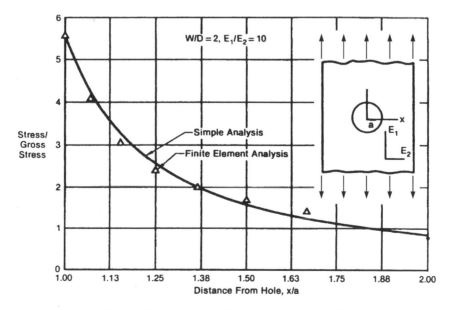

Fig. A2.7 Tangential stress distribution adjacent to a circular hole in an orthotropic sheet ($D/W = 0.5$). Comparison of the simple analysis (Eq A2.5b) and finite element analysis results. Source: Ref A2.1

- Circular hole in an orthotropic sheet where $2a/W = 0.5$

In these figures, a curve obtained using Saff's equation is compared either to an appropriate classical solution or to the result of a finite element analysis. Nearly perfect matches result in all cases.

REFERENCES

A2.1. C.R. Saff, "Durability of Continuous Fiber Reinforced Metal Matrix Composites," Report AFWAL-TR-85-3107, Air Force Flight Dynamics Laboratory, 1986

A2.2. C.E. Inglis, Stresses in a Plate due to the Presence of Cracks and Sharp Corners, *Trans. Inst. Naval Arch.*, Vol 55, 1913, p 219–230

A2.3. S. Timoshenko and J.N. Goodier, *Theory of Elasticity*, 3rd ed., McGraw-Hill, 1970

A2.4. S.G. Lekhnitskii, *Anisotropic Plates* (translated from the 2nd Russian ed.), Gordon & Breach, 1968

A2.5. R.J. Nuismer and J.M. Whitney, Uniaxial Failure of Composite Laminates Containing Stress Concentrations, *Fracture Mechanics of Composites*, STP 593, ASTM, 1975, p 117–142

A2.6. R.C. Howland, On the Stresses in the Neighborhood of a Circular Hole in a Strip Under Tension, *Philos. Trans. R. Soc. (London) A,* Vol 119, 1930, p 49–86

A2.7. R.G. Belie and F.J. Appl, Stress Concentrations in Tensile Strips with Large Circular Holes, *Exp. Mech.,* April 1972, p 190–195

A2.8. H. Fuehring, Discussion to the paper: Stress Concentrations in Tensile Strips with Large Circular Holes, *Exp. Mech.,* June 1973, p 255–256

A2.9. A.J. Durelli, V.J. Parks, V.J. Lopardo, and T.L. Chen, Normalized Stresses Around an Elliptic Hole in a Finite Plate of Linear Material Subjected to Large Uniform Inplane Loading, *Exp. Mech.,* Oct 1973, p 441–444

APPENDIX 3

Nonarbitrary Crack Size Concept for Fatigue Crack Initiation

LOW-CYCLE FATIGUE CONCEPTS based on the local stress-strain approach have been successfully used to estimate the crack initiation lives of notched structures, while analytical techniques based on fracture mechanics concepts have been developed to estimate the crack propagation lives of structures containing cracks. However, predicting the total (initiation and propagation) fatigue life of notched and cracked structures is still quite difficult, because accurate methods of combining these two concepts have not been developed.

Socie et al. (Ref A3.1, A3.2) have proposed an analytical model for combining the damage rate due to crack initiation mechanisms (from low-cycle fatigue concepts) and the damage rate due to crack propagation mechanisms (from fracture mechanics concepts). This model provides a nonarbitrary definition of fatigue crack initiation length, which serves as an analytical link between initiation and propagation analyses and appears to have considerable merit in estimating the total fatigue life of notched and cracked structures. The concept is explained here.

A fatigue crack is assumed to initiate when the fatigue damage due to crack propagation mechanisms exceeds the fatigue damage due to crack initiation or strain cycle mechanisms. To illustrate this concept, consider a center-notched component subject to remote cyclic stresses. Imagine a series of microelements ahead of the

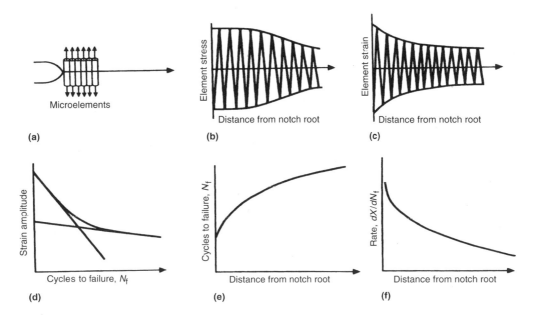

Fig. A3.1 Schematic illustration of crack initiation concept

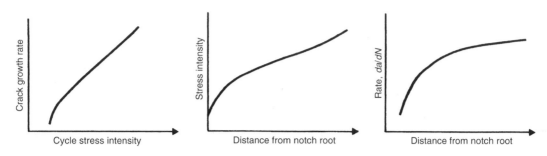

Fig. A3.2 Schematic illustration of crack propagation concept

notch, along the potential crack path, as shown in Fig. A3.1. In the absence of a primary or dominant crack, the material can be considered a continuum. Stress and strain ranges for each arbitrary element along the potential crack path can therefore be determined using an appropriate stress analysis based on the theory of elasticity. The total life of each element can be calculated from the fatigue properties (i.e., strain amplitude/cycles to failure curve) of the material and the strain ranges that occur along the potential crack path. It is then possible to construct a curve with the dimensions of distance (length) versus life (cycles) similar to that obtained in crack propagation studies (Fig. A3.1e). The reciprocal derivative of this curve represents the rate of crack initiation, as illustrated in Fig. A3.1(f). This rate can be interpreted as the rate at which the crack advances due to breaking the elements through strain-cycle fatigue mechanisms. It is important to note that this approach is independent of the selected element size, because the derivative is obtained analytically.

Now, suppose a fatigue crack has developed to a point some distance from the notch root. Since the applied loads are known, an effective stress intensity factor accounting for notch geometry and residual stress can be calculated for various crack lengths along the potential crack path. Fatigue crack growth rates can be determined from the crack growth rate of the material and the cyclic stress intensity. This procedure is illustrated in Fig. A3.2.

The crack initiation and propagation rates can be superimposed on a single plot of damage rate versus distance from the notch root, as shown in Fig. A3.3. At some distance from the notch root, the rate of damage from crack initiation mechanisms equals the rate of the damage from crack propagation mechanisms. This distance, defined as a_i, is regarded as the crack length that can be expected to develop by crack initiation mecha-

nisms. When the crack length exceeds this "nonarbitrary initiation crack length," the crack will grow by crack propagation mechanisms (i.e., fracture mechanics).

Crack initiation life and crack propagation life can now be estimated. Crack initiation life is based on the strain range occurring at a_i and is defined as the fatigue life of the element located at distance a_i. Crack propagation life is then calculated using a_i as the initial crack length. Final crack lengths can be estimated from fracture toughness or limit load calculations. The total fatigue life is obtained by summing the crack initiation and crack propagation lives.

Note that a_i is not based on actual fatigue mechanisms and, as such, may not have any physical significance; however, the concept of a nonarbitrary fatigue crack initiation length is a viable technique for predicting the total fatigue lives of notched and cracked structures. At pres-

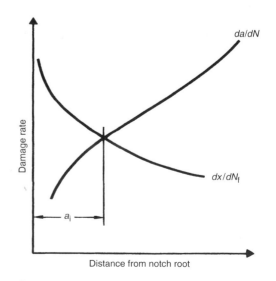

Fig. A3.3 Superposition of crack initiation and propagation rates

ent, it is probably best to consider the concept of a_i as providing a demarcation between smooth specimen and cracked specimen types of damage evaluation, and as a conceptual crack length that quantitatively indicates when a transition in the mode of damage analysis is necessary. Many variables, including geometry, material, and loading conditions, influence the value of a_i, which makes this approach rather complicated. However, this approach may be more reasonable than assuming a constant value for initiated crack size.

REFERENCES

A3.1. D.F. Socie, J. Morrow, and W.C. Chen, A Procedure for Estimating the Total Fatigue Life of Notched and Cracked Members, *Eng. Fract. Mech.*, Vol 11, 1979, p 851–860

A3.2. P. Kurath, D.F. Socie, and J. Morrow, "A Nonarbitrary Fatigue Crack Size Concept to Predict Total Fatigue Lives," Report AFFDL-TR-79–3144, Air Force Flight Dynamics Laboratory, 1979

APPENDIX 4

Fatigue Spectrum Editing

FEW STRUCTURAL APPLICATIONS involve only constant-amplitude load cycles throughout their intended service life. In a series of loading events commonly known as a "spectrum," the loading cycles can be very irregular and of random sequence. An essential element in fatigue life and crack growth life prediction is accurate interpretation of the load spectrum. Spectrum editing involves four elements:

- Deterministic arrangement of load sequences
- Realization of load characteristics
- Spectrum simplification (or condensation)
- Cycle counting

Each component in a vehicle or machine experiences different types of loads in service. Some are commonly known. For instance, pressure vessels are most likely subjected to constant-amplitude loading or predetermined stepwise loading. Load spectra for these sorts of components are defined directly in the form of a table. Otherwise, most load spectra are derived from a vast amount of strain gage data extracted from service records. The actual spectra are updated periodically after the vehicle or machine is delivered and in service. The conversion of load to stress may also be adjusted. The initially assessed fatigue lives are then recalculated.

Sometimes, a specification states only that during the service life of a given part, x number of loads occur at a certain level, x number of loads occur at another level, and so on. The so-called exceedance curve of an aircraft component spectrum is a typical example. A load exceedance curve indicates only the probability that the applied load level will equal or exceed certain values. The procedure for converting a cumulative load spectrum to a discrete load spectrum is to divide the load exceedance curve into regular segments. The difference between excessive exceedances at each increment is the number of load occurrences within that incre-

ment. There is no mention of load sequence or when these loads will occur. It is up to the engineer to make the placements. In addition, the characteristics of a part responding to applied loads must be recognized and proper adjustments made. The lug loads and bypass loads discussed in Chapter 3 are examples.

Another example is the ground-air-ground (GAG) cycle known to engineers in the aircraft industry. The variation in stress from ground taxi to takeoff, during flight, and back to ground taxi must be defined for a fatigue analysis. This stress variation, or GAG cycle, is one of the most important contributors to structural fatigue damage. Figure A4.1 illustrates two GAG cycle definitions commonly used in fatigue analysis: the mean-to-mean GAG cycle and the once-per-flight peak-to-peak GAG cycle. Of these, the peak-to-peak GAG is generally used; it provides the best correlation between spectrum fatigue testing conducted on notched coupons and fatigue analysis. Depending on use, more than one type of GAG cycle may occur during the operational life of a given vehicle. For instance, at least three kinds of GAG cycles have been identified for the Space Shuttle orbiter. These GAG cycles represent the prelaunch to docking phase, the undocking to reentry phase (all the way until touchdown) during the orbital flight, and the regular GAG cycles for the ferry mission. Generally, when fatigue cycles are counted via any of the "cycle counting" methods discussed later, insertion of GAG cycles is unnecessary.

Engineers always strive to reduce a lengthy load history to a more manageable size. Their goal is to attain a computationally efficient methodology that can reduce overall data requirements and solution times without sacrificing required accuracy. Numerous techniques have been proposed and are available in the literature. The statistic-based root mean square (rms) method, though not a cycle reduction/condensation routine, has proved effective in working with some random block spectra. This method requires no modeling other than calculating an rms value that ultimately represents all the loads in the entire spectrum. Then this rms load/stress level is directly used with the material *S-N* curve. Although the rms method is computationally efficient in making fatigue crack propagation life predictions (Ref A4.1), Fuchs and Stephens (Ref A4.2) have pointed out that this method is not suitable for general fatigue analysis. The RACETRAK method developed by Fuchs et al. (Ref A4.3) is considered most efficient and accurate for fatigue life predictions, and is discussed in Section A4.1.

After all the arrangements are made and a load history spectrum is finalized, the number of cycles in the spectrum should be counted. Because of the irregularities of load applications, recognizing a fatigue load cycle is not a straightforward matter. For example, a load may ascend, then drop slightly, then go up again, and continue this pattern several times before falling. Counting cycles has become an art form, and many methods have been proposed. Bannantine et al. (Ref A4.4) identified a dozen or so, along with various spin-off versions. Among them, the rain-flow and range-pair methods are most popular in the engineering community. The rain-flow method is widely used for identifying hysteresis loop closures, whereas the range-pair method is used for making fatigue life predictions with *S-N* curves. Each method has many versions, such as the loop closure routine described in Chapter 3. An excellent explanation

of the original rain-flow method is provided by Bannantine et al. (Ref A4.4). They also include a computer algorithm originally developed by Dowling. ASTM Standard E 1049-85 specifies the ground rules and provides counting algorithms for the rain-flow and range-pair methods, as well as four others (Ref A4.5). Section A4.2 discusses the range-pair method based on the instructions provided by Fuchs and Stephens (Ref A4.2), because their explanation of the method is much easier to understand.

A4.1 Spectrum Simplification: RACETRAK Method

The RACETRAK method, developed from a method presented in Ref A4.3, is a screening technique based on the determination of significant load reversals. The technique utilizes the analogy of a slalom course of a specified width that is represented by the load-time trace of the load spectrum. The number of direction changes required to traverse the course using the shortest route depends on the width specified for the track. Significant "corners" are identified as those involving a change in the sign of the slope of the shortest route. These may be thought of as "primary" direction changes. As the course width tends toward zero, a change in primary direction is indicated at every load level. As the course width increases toward the other limit, it becomes possible to traverse the course with very few changes in the primary direction. This

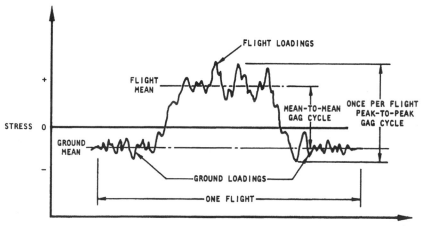

Fig. A4.1 GAG cycle

"course width" or "screening level" is given the variable name DMIN and is defined as a factor of the load used to normalize the spectrum. In the case of the spectra used for demonstration here, this normalizing load has been taken to be equal to the value of the maximum tensile spectrum load.

Figure A4.2 shows the application of the RACETRAK technique to the first few cycles of the F-18 aircraft main landing gear outboard trunnion spectrum. Note that the primary load levels are identified and stored in the order that they occur. As shown in Table A4.1, the load cycles decreased to 333 cycles from an original 727 cycles—a significant reduction. Table A4.2 summarizes the effectiveness of the RACETRAK method for condensing the number of steps in the fatigue test spectra of nine F-18 aircraft components. Two DMIN levels (0.25 and 0.5) were used in each of these exercises. Analytical predictions were made using the LOOPIN8 computer code (Ref A4.7). This code includes a subroutine for RACETRAK, which was adopted from Ref A4.3, a subroutine for tracking hysteresis loop closure, and all the elements needed to perform the ε-N analysis and summation of fatigue damage. Both the original (uncut) spectra and the post-RACETRAK spectra were used for comparison.

The effectiveness of this approach is demonstrated in Fig. A4.3 to A4.5. The data points are test results of the uncut spectra; comparisons between analytically predicted lives and fatigue test data are shown in each figure. Comparing the stress versus life curves, it is evident that the predicted results for all three versions of each spectrum are almost the same.

The RACETRAK method, which is effective and reliable, can be used to simplify fatigue test spectra and shows potential for significantly reducing the test time required to obtain fatigue data. This however, remains to be verified by a comparative test program. The applicability of the RACETRAK method to fatigue crack propagation test and life prediction should also be explored.

A4.2 Cycle Counting: Range-Pair Method

Performing a fatigue life assessment—for example, making a damage calculation using Minor's summation rule—requires reducing the complex variable loading spectrum into a number of events that can be compared to the available constant-amplitude fatigue test data. This process of reducing a complex load history into a number of constant-amplitude events involves "cycle counting." In this section, the cycle counting technique will be applied to load histories. In general, these techniques can also be applied to other loading parameters, such as strain, stress, torque, moment, and load.

The ASTM definition of the range-pair method states that this method counts a range as

Table A4.1 Comparison of original and RACETRAK-reduced load spectra for F-18 main landing gear outboard trunnion

Maximum load	Minimum load	Cycles per step	Step
Original load spectra			
0.3068	0.1534	1.0000	1
0.2784	0.0	1.0000	2
0.4602	0.3182	1.0000	3
0.5938	0.0	1.0000	4
0.3352	0.1477	1.0000	5
0.3466	0.0	1.0000	6
↓	↓	↓	↓
0.3693	0.2102	1.0000	725
0.2841	0.0	1.0000	726
0.2386	0.0	1.0000	727
Screened spectra (DMIN = 0.25)			
0.3068	0.0	1.0000	1
0.5938	0.0	1.0000	2
0.3466	0.0	1.0000	3
↓	↓	↓	↓
0.2557	0.0	1.0000	331
0.3693	0.0	1.0000	332
0.3693	0.0	1.0000	333

Source: Ref A4.6

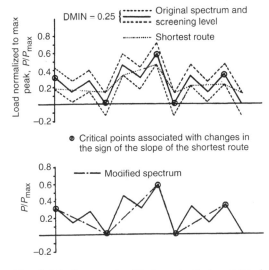

Fig. A4.2 Spectrum simplification using the RACETRAK technique. Source: Ref A4.6

a cycle if it can be paired with a subsequent loading in the opposite direction. According to the guidelines of Fuchs and Stephens (Ref A4.2), this method actually picks a pair of consecutive reversals and counts them as one cycle. It does not matter whether the reversals are counted from peak to valley or from valley to peak; they still are counted as one cycle. The procedure is:

1. First, the entire loading block is screened. All the reversal pairs with the smallest load range of the same magnitude are identified. In the artificial loading block shown in Fig. A4.6,

these pairs are BC, DE, JK, NO, and PQ. The magnitude for each pair of reversals is the same: 2 kips. Note that these pairs of reversals can be oriented in either the uploading or downloading direction. Also note that points A and V are not considered as reversals because they are the starting and ending of the block; there is no load reversal at either location.

2. The reversal points just identified are then removed from the graph (Fig A4.7). Each pair of the reversals is counted as one cycle and recorded as shown in Table A4.3. The load

Table A4.2 Comparison of steps in the original and RACETRAK-reduced spectra

Spectrum ID	Type of spectrum	Material	Steps per block		
			Original	DMIN = 0.25	DMIN = 0.5
A	Upper longeron FS 453	7079-T73511, ext, LT	4118	2332	503
B	Lower longeron FS 453	Ti-6Al-4V, ext, LT	4156	3099	1043
C20	Wing root bending moment	7050-T73651, plate, LT	3981	2255	543
D	Fin root bending moment	7050-T73651, plate, TS(a)	4553	903	81
E3	Horizontal spindle hinge moment	7050-T736, forging, LT(b)	7854	4513	1284
E4	Horizontal tail bending moment	HP9-4-.20, forging, LT	8202	4810	1853
F	Main landing gear outboard trunnion load	7050-T73651, plate, LT	727	333	87
G1	Arresting hook side load	HP9-4-.20, forging, LT	50	50	12
G2	Arresting hook drag load	Ti-6Al-4V, forging, LT(b)	300	300	100

(a) LT property was used for analysis. (b) Tests were done in forging material but analysis used plate fatigue data. Source: Ref A4.6

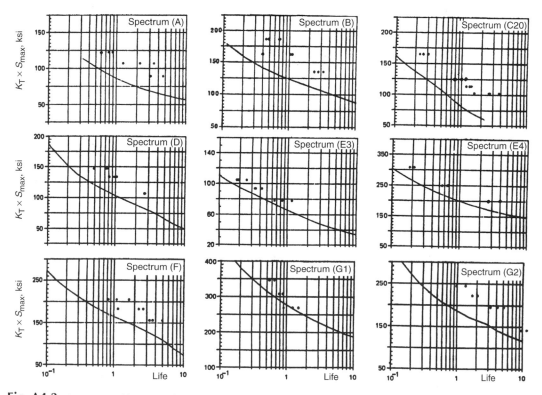

Fig. A4.3 Comparison of fatigue crack initiation test data and LOOPIN8 prediction using the uncut spectra. Source: Ref A4.6

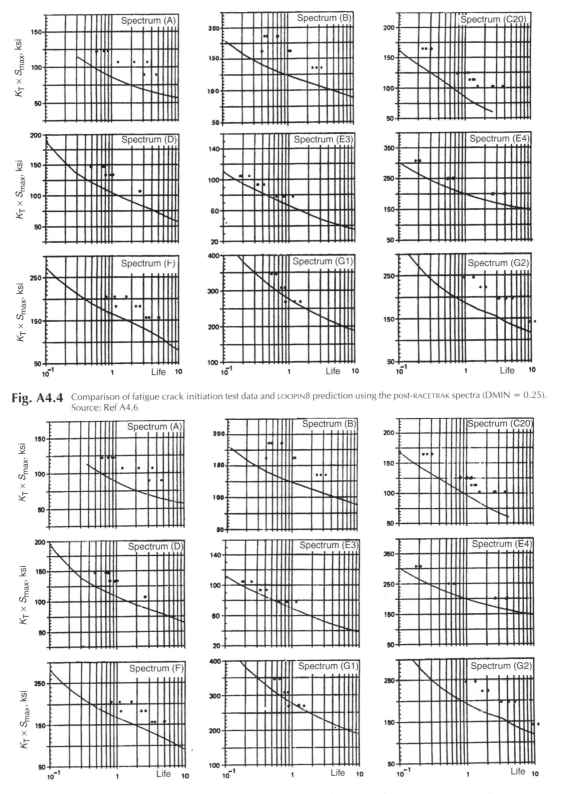

Fig. A4.4 Comparison of fatigue crack initiation test data and LOOPIN8 prediction using the post-RACETRAK spectra (DMIN = 0.25). Source: Ref A4.6

Fig. A4.5 Comparison of fatigue crack initiation test data and LOOPIN8 prediction using the post-RACETRAK spectra (DMIN = 0.50). Source: Ref A4.6

range and the mean load in conjunction with each cycle are also recorded.

3. Now look for the next smaller ranges in between the remaining reversals after connecting the open points, AF, IL, and MR (Fig. A4.8). They are ST (= 3 kips), HI (= 5 kips), and FG (= 6 kips).

4. Following removal of these six reversal points, only four reversal points remain: L, M, R, and U (Fig. A4.9).

5. After connecting AL and RU (Fig. A4.10), we find that both MR and RU have the same magnitude (i.e., 11 kips). But only three reversals are available for elimination. So, remove either MR or RU and then connect the points that are still open.

6. Shown as the last step in Fig. A4.11 and A4.12 (i.e., either option a or b of step 6), the last cycle of this spectrum will be LU or LM, having the same magnitude for their ranges (16 kips). In other words, the final result is the same whether MR or RU was removed in step 5.

As mentioned before, this procedure for range-pair counting is one of many existing versions. If it were done differently, the final spec-

trum would be different, as would the predicted fatigue lives. We also acknowledged that the range-pair method, which counts the reversals as the basis for fatigue damages, may not be suitable for low-cycle fatigue and/or fatigue crack propagation for the following reasons.

The range-pair method always retains a cycle that is equivalent to the largest bite in the spectrum block. In our example case, it is the cycle that contains the L and M (or L and U) reversals. This may be a good thing, because it is analo-

Table A4.3 Comparison of range-pair counted and real-time cycle-by-cycle spectra

	Range pair		Cycle-by-cycle		
Reversals	Load range, kips	Mean, kips	Upward half-cycle	Load range, kips	Mean, kips
BC	2	2.0	AB	3	1.5
DE	2	5.0	CD	5	3.5
JK	2	6.0	EF	5	6.5
NO	2	1.0	GH	3	4.5
PQ	2	1.0	IJ	6	4.0
ST	3	−1.5	KL	6	8.0
HI	5	3.5	MN	7	−1.5
FG	6	6.0	OP	2	1.0
MR	11	−0.5	QR	6	3.0
LU (or LM)	16	3.0	ST	3	−1.5
			UV	5	−2.5

Step 1: Original spectrum; remove reversal points
BC, DE, JK, NO, PQ

Fig. A4.6 Range-pair counting technique, step 1

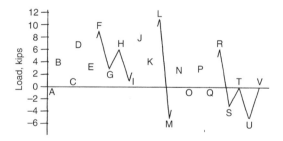

Step 2: Remaining reversal points as shown

Fig. A4.7 Range-pair counting technique, step 2

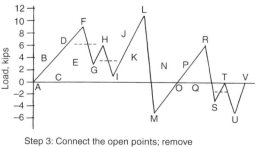

Step 3: Connect the open points; remove
range <7 kips (FG, HI, ST)

Fig. A4.8 Range-pair counting technique, step 3

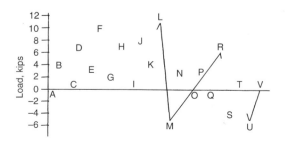

Step 4: Remaining reversal points as shown

Fig. A4.9 Range-pair counting technique, step 4

gous to the GAG cycle described earlier. However, the procedure calls for starting the counting with low-magnitude cycles. The tabulated record would show the magnitudes for all the fatigue cycles in an increasing order. The original character of the spectrum, of which the deter-

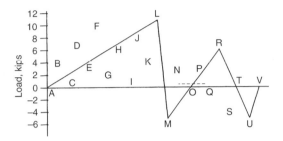

Step 5: Connect the open points AL and RU

Fig. A4.10 Range-pair counting technique, step 5

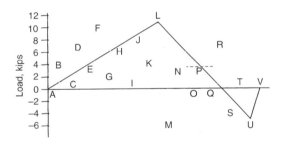

Step 6 (option a): Connect LU after removal of MR

Fig. A4.11 Range-pair counting technique, step 6(a)

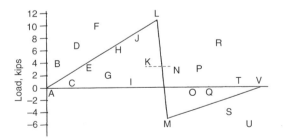

Step 6 (option b): Connect MV after removal of RU

Fig. A4.12 Range-pair counting technique, step 6(b)

ministic features are important, will be destroyed by this form of cycle ordering.

Based on the results of numerous theoretical and experimental investigations, we believe that a fatigue crack extends only while the load of a fatigue cycle is moving in the upward direction. In other words, the fatigue crack will extend while the load increases. The most logical way would be to count each uploading portion as a full cycle, and neglect the downloading events. A list of such a real-time cycle-by-cycle spectrum is placed next to the range-pair counted spectrum for comparison in Table A4.3. Clearly, these two spectra are in no way similar, even if a GAG cycle had been included in the latter.

REFERENCES

A4.1. J.M. Barsom, Fatigue Crack Growth under Variable-Amplitude Loading in Various Bridge Steels, *Fatigue Crack Growth under Spectrum Loads*, STP 595, ASTM, 1976, p 217

A4.2. H.O. Fuchs and R.I. Stephens, *Metal Fatigue in Engineering*, John Wiley & Sons, 1980

A4.3. II.O. Fuchs, D.V. Nelson, M.A. Burke, and T.L. Toomay, Shortcut in Cumulative Damage Analysis, *Advances in Engineering*, Vol 6, *Fatigue under Complex Loading: Analysis and Experiments*, R.M. Wetzel, Ed., SAE, 1977, p 145–162

A4.4. J.A. Bannantine, J.J. Comer, and J.L. Handrock, *Fundamentals of Metal Fatigue Analysis*, Prentice-Hall, 1990

A4.5. "Standard Practices for Cycle Counting in Fatigue Analysis," E 1049–85, *Annual Book of ASTM Standards*, ASTM

A4.6. P.G. Porter and A.F. Liu, "A Rapid Method to Predict Fatigue Crack Initiation, Vol I, Technical Summary," Report NADC-81010-60, Naval Air Development Center, 1983

A4.7. P.G. Porter, "A Rapid Method to Predict Fatigue Crack Initiation, Vol II, Computer Program User's Instructions," Report NADC-81010-60, Naval Air Development Center, 1983

APPENDIX 5

Stress Severity Factor

THE STRESS SEVERITY FACTOR, K_{SSF}, is an empirical factor that accounts for the geometrical stress concentration effect and additional effects such as variations in material properties, product quality, and other analytical uncertainties (Ref A5.1). It indicates the severity of stress at the potential crack initiation site. In any equation that includes K_t (Eq 3.13a, for example), K_{SSF} and K_t are interchangeable. That means the engineer may substitute K_{SSF} for K_t when the situation warrants. These two factors (K_t and K_{SSF}) both represent the ratio of the peak elastic notch stress at a critical area to the maximum nominal stress S_{max}. Actually, a K_{SSF} includes a baseline K_t in its mathematical expression. Other supplemental factors are added, depending on the situation. Thus, K_{SSF} is often used in fatigue analysis to replace the geometrical stress concentration factor K_t. This approach is equally applicable to both stress-based and strain-based analysis methods.

Stress severity factors are used in the aircraft industry for fatigue analysis of attachment joints and other areas that are prone to crack development. Attachment joint configurations are of interest to engineers because so many structural joints are distributed throughout an aircraft. The following discussion focuses on K_{SSF} in association with fastener holes in attachment joints.

As always, the local peak stress at a stress raiser point is equal to:

$$\sigma_{peak} = K_{SSF} \cdot S_{REF} \qquad \text{(Eq A5.1)}$$

where S_{REF} is the reference stress, which is usually the gross-section reference stress at the area of interest.

$$K_{SSF} = \frac{F_1 \cdot F_2 \cdot F_3 \cdot F_4 \cdot F_{FT}}{S_{REF}} [K_{SSF,P} + K_{SSF,H}]$$

$$\text{(Eq A5.2)}$$

where $K_{SSF,P}$ is associated with the pin, and $K_{SSF,H}$ is associated with the hole. Their expressions, along with definitions for all the supplemental factors, are given below:

$$K_{SSF,P} = \frac{\Delta P \cdot K_{Tg,P} \cdot F_{PS} \cdot F_{BC} \cdot F_{CS}}{W_{eff} \cdot t} \qquad \text{(Eq A5.3)}$$

$$K_{SSF,H} = \frac{P_{BP} \cdot K_{Tg,H} \cdot F_{RS} \cdot F_T \cdot F_{CS}}{W \cdot t} \qquad \text{(Eq A5.4)}$$

where

ΔP = pin load at the hole of interest (i.e., transferred local load)
P_{BP} = bypass load across the section
W = specimen or strip width in row of interest
t = specimen or strip thickness in row of interest
W_{eff} = effective width (or hole diameter in the case of bearing load)
$K_{Tg,P}$ = pin load K_t
$K_{Tg,H}$ = open hole K_t
F_1 = product form correction
F_2 = product thickness correction
F_3 = effect of grain orientation
F_4 = effect of surface finish
F_{PS} = pin load shadow factor
F_{BC} = pin bending correction
F_{CS} = countersunk correction
F_{RS} = open hole shadow factor
F_{FT} = life improvement correction
F_T = thickness correction

Depending on how the analytical K_t is derived, some of these factors—for example, the open hole shadow factor, the pin hole shadow factor, and the pin bending correction factor—may have been a part of the baseline stress concentration factor. In addition, the effects of adjacent holes scattered in various patterns, lands, hole reinforcements, faying surface friction, and other factors must be considered. The bottom line is that we can perform a stress analysis or adopt a K_t factor from a handbook. Adjustments are required if the model for the stress analysis or the configuration that was selected from the handbook does not match the exact geometry and/or the loading condition of the problem being considered. Such cases require use of the above geometry- and loading-related correction factors. Again, many of these factors are available in the literature (see the References list in Chapter 1). Equation A5.2 assumes that the loaded-hole part and the open-hole part of the fastener joint will be modeled separately, because that is how handbook solutions are presented. If both parts were modeled as a whole, Eq A5.2 would have to be rewritten. The other factors, such as those for the material product form, metal processing, hole quality, and so forth, are empirically developed correction factors.

Because every product has unique problems of its own, more correction factors of this sort may be needed. Vast amounts of such data, developed by computer or by mechanical testing, have been generated in the industry. These data usually are not available to the public, although a limited quantity has been published in the open literature (e.g., Ref A5.2, A5.3). Some of these factors are included here for the convenience of readers.

The life improvement correction, F_{FT}, is considered the product of the fastener hole quality factor α and the fastener fit factor β. The α factor accounts for the surface condition of the fastener hole (Table A5.1). The β factor accounts for the effect of fastener type and how it is fastened (Table A5.2).

The fastener bending factor (also known as the fastener tilt factor or bearing distribution factor) is defined here as the peak bearing stress

Table A5.1   Fastener quality factor α

Surface Condition	α
Standard drilled holes	
6.4 mm (0.25 in.) diam	1.0
7.9 mm (0.312 in.) diam	1.03
9.5 mm (0.375 in.) diam	1.06
Broached or reamed	0.9
Cold worked	0.7–0.8

Table A5.2   Fastener fit factor β

Fastener type and fit	β
Open holes	1.0
Clearance fit, fingertight	1.0
Threaded bolts	0.75–0.9
Hi-Lok, steel (0.025 mm, or 0.001 in., interference, full clamp-up)	0.635
Hi-Lok, steel (interference and clamp-up not specified)	0.75
Taperlocks	0.5
Hi-tigue	0.75
Rivets	0.75

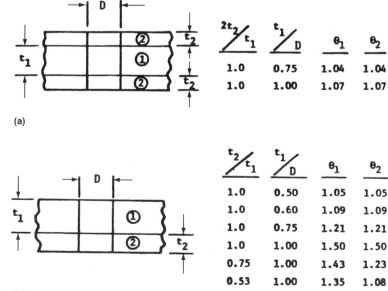

Fig. A5.1   Fastener tilt factor θ for double-shear (a) and single-shear (b) joints

divided by the average bearing stress at a hole. It is used as an amplification factor for the bearing stress of the fastener, accounting for fastener rotation and deflection. The fastener tilt factor θ that is given in Ref A5.3 is adopted here as the fastener bending factor. Shown in Fig. A5.1, the fastener bending factor is a function of the ratio of the plate thicknesses and the ratio of the plate thickness to the hole diameter. However, consideration should also be given to the ratio of the pin and the plate modulus.

Taking into account only the effects of α, β, and θ in a fastened joint, Eq A5.2 simply becomes:

$$K_{SSF} = \frac{\alpha \cdot \beta}{S_{REF}} \left[K_{tb} \cdot \frac{\Delta P}{Dt} \cdot \theta + K_t \frac{P_{BP}}{Wt} \right]$$

$$(Eq \ A5.5)$$

REFERENCES

A5.1. L.E. Jarfall, "Optimum Design of Joints: The Stress Severity Factor Concept," presented at Fifth ICAF Symposium (Melbourne), May 1967

A5.2. J.M. Waraniak and A.F. Liu, "Fatigue and Crack Propagation Analysis of Mechanically Fastened Joints," Paper 83–0839, presented at the AIAA/ASME/ASCE/AHS 24th Structures, Structural Dynamics and Materials Conference (Lake Tahoe, NV), May 1983 (synopsis appears in *J. Aircraft*, Vol 21, 1984, p 225–226)

A5.3. T.R. Brussat, S.T. Chu, and M. Creager, "Flaw Growth in Complex Structure," Report AFFDL-TR-77-79, Air Force Flight Dynamic Laboratory, Dec 1977

APPENDIX 6

Specimen Orientation and Fracture Plane Identification

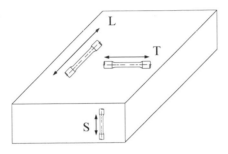

Fig. A6.1 Conventional specimen orientation code. Arrows indicate loading directions. L, longitudinal; T, transverse; S, short transverse. Note that T is the same as LT, or long-transverse; LS is also used for short-transverse. The axis of a tension test specimen is always parallel to the applied load (as well as the designated grain orientation, see figure). Crack plane of a fracture mechanics specimen (see Fig. A6.2–A6.4) is always perpendicular to the applied load.

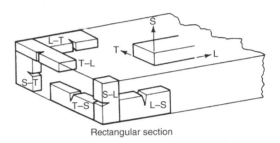

Rectangular section

Fig. A6.2 ASTM crack plane orientation identification code for rolled plate. L, length, longitudinal, principal direction of metal working (rolling, extrusion, axis of forging); T, width, long-transverse grain direction; S, thickness, short-transverse grain direction. First letter: normal to the fracture plane (loading direction); second letter: direction of crack propagation in fracture plane

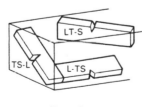

Non-primary

Fig. A6.3 ASTM crack plane orientation identification code for specimens tilted with respect to reference direction. First letter: normal to the fracture plane (loading direction); second letter: direction of crack propagation in fracture plane

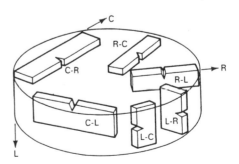

Fig. A6.4 ASTM crack plane orientation identification code for drawn bars and hollow cylinders. C, chord of cylindrical cross section; R, radius of cylindrical cross section. First letter: normal to the fracture plane (loading direction); second letter: direction of crack propagation in fracture plane

APPENDIX 7

Mechanical Properties Data for Selected Aluminum Alloys

LIMITED MECHANICAL PROPERTIES DATA for several selected aluminum alloys are compiled in this appendix. Relatively new aluminum alloys included are 7033, Al-Li 8090 and 2090, rapidly solidified power metallurgy (P/M) aluminum, and B201 and D357 aluminum castings.

A7.1 Conventional and High-Strength Aluminum Alloys

Both 2000 and 7000 series aluminum alloys are used in the aerospace/aircraft industry. Tables A7.1 and A7.2 along with Fig. A7.1 to A7.3 present tensile properties and fracture toughness test data for several of these alloys. Plane-stress fracture toughness values and crack growth resistance curves for these alloys are shown in Fig. A7.4 and A7.5, respectively. Figures A7.6 and A7.7 plot fatigue crack growth rate curves.

Tensile and fracture toughness data for the new 7033 T6 high-strength automotive alloy are presented in Table A7.3, which also compares these properties with the conventional 2014-T6 and 6061-T6 alloys. *S-N* curves for all three alloys are presented in Fig. A7.8.

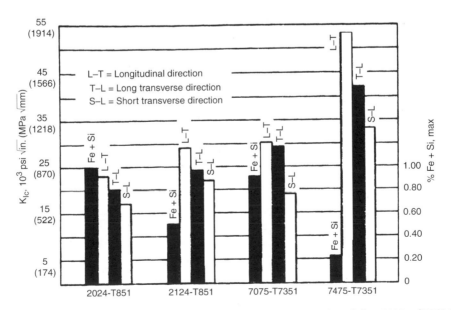

Fig. A7.1 Comparison of regular and high-purity (lower iron and silicon contents) versions of alloys 2024 and 7075. Plane-strain fracture toughness is higher in the high-purity alloys (designated as 2124 and 7475). Source: Ref A7.2

A7.2 P/M Aluminum

Mechanical properties data for several P/M aluminum forgings and extrusions are listed in Table A7.4. These data are taken from a report for the Air Force Advanced Aluminum Fighter Structures (AAFS) program (Ref A7.11).

A7.3 Aluminum-Lithium Alloys

Sheet and plate mechanical properties data for the low-density 8090 and 2090 Al-Li alloys are presented in Table A7.5. Table A7.6 lists the test results for extrusions. Fatigue and fatigue crack growth rate data for the 8090-TU51 extrusion are shown in Fig. A7.9 to A7.11.

A7.4 Aluminum Casting Alloys

This appendix includes mechanical test data for two casting materials: B201-T7 and D357-T6. Composition specifications for these castings are, respectively, AMS 4242 and AMS 4241, to which a small amount of strontium (0.014 wt% max) or sodium (0.012 wt% max) was added as a silicon modifier. Tensile properties and plane-strain fracture toughness data are listed in Tables A.7.7 to A7.10. High- and low-cycle fatigue and fatigue crack growth rate data are shown in Fig. A7.12 to A7.17.

Table A7.1 Mechanical properties of aluminum alloys at room temperature

Alloy	7150-T6E189(a)	7050-T7451	7050-T7651	7475-T7351	7475-T7651	7475-T651
Plate thickness, mm (in.)	25.4 (1)	6.4–38.1 (0.25–1.5)	6.4–25.4 (0.25–1.0)	6.4–38.1 (0.25–1.5)	12.7–25.4 (0.5–1.0)	12.7–25.4 (0.5–1.0)
Orientation	LT	T	T	T	T	T
E, MPa (ksi)	. . .	71,070 (10,300)	71,070 (10,300)	71,070 (10,300)	70,380 (10,200)	70,380 (10,200)
F_{tu}, MPa (ksi)	628 (91)	524 (76)	524 (76)	504 (73)	483 (70)	538 (78)
F_{ty}, MPa (ksi)	587 (85)	455 (66)	455 (66)	428 (62)	407 (59)	469 (68)
Elongation, %	12	9	8	9	8	9
E_c, MPa (ksi)	. . .	73,140 (10,600)	74,520 (10,800)	73,140 (10,600)	73,140 (10,600)	73,140 (10,600)
F_{cy}, MPa (ksi)	. . .	469 (68)	469 (68)	435 (63)	428 (62)	490 (7)
F_{su}, MPa (ksi)	. . .	297 (43)	297 (43)	290 (42)	269 (39)	297 (43)
F_{bru}, MPa (ksi) ($e/D = 1.5$)	. . .	759 (110)	759 (110)	725 (105)	711 (103)	780 (113)
F_{bry}, MPa (ksi) ($e/D = 1.5$)	. . .	614 (89)	600 (87)	580 (84)	559 (81)	642 (93)
K_{IC}, MPa$\sqrt{m}$ (ksi$\sqrt{in.}$)	31 (28)	. . .	. . .	(b)	. . .	. . .
K_C, MPa$\sqrt{m}$ (ksi$\sqrt{in.}$)	. . .	. . .	. . .	. . .	110 (100)(c)	99 (90)(c)

Alloy	2124-T351(a)	2124-T851	2024-T351	2024-T851	7075-T651	7075-T7351
Plate thickness, mm (in.)	25.4 (1)	25.4–38.1 (1.0–1.5)	12.7–25.4 (0.5–1.0)	12.7–25.4 (0.5–1.0)	12.7–25.4 (0.5–1.0)	6.4–12.7 (0.25–0.5)
Orientation	LT	T	T	T	T	T
E, MPa (ksi)	. . .	71,760 (10,400)	73,830 (10,700)	. . .	71,070 (10,300)	71,070 (10,300)
F_{tu}, MPa (ksi)	469 (68)	455 (66)	449 (65)	455 (66)	552 (80)	476 (69)
F_{ty}, MPa (ksi)	366 (53)	393 (57)	304 (44)	400 (58)	476 (69)	393 (57)
Elongation, %	22	5	. . .	. . .	7	7
E_c, MPa (ksi)	. . .	75,210 (10,900)	75,210 (10,900)	. . .	73,140 (10,600)	73,140 (10,600)
F_{cy}, MPa (ksi)	. . .	393 (57)	324 (47)	407 (59)	511 (74)	407 (59)
F_{su}, MPa (ksi)	. . .	. . .	262 (38)	262 (38)	311 (45)	262 (38)
F_{bru}, MPa (ksi) ($e/D = 1.5$)	. . .	. . .	676 (98)	697 (101)	828 (120)	704 (102)
F_{bry}, MPa (ksi) ($e/D = 1.5$)	. . .	. . .	524 (76)	600 (87)	711 (103)	545 (79)
K_{IC}, MPa$\sqrt{m}$ (ksi$\sqrt{in.}$)	47.9 (43.5)	(b)	. . .	(b)	30.3 (27.5)(d)	(b)
K_C, MPa$\sqrt{m}$ (ksi$\sqrt{in.}$)	. . .	. . .	105 (95)(c)	. . .	71.5 (65)(c)	. . .
Fatigue strength, MPa (ksi)	. . .	. . .	138 (20)(e)	. . .	159 (23)(e)	. . .

(a) Ref A7.1. (b) See Fig. A7.1. (c) Thin-sheet K_C value (Ref A7.2). (d) Ref A7.3. (e) At 500 million cycles, $K_t = 1$, $R = -1$ (Ref A7.4). All test data in this table are S or B values (per Ref A7.5), unless otherwise noted.

REFERENCES

A7.1. G.V. Scarich and P.E. Pretz, "Fatigue Crack-Growth Resistance of Aluminum Alloys under Spectrum Loading, Vol I—Commercial 2XXX and 7XXX Alloys," Report NOR 85-141, Northrop Corp., Aircraft Division, 1985

A7.2. R.R. Senz and E.H. Spuhler, Fracture Mechanics' Impact on Specifications and Supply, *Met. Prog.,* March 1975, p 64–66

A7.3. J.C. Evall, T.R. Brussat, A.F. Liu, and M. Creager, "Engineering Criteria and Analysis Methodology for the Appraisal of Potential Fracture Resistant Primary Aircraft Structure," Report AFFDL-TR-72-80, Wright Research and Development Center, Flight Dynamics Laboratory, Air Force Systems Command, 1972

A7.4. Guide to Engineering Materials (GEM 2002), *Adv. Mater. Process.,* Vol 159 (No. 12), 2001

A7.5. *Military Standardization Handbook: Metallic Materials and Elements for Aerospace Vehicle Structures,* MIL-HDBK-5E, U.S. Department of Defense, 1987

A7.6. J.G. Kaufman, Fracture Toughness of Aluminum Alloy Plate—Tension Test of Large Center Slotted Panels, *J. Mater.,* Vol 2, 1967, p 889–914

A7.7. J.T. Staley, *Microstructure and Toughness of Higher Strength Aluminum Alloys,* STP 605, ASTM, 1976

A7.8. R.J. Bucci, G. Nordmark, and E.A. Starke, Jr., Selecting Aluminum Alloys to Resist Failure by Fracture Mechanisms, *ASM Handbook,* Vol 19, *Fatigue and Fracture,* ASM International, 1996, p 779

A7.9. W.II. Reimann and A.W. Brisbane, *Eng. Fract. Mech.,* Vol 5, 1973, p 67

A7.10. D. Childree, High-Strength Aluminum Automotive Alloy, *Adv. Mater. Process.,* Vol 154 (No. 3), 1998, p 27–29

A7.11. P.G. Porter and D. Kane, "Advanced Aluminum Fighter Structures," Report WRDC-TR-90-3049, Wright Research and Development Center, Flight Dynamics Laboratory, Air Force Systems Command, 1990

A7.12. M.W. Ozelton, S.J. Mocarski, and P.G. Porter, "Durability and Damage Tolerance of Aluminum Castings," Materials Directorate, Wright Research and Development Center, Air Force Systems Command, 1991

A7.13. S.J. Mocarski, G.V. Scarich, and K.C. Wu, Effect of Hot Isostatic Pressure on Cast Aluminum Airframe Components, *AFS Trans.,* Vol 91, 2002, p 77–81

Table A7.2 Plane-strain fracture toughness data for aluminum alloys at various test temperatures

Alloy and condition	Room-temperature yield strength		Specimen design	Orientation	Fracture toughness, K_{1c} or $K_{1c}(J)$ at:							
					24 °C (75 °F)		−196 °C (−320 °F)		−253 °C (−423 °F)		−269 °C (−452 °F)	
	MPa	ksi			MPa√m	ksi√in.	MPa√m	ksi√in.	MPa√m	ksi√in.	MPa√m	ksi√in.
2014–T651	432	62.7	Bend	T-L	23.2	21.2	28.5	26.1	...	...	...	...
2024–T851	444	64.4	Bend	T-L	22.3	20.3	24.4	22.2	...	...	...	...
2124–T851(a)	455	66.0	CT	T-L	26.9	24.5	32.0	29.1	...	...	...	...
	435	63.1	CT	L-T	29.2	26.6	35.0	31.9	...	...	...	...
	420	60.9	CT	S-L	22.7	20.7	24.3	22.1	...	...	...	...
2219–T87	382	55.4	Bend	T-S	39.9	36.3	46.5	42.4	52.5	48.0	...	...
			CT	T-S	28.8	26.2	34.5	31.4	37.2	34.0	...	...
	412	59.6	CT	T-L	30.8	28.1	38.9	32.7	...	...	...	...
5083–O	142	20.6	CT	T-L	27.0(b)	24.6(b)	43.4(b)	39.5(b)	...	...	48.0(b)	43.7(b)
6061–T651	289	41.9	Bend	T-L	29.1	26.5	41.6	37.9	...	...	...	...
7039–T6	381	55.3	Bend	T-L	32.3	29.4	33.5	30.5	...	...	...	...
7075–T651	536	77.7	Bend	T-L	22.5	20.5	27.6	25.1	...	...	...	...
7075–T7351	403	58.5	Bend	T-L	35.9	32.7	32.1	29.2	...	...	...	...
7075–T7351	392	56.8	Bend	T-L	31.0	28.2	30.9	28.1	...	...	...	...

(a) 2124 is similar to 2024, but with higher-purity base and special processing to improve fracture toughness. (b) $K_{1c}(J)$. Source: *Metals Handbook,* 9th ed., Vol 3, American Society for Metals, 1980, p 746, compiled from several references

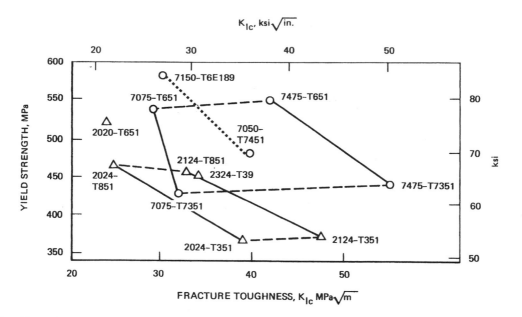

Fig. A7.2 Plane-strain fracture toughness as a function of material tensile yield strength. Comparison of several 2000 and 7000 series aluminum alloys. Source: Ref A7.1

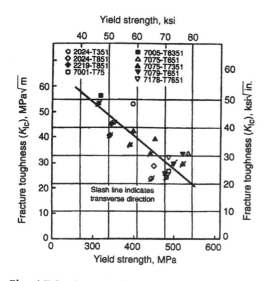

Fig. A7.3 Plane-strain fracture toughness for 25.4 to 38.1 mm (1 to 1.5 in.) thick commercial aluminum alloys. Source: Ref A7.6

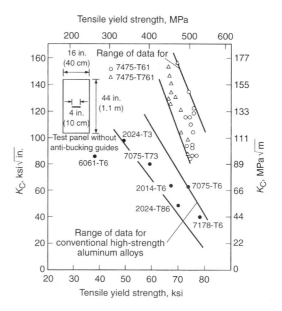

Fig. A7.4 Plane-stress fracture toughness for 1 to 4.8 mm (0.04 to 0.2 in.) thick aluminum alloy sheet. Source: Ref A7.7

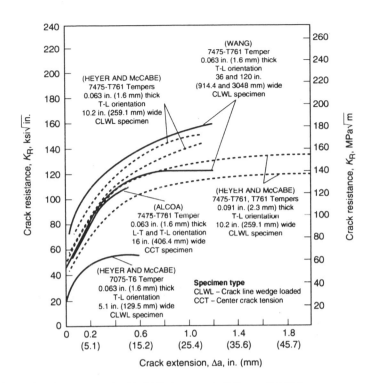

Fig. A7.5 Crack growth resistance curves for thin 7475 aluminum sheet. Source: Ref A7.8

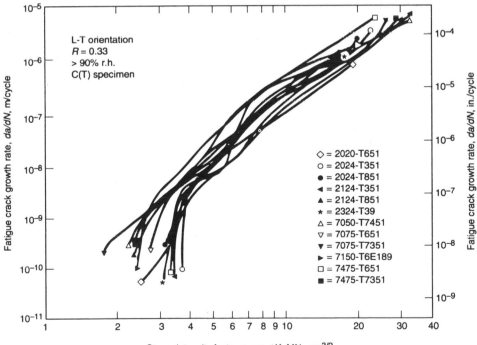

Fig. A7.6 *da/dN* curves for 12 aerospace aluminum alloys, *R* = 0.33. Source: Ref A7.9

Table A7.3 **Comparison of mechanical properties for 7033-T6, 2014-T6, and 6061-T6 aluminum alloys**

Property	7033-T6	6061-T6	2014-T6
Tensile			
Tensile strength, MPa (ksi)	518 (75)	331 (48)	449 (65)
Yield strength, MPa (ksi)	483 (70)	297 (43)	407 (59)
Elongation, % (2.0 in. gage length)	12	14	9
Elevated-temperature yield strength, MPa (ksi)			
40 °C (100 °F)	483 (70)	290 (42)	407 (59)
95 °C (200 °F)	455 (66)	283 (41)	435 (63)
150 °C (300 °F)	373 (54)	262 (38)	276 (40)
205 °C (400 °F)	248 (36)	228 (33)	110 (16)
260 °C (500 °F)	138 (20)	124 (18)	65 (9.5)
Compression			
Peak stress, MPa (ksi)	483 (70)	317 (46)	Not tested
Yield stress, MPa (ksi)	455 (66)	297 (43)	Not tested
Fracture toughness, MPa$\sqrt{m}$ (ksi$\sqrt{in.}$)			
Round bar (plate equivalent orientation)			
R-L (T-L)	41 (37)	26 (24)	21 (19)
L-R (L-T)	66 (60)	Invalid test	7.7 (7)
R-C (S-T)	39 (36)	Not tested	Not tested
Hardness, HRB	80	60	79
Electrical conductivity, % IACS	39	44	40

Source: Ref A7.10

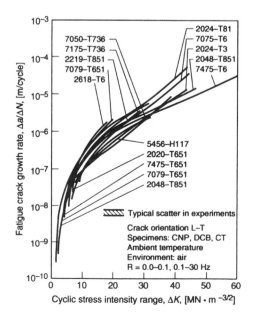

Fig. A7.7 da/dN curves for 15 aluminum alloys, R = 0 to 0.1. Source: Ref A7.8

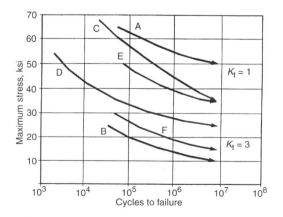

Fig. A7.8 S-N curves for three aluminum alloys, R = 0.1. Curves A and B: 7033-T6; curves C and D: 2014-T6; curves E and F: 6061-T6. Source: Ref A7.10

Table A7.4 Tensile strength and plane-strain fracture toughness of P/M (RST) aluminum alloys

Alloy and temper	CW67-T7E94 (Alcoa data)			CW67-T7E94		PM7064-TX7652 Hand forging	
Product form							
Orientation	(a) L	(a) T	(a) S	(a) L	(a) T	L	T
E, MPa (ksi)	···	···	···	71,070 (10,300)	75,210 (10,900)	72,222 (10,467)	72,222 (10,467)
F_{tu}, MPa (ksi)	631.7 (91.55)	574.4 (83.25)	573.0 (83.05)	600.6 (87.05)	567.9 (82.3)	601.2 (87.13)	596.4 (86.43)
F_{ty}, MPa (ksi)	611.3 (88.6)	534.1 (77.4)	526.8 (76.35)	572.0 (82.9)	532.3 (77.15)	554.1 (80.3)	549.0 (79.57)
Elongation, %	12.2	8	7.05	9.96	10.87	9.18	6.52
E_c, MPa (ksi)	···	···	···	···		78,660 (11,400)	77,970 (11,300)
F_{cy}, MPa (ksi)	···	···	···	···		587.4 (85.13)	576.7 (83.58)
K_{IC}, MPa√m (ksi√in.)	43.07 (39.15)	21.07 (19.15)	23.1 (21)	···		···	···

Alloy and temper	CW67-T7E94 (Alcoa data)			CW67-T7E94	PM7064-TX76510 Extrusion	
Product form						
Orientation	(b) L	(b) T	(b) S	(b) L	L	T
E, MPa (ksi)	···	···	···	75,555 (10,950)	71,298 (10,333)	74,058 (10,733)
F_{tu}, MPa (ksi)	657.2 (95.25)	610.7 (88.5)	585.1 (84.8)	631.4 (91.5)	651.4 (94.4)	623.3 (90.33)
F_{ty}, MPa (ksi)	638.3 (92.5)	579.9 (84.05)	564.1 (81.75)	651.0 (94.35)	617.1 (89.43)	586.5 (85)
Elongation, %	11.25	8.20	5.15	9.71	8.29	8.67
E_c, MPa (ksi)	76,935 (11,150)	77,453 (11,225)	···	···	76,590 (11,100)	79,578 (11,533)
F_{cy}, MPa (ksi)	599.2 (86.84)	583.1 (84.51)	···	···	629.4 (91.22)	620.2 (89.88)
K_{IC}, MPa√m (ksi√in.)	···	···	22.1 (20.1)	···	···	···

Alloy and temper	CW67-T7E94 (Alcoa data)			CW67-T7E94		
Product form						
Orientation	(c) L	(c) T	(c) S	(c) L	(c) T	(c) S
E, MPa (ksi)	···	···	···	69,462 (10,067)	70,953 (10,283)	70,842 (10,267)
F_{tu}, MPa (ksi)	572.4 (82.95)	534.6 (77.48)	559.1 (81.03)	606.3 (87.87)	583.7 (84.6)	565.3 (81.93)
F_{ty}, MPa (ksi)	535.4 (77.6)	481.1 (69.73)	508.4 (73.68)	580.3 (84.1)	548.2 (79.45)	526.1 (76.25)
Elongation, %	14.38	8.5	6.75	11.52	8.45	8.3
E_c, MPa (ksi)	···	···	···			
F_{cy}, MPa (ksi)	···	···	···			
K_{IC}, MPa√m (ksi√in.)	43.73 (39.75)	···	30.8 (28)			

(a) Specimens cut from the 58.4 mm (2.3 in.) thick flange of a die forging. (b) Specimens cut from the 12.7 mm (0.5 in.) thick flange of a die forging. (c) Specimens cut from a 114.3 mm (4.5 in.) thick hand forging. Source: Ref A7.11

Table A7.5 Tensile strength and plane-strain fracture toughness of 8090 and 2090 Al-Li sheet and plate

Alloy and temper	8090-TU51	8090-TU51	8090-TU51	8090-TU51	8090-TU51	8090-T6	8090-T6
Thickness, mm (in.)	1.6 (0.063)	1.6 (0.063)	63.5 (2.5)	63.5 (2.5)	63.5 (2.5)	1.6 (0.063)	1.6 (0.063)
Orientation	L	T	L	T	S	L	T
E, MPa (ksi)	80,502 (11,667)	80,730 (11,700)	78,432 (11,367)	79,578 (11,533)	76,818 (11,133)	80,958 (11,733)	82,800 (12,000)
F_{tu}, MPa (ksi)	529.9 (76.8)	517.3 (74.97)	492.7 (71.4)	492.7 (71.4)	451.1 (65.37)	481.1 (69.73)	463.9 (67.23)
F_{ty}, MPa (ksi)	464.4 (67.3)	438.2 (63.5)	458.0 (66.37)	405.0 (58.7)	351.0 (50.87)	380.7 (55.17)	376.9 (54.63)
Elongation, %	3.35	8.35	4.85	5.22	1.71	5.86	8.85
E_c, MPa (ksi)	…	…	81,248 (11,775)	83,490 (12,100)	82,110 (11,900)	…	…
F_{cy}, MPa (ksi)	…	…	424.0 (61.45)	442.6 (64.15)	403.3 (58.45)	…	…
F_{su}, MPa (ksi)	…	…	249.6 (36.17)	240.6 (34.87)	…	…	…
F_{bru}, MPa (ksi) ($e/D = 2$)	774.2 (112.2)	…	908.2 (131.63)	…	…	719.7 (104.3)	…
F_{bry}, MPa (ksi) ($e/D = 2$)	635.3 (92.07)	…	704.7 (102.13)	…	…	578.4 (83.83)	…
K_{Ic}, MPa$\sqrt{m}$ (ksi$\sqrt{in.}$)	…	…	37.5 (34.08)	29.1 (26.45)	…	…	…

Alloy and temper	2090-T3E27	2090-T3E27	2090-T8(a)	2090-T8(a)	2090-T8(b)	2090-T8(b)	2090-T6	2090-T6
Thickness, mm (in.)	1.6 (0.063)	1.6 (0.063)	1.6 (0.063)	1.6 (0.063)	1.6 (0.063)	1.6 (0.063)	1.6 (0.063)	1.6 (0.063)
Orientation	L	T	L	T	L	T	L	T
E, MPa (ksi)	80,040 (11,600)	80,385 (11,650)	78,315 (11,350)	77,280 (11,200)	77,970 (11,300)	77,625 (11,250)	76,935 (11,150)	75,900 (11,000)
F_{tu}, MPa (ksi)	349.5 (50.65)	349.5 (50.65)	534.8 (77.5)	540.3 (78.3)	549.6 (79.65)	533.0 (77.25)	458.9 (66.5)	476.8 (69.1)
F_{ty}, MPa (ksi)	233.2 (33.8)	231.2 (33.5)	484.4 (70.2)	509.9 (73.9)	525.4 (76.15)	479.9 (69.55)	380.5 (55.15)	411.6 (59.65)
Elongation, %	17.93	17.39	7.54	6.24	5.36	7.59	8.03	6.9
E_c, MPa (ksi)	…	…	…	…	…	…	…	…
F_{cy}, MPa (ksi)	…	…	…	…	…	…	…	…
F_{bru}, MPa (ksi) ($e/D = 1.5$)	477.5 (69.2)	…	774.5 (112.25)	…	724.2 (104.95)	…	650.7 (94.3)	…
F_{bry}, MPa (ksi) ($e/D = 1.5$)	323.6 (46.9)	…	671.7 (97.35)	…	623.4 (90.35)	…	548.6 (79.5)	…

(a) Peak age 24 h at 160 °C (325 °F). (b) Peak age 24 h at 175 °C (350 °F). Source: Ref A7.11

Table A7.6   Tensile properties of 8090 and 2090 Al-Li extrusions

Alloy and temper	8090-TU51 Al-Li		2029-T8E41 (Alcoa data)	
Product form	25.4 × 102 mm (1 × 4 in.) extrusion		10.2 × 51 mm (0.4 × 2 in.) extrusion	
Orientation	L	T	L	T
E, MPa (ksi)	80,268 (11,633)	79,350 (11,500)	81,192 (11,767)	77,508 (11,233)
F_{tu}, MPa (ksi)	589.3 (85.4)	530.1 (76.83)	595.0 (86.23)	532.9 (77.23)
F_{ty}, MPa (ksi)	555.5 (80.5)	455.2 (65.97)	574.1 (83.2)	502.8 (72.87)
Elongation, %	4.29	6.34	7.98	5.01
E_c, MPa (ksi)	81,648 (11,833)	81,993 (11,883)	...	...
F_{cy}, MPa (ksi)	511.8 (74.17)	489.7 (70.97)	...	...
F_{bru}, MPa (ksi) ($e/D = 2$)	736.2 (106.7)	...	...	...
F_{bry}, MPa (ksi) ($e/D = 2$)	549.8 (79.68)	...	...	...
Alloy and temper	8090-TU51 Al-Li			2029-T8E41 (Kaiser data)	
Product form	76.2 × 102 mm (3 × 4 in.) extrusion			T-extrusion	
Orientation	L	T	S	L	T
E, (ksi)	79,350 (11,500)	80,040 (11,600)	78,833 (11,425)	80,040 (11,600)	81,192 (11,767)
F_{tu}, (ksi)	562.7 (81.55)	497.5 (72.1)	489.6 (70.95)	600.0 (86.95)	541.0 (78.4)
F_{ty}, (ksi)	532.0 (77.1)	403.13 (58.425)	379.8 (55.05)	572.7 (83)	494.2 (71.63)
Elongation, %	5.30	4.28	3.33	3.07	1.33

Source: Ref A7.11

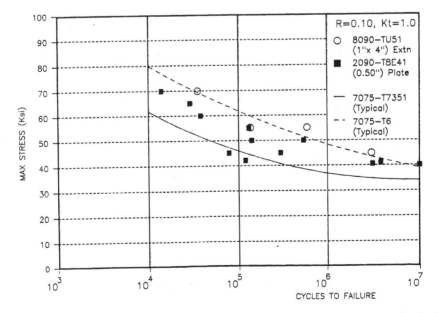

Fig. A7.9 Comparison of *S-N* curves for 8090-TU51 extrusion with 2090-T8E41 plate and typical 7075-T6/T7351 aluminum, *R* = 0.1, K_t = 1. Source: Ref A7.11

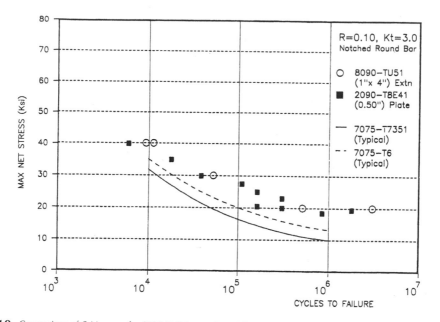

Fig. A7.10 Comparison of *S-N* curves for 8090-TU51 extrusion with 2090-T8E41 plate and typical 7075-T6/T7351 aluminum, *R* = 0.1, K_t = 3. Source: Ref A7.11

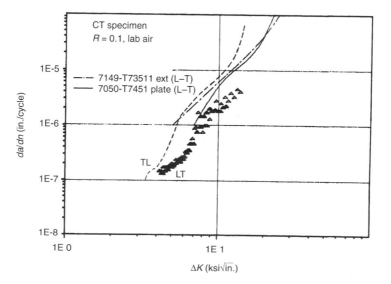

Fig. A7.11 Comparison of *da/dN* curves for 8090-TU51 extrusion (LT and TL) with 7149-T73511 extrusion and 7050-T7451 plate, *R* = 0.1. Source: Ref A7.11

Table A7.7   Tensile properties of 31.8 mm (1.25 in.) thick B201-T7 plate

Foundry	Utimate tensile strength, MPa (ksi)		Yield strength, MPa (ksi)		Elongation, %	
	Avg.	Range	Avg.	Range	Avg.	Range
A	455 (66)	449–462 (65–67)	393 (57)	373–400 (54–58)	8.7	8.3–11
B	483 (70)	469–504 (68–73)	435 (63)	414–455 (60–66)	8.2	7.5–8.8
C	462 (67)	442–483 (64–70)	414 (60)	400–428 (58–62)	8.3	6.5–9.5
Average	**469 (68)**		**414 (60)**		**8.4**	

Source: Ref A7.12

Table A7.8   Fracture toughness and notch tensile strength of 31.8 mm (1.25 in.) thick B201-T7 plate

Foundry	Fracture toughness (K_Q)		Notch tensile strength		Yield strength		
	$MPa\sqrt{m}$	$ksi\sqrt{in.}$	MPa	ksi	MPa	ksi	NTS/YS
A	51	46	639.6	92.7	398.1	57.7	1.62
	43	39	611.3	88.6	394.7	57.2	1.55
B	30	27	562.4	81 5	431.3	62.5	1.37
	35	32	585.1	84.8	454.0	65.8	1.29
C	54	49	625.8	90.7	416.1	60.3	1.50
	51	46	593.4	86.0	414.7	60.1	1.44
Average	**44**	**40**	**602.4**	**87.3**	**418.1**	**60.6**	**1.46**

Source: Ref A7.12

Table A7.9   Tensile properties of 31.8 mm (1.25 in.) thick water- and glycol-quenched D357-T6 plate

Foundry	Quench medium(a)	Ultimate tensile strength, MPa (ksi)		Yield strength, MPa (ksi)		Elongation, %	
		Avg.	Range	Avg.	Range	Avg.	Range
A	Water	373 (54)	352–380 (51–55)	311 (45)	304–324 (44–47)	5.9	3.5–8.5
	Glycol(b)	366 (53)	352–380 (51–55)	297 (43)	290–304 (42–44)	6.6	5.5–9.0
B	Water	359 (52)	345–366 (50–53)	311 (45)	304–317 (44–46)	4.5	3.0 5.6
	Glycol(b)	338 (49)	331–345 (48–50)	290 (42)	283–297 (41–43)	3.4	2.5–4.3
C	Water	373 (54)	366–380 (53–55)	324 (47)	317–331 (46–48)	5.4	3.9–8.2
	Glycol(b)	352 (51)	345–359 (50–52)	304 (44)	297–304 (43–44)	4.9	4.6–5.2
Average	**Water**	**366 (53)**		**317 (46)**		**5.2**	
	Glycol(b)	**352 (51)**		**297 (43)**		**5.0**	

(a) Room temperature. (b) 25% glycol solution. Source: Ref A7.12

Table A7.10   Fracture toughness and notch tensile strength of 31.8 mm (1.25 in.) thick D357-T6 plate

Foundry	Quench medium	Fracture toughness, $MPa\sqrt{m}$ $(ksi\sqrt{in.})$		Notch tensile strength		Yield strength		
		K_{IC}	K_Q	MPa	ksi	MPa	ksi	NTS/YS
A	Water	. . .	28 (25)	435	63	311	45	1.4
	Glycol(a)	23 (21)	. . .	400	58	297	43	1.3
B	Water	. . .	26 (24)	429	62	311	45	1.4
	Glycol(a)	24 (22)	. . .	400	58	290	42	1.4
C	Water	. . .	25 (23)	366	53	324	47	1.1
	Glycol(a)	24 (22)	. . .	304	44	304	44	1.0
Average	**Water**	**. . .**	**26 (24)**	**407**	**59**	**317**	**46**	**1.3**
	Glycol(a)	**24 (22)**	**. . .**	**366**	**53**	**297**	**43**	**1.2**

(a) 25% glycol solution. Source: Ref A7.12

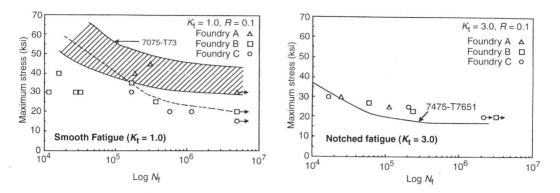

Fig. A7.12 Comparison of *S-N* curves for B201-T7 aluminum casting with HIP A201-T7 casting (dashed line, Northrop data: Ref A7.13), 7075-T73 wrought (Alcoa Green Letter GL-206, 1971), and 7475-T7651 wrought (Alcoa Green Letter GL-216, 1985). Source: Ref A7.12

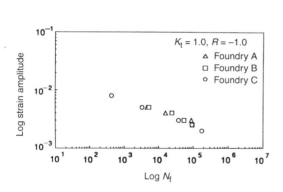

Fig. A7.13 Strain-life data for B201-T7 aluminum casting. Source: Ref A7.12

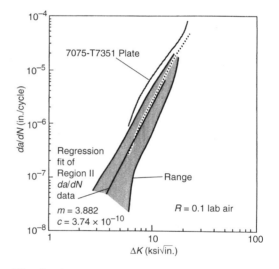

Fig. A7.14 Comparison of *da/dN* curves for B201-T7 aluminum casting with other Northrop in-house data (dotted line) and 7075-T7351 plate. Source: Ref A7.12

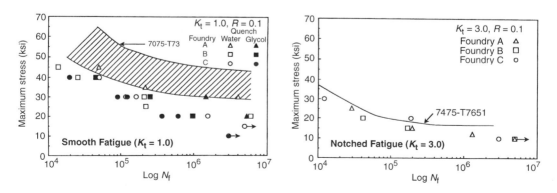

Fig. A7.15 Comparison of *S-N* curves for D357-T6 aluminum casting with 7075-T73 (Alcoa Green Letter GL-206, 1971) and 7475-T7651 (Alcoa Green Letter GL-216, 1985) wrought materials. Source: Ref A7.12

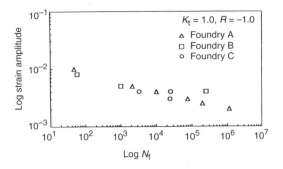

Fig. A7.16 Strain-life data for D357-T6 aluminum casting. Source: Ref A7.12

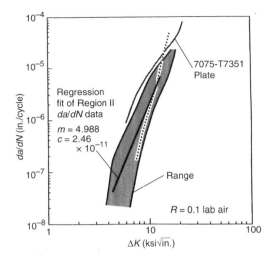

Fig. A7.17 Comparison of *da/dN* curves for D357-T6 aluminum casting with other Northrop in-house data (dotted line) and 7075-T7351 plate. Source: Ref A7.12

APPENDIX 8

Mechanical Properties Data for Selected Titanium Alloys

LIMITED MECHANICAL PROPERTIES DATA for several selected titanium alloys are compiled in this appendix. Mechanical properties for both wrought material and castings are included. Tensile properties for several $\alpha + \beta$ and β-alloys are presented in Tables A8.1 and A8.2. Figure A8.1 provides plane-strain fracture toughness properties for Ti-6Al-4V mill-annealed alloy. Fatigue crack growth rate data for Ti-6Al-4V mill-annealed plate are shown in Fig. A8.2. Tensile and fracture toughness properties of Ti-6Al-4V castings in the mill-annealed and BSTOA (beta solution treated and overaged) conditions are tabulated in Tables A8.3 and A8.4. Smooth fatigue and fatigue crack growth rate data are presented in Fig. A8.3 and A8.4, respectively.

Table A8.1 Tensile properties of some $\alpha - \beta$ titanium alloys

Alloy	Heat treatment	E MPa	E ksi	Poisson's ratio	F_{tu} MPa	F_{tu} ksi	F_{ty} MPa	F_{ty} ksi	Elongation, %
Ti-6Al-4V	Annealed	113.8	16.5	0.342	896–993	130–144	827 924	120–134	14
	STA	. . .	. . .	. . .	1172	170	1103	160	10
Ti-6Al-4V (low oxygen)	Annealed	113.8	16.5	0.342	827–896	120–130	758–827	110–120	15
Ti-6Al-6V-2Sn	Annealed	110.3	16	. . .	1034–1069	150–155	965–1000	140–145	14
	STA	. . .	. . .	. . .	1275	185	1172	170	10
Ti-6Al-2Sn-2Zr-2Mo-0.25Si	Annealed	. . .	. . .	. . .	1034	150	965	140	. . .
	STA	122	17.7	0.327	1275	185	1138	165	11

Source: Ref A8.1

Table A8.2 Tensile properties of some β-titanium alloys

Alloy	Heat treatment	E MPa	E ksi	Poisson's ratio	F_{tu} MPa	F_{tu} ksi	F_{ty} MPa	F_{ty} ksi	Elongation, %
Ti-3Al-8V-6Cr-4Mo-4Zr (beta C)	Annealed	113.8	16.5	0.342	883 min	128 min	827 min	120 min	15
	STA	105.5	15.3	. . .	1448	210	1379	200	7
Ti-11.5Mo-6Zr-4.5Sn (beta III)	Annealed		. . .	. . .	690 min	100 min	620 min	90 min	. . .
	STA	103.4	15	. . .	1448	210	1317	191	11
Ti-13V-11Cr-3Al	Annealed	101.4	14.7	0.304	1172–1220	170–177	1103–1172	160–170	8
	STA	. . .	. . .	. . .	1275	185	1207	175	8

Source: Ref A8.1

Table A8.3   Tensile properties of Ti-6Al-4V castings

Participant	Ultimate tensile strength, MPa (ksi)		Yield strength, MPa (ksi)		Modulus, MPa (ksi)		Elongation, %	
	Mill annealed	BSTOA	Mill annealed	BSTOA	Mill annealed	BSTOA	Mill annealed	BSTOA
U.S. Navy	938 (136.1)	954 (138.3)	847 (122.9)	848 (123.0)	118 (17.1)	115 (16.7)	13.4	14.2
Howmet	938 (136.1)	949 (137.7)	828 (120.1)	842 (122.1)	. . .	. . .	13.9	11.4
Lockheed-Martin	940 (136.3)	958 (139.0)	838 (121.5)	813 (123.5)	118 (17.1)	117 (17.0)	8.5	8.8
Boeing	948 (137.5)	945 (137.0)	840 (121.8)	871 (126.3)	117 (16.9)	113 (16.4)	11.5	11.5
U.S. Air Force	952 (138.0)	938 (136.0)	855 (124.0)	841 (122.0)	112 (16.2)	112 (16.2)	12.6	11.7
Average	943 (136.8)	949 (137.6)	842 (122.1)	851 (123.4)	116 (16.8)	114 (16.6)	12.0	11.5
Standard deviation	0.8	1.0	1.3	1.6	0.4	0.3	1.9	1.7

Source: Ref A8.4

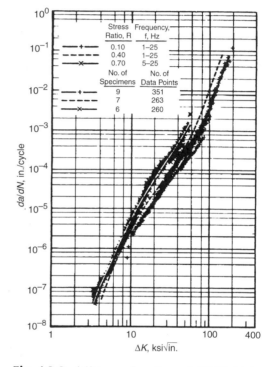

Fig. A8.2   *da/dN* curves for mill-annealed Ti-6Al-4V plate, 6.4 mm (0.25 in.) thick, with buckling restraint. Source: Ref A8.3

Table A8.4   Plane-strain fracture toughness of Ti-6Al-4V castings

Participant	MPa√m (ksi√in.)	
	Mill annealed	BSTOA
U.S. Navy	84.6 (77.0)	83.1 (75.5)
Howmet	79.6 (75.1)	82.1 (74.6)
Lockheed-Martin	72.4 (72.4)	75.5 (68.6)
Boeing	83.1 (75.5)	81.2 (73.8)
U.S. Air Force	81.7 (74.3)	92.1 (83.7)
K_Q, average	82.3 (74.8)	82.7 (75.2)
K_Q, standard deviation	1.5	4.9

Source: Ref A8.4

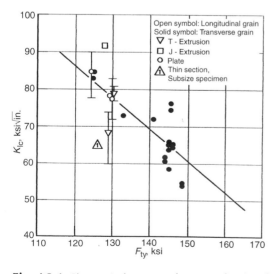

Fig. A8.1   Plane-strain fracture toughness as a function of material tensile yield strength for four-point notch-bend specimens of mill-annealed Ti-6Al-4V. Source: Ref A8.2

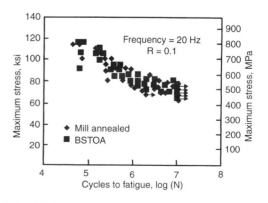

Fig. A8.3   *S-N* curves for Ti-6Al-4V castings. Source: Ref A8.4

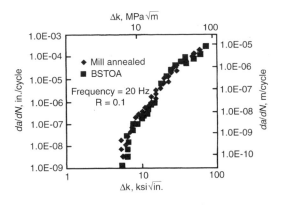

Fig. A8.4 *da/dN* curves for Ti-6Al-4V castings. Source: Ref A8.4

A8.3. *Military Standardization Handbook: Metallic Materials and Elements for Aerospace Vehicle Structures,* MIL-HDBK-5E, U.S. Department of Defense, 1986

A8.4. C.S.C. Lei, A. Davis, and E.W. Lee, Effect of BSTOA and Mill Anneal on the Mechanical Properties of Ti-6Al-4V Castings, *Adv. Mater. Process.,* Vol 157 (No. 5), 2000, p 75–77

REFERENCES

A8.1. Guide to Engineering Materials (GEM 2002), *Adv. Mater. Process.,* Vol 159 (No. 12), 2001

A8.2. J.C. Evall, T.R. Brussat, A.F. Liu, and M. Creager, "Engineering Criteria and Analysis Methodology for the Appraisal of Potential Fracture Resistant Primary Aircraft Structure," Report AFFDL-TR-72-80, Air Force Flight Dynamics Laboratory, 1972

APPENDIX 9

Mechanical Properties Data for Selected Titanium Aluminides

LIMITED MECHANICAL PROPERTIES DATA for several selected titanium aluminides are compiled in this appendix. Included are the Ti-25Al-10Nb-3V-1Mo and Ti-24Al-11Nb aluminides developed for the National Aero-Space Plane (NASP) program. Data on Ti-25Al-10Nb-3V-1Mo and Ti-24Al-11Nb beta forgings (pancake) are taken from Ref A9.1. The alloy compositions are on the basis of atomic percent. Actual weight percentages for these pancake forgings are 13.8Ti-18.4Nb-3.14V-1.95Mo and 13.5Ti-21Nb, respectively. Stabilization/age cycles of 815 °C (1500 °F) for 30 min plus 595 °C (1100 °F) for 8 h had been applied to the forgings of both alloys. Test data regarding tensile properties, plane-strain fracture toughness, high- and low-cycle fatigue, and fatigue crack growth rate are shown in Fig. A9.1 to A9.16.

REFERENCE

A9.1. D.P. DeLuca, B.A. Cowles, F.K. Haake, and K.P. Holland, "Fatigue and Fracture of Titanium Aluminides," Report WRDC-TR-89-4136, Wright Research and Development Center, 1990

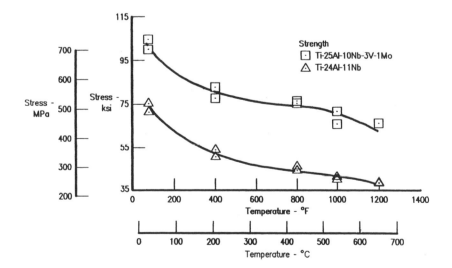

Fig. A9.1 0.2% yield strength versus temperature for Ti-25Al-10Nb-3V-1Mo and Ti-24Al-11Nb. Source: Ref A9.1

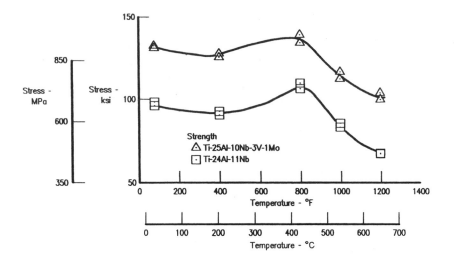

Fig. A9.2 Ultimate tensile strength versus temperature for Ti-25Al-10Nb-3V-1Mo and Ti-24Al-11Nb. Source: Ref A9.1

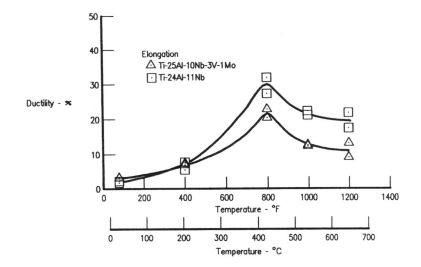

Fig. A9.3 Elongation versus temperature for Ti-25Al-10Nb-3V-1Mo and Ti-24Al-11Nb. Source: Ref A9.1

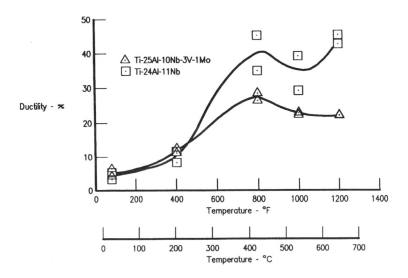

Fig. A9.4 Reduction of area versus temperature for Ti-25Al-10Nb-3V-1Mo and Ti-24Al-11Nb. Source: Ref A9.1

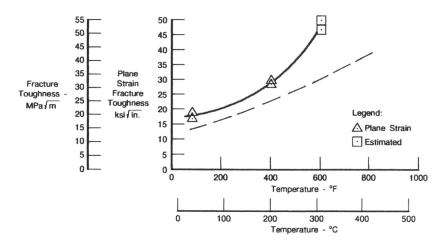

Fig. A9.5 K_{IC} versus temperature for Ti-25Al-10Nb-3V-1Mo (dotted line) and Ti-24Al-11Nb (solid line). Source: Ref A9.1

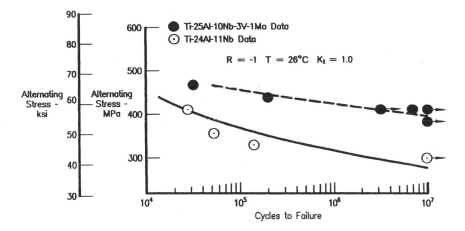

Fig. A9.6 S-N curves for Ti-25Al-10Nb-3V-1Mo and Ti-24Al-11Nb at room temperature. f = 30 Hz. Source: Ref A9.1

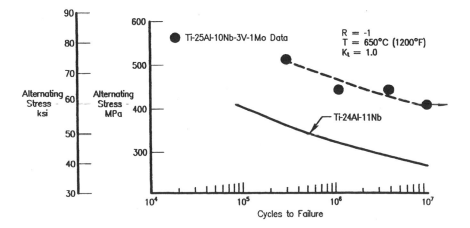

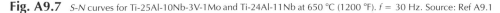

Fig. A9.7 S-N curves for Ti-25Al-10Nb-3V-1Mo and Ti-24Al-11Nb at 650 °C (1200 °F). f = 30 Hz. Source: Ref A9.1

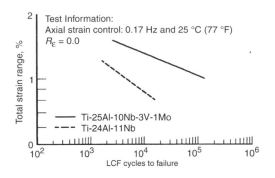

Fig. A9.8 Strain-life curves for Ti-25Al-10Nb-3V-1Mo and Ti-24Al-11Nb at room temperature. $K_t = 1$, $f = 0.17$ Hz. Source: Ref A9.1

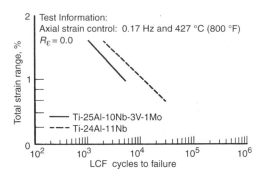

Fig. A9.9 Strain-life curves for Ti-25Al-10Nb-3V-1Mo and Ti-24Al-11Nb at 427 °C (800 °F). $K_t = 1$, $f = 0.17$ Hz. Source: Ref A9.1

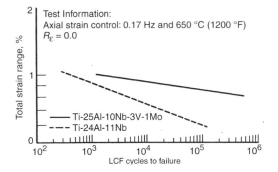

Fig. A9.10 Strain-life curves for Ti-25Al-10Nb-3V-1Mo and Ti-24Al-11Nb at 650 °C (1200 °F). $K_t = 1$, $f = 0.17$ Hz. Source: Ref A9.1

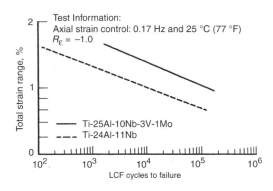

Fig. A9.11 Strain-life curves for Ti-25Al-10Nb-3V-1Mo and Ti-24Al-11Nb at room temperature. $K_t = 1$, $f = 0.17$ Hz. Source: Ref A9.1

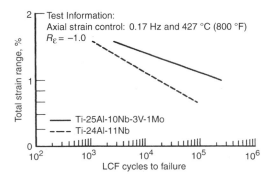

Fig. A9.12 Strain-life curves for Ti-25Al-10Nb-3V-1Mo and Ti-24Al-11Nb at 427 °C (800 °F). $K_t = 1$, $f = 0.17$ Hz. Source: Ref A9.1

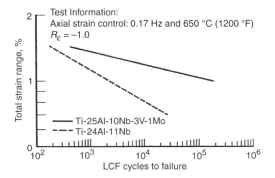

Fig. A9.13 Strain-life curves for Ti-25Al-10Nb-3V-1Mo and Ti-24Al-11Nb at 650 °C (1200 °F). $K_t = 1$, $f = 0.17$ Hz. Source: Ref A9.1

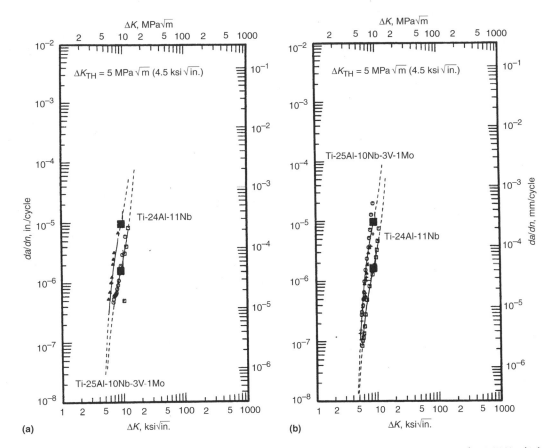

Fig. A9.14 *da/dN* curves for Ti-25Al-10Nb-3V-1Mo and Ti-24Al-11Nb at room temperature. *R* = 0.1. (a) *f* = 0.17 Hz. (b) *f* = 0.20 Hz. Source: Ref A9.1

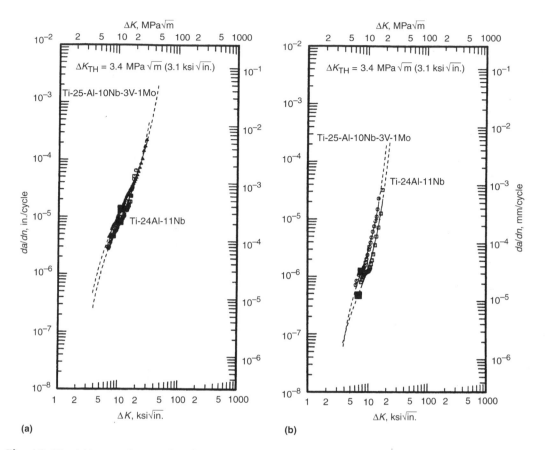

Fig. A9.15 *da/dN* curves for Ti-25Al-10Nb-3V-1Mo and Ti-24Al-11Nb at 427 °C (800 °F). *R* = 0.1. (a) *f* = 0.17 Hz. (b) *f* = 0.20 Hz. Source: Ref A9.1

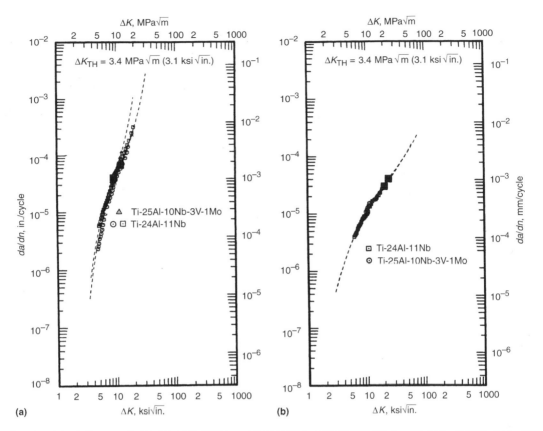

Fig. A9.16 *da/dN* curves for Ti-25Al-10Nb-3V-1Mo and Ti-24Al-11Nb at 650 °C (1200 °F). *R* = 0.1. (a) *f* = 0.17 Hz. (b) *f* = 0.20 Hz. Source: Ref A9.1

APPENDIX 10

Mechanical Properties Data for Selected Steels

LIMITED MECHANICAL PROPERTIES DATA for selected low-alloy steels and stainless steels are compiled in this appendix. Room-temperature ultimate tensile strength and plane-strain fracture toughness values for several commonly used high-strength steels are shown in Fig. A10.1 and A10.2. Plane-strain fracture toughness versus strength for 4345, 4340, and the precipitation-hardening stainless steels PH138-8Mo and Custom 465 are shown in Fig. A10.3 to A10.5. Table A10.1 tabulates the data used to plot Fig. A10.4, while Table A10.2 lists plane-strain fracture toughness data corresponding to those plotted in Fig. A10.5. Tensile strength and plane-strain fracture toughness data for several 300 series stainless steels are shown in Fig. A10.6 and Table A10.3, respectively. Also listed in Table A10.3 are tensile strength and plane-strain fracture toughness data for the Nitronic 33 and Nitronic 50 austenitic stainless steels. All data points listed in Table A10.3 are graphically presented in Fig. A10.7 in the form of K_{IC} versus tensile yield strength. Figure A10.8 replots these data, showing K_{IC} as a function of temperature.

REFERENCES

A10.1. J.M. Dahl, Ferrous-Base Aerospace Alloys, *Adv. Mater. Process.,* Vol 157 (No. 5), 2000, p 33–36

A10.2. R.P. Wei, Fracture Toughness Testing in Alloy Development, *Fracture Toughness and Its Applications,* STP 381, ASTM, 1965

A10.3. Fracture Mechanics Properties of Carbon and Alloy Steels, *ASM Handbook,* Vol 19, *Fatigue and Fracture,* ASM International, 1996, p 622

A10.4. "Fatigue Crack Growth Computer Program NASA/FLAGO, Version 2.0," Report JSC-22267A, NASA, May 1994

A10.5. S. Lampman, Fatigue and Fracture Properties of Stainless Steels, *ASM Handbook,* Vol 19, *Fatigue and Fracture,* ASM International, 1996, p 712–732

A10.6. *Military Standardization Handbook: Metallic Materials and Elements for Aerospace Vehicle Structures,* MIL-HDBK-5E, U.S. Department of Defense, 1987

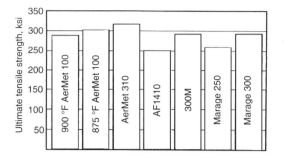

Fig. A10.1 Room-temperature ultimate tensile strengths for several high-strength steels. Source: Ref A10.1

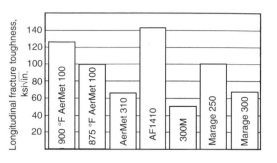

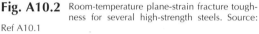

Fig. A10.2 Room-temperature plane-strain fracture toughness for several high-strength steels. Source: Ref A10.1

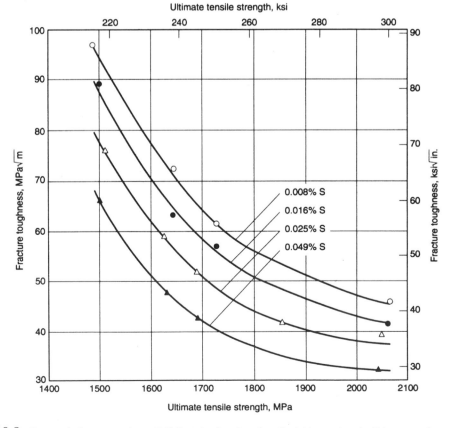

Fig. A10.3 Plane-strain fracture toughness of 4345 steel as function of tensile yield strength and sulfide content. Source: Ref A10.2

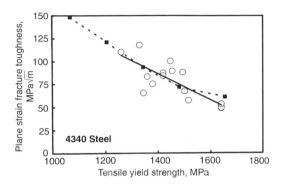

Fig. A10.4 Room-temperature plane-strain fracture toughness of 4340 steel as a function of tensile yield strength. Source: ○, Ref A10.3; ■, Ref A10.4

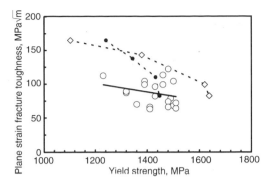

Fig. A10.5 Plane-strain fracture toughness of PH13-8Mo and Custom 465 precipitation-hardening stainless steels. ○, PH13-8Mo (Source: Ref A10.5); ●, PH13-8Mo (Source: Ref A10.1); ◇, Custom 465 (Source: Ref A10.1)

Table A10.1 Tensile strength and fracture toughness of 4340 steel for various tempers

Processing and product form	Condition or heat treatment	Ultimate tensile strength, MPa	Tensile yield strength, MPa	K_C, MPa√m	K_{IC}, MPa√m	Specimen and load type
. . .	260 °C temper	. . .	1640	. . .	48.5 ± 2	35 bend specimens
. . .	425 °C temper	. . .	1420	. . .	87 ± 4	37 bend specimens
. . .	260 °C temper	. . .	1640	. . .	50 ± 2.75	46 CT specimens
. . .	425 °C temper	. . .	1420	. . .	87 ± 3	48 CT specimens
Plate 16 mm (0.62 in.)	Heat treated to 51 HRC	. . .	1517	. . .	57	CT (L-T)
Round bar 115 mm (4.5 in.) diam	. . .	1240	1330	. . .	117.5	CT (L-T)
Plate 25 mm (1 in.) thick, oil quenched from 843 °C (1550 °F)	425 °C temper	. . .	1420	. . .	84	CT (L-T)
	260 °C temper		1640		49	
255 mm (10 in.) billet oil quenched and tempered	425 °C temper	. . .	1360–1455	. . .	83–89	CT (L-T)
Forged bar 16 mm (0.62 in.) oil quenched from 870 °C (1600 °F) and 1 h temper	205 °C temper	. . .	1345	. . .	65	CT (T-L)
	395 °C temper		1448		100	
	345 °C temper		1495		87.7	
	280 °C temper		1503		67	
. . .	426 °C temper	. . .	. . .	. . .	60	. . .
Vacuum melting with consumable electrode	149 °C temper	2332	1450	51	. . .	. . .
	204 °C temper	2038	1470	100		
	260 °C temper	1813	1540	161		
Vacuum melting with consumable electrode	316 °C temper	1764	1530	201	. . .	. . .
	371 °C temper	1636	1558	>254		
	427 °C temper	1430	1333	>225		
. . .	. . .	1920	. . .	94	. . .	. . .
		1783		218		
		1646		196		
Oil quenched bar, 25 mm (1 in.)	540 °C temper	1260	1172	. . .	110	. . .
	425 °C temper	1530	1380		75	
	205 °C temper	1950	1640		53	

Source: Ref A10.3

Table A10.2 Tensile strength and plane-strain fracture toughness of PH13-8Mo and Custom 465 precipitation-hardening stainless steels

Product form	Temper	F_{tu} (L)		F_{ty} (L)		K_{IC} (LT)		K_{IC} (TL)		Ref
		MPa	ksi	MPa	ksi	MPa$\sqrt{m}$	ksi$\sqrt{in.}$	MPa$\sqrt{m}$	ksi$\sqrt{in.}$	
PH13-8Mo										
(a)	H950	1550	225	1360	197	. . .	. . .	70	64	A10.5
(a)	H1050	1320	191	1230	178	. . .	. . .	112	102	A10.5
Forged bar	H950	1490	216	1410	204	66	60	63	57	A10.5
Rolled bar	RH950	1630	236	1500	217	68	62	. . .	. . .	A10.5
Rolled bar	RH950	1630	236	1510	219	64	58	. . .	. . .	A10.5
Rolled bar	RH975	1610	233	1490	216	79	72	. . .	. . .	A10.5
Rolled bar	RH975	1590	230	1510	219	72	65	. . .	. . .	A10.5
Forged bar	H1000	1460	212	1390	201	104	95	99	90	A10.5
Forged bar	H1000	1430	207	1320	191	87	79	89	81	A10.5
Forged bar	H1000	1510	219	1460	212	113	103	99	90	A10.5
Rolled bar	H1000	1490	216	1430	207	96	87	82	75	A10.5
Rolled bar	RH1000	1530	222	1480	214	122	111	. . .	. . .	A10.5
Rolled bar	RH1000	1560	226	1500	217	104	95	. . .	. . .	A10.5
Extruded bar	H1000	1520	220	1480	214	74	67	72	65	A10.5
(b)		. . .	. . .	1240.56	179.8	164.83	149.8	. . .	. . .	A10.1
(b)		. . .	. . .	1343.94	194.8	137.36	124.9	. . .	. . .	A10.1
(b)		. . .	. . .	1412.86	204.8	109.88	99.9	. . .	. . .	A10.1
(b)		. . .	. . .	1447.32	209.8	82.41	74.9	. . .	. . .	A10.1
Custom 465										
(b)		. . .	. . .	1102.72	159.8	164.83	149.8	. . .	. . .	A10.1
(b)		. . .	. . .	1378.40	199.8	142.85	129.9	. . .	. . .	A10.1
(b)		. . .	. . .	1619.62	234.7	98.90	89.9	. . .	. . .	A10.1
(b)		. . .	. . .	1636.85	237.2	82.41	74.9	. . .	. . .	A10.1

(a) Product form not specified. (b) Product form, temper, and gain orientation not specified. Note: Heat treatments are H950 and H1000: austenitized at 925 °C (1700 °F), AC; RH950, RH975, and RH1000: austenitized at 925 °C (1700 °F), AC, cooled to −73 °C (−100 °F) for 5 h; H950 and RH950: aged at 510 °C (950 °F) for 4 h; RH975: aged at 525 °C (975 °F) for 4 h; H1000 and RH1000: aged at 540 °C (1000 °F) for 4 h

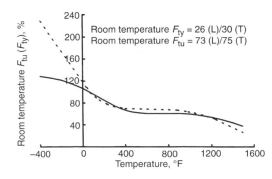

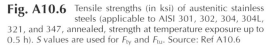

Fig. A10.6 Tensile strengths (in ksi) of austenitic stainless steels (applicable to AISI 301, 302, 304, 304L, 321, and 347, annealed, strength at temperature exposure up to 0.5 h). S values are used for F_{ty} and F_{tu}. Source: Ref A10.6

Table A10.3 Tensile strength and plane-strain fracture toughness of several 300 series (18Ni-8Cr) and Nitronic series (Fe-Cr-Ni-Mn) austenitic stainless steels(a)

Alloy	Product form/processing	F_{tu}		F_{ty}		K_{IC}	
		MPa	ksi	MPa	ksi	MPa√m	ksi√in.
301/302	Annealed plate and sheet	621	90	276	40	220	200
	Half-hard sheet	1139	165	863	125	110	100
	Full-hard sheet	1415	205	1311	190	88	80
304/304L	Annealed plate and sheet, cast	621	90	276	40	220	200
	Annealed plate and sheet, cast, 288 °C (550 °F)	442	64	166	24	165	150
	Annealed plate and sheet, cast, 427 °C (800 °F)	435	63	138	20	110	100
304LN2	Annealed plate and sheet, cast, −195 °C (−320 °F)	1415	205	690	100	220	200
316/316L	Annealed plate and sheet, cast	621	90	248	36	220	200
	Annealed plate and sheet, cast, 315 °C (600 °F)	414	60	221	32	165	150
	Annealed plate and sheet, cast, 427 °C (800 °F)	414	60	138	20	110	100
316LN2	Annealed plate and sheet, cast, −195 °C (−320 °F)	1277	185	483	70	220	200
316LHe	Annealed plate and sheet, cast, −269 °C (−452 °F)	1484	215	552	80	220	200
Nitronic 33	Annealed plate	794	115	442	64	220	200
Nitronic 33 LN2	Annealed plate, −195 °C (−320 °F)	1518	220	1139	165	165	150
Nitronic 33 LHe	Annealed plate, −269 °C (−452 °F)	1794	260	1518	220	88	80
Nitronic 50	Annealed plate	828	120	531	77	198	180
Nitronic 50 LN2	Annealed plate, −195 °C (−320 °F)	1587	230	1173	170	121	110
Nitronic 50 LHe	Annealed plate, −269 °C (−452 °F)	1898	275	1449	210	99	90

(a) All data points applicable to either L or T orientation. Source: Ref A10.4

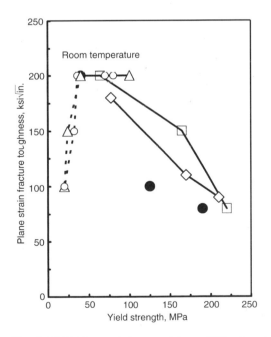

Fig. A10.7 Plane-strain fracture toughness of 300 series and nitronic series austenitic stainless steels as functions of tensile yield strength and temperature. ●, type 301/302, all three data points coming from room temperature; △, type 304; ○, type 316; □, nitronic 33; ◇, nitronic 50. Solid lines connect room temperature to data points of lower temperatures; dotted lines connect room temperature to data points of higher temperatures. Source: A10.4

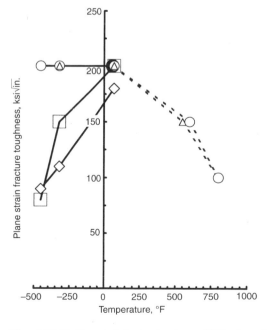

Fig. A10.8 Plane-strain fracture toughness of 300 series and nitronic series austenitic stainless steels as functions of temperature. Duplicate plot of Fig. A10.7. ●, type 301/302, annealed condition; △, type 304; ○, type 316; □, nitronic 33; ◇, nitronic 50

APPENDIX 11

Conversion Table

To convert from	to	multiply by
Density		
lb/in.^{3}	mg/m^3	27.68
lb/ft^3	kg/m^3	16.02
Length		
in.	mm	25.4
μin.	μm	0.0254
ft	m	0.3048
Area		
in.^{2}	mm^2	645.16
in.^{2}	cm^2	6.4516
in.^{2}	m^2	0.00064516
ft^2	m^2	0.09290304
Volume		
in.^{3}	m^3	0.000016387
ft^3	m^3	0.002831685
Volumetric flow		
ft^3/min	m^3/s	4.719×10^{-4}
Force, mass, weight		
lb	kg	0.4536
lb	N	4.44822
kip	N	4448.22
ton	t (metric ton)	0.9078
Moment, torque, energy		
in. · lbf	N · m	0.113
ft · lbf	N · m	1.355818
in. · lbf	J	0.113
ft · lbf	J	1.355818
Stress, pressure		
psi	MPa	0.006892
ksi	MPa	6.892
ksi	MN/m^2	6.892
ksi	N/mm^2	6.892

(continued)

Notes

giga (G) = multiple of 10^9

kilo (k) = multiple of 10^3

mega (M) = multiple of 10^6

micro (μ) = multiple of 10^{-6}

pascal (Pa) = N/m^2

newton (N) = kg · m/s^2

joule (J) = N · m

watt (W) = J/s

To convert from	to	multiply by
Modulus		
10^3 ksi	GPa	6.892
Stress intensity (stress intensity factor)		
ksi$\sqrt{\text{in.}}$	MPa$\sqrt{m}$	1.098843
ksi$\sqrt{\text{in.}}$	MN/m$^{3/2}$	1.098843
Angle		
radian	degree	57.295
Temperature		
°F	°C	5/9 · (°F − 32)
°F	°K	(°F + 459.67)/1.8
°C	°K	1 · (273.16 + °C)
°R	°K	5/9
Temperature interval		
Δ°F	Δ°C	5/9
Heating rate		
°F/s	°C/s	0.556
Quantity of heat		
cal	J	1.0
Work		
Btu	J	1054.8
Specific heat		
Btu/(lb · °F)	J/(kg · °K)	4186.8
cal/(g · °C)	J/(kg · °K)	4186.8
Thermal conductivity		
Btu/(h · ft^2 · °F)/ft	W/(m · °K)	1.7307
Thermal expansion		
in./in./°F	m/m/°K	1.8
in./in./°C	m/m/°K	1.0
Velocity		
ft/h	m/s	8.467×10^{-5}
Viscosity		
in.^{2}/s	mm^2/s	645.16
ft^2/s	m^2/s	0.0929

Notes

giga (G) = multiple of 10^9 pascal (Pa) = N/m^2
kilo (k) = multiple of 10^3 newton (N) = kg · m/s^2
mega (M) = multiple of 10^6 joule (J) = N · m
micro (μ) = multiple of 10^{-6} watt (W) = J/s

APPENDIX 12

Glossary of Terms, Symbols, and Abbreviations

A-basis. Also called the A-allowable. Mechanical property value above which at least 99% of the population of values is expected to fall, with a confidence of 95%. See also *B-basis, S-basis,* and *typical basis.*

AIAA. American Institute of Aeronautics and Astronautics.

allowable. See *material allowable* and *structural allowable.*

alloy. A metallic material with at least one metal in an amount sufficient to impart metallic characteristics to the body. An alloy also may contain one or more other elements. A binary alloy consists of two major elements, one of which is the base. A ternary alloy consists of three major elements. Brass, for example, can be defined simply as an alloy of copper and zinc, where copper is the base element. Many types of brasses also contain small amounts of other elements, such as tin, lead, or manganese. In this book, metal, metallic, and alloy are used interchangeably in general discussions when no specific type of metal or alloy is mentioned.

alternating load. See *amplitude.*

amplitude. In fatigue loading, one-half of the range of a cycle. Also known as alternating load.

anisotropic. Not isotropic; having mechanical and/or physical properties that vary with direction. In the case of composite materials, the properties of the laminate can be tailored to the exact loading and stiffness requirements of the design by orienting the fibers in a prescribed manner.

annealing twin. See Appendix 1.

aramid. A type of highly oriented organic material derived from polyamide (nylon) but incorporating an aromatic ring structure. Used primarily as high-strength, high-modulus fiber. Kelvar and Nomex are examples of aramids.

ASM. ASM International (formerly the American Society for Metal).

ASME. American Society of Mechanical Engineers.

ASTM. ASTM International (formerly the American Society for Testing and Materials).

at.%. Chemical composition (of an alloy) by atomic percentage. Contrast with *wt%.*

average linear strain. See *engineering strain.*

axial strain. The linear strain in a plane parallel to the longitudinal axis of the specimen.

batch. A definite quantity of some product or material produced under conditions that are considered uniform. A batch is usually smaller than a *lot.*

B-basis. Also called the B-allowable. Mechanical property value above which at least 90% of the population of values is expected to fall, with a confidence of 95%. See also *A-basis, S-basis,* and *typical basis.*

bcc. Abbreviation for body-centered cubic crystal structure (e.g., α-Fe, Ta).

bearing area. The product of the pin diameter and specimen thickness.

bearing strength. The maximum bearing stress that a material is capable of sustaining.

bearing stress. The force per unit of bearing area.

bend radius. (1) The inside radius of a bent section. (2) The radius of a tool around which metal is bent during testing or fabrication.

block. In fatigue loading, a specified number of *constant-amplitude loading* cycles applied consecutively, or a *spectrum loading* sequence of finite length that is repeated identically.

brittle fracture. Separation of a solid accompanied by little or no macroscopic plastic deformation. Typically, brittle fracture occurs by rapid crack propagation with less expenditure of energy than for *ductile fracture*. The fracture surface is flat and has a bright, granular appearance due to *cleavage* of individual grains. The corresponding fracture stress is below that for a net section yield.

buckling. A compression phenomenon that occurs when, after some critical level of load, a bulge, bend, bow, kink, or other wavy condition is produced in a column, bar, beam, sheet, or plate.

carburizing. Introducing carbon into the surface of solid ferrous alloy by holding above 730 °C (1350 °F) in contact with a solid, liquid, or gas containing carbon available for reaction. The carburized material is hardened by heat treatment. The carburized surface is hard and wear resistant.

Charpy test. An impact test in which a specimen (with a V-notch, U-notch, or keyhole notch), supported at both ends, is struck behind the notch by a striker mounted at the lower end of a bar that can swing as a pendulum. The energy that is absorbed in fracture is calculated from the height to which the striker would have risen had there been no specimen and the height to which it actually rises after fracture of the specimen. Contrast with *Izod test*.

chevron pattern. A fractographic pattern of radial marks (shear ledges) that look like nested letters "V"; sometimes called a herringbone pattern. Chevron patterns are typically found on brittle fracture surfaces in parts whose widths are considerably greater than their thicknesses. The points of the chevrons can be traced back to the fracture origin.

chord modulus. See *modulus of elasticity*.

cleavage. (1) Fracture of a crystal by crack propagation across a crystallographic plane of low index. (2) The tendency of a material to cleave or split along definite crystallographic planes. (3) Sometimes used to describe brittle fracture at the macro-or microscale in amorphous materials (such as glasses).

cleavage crack. In crystalline material, a transgranular fracture that extends along a *cleavage plane*, resulting in brightly reflecting facets. Contrast with *microvoid coalescence*. Sometimes used to describe the macroscale brittle fracture of amorphous materials (such as glasses and glassy polymers).

cleavage fracture. A fracture, usually of a polycrystalline metal, in which most of the grains have failed by *cleavage*, resulting in brightly reflecting facets. It is one type of *crystalline fracture* and is associated with low-energy *brittle fracture*. It exhibits a river pattern when examined under an electron microscope. Contrast with *shear fracture*.

cleavage plane. (1) In metals, a characteristic crystallographic plane or set of planes in a crystal on which a *cleavage crack* occurs easily. (2) In noncrystalline material, the plane on which *brittle fracture* occurs.

compressive modulus. The ratio of compressive stress to compressive strain below the proportional limit, theoretically equal to Young's modulus determined from tensile tests.

constant-amplitude loading. In fatigue loading, a loading in which all of the peak loads are equal and all of the valley loads are equal. Contrast with *spectrum loading*.

constant-life diagram. A plot (usually on rectangular coordinates) of a family of curves, each of which is for a single fatigue life (number of cycles), relating alternating stress, maximum stress, minimum stress, and mean stress. The constant-life (fatigue) diagram is generally derived from a family of *S-N* curves, each of which represents a different stress ratio for a 50% probability of survival.

conventional strain. See *engineering strain*.

conventional stress. See *engineering stress*.

corrosion. The deterioration of a metal by chemical or eletrochemical reaction with its environment.

corrosion fatigue. Cracking (or *fatigue failure*) produced by the combined action of repeated fluctuating stress and a corrosive environment at lower stress levels or fewer cycles than would be required in the absence of a corrosive environment.

crack extension force ($\mathcal{G}$). The elastic energy per unit of new separation area that is made available at the front of an ideal crack in an elastic solid during a virtual increment of forward crack extension.

crack growth resistance (or crack extension resistance, K_R). A measure of the resistance of a material to crack extension expressed in terms of the *stress intensity factor, K, crack extension force, $\mathcal{G}$,* or values of *J* derived using the *J-integral* concept. In the latter cases, the symbols are $\mathcal{G}_R$ and J_R, respectively.

creep. Creep, or creep deformation, is any permanent inelastic strain that occurs when a material is subjected to a sustained stress. The rate at which this deformation occurs depends not only on the magnitude of the applied stress, but also on temperature and time.

creep rate. The slope of the creep-time curve at a given time determined from a Cartesian plot.

creep-rupture strength. See *stress-rupture strength.*

creep stress. The constant load in a creep test divided by the original cross-sectional area of the specimen.

critical stress intensity factor. The critical value of *stress intensity factor.* See also *fracture toughness.*

crystalline. That form of substance comprised predominantly of one or more crystals, as opposed to glassy or amorphous.

crystalline defects. The deviations from a perfect three-dimensional atomic packing that are responsible for much of the structure-sensitive properties of materials.

crystalline fracture. A pattern of brightly reflecting crystal facets on the fracture surface of a polycrystalline metal, resulting from *cleavage fracture* of many individual crystals. See also *granular fracture.*

cup-and-cone fracture. A mixed-mode fracture often seen in tensile-test specimens of a ductile material, where the central portion undergoes plane-strain fracture and the surrounding region undergoes plane-stress fracture. One of the mating fracture surfaces looks like a miniature cup; it has a central depressed flat-face region surrounded by a shear lip. The other fracture surface looks like a miniature truncated cone.

CVN. Charpy V-notch; see *Charpy test.*

cylindrical stresses. Stresses where the magnitudes of any two of three principal stresses are equal.

d. See *denier.*

da/dN **(or** $\Delta a/\Delta N$**).** The distance that a fatigue crack propagates under one stress cycle, commonly known as the fatigue crack growth rate.

deflection. Frequently used to indicate the flexure of a beam. Also used as a general term for other types of deformations, such as axial deflection (extension) and torsion.

denier (d). A textile weight unit indicating the fineness of filaments or yarns. 1 d = 1 g/9000 m length of thread (or 1.111×10^{-7} kg/m).

design allowable. See *structural allowable.*

dimpled rupture. A fractographic term describing *ductile fracture* that occurs through the formation and coalescence of microvoids along the fracture path. The fracture surface appears dimpled when observed at high magnification and usually is most clearly resolved when viewed in a scanning electron microscope.

dislocation. The linear lattice defect that is responsible for nearly all aspects of the plastic deformation of metals. Dislocation theory is an important tool for explaining the slip of crystals, as well as other mechanical behavior in metals such as strain hardening, yield point, fatigue, fracture, creep, and even diffusion.

drop-weight tear test (DWTT). A test to determine the appearance of propagating fractures in plain-carbon or low-alloy pipe steels over the temperature range where the fracture mode changes from brittle to ductile. See Chapter 4.

drop-weight test (DWT). A test to determine the appearance of propagating fractures in plain-carbon or low-alloy pipe steels over the temperature range where the fracture mode changes from brittle to ductile. See Chapter 4.

ductile fracture. Fracture characterized by tearing of metal accompanied by appreciable gross plastic deformation and expenditure of considerable energy. Contrast with *brittle fracture.*

ductility. The ability of a material to deform plastically before fracturing, measured by elongation or reduction in area in a tensile test.

dynamic tear (DT) energy. The total energy required to fracture standard DT specimens tested in accordance with the provisions of ASTM Standard E 604. See Chapter 4.

E. See *Young's modulus.*

E_C. Compression modulus of elasticity.

edge distance. The distance from the center of a bearing hole to the edge of a plate.

effective crack size (a_e). The physical crack size augmented for the effects of cracking plastic deformation. In applied fracture mechanics the effective crack size is calculated from a measured value of a physical crack size plus a computed value of a crack-tip plastic zone adjustment.

elastic deformation. A change in dimensions directly proportional to and in phase with an increase or decrease in applied force.

elastic hysteresis. See *mechanical hysteresis.*

elastic limit. The maximum stress that a material is capable of sustaining without measurable permanent strain (deformation) remaining after the complete release of load. Compare with *proportional limit.*

elastic modulus. See *modulus of elasticity* and *Young's modulus.*

elastic-plastic fracture mechanics (EPFM). A method of fracture analysis that can determine the stress (or load) required to induce fracture instability in a structure containing a cracklike flaw of known size and shape. As opposed to *linear elastic fracture mechanics*, this method applies to very ductile material (or induced ductility as a result of stress state or temperature) that exhibits excessive yielding at the crack tip. See Chapters 4 and 5.

elastic strain. See *elastic deformation.*

elasticity. The property of a material by virtue of which deformation caused by stress disappears on removal of the stress. A perfectly elastic body completely recovers its original shape and dimensions after release of stress.

ELI. Extralow interstitial.

elongation. A term used in mechanical testing to describe the amount of extension of a testpiece when stressed. It is expressed as percentage of the original gage length, calculated using the amount of extension (the difference between the final gage length and the original gage length) divided by the original gage length. The final gage length is determined after fracture by realigning and fitting together the broken ends of the specimen.

endurance limit. The maximum stress below which a material can presumably endure an infinite number of stress cycles. The value of the maximum stress, the stress ratio, and the structural local geometry (i.e., K_t) also should be stated. There is no clear endurance limit for *fcc* and *hcp* metals. For these alloys, fatigue limit (or fatigue strength) is the maximum stress level that can be sustained for a desired number of cycles without failure. Compare with *fatigue life.*

energy release rate. See *crack extension force.*

engineering strain. A term sometimes used for average linear strain or conventional strain in order to differentiate it from *true strain*. In tension testing, it is calculated by dividing the change in gage length by the original gage length.

engineering stress. A term sometimes used for conventional stress in order to differentiate it from *true stress*. In tension testing it is calculated by dividing the breaking load applied to the specimen by the original cross-sectional area of the specimen.

environmentally assisted cracking. Brittle fracture of a normally ductile material in which the corrosive effect of the environment is a causative factor. Environmental cracking is a general term that includes *corrosion fatigue*, high-temperature hydrogen attack, hydrogen blistering, *hydrogen embrittlement*, liquid metal embrittlement, solid metal embrittlement, *stress corrosion cracking*, and sulfide stress cracking.

equiaxed grain structure. An alloy structure in which the grains have approximately the same dimensions in all directions.

exfoliation corrosion. A type of corrosion that attacks the exposed end-grain layers of an extrusion, forging, or plate parallel to the surface of the part. The corrosion products that build up in the layers are of greater volume than the metal or alloy, thus causing the layers to split apart. This is not a mechanical problem, but should be noted during design and material selection.

extensometer. An instrument for measuring changes in length over a given gage length caused by application or removal of a force. Commonly used in tension testing of metal specimens.

F_{bru}. MIL-HDBK-5 term for ultimate bearing strength.

F_{bry}. MIL-HDBK-5 term for bearing yield strength.

F_{cu}. MIL-HDBK-5 term for ultimate compression strength.

F_{su}. MIL-HDBK-5 term for ultimate shear strength.

F_{tu}. MIL-HDBK-5 term for ultimate tensile strength or *ultimate strength.*

F_{ty}. MIL-HDBK-5 term for tensile yield strength or *yield strength.*

fabric. A material constructed of interlaced yarns, fibers, or filaments, usually a planar structure. Nonwoven materials are sometimes included in this classification.

far-field stress. The stress calculated based on gross cross-sectional area (i.e., cut outs, cracks, etc., are treated as if they were not there). Gross cross-sectional area is equal to the entire width times the thickness. Sometimes called gross-area stress, or simply gross stress. Compare with *global stress*. Contrast with *nominal stress.*

fatigue. The phenomenon leading to fracture under repeated or fluctuating stresses having a maximum value less than the ultimate tensile strength of the material. *Fatigue failure* generally occurs at loads that, applied statistically, would produce little perceptible effect. Fatigue fractures are progressive, beginning as minute cracks that grow under the action of the fluctuating stresses.

fatigue allowable. The maximum stress level at which a structural member can withstand fatigue loading. It is determined by fatigue testing of a given material. Qualifiers such as stress-concentration factor and stress-ratio should be identified; otherwise, a stress ratio of -1 (i.e., fully reverse) and no stress concentrations are assumed.

fatigue failure. Failure that occurs when a specimen undergoing fatigue completely fractures into two parts.

fatigue life. The number of cycles of stress or strain of a specified character that a given specimen sustains before failure of a specified nature occurs.

fatigue limit. See *endurance limit.*

fatigue strength. See *endurance limit.*

fatigue striations. Parallel lines frequently observed in electron fractographs or fatigue-fracture surfaces. The lines are transverse to the direction of local crack propagation. The distance between successive lines represents the advance of the crack front during one cycle of stress variation.

fcc. Abbreviation for face-centered cubic crystal structure (e.g., Ag, Au, Cu).

fiber. A single homogeneous strand of material, essentially one-dimensional in the macrobehavior sense, used as a principal constituent in advanced composites because of its high axial strength and modulus.

fiber direction. The orientation or alignment of a longitudinal axis of a fiber with respect to a stated reference axis.

filamentary composites. A major form of advanced composites in which the fiber constituent consists of continuous filaments. Filamentary composites are defined here as composite materials composed of laminae in which the continuous filaments are in nonwoven, parallel, uniaxial arrays. Individual uniaxial laminae are combined into specifically oriented multiaxial laminates for application to specific envelopes of strength and stiffness requirements.

fill. Fibers running over and under the warp fibers at an angle of 90° to the warp direction. Also called woof and weft. The fill direction is the lower-strength direction, due to fiber curvature.

flow stress. The stress required to produce *plastic deformation* in a solid metal. In some literature, it is also defined as the average of the fracture stress and the yield stress.

fracture. Final failure (break) of a specimen whether it is subjected to a monotonic load or to a number of repeated (fatigue) loads. However, the latter would more appropriately be called *fatigue failure.*

fracture strength. An ultimate strength for a structural member. However, it is not a material property but rather a stress level that a structural member can withstand before fracture. Fracture strength is calculated from the load at the beginning of fracture during a tension test and the original uncracked cross-sectional area of the specimen. For estimation of design allowables its value is determined by the material ultimate strength and the original uncracked cross-sectional area. Compare with *residual strength.*

fracture stress. The gross-area stress level (fracture load divided by original cross-sectional area) that fractures a specimen in a test.

fracture toughness. A material property for measuring resistance to extension of a crack. It can be considered as the strength for a cracked structure, as opposed to the *ultimate strength* for an uncracked structure. Strictly speaking, this quantity is the *crack extension force* (also called the strain energy release rate), $\mathcal{G}$; its critical value is $\mathcal{G}_C$, that is, the energy required to initiate rapid propagation (fracture) of a given crack. Its value is determined by fracture testing of a cracked specimen. The result, computed based on fracture mechanics analysis, is conveniently presented as the *critical stress intensity factor,* K_C. Compare with *residual strength.*

frequency distribution. The way in which the frequencies of occurrence of members of a *population,* or a *sample,* are distributed in accordance with the values of the variable under consideration. For example: normal, log-normal, Weibull, etc. Generally, a normal distribution is assumed unless otherwise noted.

fretting. A type of wear that occurs between tight-fitting surfaces subjected to oscillation at very small amplitude. This type of wear can be a combination of oxidative wear and abrasive wear. See also *fretting corrosion.*

fretting corrosion. (1) Accelerated deterioration at the interface between contacting surfaces as the result of corrosion and slight oscillatory movement between the two surfaces.

(2) A form of fretting in which chemical reaction predominates. Fretting corrosion is often characterized by the removal of particles and subsequent formation of oxides, which are often abrasive and so increase the wear. Fretting corrosion can involve other chemical reaction products, which may not be abrasive.

fretting fatigue. Fatigue failure that initiates at a surface area where fretting has occurred.

G. Shear modulus. This symbol is also used for strain energy release rate, having the same meaning as $\mathcal{G}$.

$\mathcal{G}$. The original expression for strain energy release rate. See *crack extension force.*

$\mathcal{G}_C$. See *fracture toughness.*

$\mathcal{G}_R$. See *crack growth resistance.*

global stress. The stress at a point calculated on the global coordinate system by simple elasticity theory without taking into account the effect induced by stress raisers such as holes, grooves, and fillets. In this book, global stress and *far-field stress* are considered to be the same.

granular fracture. A type of irregular surface produced when metal is broken that is characterized by a rough, grainlike appearance, rather than a smooth or fibrous one. It can be subclassified as transgranular or intergranular, which means the crack path is basically running across the grain or along grain boundaries, respectively.

gross stress. See *far-field stress.*

hcp. Abbreviation for hexagonal close-packed crystal structure (e.g., Cd, Mg, Ti, Zn).

heterogeneous. Descriptive term for a material consisting of dissimilar constituents separately identifiable; a medium consisting of regions of unlike properties separated by internal boundaries. Generally taken to mean any material that is not *homogeneous.*

homogeneous. Descriptive term for a material of uniform composition throughout; a medium that has no internal physical boundaries. The properties of a homogeneous material are constant at every point; that is, for a translation of the coordinate axes there is no difference in material properties. Homogeneity does not necessarily imply that a material is *isotropic,* and vice versa.

homologous temperature. The ratio of the service temperature to the absolute melting temperature, T/T_m. This term normally appears in the literature for *creep.*

HRB (or HB). Brinell hardness of metallic materials. The test method is specified in ASTM Standard E 10.

HRC. Rockwell hardness of metallic materials. The test method is specified in ASTM Standard E 18.

hydrogen embrittlement. A condition of low toughness, low ductility, or cracking in metals, resulting from the absorption of hydrogen.

hydrogen-induced delayed cracking. A term sometimes used to identify a form of *hydrogen embrittlement* in which a metal appears to fracture spontaneously under a steady stress less than the yield stress. A delay usually occurs between the application of stress (or exposure of the stressed metal to hydrogen) and the onset of cracking. Also referred to as *static fatigue.*

hydrostatic stress. A condition where the magnitude of each principal stress component in a triaxial stress state is equal. It can be either hydrostatic tension or hydrostatic compression.

hysteresis. The phenomenon of permanently absorbed lost energy that occurs during any cycle of loading or unloading when a material is subjected to repeated loading.

impact energy. The amount of energy required to fracture a material, usually measured by means of an *Izod test* or *Charpy test.* The type of specimen and test conditions affect the values and must therefore be specified.

impact strength. The resiliency or toughness of a solid as measured by impact energy.

ISO. International Organization for Standardization.

isotropic. Having uniform properties in all directions. The measured elastic properties of an isotropic material are independent of the axis of testing. That is, they do not vary for a rotation of axes, or for a change in orientation.

Izod test. A type of impact test in which a V-notched specimen, mounted vertically, is subjected to a sudden blow delivered by the weight at the end of a pendulum arm. The energy require to break off the free end is a measure of the impact strength or toughness of the material. Contrast with *Charpy test.*

J_C. The critical value of J.

J_{IC}. Relates to K_{IC} by a factor of $(1 - \nu)/E$ when crack-tip plasticity is very low.

J_R. See *crack growth resistance.*

J-integral (J). A mathematical expression describing a line or surface integral that encloses the crack front from one crack surface to the other, used to characterize the local stress-strain field around the crack front. Its critical value is equivalent to *fracture toughness* of a material having appreciable (large-scale) plastic yielding at the crack tip before fracture. The computed values for *J* and $\mathcal{G}$ are the same for a crack tip with small-scale yielding.

K_C. Fracture toughness; plane stress is usually implied. See Chapter 4.

K_{IC}. Plane-strain fracture toughness. See Chapter 4.

K_{IE}. Plane-strain fracture toughness; its value is determined by using surface flaw specimens. See Chapter 4.

K_{IHE}. A threshold value of K_{IC} below which no failure will occur under the condition of *hydrogen embrittlement*. See Chapter 4.

K_{ISCC}. A threshold value of K_{IC} below which no failure will occur under the condition of *stress corrosion cracking*. See Chapter 4.

K_Q. K_{IC} equivalent. See Chapter 4 and ASTM Standard E 399 for details.

K_R. See *crack growth resistance*.

K_t. See *stress concentration factor*.

K-EE. K_{IC} equivalent. See Chapter 4 and ASTM Standard E 399 for details.

knitted fabrics. Fabrics produced by interlooping chains of yarn.

lamina. A single ply or layer in a multilayer *laminate*.

laminate. A product made by bonding together two or more layers, or laminae, of material or materials.

laminate orientation. The configuration of a cross-plied composite laminate with regard to the angles of cross-plying, the number of laminae at each angle, and the exact sequence of the lamina layup.

linear elastic fracture mechanics (LEFM). A method of fracture analysis that can determine the stress (or load) required to induce fracture instability in a structure containing a cracklike flaw of known size and shape. This method is applicable to situations that exhibit small-scale yielding at the crack tip. See Chapters 4 and 5. See also *stress intensity factor*. Compare with *elastic-plastic fracture mechanics*.

local stress. Compare with *global stress* and *nominal stress*.

lot. A definite quantity of a product or material accumulated under conditions that are considered uniform for sampling purposes. Compare with *batch*.

material allowable. The established material strength determined by mechanical testing. Compare with *structural allowable*.

matrix. The essentially *homogeneous* material in which the fibers or filaments of a composite are embedded.

mean stress. The algebraic average of the maximum and minimum stresses in one cycle of fatigue loading.

mechanical hysteresis. Energy absorbed in a complete cycle of loading and unloading within the *elastic limit* and represented by the closed loop of the stress-strain curves for loading and unloading. Sometimes called elastic hysteresis.

mechanical twin. See Appendix 1.

microvoid coalescence. Ductile micromechanism of fracture that occurs due to the nucleation of microscale voids, followed by their growth and eventual coalescence; initiation is caused by particle cracking or interfacial failure between an inclusion or precipitate particle and the surrounding matrix.

modulus of elasticity. The measure of rigidity or *stiffness* of a metal; the ratio of stress (up to the proportional limit) to the corresponding strain. Known as Hooke's law. This elastic stress to strain ratio (tension or compression) is commonly designated by the symbol *E*, or *Young's modulus*. Since the Young's modulus is essentially the slope of the elastic portion of the material's stress-strain curve, it is also considered the *tangent modulus*. The following moduli that are reduced from a tensile stress-strain curve are often of interest. For materials that do not conform to Hooke's law throughout the elastic range, the slope tangent to the stress-strain curve at the origin is taken as the tangent modulus. Tangent modulus also refers to the slope tangent to a given point on the stress-strain curve. The slope of a line drawn between that point and the origin of the stress-strain curve is called the secant modulus. The secant modulus is often used in connection with points in the plastic range. The chord connecting any two specific points on the stress-strain curve is called the chord modulus. Compare with *modulus of rigidity*.

modulus of resilience. The amount of energy stored in a material when loaded to its elastic limit. It is determined by measuring the area under the stress-strain curve up to the elastic limit.

modulus of rigidity. In cases of shear or torsion, the modulus of elasticity is commonly designated as modulus of rigidity, shear modulus, or torsional modulus, represented by the symbol *G*. The ratio of shear stress to the corresponding shear strain for shear stress below the proportional limit of the material is called the modulus of rigidity. Values of shear modulus are usually determined by torsion testing, represented by the symbol G. It relates to the *modulus of elasticity* and *Poisson's ratio*. An equation relating these three quantities is given in Chapter 1.

NASA. National Aeronautics and Space Administration.

net-section stress. The stress at a point calculated using the net cross-sectional area; that is, cutouts and/or cracks are not included as part of the cross section. In this book net-section stress is treated the same as *nominal stress*. However, this term is only used for parts containing a crack(s), whether or not a cutout is present. The term nominal stress is used for parts without cracks. Contrast with *far-field stress* and *global stress*.

nil-ductility transition temperature. Used along with a drop-weight test (ASTM Standard E 208) to determine the maximum temperature at which a standard drop-weight specimen breaks.

nitriding. Surface hardening of alloys by the absorption and combination of nitrogen with the metal. Nitriding is usually done in an ammonia atmosphere.

nominal stress. The stress at a point calculated on the net cross section by simple elasticity theory without taking into account the effect induced by stress raisers such as holes, grooves, and fillets. When there is no cutout, nominal stress is the same as far-field stress or gross-area stress. Compare with *far-field stress, global stress*, and *net-section stress*.

notch strength. The maximum *nominal stress* that a notched tensile specimen is capable of sustaining. Also known as notch tensile strength. A material may be classified as notch brittle if its notch strength is lower than its tensile strength (obtained from tensile testing of an unnotched specimen). Otherwise, it is said to be notch ductile.

NTS. Notch tensile strength.

orthotropic. Having three mutually perpendicular planes of symmetry. When dealing with thin laminated composites, since the stresses and strains in the thickness direction are ignored, the two planes of symmetry normal to the plane of laminate are of importance.

permanent set. The deformation or strain remaining in a previously stressed body after release of load.

plastic deformation. The permanent (inelastic) distortion of a material under applied stress that strains the material beyond its *elastic limit*.

plastic strain. Dimensional change that does not disappear when the initiating stress is removed. Usually accompanied by some *elastic deformation*.

plasticity. The property that enables a material to undergo permanent deformation without rupture.

Poisson's ratio. The absolute value of the ratio of transverse strain to the corresponding axial strain resulting from uniformly distributed axial stress below the *proportional limit* of the material.

population. In statistics, a generic term denoting any finite or infinite collection of individual samples or data points in the broadest concept; an aggregate determined by some property that distinguishes samples that do and do not belong.

precipitate-free zone (PFZ). A region adjacent to the grain boundary in which there is no (or little) precipitate, although precipitate is present in the grain interior (usually fine) and in the grain boundary. Often associated with poor corrosion resistance and/or poor fracture toughness. The presence of a PFZ often results in fracture in the region.

prepreg. Typically fabric or tape that is preimpregnated with a resin and stored for use. The resin is partially cured to a "B" stage, an intermediate stage in the polymerization reaction of thermoset resins. A prepreg must be stored in a freezer and has a limited shelf life (usually six months).

principal stress. The maximum or minimum value of the normal stress at a point in a plane considered with respect to all possible orientations. On such principal planes the shear stress is zero.

proof stress. (1) A specified stress applied to a member or structure to indicate its ability to withstand service loads. (2) The stress that will cause a specified small *permanent set* in a material.

proportional limit. The greatest stress a material is capable of developing without a deviation from straight-line proportionality between stress and strain.

quasi-isotropic. When applied to composite materials, this term refers to a particular type of laminate that is nearly *isotropic* in stiffness.

residual strength. The counterpart of *fracture strength*. This term is exclusively used for a structure or specimen that contains a crack (preexisting or otherwise). Associated with a certain crack size and configuration, its value is determined by the material *fracture toughness* and an appropriate *stress intensity factor* for a given crack.

residual stress. Stress in a body that is at rest and in equilibrium and at uniform temperature in the absence of external and mass forces.

reversal. In fatigue loading, when the first derivative of the load-time history changes sign (a peak or a valley).

rosette. A type of *strain gage,* it comes with three or four gages as a set. The gages are positioned in various arrangements for special purposes. The most common type is the rectangular rosette, consisting of three gages outlined at 45° apart to form a right angle between the outer two gages.

RST. Rapid solidification technology.

SAE. Society of Automotive Engineers.

sample. (1) One or more units of product (or a relatively small quantity of bulk material) withdrawn from a *lot* and then tested or inspected to provide information about the properties, dimensions, or other quality characteristics of the lot or process stream. Not to be confused with *specimen.* (2) A portion of a material intended to be representative of the whole. For example, the specimens selected from the population for test purposes.

sample mean. The sum of all the observed values in a *sample* divided by the sample size. It is a point estimate of the *population* mean. Also called the sample average (arithmetic average).

sample median. The middle value when all the observed values in a sample are arranged in order of magnitude if an odd number of items (units) is tested. If the sample size is even, it is the average of the two middlemost values. It is a point estimate of the *population* median, or 50% point.

sample percentage. The percentage of observed values between two stated values of the variable under consideration. It is a point estimate of the percentage of the *population* between the same two stated values. One stated value may be "minus infinity" or "plus infinity."

sample standard deviation (*s*). The square root of the *sample variance.* It is a point estimate of the population standard deviation, a measure of the "spread" of the *frequency distribution* of a *population.*

sample variance (*s²*). The sum of the squares of the differences between each observed value and the sample average divided by the sample size minus one. It is a point estimate of the *population* variance.

S-basis. Minimum property value specified by the appropriate federal, military, SAE, ASTM, or other recognized and approved specifications for the material. The statistical assurance associated with this value is not known. For certain products heat treated by the user (e.g., steels hardened and tempered to a designated strength), the S-value may reflect a specified quality-control requirement. See also *A-basis, B-basis,* and *typical basis.*

secant modulus. See *modulus of elasticity.*

SEM. Scanning electron microscopy. An electron microscope that is capable of taking a picture at a magnification from 100 to 10,000×. Compare with *TEM.*

shear fracture. A ductile fracture in which a crystal (or a polycrystalline mass) has separated by sliding or tearing under the action of shear stresses. Contrast with *cleavage fracture.*

shear modulus. See *modulus of rigidity.*

shear strength. The maximum shear stress that a material is capable of sustaining. Calculated from the maximum load during a shear or torsion test, shear strength is based on the original dimensions of the specimen cross section.

slant fracture. A type of fracture appearance, typical of plane-stress fractures, in which the plane of metal separation is inclined at an angle (usually about 45°) to the axis of the applied load.

specimen. A test object, often of standard dimensions and/or configuration, that is used for destructive or nondestructive testing. One or more specimens may be cut from each unit of a *sample.*

spectrum loading. In fatigue loading, a loading in which the magnitude and amplitude of loads are irregular and the mean level of each load cycle may vary during the lifetime of a structural/machine part. Also known as variable-amplitude loading. Contrast with *constant-amplitude loading.*

spherical state of stress. See *hydrostatic stress* and *triaxial stress.*

static fatigue. Seldom used in the literature. The term refers to a time-dependent reduction in strength with a static (noncyclic) load, such as *hydrogen-induced delayed cracking* or the effect of *creep* on the strength of plastics.

stiffness. (1) The ability of a material or shape to resist elastic deflection. (2) The rate of stress increase with respect to the rate of strain increase induced in the material or shape; the greater the stress required to produce a given strain, the stiffer the material is said to be.

strain aging. The change in ductility, hardness, yield point, and tensile strength that occurs when a metal or alloy that has been cold worked is stored for some time. In steel, strain aging is characterized by a loss of ductility and a corresponding increase in hardness, yield point, and tensile strength.

strain energy release rate. See *crack extension force.*

strain gage. A device used for measuring the strain at a point. Such gages consist of several loops of fine wire or foil of special composition, which are bonded to the surface of the body under study. When the body is deformed, the wires in the gage are strained and their electrical resistance altered. The change in resistance, which is proportional to strain, can be determined with a simple Wheatstone bridge circuit.

strain hardening. The state when a material requires higher stress to increase in strain after yielding. A material that exhibits strain-hardening behavior is called the strain-hardening material. Also known as work hardening.

stress concentration factor (K_t). The ratio of the greatest elastic stress in the region of the discontinuity to the nominal stress for the entire section.

stress corrosion. A time-dependent process in which a metallurgically susceptible material fractures prematurely under conditions of simultaneous corrosion and sustained tensile loading at lower stress levels than would be required in the absence of a corrosive environment. Tensile stress is required at the metal surface and may be a residual stress resulting from heat treatment or fabrication of the metal or the result of external sustained loading. The term *stress corrosion cracking* is also used in the literature; however, it is used more or less exclusively in fracture mechanics to describe crack growth behavior under sustained load in a corrosive environment.

stress corrosion cracking. See *stress corrosion.*

stress intensity. See *stress intensity factor.*

stress intensity factor. A scaling factor, usually denoted by the symbol K, used in *linear elastic fracture mechanics* to describe the intensification of applied stress at the tip of a crack of known size and shape. At the onset of rapid crack propagation in any structure containing a crack, the factor is called the *critical stress intensity factor,* or *fracture toughness.* Various subscripts are used to denote different loading conditions and/or combinations of structural configuration and crack morphology. Strictly speaking, stress intensity factor is not a factor and should be called stress intensity, especially when used to represent a material's fracture toughness (i.e., critical stress intensity). Because a factor (e.g., *stress concentration factor*) is unitless (or dimensionless) and thus is not a physical quantity, it is used only for magnifying the density/intensity of a given parameter. However, stress intensity has a unit (stress times the square root of crack length) that is directly related to *crack extension force* ($\mathcal{G}$). It is called a factor because it appears and acts as a factor in the solutions (equations) for crack-tip stress distributions. In this book, K is referred to as stress intensity or the stress intensity factor whenever the situation warrants. See also *stress-intensity geometry factor.*

stress-intensity geometry factor. A stress intensity factor is made up of three elements: applied stress, crack size, and a series of mathematical expressions accounting for the structural configuration and crack morphology. The last is a dimensionless component.

stress-rupture strength. The stress that will cause fracture in a creep test at a given time in a specified constant environment. Also known as creep-rupture strength or creep strength.

striation. A fracture surface marking consisting of a separation of the advancing crack front into separate fracture planes. On the microscale, striations appear as ripples on the fracture surface. A striation spacing becomes a measure of the crack rate of propagation during a given cycle.

structural allowable. The maximum load (or stress) that a part can withstand without failure; usually a static strength allowable is implied. This term can also mean *fatigue allowable,* or the residual strength of a cracked structural element/component.

structure (or structural part). Any load-carrying member, whether in a machine, a vehicle, or a building.

symmetrical laminate. A composite laminate in which the ply orientation is symmetrical about the laminate midplane.

T_g. Glass transition temperature.

T_m. Absolute melting temperature.

tangent modulus. The slope of the stress-strain curve at any specified stress or strain. See also *modulus of elasticity.*

tape. Parallel *tows* of fibers held together in a *prepreg* or cross-stitching to form standard widths of unidirectional fibers.

TEM. Transmission electron microscopy. An electron microscope that can take a picture at very high magnifications, even higher than $10,000\times$, but not lower than $2500\times$. Use of replica is required. Compare with *SEM.*

tensile strength. The maximum tensile stress that a material is capable of sustaining. Tensile strength is calculated from the maximum load during a tension test carried to rupture and the original cross-sectional area of the specimen. Also known as *ultimate strength.*

thermal fatigue. Fatigue resulting from the presence of temperature gradients that vary with time to produce cyclic stresses in a structure.

torsional modulus. See *modulus of rigidity.*

tow. An untwisted bundle of continuous fibers, commonly used in referring to manmade fibers—particularly carbon and graphite, but also glass and aramid. A tow designated at 140K has 140,000 filaments.

triaxial stress. A state of stress in which there are three principal stresses, none of which is zero. When all three stress components have an equal magnitude, the stress state is called spherical or hydrostatic. When two of three stress components have an equal magnitude, the stress state is called cylindrical. When one of the principal stresses is equal to zero, the stress state is called biaxial. When two of the principal stresses are equal to zero, the stress state is called uniaxial.

true strain. (1) The ratio of the change in dimension, resulting from a given load increment, to the magnitude of the dimension immediately prior to applying the load increment. (2) In a body subjected to axial force, the natural logarithm of the ratio of the gage length at the moment of observation to the original gage length. Also known as natural strain. Compare with *engineering strain.*

true stress. In a tension or compression test the axial stress calculated on the basis of the cross-sectional area at the moment of observation instead of the original cross-sectional area. Compare with *engineering stress.*

twinning. See Appendix 1.

typical basis. The typical value is an average value. No statistical assurance is associated with this basis. See also *A-basis, B-basis,* and *S-basis.*

ultimate strength. The same as *tensile strength,* also called ultimate tensile strength. When applied to shear and compression tests, called ultimate shear strength and ultimate compression strength, respectively.

uniform elongation. The elongation at maximum load and immediately preceding the onset of necking in a tension test.

uniform strain. The strain prior to the beginning of strain localization (necking); in tension testing, the strain to maximum load.

variable-amplitude loading. See *spectrum loading.*

variance. A measure of the squared dispersion of observed values or measurements expressed as a function of the sum of the squared deviations from the *population* mean or *sample* average.

viscoelasticity. A property involving a combination of elastic and viscous behavior. A material having this property combines the features of a perfectly elastic solid and a perfect fluid.

viscosity. The property of resistance to flow exhibited within the body of a material, expressed in terms of the relationship between applied shearing stress and resulting rate of strain in shear.

warp. Longitudinal fibers of a fabric. The warp is the high-strength direction, due to the straightness of the fibers.

work hardening. See *strain hardening.*

wt%. Chemical composition (of an alloy) by weight percentage. Contrast with *at.%.*

yielding. Evidence of *plastic deformation* in structural materials.

yield point. The first stress in a material, usually less than the maximum attainable stress, at which an increase in strain occurs without an increase in stress. Only certain metals—those of transition from elastic to plastic deformation—produce a yield point. If there is a decrease in stress after yielding, a distinction may be made between upper and lower yield points. The load at which a sudden drop in the flow curve occurs is called the upper yield point. The constant load shown on the flow curve is the lower yield point.

yield-point elongation. During discontinuous yielding, the amount of strain measured from the onset of yielding in the beginning of *strain hardening.*

yield strength. The *engineering stress* at which a material exhibits a specified limiting deviation from the proportionality of stress to strain. A 0.2% offset is used for many metals.

yield stress. The stress level of highly ductile materials, such as structural steels, at which large strains take place without further increase in stress.

Young's modulus. A term used synonymously with elastic modulus, *modulus of elasticity*, and Hooke's law. The ratio of tensile or compressive stresses to the resulting strain in the material's *elastic deformation* range.

Index

U

V

van der Waals bonds, 1
van der Waals force, 2
variable-amplitude crack growth, 260
variable-amplitude fatigue loading, 138
variable-amplitude loading, 151
viscoelastic behavior, 2, 3, 303, 303(F)
viscoelastic crack growth, 266
viscoelastic deformation, 286
viscosity behavior, 6
viscous deformation, 6(F)
viscous flow between two parallel plates, 6(F)
void coalescence, 68(F)
volume strain, 10
von Mises-Hencky criterion (VMHC), 58
von Mises relationship, 15
von Mises stress, 30
von Mises stress criterion (VMSC), 58

W

Wallner lines, 71, 73(F), 317, 317(F)
weaving, 345
Weibull's statistical theory of fracture, 92
Wei-Landes superposition model, 272

Y

yield behavior, 49(F)
yield criteria, 57–58
yielding, 90
yield phenomenon, 52–55
yield point, 52
yield point elongation, 53
yield point stress, 24
yield strength, 49, 199(T)
yield stress
    influence of strain rate on, 85(F)
    temperature dependence of, 94
yield stress frequency, 95(F)
yield zone, 30
Young's modulus, 5, 49, 339

Wei model, 187
Westergaad stress function, 233
whiskers, 319
whisker-toughened ceramics, 321
width correlation factor, 218
width-to-thickness (w/t) ratios, 81, 89
Williams stress function, 233
work hardening, 53–55
work-hardening range, 52